STUDENT'S SOLUTIO

LIAL · MILLER · HORNSBY

COLLEGE ALGEBRA

SIXTH EDITION

Prepared with the assistance of

BRIAN HAYES

Triton College

ABBY TANENBAUM

College of DuPage

HarperCollinsCollegePublishers

PREFACE

This book provides complete solutions for many of the exercises in <u>College Algebra</u>, sixth edition, by Margaret L. Lial, Charles D. Miller, and E. John Hornsby, Jr. Solutions are provided for the odd–numbered exercises as well as all the exercises in the chapter tests. Solutions are not provided for exercises that involve open–response answers.

This book should be used as an aid as you work to master your course work. Try to solve the exercises that your instructor assigns before you refer to the solutions in this book. Then, if you have difficulty, read these solutions to guide you in solving the exercises. The solutions have been written so that they are consistent with the methods used in the textbook.

You may find that some of the solutions are presented in greater detail than others. Thus, if you cannot find an explanation for a difficulty that you encountered in one exercise, you may find the explanation in the solution for a similar exercise elsewhere in the exercise set.

Solutions that require graphs will refer to the answer section of the textbook. These graphs are not included in this book.

In addition to solutions, you will find a list of suggestions on how to be successful in mathematics. A careful reading will be helpful for many students.

The following people have made valuable contributions to the production of this <u>Student's Solution Manual</u>: Abby Tanenbaum, editor; Judy Martinez, typist; Therese Brown and Charles Sullivan, artists; and Carmen Eldersveld, proofreader.

We also want to thank Tommy Thompson of Seminole Community College for his suggestions for the essay "To the Student: Success in Mathematics" that follows this preface.

TO THE STUDENT: SUCCESS IN MATHEMATICS

The main reason students have difficulty with mathematics is that they don't know how to study it. Studying mathematics *is* different from studying subjects like English or history. The key to success is regular practice.

This should not be surprising. After all, can you learn to play the piano or to ski well without a lot of regular practice? The same thing is true for learning mathematics. Working problems nearly every day is the key to becoming successful. Here is a list of things you can do to help you succeed in studying mathematics.

1. *Attend class regularly.* Pay attention in class to what your instructor says and does, and make careful notes. In particular, note the problems the instructor works on the board and copy the complete solutions. Keep these notes separate from your homework to avoid confusion when you read them over later.

2. Don't hesitate to ask questions in class. It is not a sign of weakness, but of strength. There are always other students with the same question who are too shy to ask.

3. *Read your text carefully.* Many students read only enough to get by, usually only the examples. Reading the complete section will help you to be successful with the homework problems. Most exercises are keyed to specific examples or objectives that will explain the procedures for working them.

4. Before you start on your homework assignment, rework the problems the instructor worked in class. This will reinforce what you have learned. Many students say, "I understand it perfectly when you do it, but I get stuck when I try to work the problem myself."

5. Do your homework assignment only *after* reading the text and reviewing your notes from class. Check your work with the answers in the back of the book. If you get a problem wrong and are unable to see why, mark that problem and ask your instructor about it. Then practice working additional problems of the same type to reinforce what you have learned.

6. Work as neatly as you can. Write your symbols clearly, and make sure the problems are clearly separated from each other. Working neatly will help you to think clearly and also make it easier to review the homework before a test.

7. After you have completed a homework assignment, look over the text again. Try to decide what the main ideas are in the lesson. Often they are clearly highlighted or boxed in the text.

8. Use the chapter test at the end of each chapter as a practice test. Work through the problems under test conditions, without referring to the text or the answers until you are finished. You may want to time yourself to see how long it takes you. When you have finished, check your answers against those in the back of the book and study those problems that you missed. Answers are referenced to the appropriate sections of the text.

9. Keep any quizzes and tests that are returned to you and use them when you study for future tests and the final exam. These quizzes and tests indicate what your instructor considers most important. Be sure to correct any problems on these tests that you missed, so you will have the corrected work to study.

10. Don't worry if you do not understand a new topic right away. As you read more about it and work through the problems, you will gain understanding. Each time you look back at a topic you will understand it a little better. No one understands each topic completely right from the start.

CONTENTS

CHAPTER 1   ALGEBRAIC EXPRESSIONS
Section 1.1

1. 1 and 3 are natural numbers.

3. -6, $\dfrac{-12}{4}$ (or -3), 0, 1, and 3 are integers.

5. $-\sqrt{3}$, 2π, and $\sqrt{12}$ are irrational numbers.

7. 12 is a natural number, a whole number, an integer, a rational number, and a real number.

9. $-\dfrac{3}{4}$ is a rational number and a real number.

11. $\sqrt{8}$ is an irrational number and a real number.

15. $-3^5 = -(3 \cdot 3 \cdot 3 \cdot 3 \cdot 3) = -243$

17. $(-3)^4 = (-3)(-3)(-3)(-3) = 81$

19. $(-3)^5 = (-3)(-3)(-3)(-3)(-3) = -243$

21. A negative base raised to an odd exponent is negative. A negative base raised to an even exponent is positive.

23. $8^2 - (-4) + 11 = 64 - (-4) + 11$
$$= 64 + 4 + 11$$
$$= 79$$

25. $-15 - 3(-8) = -15 + 24$
$$= 9$$

27. $9 \cdot 3 - 16 \div 4 = 27 - 4$
$$= 23$$

29. $6(-5) - (-3)(2)^4 = 6(-5) - (-3)(16)$
$$= -30 - (-48)$$
$$= -30 + 48$$
$$= 18$$

31. $[-3^2 - (-2)][\sqrt{16} - 2^3]$
$$= [-9 - (-2)](4 - 8)$$
$$= (-9 + 2)(4 - 8)$$
$$= (-7)(-4)$$
$$= 28$$

33. $\left[-\dfrac{5}{8} - \left(-\dfrac{2}{5}\right)\right] - \left(\dfrac{3}{2} - \dfrac{11}{10}\right)$
$$= \left(-\dfrac{5}{8} + \dfrac{2}{5}\right) - \left(\dfrac{3}{2} - \dfrac{11}{10}\right)$$
$$= \left(-\dfrac{25}{40} + \dfrac{16}{40}\right) - \left(\dfrac{15}{10} - \dfrac{11}{10}\right)$$
$$= \left(-\dfrac{9}{40}\right) - \left(\dfrac{4}{10}\right)$$
$$= -\dfrac{9}{40} - \dfrac{16}{40}$$
$$= -\dfrac{25}{40} = -\dfrac{5}{8}$$

35. $\dfrac{15 \div 5 \cdot 4 \div 6 - 8}{-6 - (-5) - 8 \div 2}$
$$= \dfrac{3 \cdot 4 \div 6 - 8}{-6 - (-5) - 4}$$
$$= \dfrac{12 \div 6 - 8}{-6 + 5 - 4}$$
$$= \dfrac{2 - 8}{-1 - 4}$$
$$= \dfrac{-6}{-5} = \dfrac{6}{5}$$

37. $2(q - r) = 2[4 - (-5)]$

$\qquad\qquad$ *Let $q = 4$ and $r = -5$*

$\qquad = 2(4 + 5)$

$\qquad = 2(9) = 18$

39. $\dfrac{q + r}{q + p} = \dfrac{4 + (-5)}{4 + (-2)}$ $\quad$ *Let $p = -2$, $q = 4$, and $r = -5$*

$\qquad = \dfrac{-1}{2}$

$\qquad = -\dfrac{1}{2}$

41. $\dfrac{3q}{r} - \dfrac{5}{p} = \dfrac{3 \cdot 4}{-5} - \dfrac{5}{-2}$ $\quad$ *Let $p = -2$, $q = 4$, and $r = -5$*

$\qquad = -\dfrac{12}{5} + \dfrac{5}{2}$

$\qquad = -\dfrac{24}{10} + \dfrac{25}{10}$

$\qquad = \dfrac{1}{10}$

43. $\dfrac{\dfrac{3r}{10} - \dfrac{5p}{2}}{q + \dfrac{2r}{5}}$

$= \dfrac{\dfrac{3(-5)}{10} - \dfrac{5(-2)}{2}}{4 + \dfrac{2(-5)}{5}}$ $\quad$ *Let $p = -2$, $q = 4$, and $r = -5$*

$= \dfrac{\dfrac{-15}{10} - \dfrac{-10}{2}}{4 + \dfrac{-10}{5}}$

$= \dfrac{\dfrac{-15}{10} - \dfrac{-50}{10}}{\dfrac{20}{5} + \dfrac{-10}{5}}$

$= \dfrac{\dfrac{35}{10}}{\dfrac{10}{5}}$

$= \dfrac{35}{10} \cdot \dfrac{5}{10} = \dfrac{7}{4}$

45. $8(m + 4) = 8m + 8 \cdot 4$

Distributive property

47. $\dfrac{2 + m}{2 - m} \cdot \dfrac{2 - m}{2 + m} = 1$, if $m \neq 2$ or -2

Inverse property

49. $[9 + (-9)] \cdot 5 = 0 \cdot 5$

$\qquad\qquad\qquad = 5 \cdot 0$

Inverse property and commutative property

51. $x \cdot \dfrac{1}{x} + x \cdot \dfrac{1}{x} = x\left(\dfrac{1}{x} + \dfrac{1}{x}\right)$, if $x \neq 0$

Distributive property

53. $8p - 14p = (8 - 14)p$

$\qquad\qquad = -6p$

55. $18y + 6 = 6(3y + 1)$

57. $9(r - s) = 9r - 9s$

59. $-2(m + n) = -2(m) + (-2)(n)$

$\qquad\qquad\quad = -2m - 2n$

61. $-(-2 + 5m) = -1(-2 + 5m)$

$\qquad\qquad\quad = 2 - 5m$

63. $p(q - w + x) = p \cdot q + p(-w) + p \cdot x$

$\qquad\qquad\qquad = pq - pw + px$

65. $\dfrac{10}{11}(22z) = \left(\dfrac{10}{11} \cdot 22\right)z$ $\quad$ *Associative property*

$\qquad\qquad = 20z$

67. $\left(-\frac{5}{8}p\right)(-24) = (-24)\left(-\frac{5}{8}p\right)$ *Commutative property*

$\qquad = \left[-24\left(-\frac{5}{8}\right)\right]p$ *Associative property*

$\qquad = 15p$

69. $-\frac{1}{4}(20m + 8y - 32z)$

$\qquad = -\frac{1}{4}(20m) + \left(-\frac{1}{4}\right)(8y) - \left(-\frac{1}{4}\right)(32z)$

$\qquad$ *Distributive property*

$\qquad = -5m - 2y + 8z$

71. $\frac{2}{3}\left(\frac{9}{4}a + \frac{15}{14}b - \frac{27}{20}\right)$

$\qquad = \frac{2}{3} \cdot \frac{9}{4}a + \frac{2}{3} \cdot \frac{15}{14}b - \frac{2}{3} \cdot \frac{27}{20}$

$\qquad$ *Distributive property*

$\qquad = \frac{3}{2}a + \frac{5}{7}b - \frac{9}{10}$

Section 1.2

1. Since $|-8| = 8$, $-|9| = -9$, and $-|-6| = -6$, the order is

$\qquad -|9|,\ -|-6|,\ |-8|.$

3. Since $\sqrt{8} \approx 2.83$, $-\sqrt{3} \approx -1.73$, and $\sqrt{6} \approx 2.45$, the order is

$\qquad -5,\ -4,\ -2,\ -\sqrt{3},\ \sqrt{6},\ \sqrt{8},\ 3.$

5. Since $\frac{3}{4} = .75$, $\sqrt{2} \approx 1.414$,

$\qquad \frac{7}{5} = 1.4$, $\frac{8}{5} = 1.6$, and $\frac{22}{15} \approx 1.47$,

the order is

$\qquad \frac{3}{4},\ \frac{7}{5},\ \sqrt{2},\ \frac{22}{15},\ \frac{8}{5}.$

9. $\qquad -2 < 5$

$\qquad 3(-2) < 3(5)$

$\qquad -6 < 15$

11. $\qquad -11 \geq -22$

$\left(-\frac{1}{11}\right)(-11) \leq \left(-\frac{1}{11}\right)(-22)$ *Reverse inequality symbol*

$\qquad 1 \leq 2$

13. $\qquad 3x < 9$

$\qquad \frac{1}{3} \cdot 3x < \frac{1}{3} \cdot 9$

$\qquad x < 3$

15. $\qquad -9k < 63$

$\left(-\frac{1}{9}\right)(-9k) > \left(-\frac{1}{9}\right)(63)$ *Reverse inequality symbol*

$\qquad k < -7$

17. $|2x| = |2(-4)|$ *Let $x = -4$*

$\qquad = |-8|$

$\qquad = 8$

19. $|x - y| = |-4 - 2|$ *Let $x = -4$ and $y = 2$*

$\qquad = |-6|$

$\qquad = 6$

21. $|3x + 4y| = |3(-4) + 4(2)|$

$\qquad$ *Let $x = -4$ and $y = 2$*

$\qquad = |-12 + 8|$

$\qquad = |-4|$

$\qquad = 4$

23. $|-4x + y| - |y|$

$\qquad = |(-4)(-4) + 2| - |2|$

$\qquad$ *Let $x = -4$ and $y = 2$*

$\qquad = |16 + 2| - |2|$

$\qquad = |18| - |2|$

$\qquad = 18 - 2$

$\qquad = 16$

25. $\dfrac{|x| + 2|y|}{5 + x} = \dfrac{|-4| + 2|2|}{5 + (-4)}$

$\qquad\qquad\qquad$ *Let x = -4 and y = 2*

$\qquad\qquad = \dfrac{4 + 4}{1}$

$\qquad\qquad = 8$

29. $3 - |-4| = 3 - (4) = -1$

31. $|\sqrt{7} - 5|$

Since $\sqrt{7} < 5$, $\sqrt{7} - 5 < 0$, so

$\qquad |\sqrt{7} - 5| = -(\sqrt{7} - 5)$

$\qquad\qquad\quad\; = -\sqrt{7} + 5$

$\qquad\qquad\quad\; = 5 - \sqrt{7}.$

33. $|\pi - 3|$

Since $\pi > 3$, $\pi - 3 > 0$, so

$\qquad |\pi - 3| = \pi - 3.$

35. $|x - 4|$, if $x > 4$

If $x > 4$, $x - 4 > 0$, so

$\qquad |x - 4| = x - 4.$

37. $|2k - 8|$, if $k < 4$

If $k < 4$, $2k < 8$, and $2k - 8 < 0$, so

$\qquad |2k - 8| = -(2k - 8)$

$\qquad\qquad\quad\; = -2k + 8$

$\qquad\qquad\quad\; = 8 - 2k.$

39. $|-8 - 4m|$, if $m > -2$

If $m > -2$, $4m > -8$, and $-8 - 4m < 0$, so

$\qquad |-8 - 4m| = -(-8 - 4m)$

$\qquad\qquad\qquad = 8 + 4m.$

41. $|x - y|$, if $x < y$

If $x < y$, $x - y < 0$, so

$\qquad |x - y| = -(x - y)$

$\qquad\qquad\quad = -x + y$

$\qquad\qquad\quad = y - x.$

43. $|3 + x^2|$

$(3 + x^2) > 0$, so

$\qquad |3 + x^2| = 3 + x^2.$

45. $|-1 - p^2|$

Since $p^2 \geq 0$, $-p^2 \leq 0$, and
$-1 - p^2 < 0$, so

$\qquad |-1 - p^2| = -(-1 - p^2)$

$\qquad\qquad\qquad = 1 + p^2.$

47. If $x + 8 < 15$, then $x < 7$.

$\qquad\qquad x + 8 < 15$

$\qquad x + 8 + (-8) < 15 + (-8)$

$\qquad\qquad\qquad x < 7$

Addition property of order

49. If $x < 5$ and $5 < m$, then $x < m$.

Transitive property of order

51. If $k > 0$, then $8 + k > 8$.

Addition property of order

53. $|k - m| \leq |k| + |-m|$

Triangle inequality,

$|a + b| \leq |a| + |b|$

55. $|12 + 11r| \geq 0$

Property of absolute value, $|a| \geq 0$

57. $\left|\dfrac{6}{5}\right| = \dfrac{|6|}{|5|}$

Property of absolute value,

$\left|\dfrac{a}{b}\right| = \dfrac{|a|}{|b|}$

59. $|x + y| = |x| + |y|$

This statement is true if x and y both have the same sign or if either x or y equals 0.

61. $|x| \leq 0$

This statement is true only if x = 0. For all other values of x, $|x| > 0$.

63. $||x + y|| = |x + y|$

This statement is always true.

65. $\left|\dfrac{|x|}{x}\right|$

If $x \geq 0$, $|x| = x$, so

$\left|\dfrac{|x|}{x}\right| = \left|\dfrac{x}{x}\right| = |1| = 1.$

If $x < 0$, $|x| = -x$, so

$\left|\dfrac{|x|}{x}\right| = \left|\dfrac{-x}{x}\right| = |-1| = 1.$

Thus, for all real numbers x,

$\left|\dfrac{|x|}{x}\right| = 1.$

Section 1.3

1. $(-4)^3 \cdot (-4)^2 = (-4)^{3+2}$ *Product rule*
$= (-4)^5$

3. $2^0 = 1$ *Definition of a^0*

5. $(5m)^0 = 1$ *Definition of a^0*

7. $(2^2)^5 = 2^{2 \cdot 5}$ *Power rule*
$= 2^{10}$

9. $(2x^5 y^4)^3$
$= 2^3 (x^5)^3 (y^4)^3$ *Power rule*
$= 2^3 x^{5 \cdot 3} y^{4 \cdot 3}$ *Power rule*
$= 2^3 x^{15} y^{12}$

11. $-\left(\dfrac{p^4}{q}\right)^2 = -\dfrac{(p^4)^2}{q^2}$ *Power rule*
$= -\dfrac{p^{4 \cdot 2}}{q^2}$ *Power rule*
$= -\dfrac{p^8}{q^2}$

13. $5x^{11}$ is a polynomial. It is a monomial since it has one term. It has degree 11 since 11 is the highest exponent.

15. $8p^5 q + 6pq$ is a polynomial. It is a binomial since it has two terms. It has degree 6 because 6 is the sum of the exponents in the term $8p^5 q$, and this term has a higher degree than the term 6pq.

17. $\sqrt{2}x + \sqrt{3}x^6$ is a polynomial. It is a binomial since it has two terms. It has degree 6 since 6 is the highest exponent.

19. $\frac{1}{3}r^2s - \frac{3}{5}r^4s^2 + rs^3$ is a polynomial.

It is a trinomial since it has three terms. It has degree 6 because the sum of the exponents in the term $-\frac{3}{5}r^4s^2$ is 6, and this term has the highest degree.

21. $\frac{5}{p} + \frac{2}{p^2} - \frac{5}{p^3}$ is not a polynomial since positive exponents in the denominator are equivalent to negative exponents in the numerator.

23. $5\sqrt{z} + 2\sqrt{z^3} - 5\sqrt{z^5}$

$= 5z^{1/2} + 2z^{3/2} - 5z^{5/2}$

This expression is not a polynomial since the exponents are not integers.

25. $(3x^2 - 4x + 5) + (-2x^2 + 3x - 2)$

$= (3x^2 - 2x^2) + (-4x + 3x)$

$+ (5 - 2)$ *Remove parentheses and group like terms*

$= x^2 - x + 3$ *Combine like terms*

27. $(12y^2 - 8y + 6) - (3y^2 - 4y + 2)$

$= 12y^2 - 8y + 6 - 3y^2 + 4y - 2$
    Remove parentheses

$= (12y^2 - 3y^2) + (4y - 8y)$

$+ (6 - 2)$ *Group like terms*

$= 9y^2 - 4y + 4$
    Combine like terms

29. $(6m^4 - 3m^2 + m) - (2m^3 + 5m^2 + 4m)$

$+ (m^2 - m)$

$= (6m^4 - 3m^2 + m)$

$+ (-2m^3 - 5m^2 - 4m) + (m^2 - m)$

$= 6m^4 - 2m^3 - 3m^2 - 5m^2 + m^2 + m$

$- 4m - m$

$= 6m^4 - 2m^3 + (-3 - 5 + 1)m^2$

$+ (1 - 4 - 1)m$

$= 6m^4 - 2m^3 - 7m^2 - 4m$

31. $(4r - 1)(7r + 2)$

$= 4r(7r) + 4r(2) - 1(7r) - 1(2)$
    FOIL

$= 28r^2 + 8r - 7r - 2$

$= 28r^2 + r - 2$ *Combine like terms*

33. $\left(3x - \frac{2}{3}\right)\left(5x + \frac{1}{3}\right)$

$= 3x(5x) + 3x\left(\frac{1}{3}\right) + \left(-\frac{2}{3}\right)(5x)$

$+ \left(-\frac{2}{3}\right)\left(\frac{1}{3}\right)$ *FOIL*

$= 15x^2 + \frac{3}{3}x - \frac{10}{3}x - \frac{2}{9}$

$= 15x^2 - \frac{7}{3}x - \frac{2}{9}$

35. $4x^2(3x^3 + 2x^2 - 5x + 1)$

$= (4x^2)(3x^3) + (4x^2)(2x^2)$

$- (4x^2)(5x) + (4x^2)(1)$
    Distributive property

$= 12x^5 + 8x^4 - 20x^3 + 4x^2$

37. $(2z - 1)(-z^2 + 3z - 4)$

$= 2z(-z^2 + 3z - 4) - 1(-z^2 + 3z - 4)$
    Distributive property

$= -2z^3 + 6z^2 - 8z + z^2 - 3z + 4$

$= -2z^3 + (6z^2 + z^2) + (-8z - 3z)$

$+ 4$

$= -2z^3 + 7z^2 - 11z + 4$

We may also multiply vertically.

$$
\begin{array}{r}
-z^2 + 3z - 4 \\
2z - 1 \\
\hline
z^2 - 3z + 4 \\
-2z^3 + 6z^2 - 8z \\
\hline
-2z^3 + 7z^2 - 11z + 4
\end{array}
$$

39. $(m - n + k)(m + 2n - 3k)$

$= m(m + 2n - 3k) - n(m + 2n - 3k)$

$+ k(m + 2n - 3k)$ *Distributive property*

$= m^2 + 2mn - 3km - mn - 2n^2 + 3kn$

$+ km + 2kn - 3k^2$ *Distributive property*

$= m^2 + (2mn - mn) + (-3km + km)$

$- 2n^2 + (3kn + 2kn) - 3k^2$

$= m^2 + mn - 2km - 2n^2 + 5kn - 3k^2$

41. $(a - b + 2c)^2$

$= [(a - b) + 2c]^2$ *Associative property*

$= (a - b)^2 + 2(a - b)(2c) + (2c)^2$
   Square of a binomial, treating (a − b) as one term

$= a^2 - 2ab + b^2 + 2(a - b)(2c)$

$+ (2c)^2$
   Square the binomial (a − b)

$= a^2 - 2ab + b^2 + 4ac - 4bc + 4c^2$

45. $(2m + 3)(2m - 3)$

$= (2m)^2 - 3^2$ *Difference of two squares*

$= 4m^2 - 9$

47. $(4m + 2n)^2$

$= (4m)^2 + 2(4m)(2n) + (2n)^2$
   Square of a binomial

$= 16m^2 + 16mn + 4n^2$

49. $(5r + 3t^2)^2$

$= (5r)^2 + 2(5r)(3t^2) + (3t^2)^2$
   Square of a binomial

$= 25r^2 + 30rt^2 + 9t^4$

51. $[(2p - 3) + q]^2$

$= (2p - 3)^2 + 2(2p - 3)(q) + q^2$
   Square of a binomial, treating (2p − 3) as one term

$= (2p)^2 - 2(2p)(3) + 3^2$

$+ 2(2p - 3)q + q^2$
   Square the binomial (2p − 3)

$= 4p^2 - 12p + 9 + 4pq - 6q + q^2$

53. $[(3q + 5) - p][(3q + 5) + p]$

$= (3q + 5)^2 - p^2$
   Difference of two squares

$= [(3q)^2 + 2(3q)(5) + 5^2] - p^2$
   Square of a binomial

$= 9q^2 + 30q + 25 - p^2$

55. $[a + (b + c)][a - (b + c)]$

$= a^2 - (b + c)^2$
   Difference of two squares

$= a^2 - (b^2 + 2bc + c^2)$
   Square of a binomial

$= a^2 - b^2 - 2bc - c^2$
   Remove parentheses

57. $[(3a + b) - 1]^2$

$= (3a + b)^2 - 2(3a + b)(1) + 1^2$
   Square of a binomial

$= (9a^2 + 6ab + b^2) - 2(3a + b) + 1$
   Square of a binomial

$= 9a^2 + 6ab + b^2 - 6a - 2b + 1$
   Distributive property

59. $(6p + 5q)(3p - 7q)$

$= 6p(3p - 7q) + 5q(3p - 7q)$
   Distributive property

$= 18p^2 - 42pq + 15pq - 35q^2$
   Distributive property

$= 18p^2 - 27pq - 35q^2$
   Combine like terms

61. $(3x - 4y)^3$

$$= (3x - 4y)(3x - 4y)^2$$

$$= (3x - 4y)$$

$$\bullet \; [(3x)^2 - 2(3x)(4y) + (4y)^2]$$
Square of a binomial

$$= (3x - 4y)(9x^2 - 24xy + 16y^2)$$

$$= 3x(9x^2 - 24xy + 16y^2)$$

$$- 4y(9x^2 - 24xy + 16y^2)$$
Distributive property

$$= 27x^3 - 72x^2y + 48xy^2$$

$$- 36x^2y + 96xy^2 - 64y^3$$

$$= 27x^3 - 108x^2y + 144xy^2 - 64y^3$$
Combine like terms

63. $(6k - 3)^2$

$$= (6k)^2 - 2(6k)(3) + 3^2$$
Square of a binomial

$$= 36k^2 - 36k + 9$$

65. $(p^3 - 4p^2 + p) - (3p^2 + 2p + 7)$

$$= p^3 - 4p^2 + p - 3p^2 - 2p - 7$$

$$= p^3 - 7p^2 - p - 7$$

67. $(7m + 2n)(7m - 2n)$

$$= (7m)^2 - (2n)^2 \quad \textit{Difference of}$$
two squares

$$= 49m^2 - 4n^2$$

69. $-3(4q^2 - 3q + 2) + 2(-q^2 + q - 4)$

$$= -12q^2 + 9q - 6 - 2q^2 + 2q - 8$$
Distributive property

$$= (-12q^2 - 2q^2) + (9q + 2q)$$

$$+ (-6 - 8) \quad \textit{Group like terms}$$

$$= -14q^2 + 11q - 14$$
Combine like terms

71. $p(4p - 6) + 2(3p - 8)$

$$= 4p^2 - 6p + 6p - 16$$
Distributive property

$$= 4p^2 - 16 \quad \textit{Combine like terms}$$

73. $-y(y^2 - 4) + 6y^2(2y - 3)$

$$= -y^3 + 4y + 12y^3 - 18y^2$$
Distributive property

$$= (-y^3 + 12y^3) - 18y^2 + 4y$$
Group like terms

$$= 11y^3 - 18y^2 + 4y$$
Combine like terms

75. $(x^3 + 4x)(-3x^2 + 4x - 1)$

x^3 terms come only from multiplying $(x^2)(4x)$ and $(4x)(-3x^2)$. Thus

$$4x^3 - 12x^3 = -8x^3.$$

The coefficient of x^3 is -8.

77. $(1 + x^2)(1 + x)$

The only x^3 term comes from multiplying $(x^2)(x) = x^3$. Thus the coefficient of x^3 is 1.

79. $x^2(4 - 3x)^2$

When x^2 is multiplied by the x term of $(4 - 3x)^2$, we get an x^3 term. That is, $x^2(2)(4)(-3x) = -24x^3$. The coefficient of x^3 is -24.

81. $(k^m + 2)(k^m - 2) = (k^m)^2 - 2^2$
Difference of 2 squares

$$= k^{2m} - 4$$

83. $(b^r + 3)(b^r - 2)$

$$= (b^r)^2 - 2b^r + 3b^r - 6$$

$$= b^{2r} + b^r - 6$$

85. $(3p^x + 1)(p^x - 2)$

$$= 3(p^x)^2 - 6p^x + p^x - 2$$

$$= 3p^{2x} - 5p^x - 2$$

87. $(m^x - 2)^2$

$= (m^x)^2 - 2(m^x)(2) + 2^2$

Square of a binomial

$= m^{2x} - 4m^x + 4$

89. $(q^p - 5p^q)^2$

$= (q^p)^2 - 2(q^p)(5p^q) + (5p^q)^2$

Square of a binomial

$= q^{2p} - 10q^p p^q + 25p^{2q}$

91. $(3k^a - 2)^3$

$= (3k^a - 2)(3k^a - 2)^2$

$= (3k^a - 2)$

$\quad \cdot [(3k^a)^2 - 2(3k^a)(2) + 2^2]$

Square of a binomial

$= (3k^a - 2)(9k^{2a} - 12k^a + 4)$

$= 3k^a(9k^{2a} - 12k^a + 4)$

$\quad - 2(9k^{2a} - 12k^a + 4)$

Distributive property

$= 27k^{3a} - 36k^{2a} + 12k^a$

$\quad - 18k^{2a} + 24k^a - 8$

$= 27k^{3a} - 54k^{2a} + 36k^a - 8$

93. The degree of a sum of polynomials
is the same as the degree of the
higher degree polynomial. Thus,
since m > n, the sum has degree m.

95. The product of x^m and x^n is x^{m+n}.
Thus, the degree of the product of
a polynomial of degree m and one
of degree n is m + n.

Section 1.4

1. $\dfrac{6!}{3!3!} = \dfrac{6 \cdot 5 \cdot 4 \cdot 3 \cdot 2 \cdot 1}{3 \cdot 2 \cdot 1 \cdot 3 \cdot 2 \cdot 1}$

$= \dfrac{6 \cdot 5 \cdot 4}{3 \cdot 2 \cdot 1}$

$= 20$

3. $\dfrac{7!}{3!4!} = \dfrac{7 \cdot 6 \cdot 5 \cdot 4 \cdot 3 \cdot 2 \cdot 1}{3 \cdot 2 \cdot 1 \cdot 4 \cdot 3 \cdot 2 \cdot 1}$

$= \dfrac{7 \cdot 6 \cdot 5}{3 \cdot 2 \cdot 1}$

$= 35$

5. $\binom{8}{3} = \dfrac{8!}{3!5!}$

$= \dfrac{8 \cdot 7 \cdot 6 \cdot 5 \cdot 4 \cdot 3 \cdot 2 \cdot 1}{3 \cdot 2 \cdot 1 \cdot 5 \cdot 4 \cdot 3 \cdot 2 \cdot 1}$

$= \dfrac{8 \cdot 7 \cdot 6}{3 \cdot 2 \cdot 1}$

$= 56$

7. $\binom{10}{8} = \dfrac{10!}{8!2!}$

$= \dfrac{10 \cdot 9 \cdot 8!}{8! \cdot 2 \cdot 1}$

$= \dfrac{10 \cdot 9}{2 \cdot 1}$

$= 45$

9. $\binom{n}{n-1} = \dfrac{n!}{(n-1)![n-(n-1)]!}$

$= \dfrac{n \cdot (n-1)!}{(n-1)!1!}$

$= \dfrac{n}{1} = n$

13. The signs of the terms in the
expansion of $(x - y)^n$ alternate,
starting with a positive sign
for the first term, x^n.

15. $(x + y)^6$

$$= x^6 + \binom{6}{5}x^5y + \binom{6}{4}x^4y^2 + \binom{6}{3}x^3y^3$$

$$+ \binom{6}{2}x^2y^4 + \binom{6}{1}xy^5 + y^6$$

$$= x^6 + \frac{6!}{5!1!}x^5y + \frac{6!}{4!2!}x^4y^2$$

$$+ \frac{6!}{3!3!}x^3y^3 + \frac{6!}{4!2!}x^2y^4$$

$$+ \frac{6!}{5!1!}xy^5 + y^6$$

$$= x^6 + 6x^5y + 15x^4y^2 + 20x^3y^3$$

$$+ 15x^2y^4 + 6xy^5 + y^6$$

17. $(p - q)^5$

$$= p^5 + \binom{5}{4}p^4(-q) + \binom{5}{3}p^3(-q)^2$$

$$+ \binom{5}{2}p^2(-q)^3 + \binom{5}{1}p(-q)^4 + (-q)^5$$

$$= p^5 + \frac{5!}{4!1!}p^4(-q) + \frac{5!}{3!2!}p^3q^2$$

$$+ \frac{5!}{2!3!}p^2(-q^3) + \frac{5!}{1!4!}pq^4 - q^5$$

$$= p^5 - 5p^4q + 10p^3q^2 - 10p^2q^3$$

$$+ 5pq^4 - q^5$$

19. $(r^2 + s)^5$

$$= (r^2)^5 + \binom{5}{4}(r^2)^4s + \binom{5}{3}(r^2)^3s^2$$

$$+ \binom{5}{2}(r^2)^2s^3 + \binom{5}{1}(r^2)s^4 + s^5$$

$$= r^{10} + 5r^8s + 10r^6s^2 + 10r^4s^3$$

$$+ 5r^2s^4 + s^5$$

21. $(p + 2q)^4$

$$= p^4 + \binom{4}{3}p^3(2q) + \binom{4}{2}p^2(2q)^2$$

$$+ \binom{4}{1}p(2q)^3 + (2q)^4$$

$$= p^4 + 4p^3(2q) + 6p^2(4q^2)$$

$$+ 4p(8q^3) + 16q^4$$

$$= p^4 + 8p^3q + 24p^2q^2 + 32pq^3$$

$$+ 16q^4$$

23. $(7p + 2q)^4$

$$= (7p)^4 + \binom{4}{3}(7p)^3(2q)$$

$$+ \binom{4}{2}(7p)^2(2q)^2 + \binom{4}{1}(7p)(2q)^3$$

$$+ (2q)^4$$

$$= 2401p^4 + 4(686p^3q)$$

$$+ 6(49p^2)(4q^2) + 4(7q)(8q^3)$$

$$= 2401p^4 + 2744p^3q + 1176p^2q^2$$

$$+ 224pq^3 + 16q^4$$

25. $(3x - 2y)^6$

$$= (3x)^6 + \binom{6}{5}(3x)^5(-2y)$$

$$+ \binom{6}{4}(3x)^4(-2y)^2 + \binom{6}{3}(3x)^3(-2y)^3$$

$$+ \binom{6}{2}(3x)^2(-2y)^4 + \binom{6}{1}(3x)(-2y)^5$$

$$+ (-2y)^6$$

$$= 729x^6 + 6(243x^5)(-2y)$$

$$+ 15(81x^4)(4y^2) + 20(27x^3)(-8y^3)$$

$$+ 15(9x^2)(16y^4) + 6(3x)(-32y^5)$$

$$+ 64y^6$$

$$= 729x^6 - 2916x^5y + 4860x^4y^2$$

$$- 4320x^3y^3 + 2160x^2y^4$$

$$- 576xy^5 + 64y^6$$

27. $\left(\frac{m}{2} - 1\right)^6$

$$= \left(\frac{m}{2}\right)^6 + \binom{6}{5}\left(\frac{m}{2}\right)^5(-1) + \binom{6}{4}\left(\frac{m}{2}\right)(-1)^2$$

$$+ \binom{6}{3}\left(\frac{m}{2}\right)^3(-1)^3 + \binom{6}{2}\left(\frac{m}{2}\right)^2(-1)^4$$

$$+ \binom{6}{1}\left(\frac{m}{2}\right)(-1)^5 + (-1)^6$$

$$= \frac{m^6}{64} - 6\left(\frac{m^5}{32}\right) + 15\left(\frac{m^4}{16}\right) - 20\left(\frac{m^3}{8}\right)$$

$$+ 15\left(\frac{m^2}{4}\right) - 6\left(\frac{m}{2}\right) + 1$$

$$= \frac{m^6}{64} - \frac{3m^5}{16} + \frac{15m^4}{16} - \frac{5m^3}{2} + \frac{15m^2}{4}$$

$$- 3m + 1$$

29. $\left(\sqrt{2}r + \frac{1}{m}\right)^4$

$= (\sqrt{2}r)^4 + \binom{4}{3}(\sqrt{2}r)^3\left(\frac{1}{m}\right)$

$\quad + \binom{4}{2}(\sqrt{2}r)^2\left(\frac{1}{m}\right)^2 + \binom{4}{1}(\sqrt{2}r)\left(\frac{1}{m}\right)^3$

$\quad + \left(\frac{1}{m}\right)^4$

$= (\sqrt{2})^4 r^4 + 4\left[(\sqrt{2})^3 r^3\left(\frac{1}{m}\right)\right]$

$\quad + 6\left[(\sqrt{2})^2 r^2\left(\frac{1}{m}\right)^2\right] + 4(\sqrt{2}r)\left(\frac{1}{m}\right)^3$

$\quad + \left(\frac{1}{m}\right)^4$

$= 4r^4 + \frac{8\sqrt{2}r^3}{m} + \frac{12r^2}{m^2} + \frac{4\sqrt{2}r}{m^3} + \frac{1}{m^4}$

31. To find the fifth term of $(m - 2p)^{12}$, use the formula

$$\binom{n}{n-(r-1)}x^{n-(r-1)}y^{r-1},$$

with $n = 12$ and $r = 5$,

$\quad x = m$ and $y = -2p$.

Then $r - 1 = 4$ and $n - (r - 1) = 8$. Thus, the fifth term is

$$\binom{12}{8}m^8(-2p)^4 = \frac{12!}{8!4!}m^8(16p^4)$$

$$= 495 \cdot 16m^8p^4$$

$$= 7920m^8p^4.$$

33. To find the sixth term of $(x + y)^9$, use the formula

$$\binom{n}{n-(r-1)}x^{n-(r-1)}y^{r-1},$$

with $n = 9$ and $r = 6$.

Then $r - 1 = 5$ and $n - (r - 1) = 4$. Thus, the sixth term is

$$\binom{9}{5}x^4y^5 = \frac{9!}{5!4!}x^4y^5$$

$$= 126x^4y^5.$$

35. To find the ninth term of $(2m + n)^{10}$, use the formula for the rth term with $n = 10$ and $r = 9$. Then $r - 1 = 8$ and $n - (r - 1) = 2$. The ninth term is

$$\binom{10}{8}(2m)^2n^8 = 45 \cdot 4m^2n^8$$

$$= 180m^2n^8.$$

37. To find the seventeenth term of $(p^2 + q)^{20}$, use the formula for the rth term with $n = 20$ and $r = 17$. Then $r - 1 = 16$ and $n - (r - 1) = 4$. The seventeenth term is

$$\binom{20}{16}(p^2)^4q^{16} = \frac{20!}{16!4!}p^8q^{16}$$

$$= 4845p^8q^{16}.$$

39. To find the eighth term of $(x^3 + 2y)^{14}$, use the formula for the rth term with $n = 14$ and $r = 8$. Then $r - 1 = 7$ and $n - (r - 1) = 7$. The eighth term is

$$\binom{14}{7}(x^3)(2y)^7 = \frac{14!}{7!7!}(x^3)^7(2y)^7$$

$$= 3432x^{21}(128y^7)$$

$$= 439,296x^{21}y^7.$$

41. $\sqrt[4]{630}$

Use the generalized binomial theorem.

$\left(1 + \frac{5}{625}\right)^{1/4}$

$= 1 + \frac{1}{4}\left(\frac{5}{625}\right) + \frac{\frac{1}{4}\left(\frac{1}{4} - 1\right)}{2!}\left(\frac{5}{625}\right)^2 + \cdots$

$= 1 + .002 - .000006 + \cdots$

$\approx 1.001994 \approx 1.002$

$$\sqrt[4]{630} = 625^{1/4}\left(1 + \frac{5}{625}\right)^{1/4}$$

$$\approx 5(1.002) = 5.010$$

43. $(1.02)^{-3}$

$= (1 + .02)^{-3}$

$= 1 + (-3)(.02) + \dfrac{-3(-3 - 1)}{2!}(.02)^2$

$+ \cdots$ *The generalized binomial theorem*

$= 1 - .06 + .0024 + \cdots$

$\approx .942$

Section 1.5

1. $4k^2m^3 + 8k^4m^3 - 12k^2m^4$

The greatest common factor is $4k^2m^3$.

$4k^2m^3 + 8k^4m^3 - 12k^2m^4$

$= 4k^2m^3(1) + 4k^2m^3(2k^2)$

$\quad + 4k^2m^3(-3m)$

$= 4k^2m^3(1 + 2k^2 - 3m)$

3. $2(a + b) + 4m(a + b)$

$= 2(a + b)(1 + 2m)$

$2(a + b)$ is the greatest common factor.

5. $(2y - 3)(y + 2) + (y + 5)(y + 2)$

$= (y + 2)[(2y - 3) + (y + 5)]$
   y + 2 is a common factor

$= (y + 2)[2y - 3 + y + 5]$

$= (y + 2)(3y + 2)$

7. $(5r - 6)(r + 3) - (2r - 1)(r + 3)$

$= (r + 3)[(5r - 6) - (2r - 1)]$
   r + 3 is a common factor

$= (r + 3)[5r - 6 - 2r + 1]$

$= (r + 3)(3r - 5)$

9. $2(m - 1) - 3(m - 1)^2 + 2(m - 1)^3$

$= (m - 1)[2 - 3(m - 1) + 2(m - 1)^2]$
   m - 1 is a common factor

$= (m - 1)[2 - 3m + 3]$

$\quad + 2(m^2 - 2m + 1)$

$= (m - 1)(2 - 3m + 3 + 2m^2 - 4m + 2)$

$= (m - 1)(2m^2 - 7m + 7)$

11. $6st + 9t - 10s - 15$

$= (6st + 9t) + (-10s - 15)$
   Group the terms

$= 3t(2s + 3) - 5(2s + 3)$
   Factor each group

$= (2s + 3)(3t - 5)$
   Factor out 2s + 3

13. $rt^3 + rs^2 - pt^3 - ps^2$

$= (rt^3 + rs^2) + (-pt^3 - ps^2)$
   Group the terms

$= r(t^3 + s^2) - p(t^3 + s^2)$
   Factor each group

$= (t^3 + s^2)(r - p)$
   Factor out $t^3 + s^2$

15. $6p^2 - 14p + 15p - 35$

$= (6p^2 - 14p) + (15p - 35)$
   Group the terms

$= 2p(3p - 7) + 5(3p - 7)$
   Factor each group

$= (3p - 7)(2p + 5)$
   Factor out 3p - 7

17. $20z^2 - 8zx - 45zx + 18x^2$

$= (20z^2 - 8zx) + (-45zx + 18x^2)$
   Group the terms

$= 4z(5z - 2x) - 9x(5z - 2x)$
   Factor each group

$= (5z - 2x)(4z - 9x)$
   Factor out 5z - 2x

19. $15 - 5m^2 - 3r^2 + m^2r^2$

$= (15 - 5m^2) + (-3r^2 + m^2r^2)$
Group the terms

$= 5(3 - m^2) - r^2(3 - m^2)$
Factor each group

$= (3 - m^2)(5 - r^2)$
Factor out $3 - m^2$

21. $8h^2 - 24h - 320$

First, factor out the greatest common factor, 8.

$8h^2 - 24h - 320 = 8(h^2 - 3h - 40)$

Now factor the trinomial

$h^2 - 3h - 40.$

Find two numbers whose sum is -3 and whose product is -40. The numbers are 5 and -8 since $5 + (-8) = -3$ and $5(-8) = -40$.
Thus,

$8h^2 - 24h - 320 = 8(h^2 - 3h - 40)$
$= 8(h + 5)(h - 8).$

23. $9y^4 - 54y^3 + 45y^2$

First, factor out the greatest common factor, 9.

$9y^4 - 54y^3 + 45y^2 = 9y^2(y^2 - 6y + 5)$

Now factor the trinomial $y^2 - 6y + 5$. Look for two numbers whose sum is -6 and whose product is 5. These numbers are -5 and -1.
Thus,

$9y^4 - 54y^3 + 45y^2$
$= 9y^2(y^2 - 6y + 5)$
$= 9y^2(y - 5)(y - 1).$

25. $14m^2 + 11mr - 15r^2$

The positive factors of 14 could be 2 and 7, or 1 and 14. As factors of -15, we could have 3 and -5, -3 and 5, 1 and -15, or -1 and 15. Try different combinations of these factors until the correct one is found.

$14m^2 + 11mr - 15r^2$
$= (7m - 5r)(2m + 3r)$

27. $12s^2 + 11st - 5t^2$

The positive factors of 12 could be 4 and 3, 2 and 6, or 1 and 12. As factors of -5 we could have -1 and 5 or -5 and 1. Try different combinations of these factors until the correct one is found.

$12s^2 + 11st - 5t^2 = (4s + 5t)(3s - t)$

29. $30a^2 + am - m^2$

The positive factors of 30 could be 5 and 6, 3 and 10, 2 and 15, or 1 and 30. The only factors of -1 (the coefficient of the last term) are 1 and -1. Try different combinations of these factors until the correct one is found.

$30a^2 + am - m^2 = (5a + m)(6a - m)$

31. $18x^5 + 15x^4z - 75x^3z^2$

First, factor out the greatest common factor, $3x^3$.

$18x^5 + 15x^4z - 75x^3z^2$
$= 3x^3(6x^2 + 5xz - 25z^2)$

Now factor the trinomial by trial and error.

$6x^2 + 5xz - 25z^2$

$= (2x + 5z)(3x - 5z)$

Thus,

$18x^5 + 15x^4z - 75x^3z^2$

$= 3x^3(6x^2 + 5xz - 25z^2)$

$= 3x^3(2x + 5z)(3x - 5z).$

33. $9m^2 - 12m + 4$

$= (3m)^2 - 12m + 2^2$

$= (3m)^2 - 2(3m)(2) + 2^2$
        Perfect square trinomial

$= (3m - 2)^2$

35. $32a^2 - 48ab + 18b^2$

$= 2(16a^2 - 24ab + 9b^2)$
        2 is greatest common factor

$= 2[(4a)^2 - 24ab + (3b)^2]$

$= 2[(4a)^2 - 2(4a)(3b) + (3b)^2]$
        Perfect square trinomial

$= 2(4a - 3b)^2$

37. $4x^2y^2 + 28xy + 49$

$= (2xy)^2 + 28xy + 7^2$

$= (2xy)^2 + 2(2xy)(7) + 7^2$
        Perfect square trinomial

$= (2xy + 7)^2$

39. $(a - 3b)^2 - 6(a - 3b) + 9$

Let $x = a - 3b$.
Then

$(a - 3b)^2 - 6(a - 3b) + 9$

$= x^2 - 6x + 9$

$= x^2 - 2(x)(3) + 3^2$
        Perfect square trinomial

$= (x - 3)^2.$

Replacing x with a − 3b gives

$(a - 3b)^2 - 6(a - 3b) + 9$

$= (a - 3b - 3)^2.$

41. $9a^2 - 16 = (3a)^2 - 4^2$
        Difference of two squares

$= (3a + 4)(3a - 4)$

43. $25s^4 - 9t^2 = (5s^2)^2 - (3t)^2$
        Difference of two squares

$= (5s^2 + 3t)(5s^2 - 3t)$

45. $(a + b)^2 - 16$

$= (a + b)^2 - 4^2$
        Difference of two squares

$= [(a + b) + 4][(a + b) - 4]$

$= (a + b + 4)(a + b - 4)$

47. $p^4 - 625$

$= (p^2)^2 - 25^2$
        Difference of two squares

$= (p^2 + 25)(p^2 - 25)$

$= (p^2 + 25)(p^2 - 5^2)$
        Difference of two squares

$= (p^2 + 25)(p + 5)(p - 5)$

49. The correct complete factorization of $x^4 - 1$ is (b): $(x^2 + 1)(x + 1)$ $\cdot (x - 1)$. Choice (a) is not a complete factorization, since $x^2 - 1$ can be factored as $(x + 1)(x - 1)$. The other choices are not correct factorizations of $x^4 - 1$.

51. $8 - a^3$

$= 2^3 - a^3$ *Difference of two cubes*

$= (2 - a)(a^2 + 2 \cdot a + a^2)$

$= (2 - a)(4 + 2a + a^2)$

53. $125x^3 - 27$

$= (5x)^3 - 3^3$ *Difference of two cubes*

$= (5x - 3)[(5x)^2 + 5x \cdot 3 + 3^2]$

$= (5x - 3)(25x^2 + 15x + 9)$

55. $27y^9 + 125z^6$

$= (3y^3)^3 + (5z^2)^3$ *Sum of two cubes*

$= (3y^3 + 5z^2)$

$\cdot [(3y^3)^2 - (3y^3)(5z^2) + (5z^2)^2]$

$= (3y^3 + 5z^2)(9y^6 - 15y^3z^2 + 25z^4)$

57. $(r + 6)^3 - 216$

Let $x = r + 6$. Then

$(r + 6)^3 - 216$

$= x^3 - 216$

$= x^3 - 6^3$ *Difference of two cubes*

$= (x - 6)(x^2 + 6x + 6^2)$

$= (x - 6)(x^2 + 6x + 36)$.

Replacing x with $(r + 6)$ gives

$(r + 6)^3 - 216$

$= ((r + 6) - 6)$

$\cdot [(r + 6)^2 + 6(r + 6) + 36)]$

$= r(r^2 + 12r + 36 + 6r + 36 + 36)$

$= r(r^2 + 18r + 108)$.

59. $27 - (m + 2n)^3$

Let $x = m + 2n$. Then

$27 - (m + 2n)^3$

$= 27 - x^3$

$= 3^3 - x^3$ *Difference of two cubes*

$= (3 - x)(3^2 + 3x + x^2)$

$= (3 - x)(9 + 3x + x^2)$.

Replacing x with $m + 2n$ gives

$27 - (m + 2n)^3$

$= [3 - (m + 2n)]$

$\cdot [9 + 3(m + 2n) + (m + 2n)^2]$

$= (3 - m - 2n)$

$\cdot (9 + 3m + 6n + m^2 + 4mn + 4n^2)$.

63. $a^4 - 2a^2 - 48$

Let $z = y^2$, so that $z^2 = (y^2)^2 = y^4$.

$z^2 - z - 48 = (z - 8)(z + 6)$

Replacing z with y^2 gives

$a^4 - 2a^2 - 48 = (a^2 - 8)(a^2 + 6)$.

65. $6(4z - 3)^2 + 7(4z - 3) - 3$

Let $a = 4z - 3$.
Substituting a for $4z - 3$, we have

$6a^2 + 7a - 3$.

Factor this trinomial as

$6a^2 + 7a - 3 = (2a + 3)(3a - 1)$.

Replacing a with $4z - 3$ gives

$6(4z - 3)^2 + 7(4z - 3) - 3$

$= [2(4z - 3) + 3][3(4z - 3) - 1]$

$= (8z - 6 + 3)(12z - 9 - 1)$

$= (8z - 3)(12z - 10)$

$= 2(8z - 3)(6z - 5)$.

67. $20(4 - p)^2 - 3(4 - p)^2$

Let $x = 4 - p$.
Substituting x for $4 - p$, we have

$20x^2 - 3x - 2$.

Factor this trinomial as

$20x^2 - 3x - 2 = (5x - 2)(4x + 1)$.

Replacing x with 4 − p gives

$20(4 - p)^2 - 3(4 - p) - 2$

$= [5(4 - p) - 2][4(4 - p) + 1]$

$= [20 - 5p - 2][16 - 4p + 1]$

$= (-5p + 18)(-4p + 17)$

$= (18 - 5p)(17 - 4p).$

69. $4b^2 + 4bc + c^2 - 16$

$= (4b^2 + 4bc + c^2) - 16$

$= (2b + c)^2 - 4^2$ *Difference of*
two squares

$= [(2b + c) + 4][(2b + c) - 4]$

$= (2b + c + 4)(2b + c - 4)$

71. $x^2 + xy - 5x - 5y$

$= (x^2 + xy) + (-5x - 5y)$
Group the terms

$= x(x + y) - 5(x + y)$
Factor each group

$= (x + y)(x - 5)$ *Factor out x + y*

73. $p^4(m - 2n) + q(m - 2n)$

$= (m - 2n)(p^4 + q)$ *Factor out*
m − 2n

75. $4z^2 + 28z + 49$

$= (2z)^2 + 2(2z)(7) + 7^2$
Perfect square trinomial

$= (2z + 7)^2$

77. $1000x^3 + 343y^3$

$= (10x)^3 + (7y)^3$ *Sum of two cubes*

$= (10x + 7y)$

$\quad \cdot [(10x^2)^2 - (10x)(7y) + (7y)^2]$

$= (10x + 7y)(100x^2 - 70xy + 49y^2)$

79. $125m^6 - 216$

$= (5m^2)^3 - 6^3$ *Difference of*
two cubes

$= (5m^2 - 6)[(5m^2)^2 + 5m^2 \cdot 6 + 6^2]$

$= (5m^2 - 6)(25m^4 + 30m^2 + 36)$

81. $12m^2 + 16mn - 35n^2$

Try different combinations of the
factors of 12 and −35 until the
correct one is found.

$12m^2 + 16mn - 35n^2$

$= (6m - 7n)(2m + 5n)$

83. $4p^2 + 3p - 1$

The positive factors of 4 could be 2
and 2 or 1 and 4. The factors of −1
can only be 1 and −1. Try different
combinations of these factors until
the correct one is found.

$4p^2 + 3p - 1 = (4p - 1)(p + 1)$

85. $144z^2 + 121$

The sum of two squares cannot be
factored. $144z^2 + 121$ is prime.

87. $(x + y)^2 - (x - y)^2$

Factor this expression as the
difference of two squares.

$= [(x + y) - (x - y)]$

$\quad \cdot [(x + y) + (x - y)]$

$= (x + y - x + y)(x + y + x - y)$

$= (2y)(2x)$

$= 4xy$

89. $r^2 + rs^q - 6s^{2q} = (r + 3s^q)(r - 2s^q)$

91. $9a^{4k} - b^{8k}$

  $= (3a^{2k})^2 - (b^{4k})^2$ *Difference of*
                                  two squares

  $= (3a^{2k} + b^{4k})(3a^{2k} - b^{4k})$

93. $4y^{2a} - 12y^a + 9$

  $= (2y^a)^2 - 12y^a + 3^2$

  $= (2y^a)^2 - 2(2y^a)(3) + 3^2$
           Perfect square trinomial

  $= (2y^a - 3)^2$

95. $6(m + p)^{2k} + (m + p)^k - 15$

  Let $u = (m + p)^k$, so that

  $u^2 = [(m + p)^k]^2 = (m + p)^{2k}.$

  With this substitution, the expression becomes

  $6u^2 + u - 15.$

  Factor this trinomial as

  $6u^2 + u - 15 = (3u + 5)(2u - 3).$

  Replacing u with $(m + p)^k$ gives

  $6(m + p)^{2k} + (m + p)^k - 15$

  $= [3(m + p)^k + 5][2(m + p)^k - 3].$

97. $4z^2 + bz + 81 = (2z)^2 + bz + 9^2$

  will be a perfect trinomial if

  $bz = \pm 2(2z)9$

  $bz = \pm 36z$

  $b = \pm 36.$

  If $b = 36$,

  $4z^2 + 36z + 81 = (2z + 9)^2.$

  If $b = -36$,

  $4z^2 - 36z + 81 = (2z - 9)^2.$

99. $100r^2 - 60r + c$

  The perfect square form is

  $(10x)^2 - \underbrace{2(10r)(3)}_{-60r} + 3^2.$

  Therefore, $c = 9$.

Section 1.6

1. Give restrictions on the variable in the expression

  $$\frac{x - 2}{x + 6}.$$

  Replacing x with −6 makes the denominator equal 0. Therefore, the restriction for this expression is $x \neq -6$.

3. Give restrictions on the variable in the expression

  $$\frac{2x}{5x - 3}.$$

  Replacing x with 3/5 makes the denominator equal 0. This can be found by solving the equation $5x - 3 = 0$. Therefore, for this expression the restriction is $x \neq \dfrac{3}{5}$.

5. Give restrictions on the variable in the expression

  $$\frac{-8}{x^2 + 1}.$$

  We must exclude any real number x which makes the denominator equal 0. However, the equation $x^2 + 1 = 0$

has no real solution. Therefore, there are no restrictions on the variable for this expression.

7. Write the given expression in lowest terms.

$$\frac{x^2 + 4x + 3}{x + 1}$$

$$= \frac{(x + 1)(x + 3)}{x + 1} \quad \text{Factor the numerator}$$

$$= x + 3 \quad \text{Fundmental principle}$$

Therefore, the expression which is equivalent to $\frac{x^2 + 4x + 3}{x + 1}$ is

(a): $x + 3$.

9. $\frac{25p^3}{10p^2} = \frac{5p \cdot 5p^2}{2 \cdot 5p^2} \quad \begin{array}{l}\text{Factor numerator} \\ \text{and denominator}\end{array}$

$$= \frac{5p}{2} \quad \begin{array}{l}\text{Fundamental} \\ \text{principle}\end{array}$$

11. $\frac{8k + 16}{9k + 18} = \frac{8(k + 2)}{9(k + 2)} \quad \begin{array}{l}\text{Factor numerator} \\ \text{and denominator}\end{array}$

$$= \frac{8}{9} \quad \begin{array}{l}\text{Fundamental} \\ \text{principle}\end{array}$$

13. $\frac{3(t + 5)}{(t + 5)(t - 3)} = \frac{3}{t - 3} \quad \begin{array}{l}\text{Fundamental} \\ \text{priniciple}\end{array}$

15. $\frac{8x^2 + 16x}{4x^2}$

$$= \frac{8x(x + 2)}{4x^2} \quad \text{Factor}$$

$$= \frac{2 \cdot 4x(x + 2)}{x \cdot 4x} \quad \text{Factor}$$

$$= \frac{2(x + 2)}{x} \quad \begin{array}{l}\text{Fundamental} \\ \text{principle}\end{array}$$

$$\text{or} \quad \frac{2x + 4}{x}$$

17. $\frac{m^2 - 4m + 4}{m^2 + m - 6}$

$$= \frac{(m - 2)(m - 2)}{(m - 2)(m + 3)} \quad \text{Factor}$$

$$= \frac{m - 2}{m + 3} \quad \begin{array}{l}\text{Use the fundamental} \\ \text{principle to write} \\ \text{the expression in} \\ \text{lowest terms}\end{array}$$

19. $\frac{8m^2 + 6m - 9}{16m^2 - 9}$

$$= \frac{(2m + 3)(4m - 3)}{(4m + 3)(4m - 3)} \quad \text{Factor}$$

$$= \frac{2m + 3}{4m + 3} \quad \begin{array}{l}\text{Use the fundamental} \\ \text{principle to write} \\ \text{the expression in} \\ \text{lowest terms}\end{array}$$

21. $\frac{15p^3}{9p^2} \div \frac{6p}{10p^2}$

$$= \frac{15p^3}{9p^2} \cdot \frac{10p^2}{6p} \quad \begin{array}{l}\text{Definition} \\ \text{of division}\end{array}$$

$$= \frac{150p^5}{54p^3} \quad \text{Multiply}$$

$$= \frac{25 \cdot 6p^5}{9 \cdot 6p^3} \quad \text{Factor}$$

$$= \frac{25p^2}{9} \quad \begin{array}{l}\text{Fundamental} \\ \text{principle}\end{array}$$

23. $\frac{2k + 8}{6} \div \frac{3k + 12}{2}$

$$= \frac{2k + 8}{6} \cdot \frac{2}{3k + 12} \quad \begin{array}{l}\text{Definition} \\ \text{of division}\end{array}$$

$$= \frac{2(k + 4)^2}{6(3)(k + 4)} \quad \begin{array}{l}\text{Multiply} \\ \text{and factor}\end{array}$$

$$= \frac{4}{18} = \frac{2}{9} \quad \begin{array}{l}\text{Fundamental} \\ \text{principle}\end{array}$$

25. $\frac{x^2 + x}{5} \cdot \frac{25}{xy + y}$

$$= \frac{x(x + 1)}{5} \cdot \frac{25}{y(x + 1)} \quad \text{Factor}$$

$$= \frac{25x(x + 1)}{5y(x + 1)} \quad \text{Multiply}$$

$$= \frac{5x}{y} \quad \begin{array}{l}\text{Fundamental} \\ \text{principle}\end{array}$$

27. $\dfrac{4a + 12}{2a - 10} \div \dfrac{a^2 - 9}{a^2 - a - 20}$

$= \dfrac{4a + 12}{2a - 10} \cdot \dfrac{a^2 - a - 20}{a^2 - 9}$

    Definition of division

$= \dfrac{4(a + 3)(a - 5)(a + 4)}{2(a - 5)(a + 3)(a - 3)}$

    Multiply and factor

$= \dfrac{2(a + 4)}{(a - 3)}$ *Fundamental principle*

29. $\dfrac{p^2 - p - 12}{p^2 - 2p - 15} \cdot \dfrac{p^2 - 9p + 20}{p^2 - 8p + 16}$

$= \dfrac{(p - 4)(p + 3)(p - 5)(p - 4)}{(p - 5)(p + 3)(p - 4)(p - 4)}$

    Multiply and factor

$= 1$ *Fundamental principle*

31. $\dfrac{m^2 + 3m + 2}{m^2 + 5m + 4} \div \dfrac{m^2 + 5m + 6}{m^2 + 10m + 24}$

$= \dfrac{m^2 + 3m + 2}{m^2 + 5m + 4} \cdot \dfrac{m^2 + 10m + 24}{m^2 + 5m + 6}$

    Definition of division

$= \dfrac{(m + 2)(m + 1)(m + 6)(m + 4)}{(m + 4)(m + 1)(m + 3)(m + 2)}$

    Multiply and factor

$= \dfrac{m + 6}{m + 3}$ *Fundamental principle*

33. $\dfrac{2m^2 - 5m - 12}{m^2 - 10m + 24} \div \dfrac{4m^2 - 9}{m^2 - 9m + 18}$

$= \dfrac{2m^2 - 5m - 12}{m^2 - 10m + 24} \cdot \dfrac{m^2 - 9m + 18}{4m^2 - 9}$

    Definition of division

$= \dfrac{(2m + 3)(m - 4)(m - 6)(m - 3)}{(m - 6)(m - 4)(2m + 3)(2m - 3)}$

    Multiply and factor

$= \dfrac{m - 3}{2m - 3}$ *Fundamental principle*

35. $\left(1 + \dfrac{1}{x}\right)\left(1 - \dfrac{1}{x}\right)$

$= 1^2 - \left(\dfrac{1}{x}\right)^2$ *Difference of two squares*

$= 1 - \dfrac{1}{x^2}$

$= \dfrac{x^2}{x^2} - \dfrac{1}{x^2}$ *Common denominator*

$= \dfrac{x^2 - 1}{x^2}$ *Write as a single fraction*

37. $\dfrac{x^3 + y^3}{x^2 - y^2} \cdot \dfrac{x + y}{x^2 - xy + y^2}$

$= \dfrac{(x + y)(x^2 - xy + y^2)(x + y)}{(x + y)(x - y)(x^2 - xy + y^2)}$

    Factor, using sum of two cubes and difference of two squares

$= \dfrac{x + y}{x - y}$ *Fundamental principle*

39. $\dfrac{x^3 + y^3}{x^3 - y^3} \cdot \dfrac{x^2 - y^2}{x^2 + 2xy + y^2}$

$= \dfrac{(x + y)(x^2 - xy + y^2)(x - y)(x + y)}{(x - y)(x^2 - xy + y^2)(x + y)(x + y)}$

    Factor, using sum of two cubes and difference of two squares

$= \dfrac{x^2 - xy + y^2}{x^2 + xy + y^2}$ *Fundamental principle*

41. Expression (b) and (c) are both equal to -1, since the numerator and denominator are additive inverses.

(b) $\dfrac{-x - 4}{x + 4} = \dfrac{-1(x + 4)}{x + 4} = -1$

(c) $\dfrac{x - 4}{4 - x} = \dfrac{-1(4 - x)}{4 - x} = -1$

43. $\dfrac{3}{2k} + \dfrac{5}{3k} = \dfrac{3 \cdot 3}{2k \cdot 3} + \dfrac{5 \cdot 2}{3k \cdot 2}$ *Fundamental principle*

$= \dfrac{9}{6k} + \dfrac{10}{6k}$ *Common denominator*

$= \dfrac{19}{6k}$ *Add numerators*

45. $\dfrac{a + 1}{2} - \dfrac{a - 1}{2}$

$= \dfrac{(a + 1) - (a - 1)}{2}$ *Subtract numerators*

$= \dfrac{a + 1 - a + 1}{2}$ *Remove parentheses*

$= \dfrac{2}{2} = 1$

47. $\dfrac{3}{p} + \dfrac{1}{2} = \dfrac{3 \cdot 2}{p \cdot 2} + \dfrac{1 \cdot p}{2 \cdot p}$ *Fundamental principle*

$= \dfrac{6}{2p} + \dfrac{p}{2p}$ *Common denominator*

$= \dfrac{6 + p}{2p}$ *Add numerators*

49. $\dfrac{1}{6m} + \dfrac{2}{5m} + \dfrac{4}{m}$

$= \dfrac{1 \cdot 5}{6m \cdot 5} + \dfrac{2 \cdot 6}{5m \cdot 6} + \dfrac{4 \cdot 6 \cdot 5}{m \cdot 6 \cdot 5}$ *Fundamental principle*

$= \dfrac{5}{30m} + \dfrac{12}{30m} + \dfrac{120}{30m}$ *Common denominator*

$= \dfrac{137}{30m}$ *Add numerators*

51. $\dfrac{1}{a + 1} - \dfrac{1}{a - 1}$

$= \dfrac{1(a - 1)}{(a + 1)(a - 1)} - \dfrac{1(a + 1)}{(a - 1)(a + 1)}$ *Fundamental principle*

$= \dfrac{a - 1}{(a - 1)(a + 1)} - \dfrac{a + 1}{(a + 1)(a - 1)}$ *Common denominator*

$= \dfrac{(a - 1) - (a + 1)}{(a - 1)(a + 1)}$ *Subtract numerators*

$= \dfrac{a - 1 - a - 1}{(a - 1)(a + 1)}$ *Remove parentheses*

$= \dfrac{-2}{(a - 1)(a + 1)}$

53. $\dfrac{m + 1}{m - 1} + \dfrac{m - 1}{m + 1}$

$= \dfrac{(m + 1)(m + 1)}{(m - 1)(m + 1)} + \dfrac{(m - 1)(m - 1)}{(m + 1)(m - 1)}$

$= \dfrac{m^2 + 2m + 1}{(m + 1)(m - 1)} + \dfrac{m^2 - 2m + 1}{(m - 1)(m + 1)}$

$= \dfrac{m^2 + 2m + 1 + m^2 - 2m + 1}{(m + 1)(m - 1)}$

$= \dfrac{2m^2 + 2}{(m + 1)(m - 1)}$

55. $\dfrac{3}{a - 2} - \dfrac{1}{2 - a}$

$= \dfrac{3}{a - 2} - \dfrac{1(-1)}{(2 - a)(-1)}$

$\qquad\qquad a - 2 = (-1)(2 - a)$

$= \dfrac{3}{a - 2} - \dfrac{-1}{a - 2}$

$= \dfrac{3 + 1}{a - 2} = \dfrac{4}{a - 2}$

We may also use $2 - a$ as the common denominator.

$\dfrac{3(-1)}{(a - 2)(-1)} - \dfrac{1}{2 - a}$

$= \dfrac{-3}{2 - a} - \dfrac{1}{2 - a}$

$= \dfrac{-4}{2 - a}$

The two results, $\dfrac{4}{a - 2}$ and $\dfrac{-4}{2 - a}$, are equivalent rational expressions.

57. $\dfrac{x + y}{2x - y} - \dfrac{2x}{y - 2x}$

$= \dfrac{x + y}{2x - y} - \dfrac{2x(-1)}{(y - 2x)(-1)}$

$\qquad\qquad 2x - y = (-1)(y - 2x)$

$= \dfrac{x + y}{2x - y} - \dfrac{-2x}{2x - y}$

$= \dfrac{x + y + 2x}{2x - y} = \dfrac{3x + y}{2x - y}$

We may also use $y - 2x$ as the common denominator. In this case, our result will be

$$\dfrac{-3x - y}{y - 2x}.$$

The two results are equivalent rational expressions.

59. $\dfrac{1}{a^2 - 5a + 6} - \dfrac{1}{a^2 - 4}$

$= \dfrac{1}{(a - 3)(a - 2)} - \dfrac{1}{(a + 2)(a - 2)}$

$\qquad\qquad$ *Factor denominators*

The least common denominator is $(a - 3)(a - 2)(a + 2)$.

$= \dfrac{1(a + 2)}{(a - 3)(a - 2)(a + 2)}$

$\quad - \dfrac{1(a - 3)}{(a + 2)(a - 2)(a - 3)}$

$\qquad\qquad$ *Fundamental principle*

$= \dfrac{(a + 2) - (a - 3)}{(a - 3)(a - 2)(a + 2)}$

$\qquad\qquad$ *Subtract numerators*

$= \dfrac{a + 2 - a + 3}{(a - 3)(a - 2)(a + 2)}$

$\qquad\qquad$ *Remove parentheses*

$= \dfrac{5}{(a - 3)(a - 2)(a + 2)}$

61. $\dfrac{1}{x^2 + x - 12} - \dfrac{1}{x^2 - 7x + 12} + \dfrac{1}{x^2 - 16}$

$= \dfrac{1}{(x + 4)(x - 3)} - \dfrac{1}{(x - 4)(x - 3)}$

$\quad + \dfrac{1}{(x - 4)(x + 4)}$ $\quad$ *Factor denominators*

The least common denominator is $(x + 4)(x - 3)(x - 4)$.

$\dfrac{1(x - 4)}{(x + 4)(x - 3)(x - 4)} - \dfrac{1(x + 4)}{(x - 4)(x - 3)(x + 4)}$

$+ \dfrac{1(x - 3)}{(x - 4)(x + 4)(x - 3)}$ $\quad$ *Fundamental principle*

$= \dfrac{(x - 4) - (x + 4) + (x - 3)}{(x - 4)(x + 4)(x - 3)}$

$\qquad\qquad$ *Subtract and add numerators*

$= \dfrac{x - 4 - x - 4 + x - 3}{(x - 4)(x + 4)(x - 3)}$ $\quad$ *Remove parentheses*

$= \dfrac{x - 11}{(x - 4)(x + 4)(x - 3)}$

63. $\dfrac{3a}{a^2 + 5a - 6} - \dfrac{2a}{a^2 + 7a + 6}$

$= \dfrac{3a}{(a + 6)(a - 1)} - \dfrac{2a}{(a + 6)(a + 1)}$

$\qquad\qquad$ *Factor denominator*

The least common denominator is $(a - 1)(a + 6)(a + 1)$.

$\dfrac{3a(a + 1)}{(a + 6)(a - 1)(a + 1)} - \dfrac{2a(a - 1)}{(a + 6)(a + 1)(a - 1)}$

$\qquad\qquad$ *Fundamental principle*

$= \dfrac{3a(a + 1) - 2a(a - 1)}{(a + 1)(a - 1)(a + 6)}$ $\quad$ *Subtract numerators*

$= \dfrac{3a^2 + 3a - 2a^2 + 2a}{(a + 1)(a - 1)(a + 6)}$ $\quad$ *Distributive property*

$= \dfrac{a^2 + 5a}{(a + 1)(a - 1)(a + 6)}$ $\quad$ *Combine like terms*

65. $\dfrac{1 + \dfrac{1}{x}}{1 - \dfrac{1}{x}}$

Multiply both numerator and denominator by the least common denominator of all the fractions, x.

$$\frac{1 + \frac{1}{x}}{1 - \frac{1}{x}} = \frac{x\left(1 + \frac{1}{x}\right)}{x\left(1 - \frac{1}{x}\right)}$$

$$= \frac{x \cdot 1 + x\left(\frac{1}{x}\right)}{x \cdot 1 - x\left(\frac{1}{x}\right)}$$

Distributive property

$$= \frac{x + 1}{x - 1}$$

67. $\dfrac{\dfrac{1}{x + 1} - \dfrac{1}{x}}{\dfrac{1}{x}}$

Multiply both numerator and denominator by the least common denominator of all the fractions, $x(x + 1)$.

$$\frac{\dfrac{1}{x + 1} - \dfrac{1}{x}}{\dfrac{1}{x}}$$

$$= \frac{x(x + 1)\left(\dfrac{1}{x + 1} - \dfrac{1}{x}\right)}{x(x + 1)\left(\dfrac{1}{x}\right)}$$

$$= \frac{x(x + 1)\left(\dfrac{1}{x + 1}\right) - x(x + 1)\left(\dfrac{1}{x}\right)}{x(x + 1)\left(\dfrac{1}{x}\right)}$$

Distributive property

$$= \frac{x - (x + 1)}{x + 1}$$

$$= \frac{x - x - 1}{x + 1}$$

$$= \frac{-1}{x + 1}$$

69. $\dfrac{1 + \dfrac{1}{1 - b}}{1 - \dfrac{1}{1 + b}}$

Multiply both numerator and denominator by the least common denominator of all the fractions, $(1 - b)(1 + b)$.

$$\frac{1 + \dfrac{1}{1 - b}}{1 - \dfrac{1}{1 + b}}$$

$$= \frac{(1 - b)(1 + b)\left(1 + \dfrac{1}{1 - b}\right)}{(1 - b)(1 + b)\left(1 - \dfrac{1}{1 + b}\right)}$$

$$= \frac{(1 - b)(1 + b) + (1 + b)}{(1 - b)(1 + b) - (1 - b)}$$

$$= \frac{(1 + b)[(1 - b) + 1]}{(1 - b)[(1 + b) - 1]}$$

Factor out common factors in numerator and denominator

$$= \frac{(1 + b)(2 - b)}{(1 - b)b} \quad \text{or} \quad \frac{(2 - b)(1 + b)}{b(1 - b)}$$

Remove parentheses and simplify

71. $\dfrac{m - \dfrac{1}{m^2 - 4}}{\dfrac{1}{m + 2}} = \dfrac{m - \dfrac{1}{(m + 2)(m - 2)}}{\dfrac{1}{m + 2}}$

Multiply both numerator and denominator by the least common denominator of all the fractions,

$(m + 2)(m - 2)$.

$$\frac{m - \dfrac{1}{m^2 - 4}}{\dfrac{1}{m + 2}}$$

$$= \frac{(m + 1)(m - 2)\left(m - \dfrac{1}{(m + 2)(m - 2)}\right)}{(m + 1)(m - 2)\left(\dfrac{1}{m + 2}\right)}$$

$$= \frac{(m+1)(m-2)(m) - (m+1)(m-2)\left(\dfrac{1}{(m+2)(m-2)}\right)}{(m+2)(m-2)\left(\dfrac{1}{m+2}\right)}$$

Distributive property

$$= \frac{m(m^2 - 4) - 1}{m - 2}$$

$$= \frac{m^3 - 4m - 1}{m - 2}$$

73. $\left(\dfrac{3}{p-1} - \dfrac{2}{p+1}\right)\left(\dfrac{p-1}{p}\right)$

$= \left[\dfrac{3(p+1)}{(p-1)(p+1)} - \dfrac{2(p-1)}{(p+1)(p-1)}\right]$

$\qquad \cdot \left(\dfrac{p-1}{p}\right)$

$= \left[\dfrac{3(p+1) - 2(p-1)}{(p-1)(p+1)}\right]\left(\dfrac{p-1}{p}\right)$

$= \dfrac{(3p+3-2p+2)}{(p+1)(p-1)}\left(\dfrac{p-1}{p}\right)$

$= \dfrac{(p-5)}{(p+1)(p-1)} \cdot \dfrac{(p-1)}{p}$

$= \dfrac{p+5}{(p+1)p} \quad \text{or} \quad \dfrac{p+5}{p(p+1)}$

75. $\dfrac{\dfrac{1}{x+h} - \dfrac{1}{x}}{h}$

To simplify this complex fraction, multiply both numerator and denominator by the least common denominator of all the fractions, $x(x+h)$.

$\dfrac{\dfrac{1}{x+h} - \dfrac{1}{x}}{h}$

$= \dfrac{x(x+h)\left(\dfrac{1}{x+h} - \dfrac{1}{x}\right)}{x(x+h)(h)}$

$= \dfrac{x(x+h)\left(\dfrac{1}{x+h}\right) - x(x+h)\left(\dfrac{1}{x}\right)}{x(x+h)(h)}$

$\qquad\qquad$ *Distributive property*

$= \dfrac{x - (x+h)}{xh(x+h)}$

$= \dfrac{-h}{xh(x+h)}$

$= \dfrac{-1}{x(x+h)} \quad$ *Fundamental principle*

Section 1.7

1. $(-4)^{-3} = \dfrac{1}{(-4)^3}$

$\qquad\quad = \dfrac{1}{(-4)(-4)(-4)}$

$\qquad\quad = \dfrac{1}{-64}$

$\qquad\quad = -\dfrac{1}{64}$

3. $\left(\dfrac{1}{2}\right)^{-3} = \dfrac{1}{\left(\dfrac{1}{2}\right)^3}$

$\qquad\quad = \dfrac{1}{\dfrac{1}{8}}$

$\qquad\quad = \dfrac{8}{1} = 8$

5. $-4^{1/2}$

$4^{1/2} = 2$ because $2^2 = 4$.
Thus,

$$-4^{1/2} = -2.$$

7. $8^{2/3} = (8^{1/3})^2 = 2^2 = 4$

This expression can also be evaluated as

$$8^{2/3} = (8^2)^{1/3} = 64^{1/3} = 4.$$

9. $27^{-2/3} = \dfrac{1}{(27)^{2/3}}$

$\qquad\qquad = \dfrac{1}{(27^{1/3})^2}$

$\qquad\qquad = \dfrac{1}{3^2} = \dfrac{1}{9}$

11. $\left(-\dfrac{4}{9}\right)^{-3/2}$ is not defined because the base, $-4/9$, is negative and the exponent, $-3/2$, has an even denominator.

13. $\left(\frac{27}{64}\right)^{-4/3} = \left[\left(\frac{27}{64}\right)^{1/3}\right]^{-4}$

$\qquad = \left(\frac{27^{1/3}}{64^{1/3}}\right)^{-4}$

$\qquad = \left(\frac{3}{4}\right)^{-4} = \frac{1}{\left(\frac{3}{4}\right)^4}$

$\qquad = \frac{1}{\frac{81}{256}} = \frac{256}{81}$

15. $7^{-3.1} = \frac{1}{7^{3.1}} \approx .0024$

17. $(16p^4)^{1/2} = 16^{1/2}(p^4)^{1/2} = 4p^2$

19. $(27x^6)^{2/3} = 27^{2/3}(x^6)^{2/3}$

$\qquad = (27^{1/3})^2 x^4$

$\qquad = 3^2 x^4 = 9x^4$

23. $(2x^{-3/2})^2 = 2^2(x^{-3/2})^2$

$\qquad = 2^2 \cdot x^{-3}$

$\qquad = \frac{2^2}{x^3}$

Therefore, expression (d) is equivalent to $(2x^{-3/2})^2$.

25. $2^{-3} \cdot 2^{-4} = 2^{-3+(-4)}$ *Product rule*

$\qquad = 2^{-7}$

$\qquad = \frac{1}{2^7}$ *Definition of negative exponent*

27. $27^{-2} \cdot 27^{-1} = 27^{-3}$ *Product rule*

$\qquad = \frac{1}{27^3}$ *Definition of negative exponent*

29. $\frac{4^{-2} \cdot 4^{-1}}{4^{-3}} = \frac{4^{-3}}{4^{-3}} = 1$

31. $(m^{2/3})(m^{5/3})$

$\qquad = m^{2/3 + 5/3}$ *Product rule*

$\qquad = m^{7/3}$

33. $(1 + n)^{1/2}(1 + n)^{3/4}$

$\qquad = (1 + n)^{1/2 + 3/4}$

$\qquad = (1 + n)^{5/4}$ *Add exponents*

35. $(2y^{3/4} z)(3y^{-2} z^{-1/3})$

$\qquad = 6y^{3/4 + (-2)} z^{1 + (-1/3)}$

$\qquad = 6y^{3/4 - 8/4} z^{3/3 - 1/3}$

$\qquad = 6y^{-5/4} z^{2/3}$

$\qquad = \frac{6z^{2/3}}{y^{5/4}}$

37. $(4a^{-2}b^7)^{1/2} \cdot (2a^{1/4} b^3)^5$

$\qquad = (4^{1/2} s^{-1} b^{7/2})(2^5 a^{5/4} b^{15})$

$\qquad = 2 \cdot 2^5 \cdot a^{-1} \cdot a^{5/4} \cdot b^{7/2} \cdot b^{15}$

$\qquad = 2^6 a^{-4/4 + 5/4} b^{7/2 + 30/2}$

$\qquad = 2^6 a^{1/4} b^{37/2}$

39. $\left(\frac{r^{-2}}{s^{-5}}\right)^{-3} = \frac{(r^{-2})^{-3}}{(s^{-5})^{-3}}$

$\qquad = \frac{r^{(-2)(-3)}}{s^{(-5)(-3)}} = \frac{r^6}{s^{15}}$

41. $\left(\frac{-a}{b^{-3}}\right)^{-1} = \frac{(-a)^{-1}}{(b^{-3})^{-1}}$

$\qquad = \frac{1}{-ab^3}$

$\qquad = -\frac{1}{ab^3}$

43. $\frac{12^{5/4} y^{-2}}{12^{-1} y^{-3}} = 12^{5/4 - (-1/4)} y^{-2 - (-3)}$

$\qquad\qquad\qquad\qquad$ *Quotient rule*

$\qquad = 12^{9/4} y$

45. $\dfrac{8p^{-3}(4p^2)^{-2}}{p^{-5}} = \dfrac{8p^{-3} \cdot 4^{-2}p^{-4}}{p^{-5}}$

$\qquad\qquad\qquad = \dfrac{8 \cdot 4^{-2} \cdot p^{-7}}{p^{-5}}$

$\qquad\qquad\qquad = \dfrac{8}{4^2 p^{-5}p^7}$

$\qquad\qquad\qquad = \dfrac{8}{16p^2}$

$\qquad\qquad\qquad = \dfrac{1}{2p^2}$

47. $\dfrac{m^{7/3}\,n^{-2/5}\,p^{3/8}}{m^{-2/3}\,n^{3/5}\,p^{-5/8}}$

$\qquad = \dfrac{m^{7/3}\,m^{2/3}\,p^{3/8}\,p^{5/8}}{n^{3/5}\,n^{2/5}}$

$\qquad = \dfrac{m^{9/3}\,p^{8/8}}{n^{5/5}} = \dfrac{m^3 p}{n}$

49. $\dfrac{-4a^{-1}a^{2/3}}{a^{-2}} = \dfrac{-4a^{-3/3}\,a^{-2/3}}{a^{-2}}$

$\qquad\qquad = \dfrac{-4a^{-1/3}}{a^{-2}}$

$\qquad\qquad = \dfrac{-4a^{-1/3}\,a^2}{1}$

$\qquad\qquad = -4a^{5/3}$

51. $\dfrac{(k+5)^{1/2}\,(k+5)^{-1/4}}{(k+5)^{3/4}}$

$\qquad = (k+5)^{1/2\,-1/4\,-3/4}$

$\qquad = (k+5)^{-1/2}$

$\qquad = \dfrac{1}{(k+5)^{1/2}}$

53. $\left(\dfrac{x^4 y^3 z}{16x^{-6}yz^5}\right)^{-1/2}$

$\qquad = \dfrac{x^{-2}y^{-3/2}z^{-1/2}}{16^{-1/2}x^3 y^{-1/2}z^{-5/2}}$

$\qquad = 16^{1/2}\,x^{-2-3}y^{-3/2+1/2}z^{-1/2+5/2}$

$\qquad = 4x^{-5}y^{-1}z^2$

$\qquad = \dfrac{4z^2}{x^5 y}$

55. $(r^{3/p})^{2p}(r^{1/p})p^2$

$\qquad = r^{(3/p)(2p)}\,r^{(1/p)(p^2)}$
$\qquad\qquad\qquad$ *Power rule*

$\qquad = r^6 r^p \quad$ *Product rule*

$\qquad = r^{6+p}$

57. $\dfrac{m^{1-a}m^a}{m^{-1/2}}$

$\qquad = \dfrac{m^{1-a+a}}{m^{-1/2}} \qquad$ *Product rule*

$\qquad = \dfrac{m^1}{m^{-1/2}}$

$\qquad = m^{1-(-1/2)} \qquad$ *Quotient rule*

$\qquad = m^{3/2}$

59. $\dfrac{(x^{n/2})(x^{3n})^{1/2}}{x^{1/n}}$

$\qquad = \dfrac{x^{n/2}\,x^{3n/2}}{x^{1/n}}$

$\qquad = x^{n/2\,+3n/2\,-1/n} \quad$ *Product and*
$\qquad\qquad\qquad\qquad\qquad$ *quotient rules*

$\qquad = x^{2n-1/n}$

The exponent $2n - \dfrac{1}{n}$ can be written

as a single fraction:

$$2n - \dfrac{1}{n} = \dfrac{n}{n} \cdot \dfrac{2n}{1} - \dfrac{1}{n}$$

$$= \dfrac{2n^2 - 1}{n}.$$

Therefore, the result can be written

$\qquad x^{(2n^2-1)/n}$.

61. $\dfrac{(p^{1/n})(p^{1/m})}{p^{-m/n}}$

$\qquad = p^{1/n\,+1/m\,-(-m/n)}$

The exponent in the above expression

can be simplified as follows:

$$\dfrac{1}{n} + \dfrac{1}{m} + \dfrac{m}{n} - \dfrac{m+n+m^2}{mn}.$$

Therefore, the result is

$\qquad p^{(m+n+m^2)/(mn)}$.

63. $y^{5/8}(y^{3/8} - 10y^{11/8})$

$= y^{5/8}y^{3/8} - 10y^{5/8}y^{11/8}$

$= y^{5/8 + 3/8} - 10y^{5/8 + 11/8}$

$= y - 10y^2$

65. $-4k(k^{7/3} - 6k^{1/3})$

$= -4k^1k^{7/3} + 24k^1k^{1/3}$

$= -4k^{10/3} + 24k^{4/3}$

67. $(x + x^{1/2})(x - x^{1/2})$

$= x^2 - (x^{1/2})^2$ *Difference of two squares*

$= x^2 - x$

69. $(r^{1/2} - r^{-1/2})^2$

$= (r^{1/2})^2 - 2(r^{1/2})(r^{-1/2})$

$\quad + (r^{-1/2})^2$ *Square of a binomial*

$= r - 2r^0 + r^{-1}$

$= r - 2 + r^{-1}$ or $r - 2 + \dfrac{1}{r}$

71. Factor $4k^{-1} + k^{-2}$, using the common factor k^{-2}.

$4k^{-1} + k^{-2} = k^{-2}(4k + 1)$

$\quad\quad\quad\quad\text{or}\quad \dfrac{4k + 1}{k^2}$

73. Factor $9z^{-1/2} + 2z^{1/2}$, using the common factor $z^{-1/2}$.

$9z^{-1/2} + 2z^{1/2} = z^{-1/2}(9 + z)$

$\quad\quad\quad\quad\text{or}\quad \dfrac{9 + z}{z^{1/2}}$

75. Factor $p^{-3/4} - 2p^{-7/4}$, using the common factor $p^{-7/4}$.

$p^{-3/4} - 2p^{-7/4} = p^{-7/4}(p - 2)$

$\quad\quad\quad\quad\text{or}\quad \dfrac{p - 2}{p^{7/4}}$

77. Factor $(p + 4)^{-3/2} + (p + 4)^{-1/2} + (p + 4)^{1/2}$, using the common factor $(p + 4)^{-3/2}$.

$= (p + 4)^{-3/2}$

$\quad \cdot [1 + (p + 4) + (p + 4)^2]$

$= (p + 4)^{-3/2}$

$\quad \cdot (1 + p + 4 + p^2 + 8p + 16)$

$= (p + 4)^{-3/2}(p^2 + 9p + 21)$

or $\dfrac{p^2 + 9p + 21}{(p + 4)^{3/2}}$

79. $p = 2x^{1/2} + 3x^{2/3}$

If $x = 64$,

$p = 2(64)^{1/2} + 3(64)^{2/3}$

$= 2(8) + 3(64^{1/3})^2$

$= 16 + 3(4)^2$

$= 16 + 48 = 64.$

The price is \$64 when the supply is 64 units.

81. $E_{large} = 48$; $E_{small} = 3$

Amount for small state = \$1,000,000

Amount for large state

$= \left(\dfrac{E_{large}}{E_{small}}\right)^{3/2} \times \begin{array}{l}\text{amount for}\\\text{small state}\end{array}$

$= \left(\dfrac{48}{3}\right)^{3/2} \times 1{,}000{,}000$

$= (16)^{3/2} \times 1{,}000{,}000$

$= (16^{1/2})^3 \times 1{,}000{,}000$

$= 4^3 \times 1{,}000{,}000$

$= 64 \times 1{,}000{,}000$

$= 64{,}000{,}000$

If \$1,000,000 is spent in the small state, \$64,000,000 should be spent in the large state.

83. $E_{large} = 28$; $E_{small} = 6$

Amount for large state

$= \left(\dfrac{28}{6}\right)^{3/2} \times 1{,}000{,}000$

$\approx 10 \times 1{,}000{,}000$

About \$10,000,000 should be spent in the large state.

85. $S = 28.6\,A^{.32}$

If $A = 1$,

$$S = (28.6)(1^{.32})$$
$$= 28.6$$
$$\approx 29.$$

If the area of an island is 1 sq mi, the number of land–plant species is approximately 29.

87. $S = 28.6\,A^{.32}$

If $A = 300$,

$$S = 28.6(300)^{.32}$$
$$\approx 28.6(6.204)$$
$$\approx 177.$$

If the area of an island is 300 sq mi, the number of land–plant species is approximately 177.

89. $\dfrac{a^{-1} + b^{-1}}{(ab)^{-1}}$

$= \dfrac{\dfrac{1}{a} + \dfrac{1}{b}}{\dfrac{1}{ab}}$ *Definition of negative integer exponent*

$= \dfrac{\dfrac{1 \cdot b}{a \cdot b} + \dfrac{1 \cdot a}{b \cdot a}}{\dfrac{1}{ab}}$

$= \dfrac{\dfrac{b + a}{ab}}{\dfrac{1}{ab}}$

$= \dfrac{b + a}{ab} \cdot \dfrac{ab}{1}$ *Definition of division*

$= b + a$

91. $\dfrac{r^{-1} + q^{-1}}{r^{-1} - q^{-1}} \cdot \dfrac{r - q}{r + q}$

$= \dfrac{\dfrac{1}{r} + \dfrac{1}{q}}{\dfrac{1}{r} - \dfrac{1}{q}} \cdot \dfrac{r - q}{r + q}$

$= \dfrac{rq\left(\dfrac{1}{r} + \dfrac{1}{q}\right)}{rq\left(\dfrac{1}{r} - \dfrac{1}{q}\right)} \cdot \dfrac{r - q}{r + q}$

Multiply numerator and denominator of first fraction by common denominator, rq

$= \dfrac{q + r}{q - r} \cdot \dfrac{r - q}{r + q}$

$= \dfrac{r - q}{q - r} = \dfrac{-1(r - q)}{-1(q - r)}$

$= \dfrac{-1(r - q)}{r - q} = -1$

93. $\dfrac{x - 9y^{-1}}{(x - 3y^{-1})(x + 3y^{-1})}$

$= \dfrac{x - \dfrac{9}{y}}{\left(x - \dfrac{3}{y}\right)\left(x + \dfrac{3}{y}\right)}$ *Definition of negative integer exponent*

$= \dfrac{x - \dfrac{9}{y}}{x^2 - \dfrac{9}{y^2}}$ *Multiply in denominator*

$= \dfrac{y^2\left(x - \dfrac{9}{y}\right)}{y^2\left(x^2 - \dfrac{9}{y^2}\right)}$

Multiply numerator and denominator by least common denominator, y

$= \dfrac{y^2 x - 9y}{y^2 x^2 - 9}$ *Distributive property*

or $\dfrac{y(xy - 9)}{x^2 y^2 - 9}$ *Factor numerator*

Section 1.8

1. $(-m)^{2/3} = \sqrt[3]{(-m)^2}$ or $(\sqrt[3]{-m})^2$

3. $5m^{4/5} = 5\sqrt[5]{m^4}$ or $5(\sqrt[5]{m})^4$

5. $-4z^{-1/3} = -\dfrac{4}{\sqrt[3]{z}}$

7. $(2m + p)^{2/3}$

    $= \sqrt[3]{(2m + p)^2}$ or $(\sqrt[3]{2m + p})^2$

9. $\sqrt[5]{k^2} = k^{2/5}$

11. $-\sqrt[3]{a^2} = -a^{2/3}$

13. $-3\sqrt{5p^3} = -3(5p^3)^{1/2}$

    $= -3 \cdot 5^{1/2} p^{3/2}$

15. $18\sqrt{m^2n^3p} = 18m^{2/2} n^{3/2} p^{1/2}$

    $= 18mn^{3/2} p^{1/2}$

21. $\sqrt[3]{125} = 5$

23. $\sqrt[5]{-3125} = -5$

25. $\sqrt{50} = \sqrt{25 \cdot 2} = \sqrt{25} \cdot \sqrt{2} = 5\sqrt{2}$

27. $\sqrt[3]{81} = \sqrt[3]{27 \cdot 3} = \sqrt[3]{27} \cdot \sqrt[3]{3} = 3\sqrt[3]{3}$

29. $-\sqrt[4]{32} = -\sqrt[4]{16 \cdot 2} = -\sqrt[4]{16} \cdot \sqrt[4]{2}$

    $= -2\sqrt[4]{2}$

31. $-\sqrt{\dfrac{9}{5}} = \dfrac{-3}{\sqrt{5}} \cdot \dfrac{\sqrt{5}}{\sqrt{5}} = \dfrac{-3\sqrt{5}}{5}$

33. $-\sqrt[3]{\dfrac{4}{5}} = -\dfrac{\sqrt[3]{4}}{\sqrt[3]{5}} \cdot \dfrac{\sqrt[3]{5^2}}{\sqrt[3]{5^2}}$

    $= -\dfrac{\sqrt[3]{4} \cdot \sqrt[3]{5}}{\sqrt[3]{5^3}}$

    $= -\dfrac{\sqrt[3]{4 \cdot 25}}{5}$

    $= -\dfrac{\sqrt[3]{100}}{5}$

35. $\sqrt[3]{16(-2)^4(2)^8} = \sqrt[3]{2^4 \cdot (-2)^4 2^8}$

    $= \sqrt[3]{2^4 \cdot 2^4 \cdot 2^8}$

    $= \sqrt[3]{2^{16}}$

    $= \sqrt[3]{2^{15} \cdot 2}$

    $= \sqrt[3]{2^{15}} \cdot \sqrt[3]{2}$

    $= 2^5 \cdot \sqrt[3]{2}$

    $= 32\sqrt[3]{2}$

37. $\sqrt{8x^5z^8} = \sqrt{2 \cdot 4 \cdot x^4 \cdot x \cdot z^8}$

    $= \sqrt{4x^4z^8} \cdot \sqrt{2x}$

    $= \sqrt{2x} \cdot 2x^2z^4$

    $= 2x^2z^4\sqrt{2x}$

39. $\sqrt[3]{16z^5x^8y^4} = \sqrt[3]{8 \cdot 2 \cdot z^3z^2x^6x^2y^3y}$

    $= \sqrt[3]{(8z^3x^6y^3)(2z^2x^2y)}$
    Group all perfect cubes

    $= \sqrt[3]{8z^3x^6y^3} \cdot \sqrt[3]{2z^2x^2y}$

    $= 2zx^2y\sqrt[3]{2z^2x^2y}$

41. $\sqrt[4]{m^2n^7p^8} = \sqrt[4]{m^2n^4n^3p^8}$

    $= \sqrt[4]{n^4p^8} \cdot \sqrt{m^2n^3}$

    $= np^2\sqrt[4]{m^2n^3}$

43. $\sqrt[4]{x^4 + y^4}$ cannot be simplified further.

45. $\sqrt{\dfrac{2}{3x}} = \dfrac{\sqrt{2}}{\sqrt{3x}}$

$= \dfrac{\sqrt{2}}{\sqrt{3x}} \cdot \dfrac{\sqrt{3x}}{\sqrt{3x}} = \dfrac{\sqrt{6x}}{3x}$

47. $\sqrt{\dfrac{x^5 y^3}{z^2}} = \dfrac{\sqrt{x^5 y^3}}{\sqrt{z^2}} = \dfrac{\sqrt{x^4 xy^2 y}}{z}$

$= \dfrac{\sqrt{x^4 y^2} \cdot \sqrt{xy}}{z} = \dfrac{x^2 y \sqrt{xy}}{z}$

49. $\sqrt[3]{\dfrac{8}{x^2}} = \dfrac{\sqrt[3]{8}}{\sqrt[3]{x^2}}$

$= \dfrac{2}{\sqrt[3]{x^2}} \cdot \dfrac{\sqrt[3]{x}}{\sqrt[3]{x}}$

$= \dfrac{2\sqrt[3]{x}}{x}$

51. $\sqrt[4]{\dfrac{g^3 h^5}{9r^6}} = \dfrac{\sqrt[4]{g^3 h^5}}{\sqrt[4]{9r^6}}$

$= \dfrac{h\sqrt[4]{g^3 h}}{\sqrt[4]{9r^6}} \cdot \dfrac{\sqrt[4]{9r^2}}{\sqrt[4]{9r^2}}$

$= \dfrac{h\sqrt[4]{9g^3 hr^2}}{\sqrt[4]{81r^8}}$

$= \dfrac{h\sqrt[4]{9g^3 hr^2}}{3r^2}$

53. $\dfrac{\sqrt[3]{mn} \cdot \sqrt[3]{m^2}}{\sqrt[3]{n^2}} = \sqrt[3]{\dfrac{mnm^2}{n^2}} = \sqrt[3]{\dfrac{m^3}{n}}$

$= \dfrac{\sqrt[3]{m^3}}{\sqrt[3]{n}} \cdot \dfrac{\sqrt[3]{n^2}}{\sqrt[3]{n^2}} = \dfrac{m\sqrt[3]{n^2}}{n}$

55. $\dfrac{\sqrt[4]{32x^5 y} \cdot \sqrt[4]{2xy^4}}{\sqrt[4]{4x^3 y^2}} = \sqrt[4]{\dfrac{64x^6 y^5}{4x^3 y^2}}$

$= \sqrt[4]{16x^3 y^3}$

$= 2\sqrt[4]{x^3 y^3}$

57. $\sqrt[3]{\sqrt{4}} = \sqrt[3]{4^{1/2}} = (4^{1/2})^{1/3} = 4^{1/6}$

$= (2^2)^{1/6} = 2^{2/6} = 2^{1/3} = \sqrt[3]{2}$

59. $\sqrt[6]{\sqrt[3]{x}} = \sqrt[6]{x^{1/3}} = (x^{1/3})^{1/6}$

$= x^{1/18} = \sqrt[18]{x}$

61. $4\sqrt{3} - 5\sqrt{12} + 3\sqrt{75}$

$= 4\sqrt{3} - 5\sqrt{4 \cdot 3} + 3\sqrt{25 \cdot 3}$

$= 4\sqrt{3} - 5 \cdot 2\sqrt{3} + 3 \cdot 5\sqrt{3}$

$= 4\sqrt{3} - 10\sqrt{3} + 15\sqrt{3}$

$= 9\sqrt{3}$ *Combine like radicals*

63. $3\sqrt{28p} - 4\sqrt{63p} + \sqrt{112p}$

$= 3\sqrt{4 \cdot 7p} - 4\sqrt{9 \cdot 7p} + \sqrt{16 \cdot 7p}$

$= 3(2\sqrt{7p}) - 4(3\sqrt{7p}) + 4\sqrt{7p}$

$= 6\sqrt{7p} - 12\sqrt{7p} + 4\sqrt{7p}$

$= -2\sqrt{7p}$ *Combine like radicals*

65. $2\sqrt[3]{3} + 4\sqrt[3]{24} - \sqrt[3]{81}$

$= 2\sqrt[3]{3} + 4\sqrt[3]{8 \cdot 3} - \sqrt[3]{27 \cdot 3}$

$= 2\sqrt[3]{3} + 4(2\sqrt[3]{3}) - 3\sqrt[3]{3}$

$= 2\sqrt[3]{3} + 8\sqrt[3]{3} - 3\sqrt[3]{3}$

$= 7\sqrt[3]{3}$ *Combine like radicals*

67. $\dfrac{1}{\sqrt{3}} - \dfrac{2}{\sqrt{12}} + 2\sqrt{3}$

$= \dfrac{1}{\sqrt{3}} - \dfrac{2}{\sqrt{4 \cdot 3}} + 2\sqrt{3}$

$= \dfrac{1}{\sqrt{3}} - \dfrac{2}{2\sqrt{3}} + 2\sqrt{3}$

$= \dfrac{1}{\sqrt{3}} - \dfrac{1}{\sqrt{3}} + 2\sqrt{3}$

$= 2\sqrt{3}$

69. $\dfrac{5}{\sqrt[3]{2}} - \dfrac{2}{\sqrt[3]{16}} + \dfrac{1}{\sqrt[3]{54}}$

$= \dfrac{5}{\sqrt[3]{2}} - \dfrac{2}{\sqrt[3]{8 \cdot 2}} + \dfrac{1}{\sqrt[3]{27 \cdot 2}}$

$= \dfrac{5}{\sqrt[3]{2}} - \dfrac{2}{2\sqrt[3]{2}} + \dfrac{1}{3\sqrt[3]{2}}$

$= \dfrac{5}{\sqrt[3]{2}} \cdot \dfrac{3}{3} - \dfrac{1}{\sqrt[3]{2}} \cdot \dfrac{3}{3} + \dfrac{1}{3\sqrt[3]{2}}$

$= \dfrac{15 - 3 + 1}{3\sqrt[3]{2}} = \dfrac{13}{3\sqrt[3]{2}} = \dfrac{13}{3\sqrt[3]{2}} \cdot \dfrac{\sqrt[3]{4}}{\sqrt[3]{4}}$

$= \dfrac{13\sqrt[3]{4}}{3\sqrt[3]{8}} = \dfrac{13\sqrt[3]{4}}{6}$

71. $(\sqrt{2} + 3)(\sqrt{2} - 3)$

$\quad = (\sqrt{2})^2 - 3^2$ *Product of the sum and difference of two terms*

$\quad = 2 - 9 = -7$

73. $(\sqrt[3]{11} - 1)(\sqrt[3]{11^2} + \sqrt[3]{11} + 1)$

This product has the pattern

$(a - b)(a^2 + 2ab + b^2) = a^3 - b^3$,

the difference of two cubes.

Thus,

$(\sqrt[3]{11^2} - 1)(\sqrt[3]{11^2} + \sqrt[3]{11} + 1)$

$\quad = (\sqrt[3]{11})^3 - 1^3$

$\quad = 11 - 1 = 10.$

75. $(\sqrt{3} + \sqrt{8})^2$

$\quad = (\sqrt{3} + 2\sqrt{2})^2$ *Simplify $\sqrt{8}$*

$\quad = (\sqrt{3})^2 + 2(\sqrt{3})(2\sqrt{2}) + (2\sqrt{2})^2$

$\qquad\qquad\qquad$ *Square of a binomial*

$\quad = 3 + 4\sqrt{6} + 8$

$\quad = 11 + 4\sqrt{6}$

77. $(3\sqrt{2} + \sqrt{3})(2\sqrt{3} - \sqrt{2})$

$= 3\sqrt{2}(2\sqrt{3}) - 3\sqrt{2}(\sqrt{2}) + \sqrt{3}(2\sqrt{3})$

$\quad - \sqrt{3}\sqrt{2}$ *FOIL*

$= 6\sqrt{6} - 3 \cdot 2 + 2 \cdot 3 - \sqrt{6}$

$= 6\sqrt{6} - 6 + 6 - \sqrt{6}$

$= 5\sqrt{6}$

79. $(2\sqrt[3]{3} + 1)(\sqrt[3]{3} - 4)$

$= 2\sqrt[3]{3} \cdot \sqrt[3]{3} - 2\sqrt[3]{3} \cdot 4 + 1 \cdot \sqrt[3]{3} - 4$

$\qquad\qquad\qquad\qquad$ *FOIL*

$= 2\sqrt[3]{9} - 8\sqrt[3]{3} + \sqrt[3]{3} - 4$

$= 2\sqrt[3]{9} - 7\sqrt[3]{3} - 4$ *Combine like radicals*

81. $\dfrac{\sqrt{3}}{\sqrt{5} + \sqrt{3}}$

$= \dfrac{\sqrt{3}}{\sqrt{5} + \sqrt{3}} \cdot \dfrac{\sqrt{5} - \sqrt{3}}{\sqrt{5} - \sqrt{3}}$

$\quad$ *Multiply numerator and denominator by conjugate of denominator*

$= \dfrac{\sqrt{3}(\sqrt{5} - \sqrt{3})}{(\sqrt{5})^2 - (\sqrt{3})^2}$ *Difference of two squares*

$= \dfrac{\sqrt{3}\sqrt{5} - \sqrt{3}\sqrt{3}}{5 - 3}$ *Distributive property*

$= \dfrac{\sqrt{15} - 3}{2}$

83. $\dfrac{1 + \sqrt{3}}{3\sqrt{5} + 2\sqrt{3}}$

$= \dfrac{1 + \sqrt{3}}{3\sqrt{5} + 2\sqrt{3}} \cdot \dfrac{3\sqrt{5} - 2\sqrt{3}}{3\sqrt{5} - 2\sqrt{3}}$

$\quad$ *Multiply numerator and denominator by conjugate of denominator*

$= \dfrac{(1 + \sqrt{3})(3\sqrt{5} - 2\sqrt{3})}{(3\sqrt{5})^2 - (2\sqrt{3})^2}$

$\quad$ *Difference of two squares*

$= \dfrac{3\sqrt{5} - 2\sqrt{3} + 3\sqrt{15} - 6}{45 - 12}$

$= \dfrac{3\sqrt{5} - 2\sqrt{3} + 3\sqrt{15} - 6}{33}$

85. $\dfrac{p}{\sqrt{p} + 2} = \dfrac{p}{\sqrt{p} + 2} \cdot \dfrac{\sqrt{p} - 2}{\sqrt{p} - 2}$

Multiply numerator and denominator by conjugate of denominator

$= \dfrac{p(\sqrt{p} - 2)}{(\sqrt{p})^2 - 2^2}$ *Difference of two squares*

$= \dfrac{p(\sqrt{p} - 2)}{p - 4}$

87. $\dfrac{a}{\sqrt{a + b} - 1}$

$= \dfrac{a}{\sqrt{a + b} - 1} \cdot \dfrac{\sqrt{a + b} + 1}{\sqrt{a + b} + 1}$

$= \dfrac{a(\sqrt{a + b} + 1)}{(\sqrt{a + b})^2 - 1^2}$

$= \dfrac{a(\sqrt{a + b} + 1)}{a + b - 1}$

89. $\dfrac{1 + \sqrt{2}}{2}$

$= \dfrac{1 + \sqrt{2}}{2} \cdot \dfrac{1 - \sqrt{2}}{1 - \sqrt{2}}$

Multiply numerator and denominator by conjugate of numerator

$= \dfrac{1^2 - (\sqrt{2})^2}{2(1 - \sqrt{2})}$ *Difference of two squares*

$= \dfrac{1 - 2}{2(1 - \sqrt{2})}$

$= \dfrac{-1}{2(1 - \sqrt{2})}$

91. $\dfrac{\sqrt{x}}{1 + \sqrt{x}} = \dfrac{\sqrt{x}}{1 + \sqrt{x}} \cdot \dfrac{\sqrt{x}}{\sqrt{x}}$

$= \dfrac{x}{(1 + \sqrt{x})\sqrt{x}}$

$= \dfrac{x}{\sqrt{x} + x}$

93. $\dfrac{\sqrt{x} + \sqrt{x + 1}}{\sqrt{x} - \sqrt{x + 1}}$

$= \dfrac{\sqrt{x} + \sqrt{x + 1}}{\sqrt{x} - \sqrt{x + 1}} \cdot \dfrac{\sqrt{x} - \sqrt{x + 1}}{\sqrt{x} - \sqrt{x + 1}}$

Multiply numerator and denominator by conjugate of numerator

$= \dfrac{(\sqrt{x})^2 - (\sqrt{x + 1})^2}{(\sqrt{x} - \sqrt{x + 1})^2}$ *Difference of two squares*

$= \dfrac{x - (x + 1)}{(\sqrt{x} - \sqrt{x + 1})^2}$

$= \dfrac{-1}{(\sqrt{x} - \sqrt{x + 1})^2}$

$= \dfrac{-1}{x - 2\sqrt{x}\sqrt{x + 1} + x + 1}$ *Square of a binomial*

$= \dfrac{-1}{2x - 2\sqrt{x(x + 1)} + 1}$

95. $\sqrt{(m + n)^2} = |m + n|$ since $\sqrt{x^2} = |x|$.

97. $\sqrt{z^2 - 6zx + 9x^2}$

$= \sqrt{(z - 3x)^2}$ *Factor*

$= |z - 3x|$

since

$\sqrt{x} = |x|$.

Chapter 1 Review Exercises

1. $-12, -6, -\sqrt{4}$ (or -2), 0, and 6 are integers.

3. $-\sqrt{7}, \dfrac{\pi}{4}$, and $\sqrt{11}$ are irrational numbers.

5. $-\sqrt{25} = -5$ is an integer, a rational number, and a real number.

7. $\frac{3\pi}{4}$ is an irrational number and a

    real number.

9. $[2^3 - (-5)] - 2^2 = (8 + 5) - 4$
    $$= 13 - 4$$
    $$= 9$$

11. $(6 - 9)(-2 - 7) - (-4)$
    $$= (-3)(-9) - (-4)$$
    $$= 27 + 4$$
    $$= 31$$

13. $\left(-\frac{2^3}{5} - \frac{3}{4}\right) - \left(-\frac{1}{2}\right)$
    $$= \left(-\frac{32}{20} - \frac{15}{20}\right) - \left(-\frac{1}{2}\right)$$
    $$= \frac{-47}{20} + \frac{1}{2}$$
    $$= \frac{-47}{20} + \frac{10}{20}$$
    $$= -\frac{37}{20}$$

15. $\dfrac{(-7)(-3) - (-2^3)(-5)}{(-2^2 - 2)(-1 - 6)} = \dfrac{21 - 40}{(-6)(-7)}$
    $$= -\frac{19}{42}$$

17. $-4(2a - 5b)$
    $$= -4[2(-1) - 5(-2)] \quad Let\ a = -1,$$
    $$\qquad\qquad\qquad\qquad\qquad b = -2$$
    $$= -4(-2 + 10)$$
    $$= -4(8) = -32$$

21. $4 \cdot 6 + 4 \cdot 12 = 4(6 + 12)$

    Distributive property

23. $-(r - 2) = -r + 2$

    Distributive property

25. $[6 \cdot 5 + 8 \cdot 5] \cdot 2 = [(6 + 8) \cdot 5] \cdot 2$

    Distributive property

27. $k(r + s - t) = kr + ks - kt$

29. Simplify each number.

    $|6 - 4| = 2, \ -|-2| = -2,$
    $|8 + 1| = 9, \ -|3 - (-2)| = -5$

    The correct order is

    $-|3 - (-2)|, \ -|-2|, \ |6-4|, \ |8+1|.$

31. $-|-6| + |3| = -6 + 3 = -3$

33. $|\sqrt{8} - 3|$

    Since $\sqrt{8} < 3$, $\sqrt{8} - 3 < 0$, so
    $$|\sqrt{8} - 3| = -(\sqrt{8} - 3)$$
    $$= -\sqrt{8} + 3$$
    $$= 3 - \sqrt{8}.$$

35. $|m - 3|$ if $m > 3$

    If $m > 3$, $m - 3 > 0$, so
    $$|m - 3| = m - 3.$$

37. $|\pi - 4|$

    Since $\pi < 4$, $\pi - 4 < 0$, so
    $$|\pi - 4| = -(\pi - 4)$$
    $$= -\pi + 4$$
    $$= 4 - \pi.$$

39. $(3q^3 - 9q^2 + 6) + (4q^3 - 8q + 3)$
    $$= 3q^3 + 4q^3 - 9q^2 - 8q + 6 + 3$$
    $$= 7q^3 - 9q^2 - 8q + 9$$

41. $(8y - 7)(2y + 7)$

$\quad = 16y^2 + 56y - 14y - 49 \quad$ *FOIL*

$\quad = 16y^2 + 42y - 49$

43. $(3k - 5m)^2$

$\quad = (3k)^2 - 2(3k)(5m) + (5m)^2$
$\qquad\qquad$ *Square of a binomial*

$\quad = 9k^2 - 30km + 25m^2$

45. $(3w - 2)(5w^2 - 4w + 1)$

$\quad = 3w(5w^2 - 4w + 1)$

$\quad - 2(5w^2 - 4w + 1)$
$\qquad\qquad$ *Distributive property*

$\quad = 15w^3 - 12w^2 + 3w - 10w^2 + 8w - 2$
$\qquad\qquad$ *Distributive property*

$\quad = 15w^3 - 22w^2 + 11w - 2$
$\qquad\qquad$ *Combine like terms*

47. $(p^q + 1)(p^q - 3)$

$\quad = p^{2q} - 3p^q + p^q - 3 \quad$ *FOIL*

$\quad = p^{2q} - 2p^q - 3 \qquad$ *Combine like terms*

49. $(x + 2y)^4$

$\quad = x^4 + \binom{4}{3}x^3(2y) + \binom{4}{2}x^2(2y)^2$

$\quad + \binom{4}{1}x(2y)^3 + (2y)^4$

$\quad = x^4 + 4x^3(2y) + 6x^2(4y^2)$

$\quad + 4x(8y^3) + 16y^4$

$\quad = x^4 + 8x^3y + 24x^2y^2 + 32xy^3$

$\quad + 16y^4$

51. To find the fifth term of $(3x - 2y)^6$, use the formula

$$\binom{n}{n - (r - 1)}x^{n-(r-1)}y^{r-1}$$

with $n = 6$ and $r = 5$.

Then $r - 1 = 4$ and $x - (r - 1) = 2$.

$\binom{6}{2}(3x)^2(-2y)^4 = \dfrac{6!}{2!4!}(9x^2)(16y^4)$

$\qquad\qquad\qquad = 15(9x^2)(16y^4)$

$\qquad\qquad\qquad = 2160x^2y^4$

53. First four terms of $(3 + x)^{16}$

$3^{16} + \binom{16}{15}3^{15}x + \binom{16}{14}3^{14}x^2 + \binom{16}{13}3^{13}x^3$

$\quad = 3^{16} + 16 \cdot 3^{15}x + 120 \cdot 3^{14}x^2$

$\quad + 560 \cdot 3^{13}x^3$

55. $7z^2 - 9z^3 + z = z(7z - 9z^2 + 1)$
$\qquad\qquad\qquad$ *z is greatest*
$\qquad\qquad\qquad$ *common factor*

57. $r^2 + rp - 42p^2$

Find two numbers whose product is -42 and whose sum is 1. They are 7 and -6. Thus,

$\qquad r^2 + rp - 42p^2$

$\qquad = (r + 7p)(r - 6p).$

59. $6m^2 - 13m - 5$

The positive factors of 6 could be 2 and 3 or 1 and 6. As factors of -5, we could have -1 and 5 or -5 and 1. Try different combinations of these factors until the correct one is found.

$\quad 6m^2 - 13m - 5 = (3m + 1)(2m - 5)$

61. $169y^4 - 1$

$\quad = (13y^2)^2 - 1^2 \quad$ *Difference of*
$\qquad\qquad\qquad$ *two squares*

$\quad = (13y^2 + 1)(13y^2 - 1)$

63. $8y^3 - 1000z^6$

$= 8(y^3 - 125z^6)$ *Factor out 8*

$= 8[y^3 - (5z^2)^3]$
   Difference of two cubes

$= 8(y - 5z^2)$
   $\cdot [y^2 + y(5z^2) + (5z^2)^2]$

$= 8(y - 5z^2)(y^2 + 5yz^2 + 25z^4)$

65. $ar - 3as + 5rb - 15sb$

$= (ar - 3as) + (5rb - 15sb)$
   Group the terms

$= a(r - 3s) + 5b(r - 3s)$
   Factor each group

$= (r - 3s)(a + 5b)$
   Factor our r - 3s

67. $(16m^2 - 56m + 49) - 25a^2$

$= (4m - 7)^2 - (5a)^2$
   Difference of two squares

$= [(4m - 7) + 5a][(4m - 7) - 5a]$

$= (4m - 7 + 5a)(4m - 7 - 5a)$

69. $\dfrac{2a + b}{4a^2 - b^2}$

$= \dfrac{2a + b}{(2a + b)(2a - b)}$
   Factor denominator as difference of two squares

$= \dfrac{1}{2a - b}$ *Use fundamental principle to write expression in lowest terms*

Thus, expression (a) is equal to

$\dfrac{2a + b}{4a^2 - b^2}$.

71. $\dfrac{3r^3 - 9r^2}{r^2 - 9} \div \dfrac{8r^3}{r + 3}$

$= \dfrac{3r^3 - 9r^2}{r^2 - 9} \cdot \dfrac{r + 3}{8r^3}$
   Definition of division

$= \dfrac{3r^2(r - 3)}{(r + 3)(r - 3)} \cdot \dfrac{(r + 3)}{8r^3}$ *Factor*

$= \dfrac{3}{8r}$ *Use fundamental principle to write expression in lowest terms*

73. $\dfrac{27m^3 - n^3}{3m - n} \div \dfrac{9m^2 + 3mn + n^2}{9m^2 - n^2}$

$= \dfrac{27m^3 - n^3}{3m - n} \cdot \dfrac{9m^2 - n^2}{9m^2 + 3mn + n^2}$
   Definition of division

$= \dfrac{(3m)^3 - n^3}{3m - n} \cdot \dfrac{(3m)^2 - n^2}{9m^2 + 3mn + n^2}$
   Difference of two cubes and difference of two squares

$= \dfrac{(3m - n)[(3m)^2 + 3mn + n^2]}{3m - n}$
   $\cdot \dfrac{(3m + n)(3m - n)}{9m^2 + 3mn + n^2}$ *Factor numerators*

$= \dfrac{(3m-n)(9m^2+3mn+n^2)(3m+n)(3m-n)}{(3m-n)(9m^2+3mn+n^2)}$
   Multiply

$= (3m + n)(3m - n)$
   Use fundmental principle to write expressions in lowest terms

75. $\dfrac{1}{4y} + \dfrac{8}{5y} = \dfrac{1 \cdot 5}{4y \cdot 5} + \dfrac{8 \cdot 4}{5y \cdot 4}$

$= \dfrac{5}{20y} + \dfrac{32}{20y}$

$= \dfrac{37}{20y}$

77. $\dfrac{3}{x^2 - 4x + 3} - \dfrac{2}{x^2 - 1}$

$= \dfrac{3}{(x - 3)(x - 1)} - \dfrac{2}{(x + 1)(x - 1)}$

The least common denominator is
$(x - 3)(x - 1)(x + 1)$.

$= \dfrac{3(x + 1)}{(x - 3)(x - 1)(x + 1)}$

$\quad - \dfrac{2(x - 3)}{(x + 1)(x - 1)(x - 3)}$

$= \dfrac{3(x + 1) - 2(x - 3)}{(x - 3)(x - 1)(x + 1)}$

$= \dfrac{3x + 3 - 2x + 6}{(x - 3)(x - 1)(x + 1)}$

$= \dfrac{x + 9}{(x - 3)(x - 1)(x + 1)}$

79. $\dfrac{\dfrac{1}{p} + \dfrac{1}{q}}{1 - \dfrac{1}{pq}}$

Multiply both numerator and denominator by the least common denominator of all the fractions, pq.

$\dfrac{\dfrac{1}{p} + \dfrac{1}{q}}{1 - \dfrac{1}{pq}} = \dfrac{pq\left(\dfrac{1}{p} + \dfrac{1}{q}\right)}{pq\left(1 - \dfrac{1}{pq}\right)}$

$= \dfrac{pq\left(\dfrac{1}{p}\right) + pq\left(\dfrac{1}{q}\right)}{pq(1) - pq\left(\dfrac{1}{pq}\right)}$

Distributive property

$= \dfrac{q + p}{pq - 1}$

81. $2^{-6} = \dfrac{1}{2^6} = \dfrac{1}{64}$

83. $\left(-\dfrac{5}{4}\right)^{-2} = \dfrac{1}{\left(-\dfrac{5}{4}\right)^2}$

$= \dfrac{1}{\dfrac{25}{16}}$

$= \dfrac{16}{25}$

85. $(5z^3)(-2z^5) = -10z^{3+5}$ *Product rule*

$= -10z^8$

87. $(-6p^5w^4m^{12})^0 = 1$ *Definition of a^0*

89. $\dfrac{-8y^7p^{-2}}{y^{-4}p^{-3}} = -8y^{7-(-4)}p^{(-2)-(-3)}$

Quotient rule

$= -8y^{11}p$

91. $\dfrac{(p + q)^4(p + q)^{-3}}{(p + q)^6}$

$= (p + q)^{4+(-3)-6}$ *Product and quotient rules*

$= (p + q)^{-5}$

$= \dfrac{1}{(p + q)^5}$

93. $\left(\dfrac{r^{-1}s^2}{t^{-2}}\right)^{-2} = \dfrac{(r^{-1})^{-2}(s^2)^{-2}}{(t^{-2})^{-2}}$

Power rule

$= \dfrac{r^2s^{-4}}{t^4}$ *Power rule*

$= \dfrac{r^2}{s^4t^4}$

95. $\dfrac{p^4(p^{-2})}{p^{5/3}} = \dfrac{p^2}{p^{5/3}}$ *Product rule*

$= p^{2-5/3}$ *Quotient rule*

$= p^{6/3 - 5/3}$ *Common denominator*

$= p^{1/3}$

97. $(7r^{1/2})(2r^{3/4})(-r^{1/6})$

$= -14r^{1/2\,+3/4\,+1/6}$ *Product rule*

$= -14r^{17/12}$

99. $\dfrac{y^{5/3} \cdot y^{-2}}{y^{-5/6}} = y^{5/3\,+(-2)\,-(-5/6)}$

$\quad$ *Product and quotient rules*

$= y^{10/6\,-12/6\,+5/6}$

$\quad$ *Common denominator*

$= y^{3/6} = y^{1/2}$

101. $\dfrac{k^{2+p} \cdot k^{-4p}}{k^{6p}} = k^{2+p-4p-6p}$

$\quad$ *Product and quotient rules*

$= k^{2-9p}$

103. $2z^{1/3}(5z^2 - 2)$

$= 2z^{1/3}(5z^2) - 2z^{1/3}(2)$

$\quad\quad\quad$ *Distributive property*

$= 10z^{7/3} - 4z^{1/3}$

105. $(p + p^{1/2})(3p - 5)$

$= 3p^2 - 5p + 3p^{3/2} - 5p^{1/2}$ *FOIL*

$= 3p^2 + 3p^{3/2} - 5p - 5p^{1/2}$

107. $\sqrt{200} = \sqrt{100 \cdot 2}$

$= \sqrt{100} \cdot \sqrt{2}$

$= 10\sqrt{2}$

109. $\sqrt[4]{1250} = \sqrt[4]{625 \cdot 2}$

$= \sqrt[4]{625} \cdot \sqrt[4]{2}$

$= 5\sqrt[4]{2}$

111. $-\sqrt[3]{\dfrac{2}{5p^2}} = -\dfrac{\sqrt[3]{2}}{\sqrt[3]{5p^2}}$

$= -\dfrac{\sqrt[3]{2}}{\sqrt[3]{5p^2}} \cdot \dfrac{\sqrt[3]{25p}}{\sqrt[3]{25p}}$ *Rationalize denominator*

$= -\dfrac{\sqrt[3]{50p}}{\sqrt[3]{125p^3}}$

$= -\dfrac{\sqrt[3]{50p}}{5p}$

113. $\sqrt[4]{\sqrt[3]{m}} = (\sqrt[3]{m})^{1/4} = (m^{1/3})^{1/4}$

$= m^{1/3\,\cdot 1/4} = m^{1/12} = \sqrt[12]{m}$

115. $(\sqrt[3]{2} + 4)(\sqrt[3]{2^2} - 4\sqrt[3]{2} + 16)$

$= \sqrt[3]{2}(\sqrt[3]{2^2} - 4\sqrt[3]{2} + 16)$

$\quad + 4(\sqrt[3]{2^2} - 4\sqrt[3]{2} + 16)$

$\quad\quad\quad$ *Distributive property*

$= \sqrt[3]{2^3} - 4\sqrt[3]{2^2} + 16\sqrt[3]{2} + 4\sqrt[3]{2^2}$

$\quad - 16\sqrt[3]{2} + 64$

$\quad\quad\quad$ *Distributive property*

$= 2 + 64 = 66$

Alternate solution:

$(\sqrt[3]{2} + 4)(\sqrt[3]{2} - 4\sqrt[3]{2} + 16)$

$= (\sqrt[3]{2} + 4)[(\sqrt[3]{2})^2 - \sqrt[3]{2} \cdot 4 + 4^2]$

$= (\sqrt[3]{2})^3 + 4^3$ *Sum of two cubes*

$= 2 + 64 = 66$

117. $\sqrt{18m^3} - 3m\sqrt{32m} + 5\sqrt{m^3}$

$= \sqrt{9m^2 \cdot 2m} - 3m\sqrt{16 \cdot 2m} + 5\sqrt{m^2m}$

$= 3m\sqrt{3m} - 12m\sqrt{2m} + 5m\sqrt{m}$

$= -9m\sqrt{2m} + 5m\sqrt{m}$

or $\quad m(-9\sqrt{2m} + 5\sqrt{m})$

119. $\dfrac{6}{3 - \sqrt{2}}$

$= \dfrac{6}{3 - \sqrt{2}} \cdot \dfrac{3 + \sqrt{2}}{3 + \sqrt{2}}$

Multiply numerator and denominator by conjugate of denominator

$= \dfrac{6(3 + \sqrt{2})}{9 - 2}$

$= \dfrac{6(3 + \sqrt{2})}{7}$

121. $\dfrac{k}{\sqrt{k} - 3}$

$= \dfrac{k}{\sqrt{k} - 3} \cdot \dfrac{\sqrt{k} + 3}{\sqrt{k} + 3}$

Multiply numerator and denominator by conjugate of denominator

$= \dfrac{k(\sqrt{k} + 3)}{k - 9}$

Chapter 1 Test

1. $\dfrac{4}{\pi}$ is an irrational number.

2. -10, $-\dfrac{9}{3}$ (or -3), 0, and $\sqrt{25}$ (or 5) are integers.

3. $\dfrac{2x - |3y| - |z|}{-|x + y| + z}$

$= \dfrac{2(-2) - |3(-3)| - |-4|}{-|(-2) + (-3)| + (-4)}$

Let $x = -2$, $y = -3$, and $z = -4$

$= \dfrac{-4 - |-9| - 4}{-|-5| + (-4)}$

$= \dfrac{-4 - 9 - 4}{-5 + (-4)}$

$= \dfrac{-17}{-9} = \dfrac{17}{9}$

4. $2(x + 3) = 2(3 + x)$

Commutative property

5. $3(x + 4) = 3x + 12$

Distributive property

6. $(a^2 - 3a + 1) - (a - 2a^2)$
   $+ (-4 + 3a - a^2)$

$= (1 + 2 - 1)a^2 + (-3 - 1 + 3)a$
   $+ (1 - 4)$

$= 2a^2 - a - 3$

7. $(2x^n - 3)^2$

$= (2x^n)^2 - 2(2x^n)(3) + 3^2$

$= 4x^{2n} - 12x^n + 9$

8. $(3y^2 - y + 4)(y + 2)$

$= (3y^2 - y + 4)(y)$
   $+ (3y^2 - y + 4)(2)$

$= 3y^2(y) - y(y) + 4(y)$
   $+ 3y^2(2) - y(2) + 4(2)$

$= 3y^3 - y^2 + 4y + 6y^2 - 2y + 8$

$= 3y^3 + 5y^2 + 2y + 8$

9. $(2x - 3y)^4$

$= (2x)^4 + \binom{4}{3}(2x)^3(-3y)$

$+ \binom{4}{2}(2x)^2(-3y)^2 + \binom{4}{1}(2x)(-3y)^3$
   $+ (-3y)^4$

$= 16x^4 + 4(8x^3)(-3y) + 6(4x^2)(9x^2)$
   $+ 4(2x)(-27y^3) + 81y^4$

$= 16x^4 - 96x^3y + 216x^2y^2 - 216xy^3$
   $+ 81y^4$

10. To find the third term in the expansion of $(w - 2y)^6$, use the formula

$$\binom{n}{n - (r - 1)} x^{n - (r - 1)} y^{r - 1}$$

with $n = 6$ and $r = 3$.
Then $r - 1 = 2$ and $n - (r - 1) = 4$.
Thus, the third term is

$$\binom{6}{4} w^4 (-2y)^2 = 15 w^4 (4y^2)$$

$$= 60 w^4 y^2.$$

11. $x^4 - 16$

$$= (x^2)^2 - 4^2$$

$$= (x^2 + 4)(x^2 - 4)$$

$$= (x^2 + 4)(x^2 - 2^2)$$

$$= (x^2 + 4)(x + 2)(x - 2)$$

12. $24m^3 - 14m^2 - 24m$

$$= 2m(12m^2 - 7m - 12)$$

$$= 2m(4m + 3)(3m - 4)$$

13. $x^3 y^2 - 9x^3 - 8y^2 + 72$

$$= (x^3 y^2 - 9x^3) + (-8y^2 + 72)$$

$$= x^3(y^2 - 9) - 8(y^2 - 9)$$

$$= (x^3 - 8)(y^2 - 9)$$

$$= (x - 2)(x^2 + 2x + 4)(y + 3)$$

$$\cdot (y - 3)$$

14. The first step is wrong. The numerator must be in factored form before cancelling factors of $x - 1$. As shown, the numerator is a difference, not a product.

15. $\dfrac{5x^2 - 9x - 2}{30x^3 + 6x^2} \cdot \dfrac{2x^8 + 6x^7 + 4x^6}{x^4 - 3x^2 - 4}$

$$= \frac{(5x + 1)(x - 2)}{6x^2(5x + 1)}$$

$$\cdot \frac{2x^6(x^2 + 3x + 2)}{(x^2 - 4)(x^2 + 1)}$$

$$= \frac{(5x + 1)(x - 2)(2x^6)(x + 2)(x + 1)}{6x^2(5x + 1)(x + 2)(x - 2)(x^2 + 1)}$$

$$= \frac{2x^6(x + 1)}{6x^2(x^2 + 1)}$$

$$= \frac{x^4(x + 1)}{3(x^2 + 1)}$$

16. $\dfrac{x}{x^2 + 3x + 2} + \dfrac{2x}{2x^2 - x - 3}$

$$= \frac{x}{(x + 2)(x + 1)} + \frac{2x}{(2x - 3)(x + 1)}$$

The least common denominator is
$(x + 2)(x + 1)(2x - 3)$.

$$= \frac{x(2x - 3)}{(x + 2)(x + 1)(2x - 3)}$$

$$+ \frac{2x(x + 2)}{(2x - 3)(x + 1)(x + 2)}$$

$$= \frac{2x^2 - 3x}{(x + 2)(x + 1)(2x - 3)}$$

$$+ \frac{2x^2 + 4x}{(x + 2)(x + 1)(2x - 3)}$$

$$= \frac{4x^2 + x}{(x + 2)(x + 1)(2x - 3)}$$

17. $\dfrac{a + b}{2a - 3} - \dfrac{a - b}{3 - 2a}$

$$= \frac{a + b}{2a - 3} - \frac{(a - b)(-1)}{(3 - 2a)(-1)}$$

$$= \frac{a + b}{2a - 3} + \frac{a - b}{2a - 3}$$

$$= \frac{2a}{2a - 3}$$

If $3 - 2a$ is used as the common denominator, the result will be

$\dfrac{-2a}{3 - 2a}$. The rational expressions

$\dfrac{2a}{2a - 3}$ and $\dfrac{-2a}{3 - 2a}$ are equivalent.

18. $\dfrac{y - 2}{y - \dfrac{4}{y}} = \dfrac{y(y - 2)}{y\left(y - \dfrac{4}{y}\right)}$

$\qquad = \dfrac{y^2 - 2y}{y^2 - 4}$

$\qquad = \dfrac{y(y - 2)}{(y + 2)(y - 2)}$

$\qquad = \dfrac{y}{y + 2}$

19. $\left(\dfrac{x^{-1}x^3}{2x^0x^{-5}}\right)^{-2} = \dfrac{(x^{-1})^{-2}(x^3)^{-2}}{2^{-2}(x^0)^{-2}(x^{-5})^{-2}}$

$\qquad = \dfrac{x^2 \cdot x^{-6}}{\dfrac{1}{4} \cdot 1 \cdot x^{10}}$

$\qquad = 4x^{2-6-10}$

$\qquad = 4x^{-14}$

$\qquad = \dfrac{4}{x^{14}}$

20. $\left(\dfrac{x^{-2}y^{-1/3}\,z}{x^{-5/3}\,y^{-2/3}\,z^{2/3}}\right)^3$

$\qquad = \dfrac{x^{-6}y^{-1}z^3}{x^{-5}y^{-2}z^2}$

$\qquad = x^{-6-(-5)}\,y^{-1-(-2)}\,z^{3-2}$

$\qquad = x^{-1}y^1z^1$

$\qquad = \dfrac{yz}{x}$

21. $\sqrt{18x^5y^8} = \sqrt{(9x^4y^8)(2x)}$

$\qquad = \sqrt{9x^4y^8} \cdot \sqrt{2x}$

$\qquad = 3x^2y^4\sqrt{2x}$

22. $\sqrt[5]{\dfrac{x}{4y^2}} = \dfrac{\sqrt[5]{x}}{\sqrt[5]{4y^2}}$

$\qquad = \dfrac{\sqrt[5]{x} \cdot \sqrt[5]{8y^3}}{\sqrt[5]{4y^2} \cdot \sqrt[5]{8y^3}}$

$\qquad = \dfrac{\sqrt[5]{8xy^3}}{\sqrt[5]{32y^5}}$

$\qquad = \dfrac{\sqrt[5]{8xy^3}}{2y}$

23. $\sqrt{32x} + \sqrt{2x} - \sqrt{18x}$

$\qquad = \sqrt{16 \cdot 2x} + \sqrt{2x} - \sqrt{9 \cdot 2x}$

$\qquad = 4\sqrt{2x} + \sqrt{2x} - 3\sqrt{2x}$

$\qquad = 2\sqrt{2x}$

24. $\dfrac{3}{\sqrt{2} - 1} = \dfrac{3}{\sqrt{2} - 1} \cdot \dfrac{\sqrt{2} + 1}{\sqrt{2} + 1}$

$\qquad = \dfrac{3(\sqrt{2} + 1)}{2 - 1}$

$\qquad = 3(\sqrt{2} + 1) \quad \text{or} \quad 3\sqrt{2} + 3$

25. $(\sqrt{x} - \sqrt{y})(\sqrt{x} + \sqrt{y})$

$\qquad = (\sqrt{x})^2 - (\sqrt{y})^2$

$\qquad = x - y$

CHAPTER 2   EQUATIONS AND INEQUALITIES

Section 2.1

1. $x^2 + 5x = x(x + 5)$

    Since the product of x and x + 5 is $x^2 + 5x$, the given equation is true for every value of x, and is an identity. The solution set is {all real numbers}.

3. $2(x - 7) = 5x + 3 - x$

    Simplified, the given equation becomes $2x - 14 = 4x + 3$. Replacing x with $-\frac{17}{2}$ gives $-31 = -31$, a true statement. However, x = 1 yields $-12 = 7$, a false statement. The equation $2(x - 7) = 5x + 3 - x$ is true for some values of x, but not all values of x, and thus is a conditional equation. The solution set is $\left\{-\frac{17}{2}\right\}$.

5. $\frac{m + 3}{m} = 1 + \frac{3}{m}$

    Since $1 + \frac{3}{m} = \frac{m}{m} + \frac{3}{m} = \frac{m + 3}{m}$, the given equation is true for every number that is a meaningful replacement for m and is an identity. Notice that m cannot be zero, since $\frac{m + 3}{m}$ and $\frac{3}{m}$ are not defined when m = 0. The solution set is {all real numbers except 0}.

7. $4q + 20q - 25 = 24q + 5$
    $24q - 25 = 24q + 5$

    Since this equation is false for all values of q, the original equation is a contradiction. The solution set is ∅.

9. $\frac{3x}{x - 1} = \frac{2}{x - 1}$
    $3x = 2$

    The solution set of $\frac{3x}{x - 1} = \frac{2}{x - 1}$ is $\left\{\frac{2}{3}\right\}$, and this is also the solution set of 3x = 2. Since the solution sets are equal, the equations are equivalent.

11. $\frac{x}{x - 2} = \frac{2}{x - 2}$
    $x = 2$

    The solution set of $\frac{x}{x - 2} = \frac{2}{x - 2}$ is ∅, while the solution set of x = 2 is {2}. Since the solution sets are not equal, the equations are not equivalent.

13. $x = 4$
    $x^2 = 16$

    The solution set for x = 4 is {4}, while the solution set for $x^2 = 16$ is {4, -4}. Since the equation sets are not equal, the equations are not equivalent.

15. (b) $8x^2 - 4x + 3 = 0$

Because of the x^2 that appears, this equation cannot be written in the form $ax + b = 0$. It is not a linear equation. All of the other choices are equations that can be written in the form $ax + b = 0$ and therefore are linear equations.

19.

$$2m - 5 = m + 7$$

$2m - 5 - m = m + 7 - m$

Subtract m from both sides

$$m - 5 = 7$$

$m - 5 + 5 = 7 + 5$ *Add 5 to both sides*

$$m = 12$$

Solution set: $\{12\}$

21. $\frac{5}{6}k - 2k + \frac{1}{3} = \frac{2}{3}$

Multiply both sides of the equation by the least common denominator, 6.

$6\left(\frac{5}{6}k - 2k + \frac{1}{3}\right) = 6\left(\frac{2}{3}\right)$

$5k - 12k + 2 = 4$ *Distributive property*

$-7k + 2 = 4$

$-7k = 2$ *Subtract 2*

$k = -\frac{2}{7}$ *Divide by 7*

Solution set: $\left\{-\frac{2}{7}\right\}$

23. $3r + 2 - 5(r + 1) = 6r + 4$

$3r + 2 - 5r - 5 = 6r + 4$

Distributive property

$-2r - 3 = 6r + 4$

Combine terms

$-2r - 3 + 2r = 6r + 4 + 2r$

Add 2r to both sides

$-3 = 8r + 4$

Combine terms

$-3 - 4 = 8r + 4 - 4$

Subtract 4 from both sides

$-7 = 8r$

Combine terms

$-\frac{7}{8} = r$

Divide both sides by 8

Solution set: $\left\{-\frac{7}{8}\right\}$

25. $2[m - (4 + 2m) + 3] = 2m + 2$

$2[m - 4 - 2m + 3] = 2m + 2$

$2[-m - 1] = 2m + 2$

$-2m - 2 = 2m + 2$

$-2m - 2 + 2m = 2m + 2 + 2m$

$-2 = 4m + 2$

$-2 - 2 = 4m + 2 - 2$

$-4 = 4m$

$-1 = m$

Solution set: $\{-1\}$

27. $\frac{3x - 2}{7} = \frac{x + 2}{5}$

Multiply both sides of the equation by the least common denominator, 35.

$35\left(\frac{3x - 2}{7}\right) = 35\left(\frac{x + 2}{5}\right)$

$5(3x - 2) = 7(x + 2)$

$15x - 10 = 7x + 14$

$15x - 10 - 7x = 7x + 14 - 7x$

$$8x - 10 = 14$$
$$8x - 10 + 10 = 14 + 10$$
$$8x = 24$$
$$x = 3$$

Solution set: $\{3\}$

29.
$$\frac{1}{4p} + \frac{2}{p} = 3$$

Multiply both sides of the equation by the least common denominator, 4p, assuming $p \neq 0$.

$$4p\left(\frac{1}{4p}\right) + 4p\left(\frac{2}{p}\right) = 4p \cdot 3$$
$$1 + 8 = 12p$$
$$9 = 12p$$
$$\frac{1}{12} \cdot 9 = \frac{1}{12} \cdot 12p$$
$$\frac{3}{4} = p$$

Solution set: $\left\{\frac{3}{4}\right\}$

31.
$$\frac{m}{2} - \frac{1}{m} = \frac{6m + 5}{12}$$

Multiply both sides of the equation by the least common denominator, 12m, assuming $m \neq 0$.

$$12m\left(\frac{m}{2} - \frac{1}{m}\right) = 12m\left(\frac{6m + 5}{12}\right)$$
$$6m^2 - 12 = 6m^2 + 5m$$
$$6m^2 - 12 - 6m^2 = 6m^2 + 5m - 6m^2$$
$$-12 = 5m$$
$$\frac{1}{5}(-12) = \frac{1}{5} \cdot 5m$$
$$-\frac{12}{5} = m$$

Solution set: $\left\{-\frac{12}{5}\right\}$

33.
$$\frac{2r}{r - 1} = 5 + \frac{2}{r - 1}$$

Multiply both sides of the equation by the least common denominator, $r - 1$, assuming $r \neq 1$.

$$(r - 1)\left(\frac{2r}{r - 1}\right) = (r - 1)\left(5 + \frac{2}{r - 1}\right)$$
$$2r = 5(r - 1) + 2$$
$$2r = 5r - 5 + 2$$
$$2r = 5r - 3$$
$$2r + 3 = 5r$$
$$3 = 3r$$
$$1 = r$$

Substituting 1 for r in the original equation would result in a denominator of 0, so 1 is not a solution. Solution set: Ø

35.
$$\frac{5}{2a + 3} + \frac{1}{a - 6} = 0$$

Multiply both sides by the least common denominator, $(2a + 3)(a - 6)$, assuming $a \neq -3/2$ and $a \neq 6$.

$$(2a + 3)(a - 6)\left(\frac{5}{2a + 3} + \frac{1}{a - 6}\right)$$
$$= (2a + 3)(a - 6) \cdot 0$$
$$5(a - 6) + (2a + 3) = 0$$
$$5a - 30 + 2a + 3 = 0$$
$$7a - 27 = 0$$
$$7a = 27$$
$$a = \frac{27}{7}$$

Solution set: $\left\{\frac{27}{7}\right\}$

37. $\dfrac{4}{x - 3} - \dfrac{8}{2x + 5} + \dfrac{3}{x - 3} = 0$

$\dfrac{7}{x - 3} - \dfrac{8}{2x + 5} = 0$

Multiply both sides by the least common denominator, $(x - 3)(2x + 5)$, assuming $x \neq 3$ and $x \neq -5/2$.

$(x - 3)(2x + 5)\left(\dfrac{7}{x - 3} - \dfrac{8}{2x + 5}\right)$

$= (x - 3)(2x + 5) \cdot 0$

$7(2x + 5) - 8(x - 3) = 0$

$14x + 35 - 8x + 24 = 0$

$6x + 59 = 0$

$6x = -59$

$x = -\dfrac{59}{6}$

Solution set: $\left\{-\dfrac{59}{6}\right\}$

39. $\dfrac{2p}{p - 2} = 3 + \dfrac{4}{p - 2}$

Multiply both sides by the least common denominator, $p - 2$, assuming $p \neq 2$.

$(p - 2)\left(\dfrac{2p}{p - 2}\right) = (p - 2)\left(3 + \dfrac{4}{p - 2}\right)$

$2p = (p - 2) \cdot 3 + 4$

$2p = 3p - 6 + 4$

$2p = 3p - 2$

$0 = p - 2$

$2 = p$

However, substituting 2 for p in the original equation would result in a denominator at 0, so 2 is not a solution.

Solution set: $\emptyset$

41. $\dfrac{3}{y - 2} + \dfrac{1}{y + 1} = \dfrac{1}{y^2 - y - 2}$

$\dfrac{3}{y - 2} + \dfrac{1}{y + 1} = \dfrac{1}{(y - 2)(y + 1)}$

Multiply both sides by the least common denominator, $(y - 2)(y + 1)$, assuming $y \neq 2$ and $y \neq -1$.

$(y - 2)(y + 1)\left(\dfrac{3}{y - 2}\right)$

$+ (y - 2)(y + 1)\left(\dfrac{1}{y + 1}\right)$

$= (y - 2)(y + 1)\left(\dfrac{1}{(y - 2)(y + 1)}\right)$

$3(y + 1) + (y - 2) = 1$

$3y + 3 + y - 2 = 1$

$4y + 1 = 1$

$4y = 0$

$y = 0$

Solution set: $\{0\}$

43. $.08w + .06(w + 12) = 7.72$

$.08w + .06w + .72 = 7.72$ *Distributive property*

$.14w + .72 = 7.72$

$.14w + .72 - .72 = 7.72 - .72$ *Subtract .72*

$.14w = 7$

$\dfrac{.14w}{.14} = \dfrac{7}{.14}$ *Divide by .14*

$w = 50$

Solution set: $\{50\}$

45. $(3x - 4)^2 - 5 = 3(x + 5)(3x + 2)$

$9x^2 - 24x + 16 - 5 = 3(3x^2 + 17x + 10)$

$9x^2 - 24x + 11 = 9x^2 + 51x + 30$

$9x^2 - 24x + 11 - 9x^2 = 9x^2 + 51x + 30 - 9x^2$

$-24x + 11 = 51x + 30$

$-24x + 11 + 24x = 51x + 30 + 24x$

$11 = 75x + 30$

$11 - 30 = 75x + 30 - 30$

$$-19 = 75x$$

$$\frac{1}{75}(-19) = \frac{1}{75} \cdot 75x$$

$$-\frac{19}{75} = x$$

Solution set: $\left\{-\dfrac{19}{75}\right\}$

47.
$$2(x - a) + b = 3x + a$$
$$2x - 2a + b = 3x + a$$
$$2x - 2a + b - 2x = 3x + a - 2x$$
$$-2a + b = x + a$$
$$-2a + b - a = x + a - a$$
$$-3a + b = x$$

49.
$$ax + b = 3(x - a)$$
$$ax + b = 3x - 3a$$
$$ax + b - ax = 3x - 3a - ax$$
$$b = 3x - ax - 3a$$
$$3a + b = 3x - ax - 3a + 3a$$
$$3a + b = 3x - ax$$
$$3a + b = (3 - a)x \quad \textit{Factor out } x$$
$$\frac{3a + b}{3 - a} = x \qquad \begin{array}{l}\textit{Divide by}\\ 3 - a\end{array}$$

51.
$$\frac{x}{a - 1} = ax + 3$$
$$(a - 1) \cdot \frac{x}{a - 1} = (a - 1)(ax + 3)$$
$$x = a^2x + 3a - ax - 3$$
$$0 = a^2x - ax - x + 3a - 3$$
$$3 - 3a = a^2x - ax - x$$
$$3 - 3a = x(a^2 - a - 1)$$
$$\frac{3 - 3a}{a^2 - a - 1} = x$$

53.
$$a^2x + 3x = 2a^2$$
$$(a^2 + 3)x = 2a^2$$
$$x = \frac{2a^2}{a^2 + 3}$$

55.
$$y = \frac{ax + b}{cx + d}$$
$$y(cx + d) = \frac{ax + b}{cx + d}(cx + d)$$
$$\textit{Multiply by } cx + d$$
$$ycx + yd = ax + b$$
$$\textit{Distributive property}$$
$$ycx - ax = b - yd$$
$$\textit{Subtract } ax \textit{ and } yd$$
$$x(yc - a) = b - yd \quad \textit{Factor out } x$$
$$x = \frac{b - yd}{yc - a} \quad \begin{array}{l}\textit{Divide by}\\ yc - a\end{array}$$

57. Let x = the number of earned runs Clemens allowed.

$$3.13 = \frac{9x}{253\frac{1}{3}}$$

Multiply numerator and denominator of the fraction by 3.

$$3.13 = \frac{27x}{760}$$

Multiply both sides of the equation by 760.

$$760(3.13) = 760 \cdot \frac{27x}{760}$$
$$2378.8 = 27x$$

Divide both sides by 27.

$$\frac{2378.8}{27} = x$$
$$88.10 = x$$

Roger Clemens allowed 88 runs.

59. Let x = Swindell's E.R.A.

$$x = \frac{9(69)}{184\frac{1}{3}}$$

$$= \frac{621}{184\frac{1}{3}}$$

Multiply numerator and denominator of the fraction by 3.

$$x = \frac{1863}{553}$$

$$= 3.37$$

Greg Swindell had an E.R.A. of 3.37.

61. Let x = the number of innings pitched by Blyleven.

$$2.73 = \frac{9(73)}{x}$$

$$2.73 = \frac{657}{x}$$

$$2.73x = x \cdot \frac{657}{x}$$

$$2.73x = 657$$

$$x = \frac{657}{2.73}$$

$$x = 240.66$$

Bert Blyleven pitched 241 innings.

63. 20°C

$$F = \frac{9}{5}C + 32$$

$$= \frac{9}{5}(20) + 32$$

$$= 36 + 32 = 68$$

Therefore, 20°C = 68°F.

65. 59°F

$$C = \frac{5(F - 32)}{9}$$

$$= \frac{5(59 - 32)}{9}$$

$$= \frac{5(27)}{9} = 15$$

Therefore, 59°F = 15°C.

67. 100°F

$$C = \frac{5(F - 32)}{9}$$

$$= \frac{5(100 - 32)}{9}$$

$$= \frac{5}{9}(68) = \frac{340}{9}$$

$$= 37.8$$

Therefore, 100°F = 37.8°C.

69. p = 12, f = $800, b = $4000, q = 36; find A

$$A = \frac{2pf}{b(q + 1)}$$

$$= \frac{2(12)(800)}{4000(36 + 1)}$$

$$= \frac{19,200}{148,000}$$

$$\approx .13$$

The annual interest rate, to the nearest percent, is 13%.

71. A = 14% (or .14), p = 12, b = $2000, q = 36, find f

$$A = \frac{2pf}{b(q + 1)}$$

$$.14 = \frac{2(12)f}{2000(36 + 1)}$$

$$.14(2000)(37) = 24f$$

$$\frac{.14(2000)(37)}{24} = f$$

$$431.67 = f$$

The finance charge, to the nearest dollar, is $432.

73. $A = 16\%$, $p = 12$, $f = \$370$, $q = 36$; find b

$$A = \frac{2pf}{b(q + 1)}$$

$$.16 = \frac{2(12)(370)}{b(36 + 1)}$$

$$37(.16)b = 2(12)(370)$$

$$b = \frac{2(12)(370)}{37(.16)}$$

$$= 1500$$

The balance owed is $1500.

75. $f = \$800$, $q = 36$, $n = 18$

$$u = f \cdot \frac{n(n + 1)}{q(q + 1)}$$

$$= 800 \cdot \frac{18(18 + 1)}{36(36 + 1)}$$

$$= \frac{273,600}{1332}$$

$$= 205.41$$

The amount of unearned interest is $205.41.

77. $f = \$950$, $q = 24$, $n = 6$

$$u = f \cdot \frac{n(n + 1)}{q(q + 1)}$$

$$u = 950 \cdot \frac{6(6 + 1)}{24(24 + 1)}$$

$$= 66.50$$

The amount of unearned interest is $66.50.

79. Since 3 is the solution of the equation, dividing by x − 3 is dividing by zero, which is not allowed. Many such errors result from dividing by zero.

81. $-5x + 11x - 2 = k + 4$

Substitute 2 for x and solve for k.

$$-5(2) + 11(2) - 2 = k + 4$$

$$-10 + 22 - 2 = k + 4$$

$$12 - 2 = k + 4$$

$$10 = k + 4$$

$$k = 6$$

83. $\sqrt{x + k} = 0$

Substitute 2 for x and solve for k.

$$\sqrt{2 + k} = 0$$

$$2 + k = 0 \quad \textit{Square both sides}$$

$$k = -2$$

Substitute 2 for x and −2 for k in the original equation to check this solution.

Section 2.2

1. $i = prt$ for p (simple interest)

$$\frac{i}{rt} = p \quad \textit{Divide by rt}$$

3. $P = 2\ell + 2w$ for w (perimeter of a rectangle)

$$P - 2\ell = 2w \quad \textit{Subtract } 2\ell$$

$$\frac{P - 2\ell}{2} = w \quad \textit{Divide by 2}$$

or $\dfrac{P}{2} - \ell = w$

5. $A = \frac{1}{2}(B + b)h$ for h (area of a trapezoid)

$2A = (B + b)h$ *Multiply by 2*

$\frac{2A}{B + b} = h$ *Divide by B + b*

7. $S = 2\ell w + 2wh + 2h\ell$ for h (surface area of a rectangular box)

$S - 2\ell w = 2\ell w + 2wh$ *Subtract ℓw*

$S - 2\ell w = h(2w + 2\ell)$ *Factor out h*

$\frac{S - 2\ell w}{2w + 2\ell} = h$ *Divide by 2w + 2ℓ*

9. $S = \frac{1}{2}gt^2$ for g (distance traveled by a falling object)

$2s = gt^2$ *Multiply by 2*

$\frac{2s}{t^2} = g$ *Divide by t^2*

11. $\frac{1}{R} = \frac{1}{r_1} + \frac{1}{r_2}$ for R (electricity)

Multiply both sides of the equation by the least common denominator Rr_1r_2.

$Rr_1r_2\left(\frac{1}{R}\right) = Rr_1r_2\left(\frac{1}{r_1}\right) + Rr_1r_2\left(\frac{1}{r_2}\right)$

$r_1r_2 = Rr_2 + Rr_1$ *Distributive property*

$r_1r_2 = R(r_2 + r_1)$ *Factor out R*

$\frac{r_1r_2}{r_2 + r_1} = R$ *Divide by $r_2 + r_1$*

13. $A = \frac{24f}{B(p + 1)}$ for f (approximate annual interest rate)

$AB(p + 1) = \frac{24f}{B(p + 1)} \cdot B(p + 1)$

$\qquad\qquad$ *Multiply by B(p + 1)*

$AB(p + 1) = 24f$

$\frac{AB(p + 1)}{24} = f$ *Divide by 24*

15. $A = P\left(1 + \frac{i}{m}\right)$ for m (compound interest)

$A = P + \frac{Pi}{m}$ *Distributive property*

$A - P = \frac{Pi}{m}$

$m(A - P) = Pi$ *Multiply by m*

$m = \frac{Pi}{A - P}$ *Divide by A − P*

19. The concentration of the mixture cannot possibly be (a) 36%, since this is higher than the concentrations of either of the two solutions being mixed. The concentration of the mixture must be between 26% and 32%, so any of the other choices are possible.

21. Let x = length of shortest side.

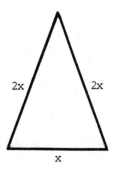

Then 2x = length of one side;

2x = length of other side.

$$x + 2x + 2x = 30$$
$$5x = 30$$
$$x = 6$$

The length of the shortest side is 6 cm.

23. Let x = Hien's grade on the fourth test.

$$\frac{84 + 88 + 92 + x}{4} = 90$$
$$\frac{264 + x}{4} = 90$$
$$264 + x = 360$$
$$x = 96$$

For Hein to have an average of 90, his grade on the fourth test must be 96.

25. Let x = number of liters of pure alcohol to be added.

Strength	Liters of solution	Liters of pure alcohol
10%	7	.10(7)
100%	x	1(x)
30%	7 + x	.30(7 + x)

Liters of alcohol in 10% solution	+	Liters of alcohol in 100% solution	=	Liters of alcohol in 30% solution
.10(7)	+	1(x)	=	.30(7 + x)

$$.10(7) + 1(x) = .30(7 + x)$$
$$.7 + x = 2.1 + .3x$$
$$.7x = 1.4$$
$$x = 2$$

He should add 2 liters of pure alcohol.

27. Let h = the height of the box. Use the formula for the surface area of a rectangular box.

$$S = 2\ell w + 2wh + 2h\ell$$
$$496 = 2 \cdot 18 \cdot 8 + 2 \cdot 8 \cdot h + 2 \cdot h \cdot 18$$
$$\text{Let } S = 496,\ \ell = 18,\ w = 8$$
$$496 = 288 + 16h + 36h$$
$$496 = 288 + 52h$$
$$208 = 52h$$
$$4 = h$$

The height of the box is 4 ft.

29. Let x = the number of liters of 94 octane gasoline.

Strength	Liters of solution	Amount of pure isooctane
94%	x	.94x
99%	200	.99(200)
97%	x + 200	.97(x + 200)

$$.94x + .99(220) = .97(x + 200)$$
$$.94x + 198 = .97x + 194$$
$$4 = .03x$$
$$133\frac{1}{3} = x$$

$133\frac{1}{3}$ (or $\frac{400}{3}$) liters of 94−octane gasoline should be used in the mixture.

31. Let x = time on trip from Denver to Minneapolis.

Set up a chart, using the relationship d = rt.

	d	r	t
Denver to Minneapolis	50x	50	x
Minneapolis to Denver	55(32 − x)	55	32 − x

$$\text{Distance to Minneapolis} = \text{Return distance}$$

$$50x = 55(32 - x)$$
$$50x = 1760 - 55x$$
$$105x = 1760$$
$$x = 16.76$$
$$d = rt$$
$$= 50(16.76)$$
$$\approx 840$$

The distance between the two cities is about 840 mi.

33.   Let d = distance Janet runs.

Then $d + \dfrac{1}{2}$ = distance Russ runs.

Set up a chart, using d = rt.
This is equivalent to t = d/r.

	d	r	t
Russ	$d + \dfrac{1}{2}$	7	$\dfrac{d + \dfrac{1}{2}}{7}$
Janet	d	5	$\dfrac{d}{5}$

Since they both traveled for the same time, we have the equation

$$\frac{d + \frac{1}{2}}{7} = \frac{d}{5}.$$

Multiply both sides by the least common denominator, 35.

$$35\left(\frac{d + \frac{1}{2}}{7}\right) = 35\left(\frac{d}{5}\right)$$

$$5\left(d + \frac{1}{2}\right) = 35\left(\frac{d}{5}\right)$$

$$5d + \frac{5}{2} = 7d$$

$$\frac{5}{2} = 2d$$

$$\frac{5}{4} = d$$

To find t, use either

$$t = \frac{d + \frac{1}{2}}{7}$$

$$\text{or} \quad t = \frac{d}{5}.$$

Then $t = \dfrac{d}{5} = \dfrac{\frac{5}{4}}{5} = \dfrac{1}{4}.$

It will take $\dfrac{1}{4}$ hr or 15 min until they are $\dfrac{1}{2}$ mi apart.

35. Let b = boat speed.

speed upstream
   = boat speed − current
   = b − 5
speed downstream
   = boat speed + current
   = b + 5

Set up a chart, using the relation—ship $d = rt$.

	d	r	t
Upstream	$\frac{1}{3}(b - 5)$	$b - 5$	$\frac{1}{3}$
Downstream	$\frac{1}{6}(b + 5)$	$b + 5$	$\frac{1}{4}$

Since the speeds are given in kilo-meters per hour, the times must be given in hours.

$$\frac{\text{Distance}}{\text{upstream}} = \frac{\text{Distance}}{\text{downsteam}}$$

$$\frac{1}{3}(b - 5) = \frac{1}{4}(b + 5)$$

$$12 \cdot \frac{1}{3}(b - 5) = 12 \cdot \frac{1}{4}(b + 5)$$

$$4(b - 5) = 3(b + 5)$$

$$4b - 20 = 3b + 15$$

$$b = 35$$

The boat speed was 35 km per hr.

37.   Let x = biking speed;

$x + 4.5$ = driving speed.

Set up a chart, using $d = rt$.

	d	r	t
Car	$\frac{1}{3}(x + 4.5)$	$x + 4.5$	$\frac{1}{3}$
Bike	$\frac{3}{4}x$	x	$\frac{3}{4}$

Since the speeds are given in miles per hour, the times must be changed from minutes to hours.

$$\frac{\text{Distance}}{\text{driving}} = \frac{\text{Distance}}{\text{biking}}$$

$$\frac{1}{3}(x + 4.5) = \frac{3}{4}x$$

$$12 \cdot \frac{1}{3}(x + 4.5) = 12 \cdot \frac{3}{4}x$$

$$4(x + 4.5) = 9x$$

$$4x + 18 = 9x$$

$$18 = 5x$$

$$\frac{18}{5} = x$$

To find the distance, use

$$d = \frac{3}{4}x = \frac{3}{4}\left(\frac{18}{5}\right) = \frac{27}{10} = 2.7.$$

Johnny travels 2.7 mi to work.

39.   Let x = the number of hours it takes Le and Tran to clean the house working together.

Their individual rates are

$$\frac{1}{9} = \text{Le's rate (job per hour)};$$

$$\frac{1}{6} = \text{Tran's rate (job per hour)}.$$

Use the formula
rate × time
= portion of the job completed
to complete the chart.

	Rate	Time	Part of the job accomplished
Le	$\frac{1}{9}$	x	$\frac{1}{9}x$
Tran	$\frac{1}{6}$	x	$\frac{1}{6}x$

The sum of the two parts of the job is accomplished is 1, since one whole job is done.

$$\frac{1}{9}x + \frac{1}{6}x = 1$$

Multiply both sides by the least common denominator, 18.

$$18\left(\frac{1}{9}x + \frac{1}{6}x\right) = 18 \cdot 1$$

$$2x + 3x = 18$$

$$5x = 18$$

$$x = \frac{18}{5}$$

It will take them $\frac{18}{5}$ hr or 3.6 hr to clean the house working together.

41. Let   x = the number of hours it takes Plant A to produce the maximum amount of pollutant.

Then 2x = the number of hours it takes Plant B to produce the maximum amount of pollutant.

	Rate	Time	Part of the job accomplished
Plant B	$\frac{1}{2x}$	26	$\frac{1}{2x}(26) = \frac{13}{x}$
Plant A	$\frac{1}{x}$	26	$\frac{1}{x}(26) = \frac{26}{x}$

$$\begin{array}{ccccc} \text{Part done by} & + & \text{Part done by} & = & \text{1 whole} \\ \text{Plant B} & & \text{Plant A} & & \text{job} \\ \downarrow & \downarrow & \downarrow & \downarrow & \downarrow \\ \frac{13}{x} & + & \frac{26}{x} & = & 1 \end{array}$$

Multiply both sides by the least common denominator, x.

$$x\left(\frac{13}{x} + \frac{26}{x}\right) = x \cdot 1$$

$$13 + 26 = x$$

$$39 = x$$

$$78 = 2x$$

It will take plant B 78 hr to produce the maximum pollutant alone.

43. Let x = the number of hours to fill the pool with both pipes open.

	Rate	Time	Part of the job accomplished
Inlet pipe	$\frac{1}{5}$	x	$\frac{1}{5}x$
Outlet pipe	$\frac{1}{8}$	x	$\frac{1}{8}x$

$$\begin{array}{ccccc} \text{Part done by} & & \text{Part done by} & = & \text{Full} \\ \text{inlet pipe} & - & \text{outlet pipe} & & \text{pool} \\ \downarrow & \downarrow & \downarrow & \downarrow & \downarrow \\ \frac{1}{5}x & - & \frac{1}{8}x & = & 1 \end{array}$$

Multiply both sides by the least common denominator, 40.

$$40\left(\frac{1}{5}x - \frac{1}{8}x\right) = 40 \cdot 1$$

$$8x - 5x = 40$$

$$3x = 40$$

$$x = \frac{40}{3}$$

It took $\frac{40}{3}$ hr to fill the pool.

45. Let   x = the regular price.

Then .30x = 30% of the regular price.

$$x - .30x = 245$$

$$.70x = 245$$

$$x = \frac{245}{.70}$$

$$x = 350$$

The regular price was $350.

47. Let x = Jack's mental age.

$$130 = \frac{100 \cdot x}{7}$$

$$910 = 100x$$

$$9.1 = x$$

Jack's mental age is 9.1 yr.

49. 30% of $200,000 is $60,000, so after paying his income tax, Bill had $140,000 left to invest.

Let x = amount invested at 8.5%.
Then 140,000 − x

= amount invested at 7%.

.085x + .07(140,000 − x) = 10,700

Multiply by 1000 to clear decimals.

$$85x + 70(140,000 - x) = 10,700,000$$

$$85x + 9,800,000 - 70x = 10,700,000$$

$$15x = 900,000$$

$$x = 60,000$$

Bill invested $60,000 at 8.5% and
$140,000 − $60,000 = $80,000 at 7%.

51. Let x = amount paid for first plot.
Then 120,000 − x

= amount paid for second plot.

Profit on 1st plot − Loss on 2nd plot = Total profit

$$.15x - .10(120,000 - x) = 5500$$

$$.15x - 12,000 + .10x = 5500$$

$$.25x = 17,500$$

$$x = 70,000$$

$$120,000 - x = 50,000$$

She paid $70,000 for the first plot
and $50,000 for the second plot.

53. Let   x = Cathy's gross weekly pay.

Then .26x = Cathy's weekly deductions.

$$x - .26x = 592$$

$$.74x = 592$$

$$x = 800$$

Cathy's weekly pay is $800 before deductions.

55. Let x = amount invested at 5%.
Then 20,000 − x

= amount invested at 7%.

Interest earned at 5% + Interest earned at 7% = Total interest

$$.05x + .07(20,000 - x) = 1340$$

$$.05x + 1400 - .07x = 1340$$

$$-.02x = -60$$

$$x = 3000$$

$$20,000 - x = 17,000$$

$3000 is invested at 5% (passbook account) and $17,000 is invested at 7% (long-term deposit).

57. Let   x = amount invested at 4.5%.
Then 2x = amount invested at 5%.

$$.045x + .05(2x) = 2900$$

$$45x + 50(2x) = 2,900,000$$
Multiply by 1000

$$45x + 100x = 2,900,000$$

$$145x = 2,900,000$$

$$x = 20,000$$

Louise deposited $20,000 at 4.5% and
2($20,000) = $40,000 at 5%.

59. (b) and (c)

In (b), $-2x + 7(5 - x) = 62$

$$-2x + 35 - 7x = 62$$
$$35 - 9x = 62$$
$$-9x = 27$$
$$x = -3,$$

but the length of a rectangle cannot be negative.

In (c), $4(x + 2) + 4x = 8$

$$4x + 8 + 4x = 8$$
$$8 + 8x = 8$$
$$8x = 0$$
$$x = 0,$$

but the length of a rectangle cannot be zero.

Section 2.3

1. $-9i$ is an imaginary number.

3. π is a real number.

5. $i\sqrt{6}$ is an imaginary number.

7. $2 + 5i$ is an imaginary number.

9. $\sqrt{-100} = i\sqrt{100} = 10i$

11. $-\sqrt{-400} = -i\sqrt{400} = -20i$

13. $-\sqrt{-39} = -i\sqrt{39}$

15. $5 + \sqrt{-4} = 5 + i\sqrt{4} = 5 + 2i$

17. $9 - \sqrt{-50} = 9 - i\sqrt{50}$
$$= 9 - 5i\sqrt{2}$$

19. $\sqrt{-5} \cdot \sqrt{-5} = i\sqrt{5} \cdot i\sqrt{5}$
$$= i^2 \cdot (\sqrt{5})^2$$
$$= (-1)(5) \quad i^2 = -1$$
$$= -5$$

21. $\dfrac{\sqrt{-40}}{\sqrt{-10}} = \dfrac{i\sqrt{40}}{i\sqrt{10}}$
$$= \sqrt{\dfrac{40}{10}}$$
$$= \sqrt{4} = 2$$

23. $\dfrac{\sqrt{-6} \cdot \sqrt{-2}}{\sqrt{3}} = \dfrac{i\sqrt{6} \cdot i\sqrt{2}}{\sqrt{3}}$
$$= \dfrac{i^2 \cdot \sqrt{12}}{\sqrt{3}}$$
$$= -1\sqrt{4} = -2$$

25. $(3 + 2i) + (4 - 3i)$
$$= (3 + 4) + [2 + (-3)]i$$
$$= 7 - i$$

27. $(-2 + 3i) - (-4 + 3i)$
$$= [(-2) - (-4)] + (3 - 3)i$$
$$= 2 + 0i$$

29. $(2 - 5i) - (3 + 4i) - (-2 + i)$
$$= [2 - 3 - (-2)] + [-5 - 4 - 1]i$$
$$= 1 - 10i$$

31. $(2 + i)(3 - 2i)$
$$= 2(3) + 2(-2i) + i(3) + i(-2i)$$
$$= 6 - 4i + 3i - 2i^2$$
$$= 6 - i - 2(-1) \quad i^2 = -1$$
$$= 8 - i$$

33. $(2 + 4i)(-1 + 3i)$

$\quad = 2(-1) + 2(3i) + 4i(-1) + 4i(3i)$

$\quad = -2 + 6i - 4i + 12i^2$

$\quad = -2 + 2i + 12(-1) \quad i^2 = -1$

$\quad = -14 + 2i$

35. $(-3 + 2i)^2$

$\quad = (-3)^2 + 2(-3)(2i) + (2i)^2$
$\qquad\qquad$ *Square of a binomial*

$\quad = 9 - 12i + 4i^2$

$\quad = 9 - 12i + 4(-1)$

$\quad = 5 - 12i$

37. $(2 + 3i)(2 - 3i)$

$\quad = 2^2 - (3i)^2 \quad$ *Difference of*
$\qquad\qquad\qquad$ *two squares*

$\quad = 4 - 9i^2$

$\quad = 4 - 9(-1)$

$\quad = 13 = 13 + 0i \quad$ *Standard form*

39. $(\sqrt{6} + i)(\sqrt{6} - i)$

$\quad = (\sqrt{6})^2 - i^2 \quad$ *Difference of*
$\qquad\qquad\qquad$ *two squares*

$\quad = 6 - i^2$

$\quad = 6 - (-1)$

$\quad = 7 = 7 + 0i \quad$ *Standard form*

41. $i(3 - 4i)(3 + 4i)$

$\quad = i[3^2 - (4i)^2] \quad$ *Difference of*
$\qquad\qquad\qquad\quad$ *two squares*

$\quad = i[9 - (-16)]$

$\quad = 25i = 0 + 25i \quad$ *Standard form*

43. $3i(2 - i)^2$

$\quad = 3i[2^2 - 2(2)(i) + i^2]$
$\qquad\qquad\qquad$ *Square of a binomial*

$\quad = 3i(4 - 4i - 1)$

$\quad = 3i(3 - 4i)$

$\quad = 9i - 12i^2$

$\quad = 9i - 12(-1)$

$\quad = 12 + 9i$

45. $i^5 = i^4 \cdot i = 1 \cdot i = i$

47. $i^9 = i^8 \cdot i$

$\quad = (i^4)^2 \cdot i$

$\quad = 1^2 \cdot i = i$

49. $i^{12} = (i^4)^3 = 1^3 = 1$

51. $i^{43} = i^{40} \cdot i^3$

$\quad = (i^4)^{10} \cdot i^3$

$\quad = 1^{10} \cdot i^3 = -i$

53. $\dfrac{1}{i^{12}} = \dfrac{1}{(i^4)^3} = \dfrac{1}{1^3} = 1$

55. $i^{-15} = i^{-16} \cdot i$

$\quad = (i^4)^{-4} \cdot i$

$\quad = 1^{-4} \cdot i = i$

59. $\dfrac{1 + i}{1 - i}$

Multiply numerator and denominator by $1 + i$, the conjugate of the denominator.

$\dfrac{1 + i}{1 - i} = \dfrac{(1 + i)(1 + i)}{(1 - i)(1 + i)}$

$\quad = \dfrac{1 + 2i + i^2}{1 - i^2} \quad$ *Multiply*

$\quad = \dfrac{1 + 2i - 1}{1 + 1} \quad i^2 = -1$

$\quad = \dfrac{2i}{2}$

$\quad = i = 0 + i \quad$ *Standard form*

61. $\dfrac{4 - 3i}{4 + 3i}$

Multiply numerator and denominator by $4 - 3i$, the conjugate of the denominator.

$$\frac{4 - 3i}{4 + 3i} = \frac{(4 - 3i)(4 - 3i)}{(4 + 3i)(4 - 3i)}$$

$$= \frac{16 - 24i + 9i^2}{16 - 9i^2} \quad \textit{Multiply}$$

$$= \frac{16 - 24i - 9}{16 + 9} \quad i^2 = -1$$

$$= \frac{7 - 24i}{25}$$

$$= \frac{7}{25} - \frac{24}{25}i \qquad \begin{array}{l}\textit{Standard}\\ \textit{form}\end{array}$$

63. $\dfrac{3 - 4i}{2 - 5i} = \dfrac{(3 - 4i)(2 + 5i)}{(2 - 5i)(2 + 5i)}$

$$= \frac{6 + 15i - 8i - 20i^2}{4 - 25i^2}$$
$$\qquad\qquad\qquad \textit{Multiply}$$

$$= \frac{26 + 7i}{29}$$

$$= \frac{26}{29} + \frac{7}{29}i \quad \begin{array}{l}\textit{Standard}\\ \textit{form}\end{array}$$

65. $\dfrac{-3 + 4i}{2 - i} = \dfrac{(-3 + 4i)(2 + i)}{(2 - i)(2 + i)}$

$$= \frac{-6 - 3i + 8i + 4i^2}{4 - i^2}$$

$$= \frac{-10 + 5i}{5}$$

$$= -\frac{10}{5} + \frac{5}{5}i$$

$$= -2 + i \quad \textit{Lowest terms}$$

67. $\dfrac{2}{i} = \dfrac{2(-i)}{i(-i)}$ $-i$ *is the conjugate of* i

$$= \frac{-2i}{-i^2} = \frac{-2i}{1} = -2i = 0 - 2i$$

69. $\dfrac{1 - \sqrt{-5}}{3 + \sqrt{-4}} = \dfrac{1 - i\sqrt{5}}{3 + 2i}$

$$= \frac{(1 - i\sqrt{5})(3 - 2i)}{(3 + 2i)(3 - 2i)}$$

$$= \frac{3 - 2i - 3i\sqrt{5} + 2i^2\sqrt{5}}{9 - 4i^2}$$

$$= \frac{3 - 2\sqrt{5} - (2 + 3\sqrt{5})i}{13}$$

$$= \frac{3 - 2\sqrt{5}}{13} + \frac{-2 - 3\sqrt{5}}{13}i$$
$$\qquad\qquad\qquad \textit{Standard form}$$

71. $\dfrac{2 + i}{3 - i} \cdot \dfrac{5 + 2i}{1 + i}$

$$= \frac{10 + 4i + 5i + 2i^2}{3 - i + 3i - i^2}$$

$$= \frac{8 + 9i}{4 + 2i}$$

$$= \frac{(8 + 9i)(4 - 2i)}{(4 + 2i)(4 - 2i)}$$

$$= \frac{32 - 16i + 36i - 18i^2}{16 - 4i^2}$$

$$= \frac{50 + 20i}{20} = \frac{50}{20} + \frac{20}{20}i$$

$$= \frac{5}{2} + i \quad \textit{Lowest terms}$$

73. $\dfrac{6 + 2i}{5 - i} \cdot \dfrac{1 - 3i}{2 + 6i}$

$$= \frac{2(3 + i)}{5 - i} \cdot \frac{1 - 3i}{2(1 + 3i)}$$

$$= \frac{(3 + i)}{(5 - i)} \cdot \frac{(1 - 3i)}{(1 + 3i)} \quad \begin{array}{l}\textit{Lowest}\\ \textit{terms}\end{array}$$

$$= \frac{3 - 9i + i - 3i^2}{5 - i + 15i - 3i^2}$$

$$= \frac{6 - 8i}{8 + 14i} = \frac{2(3 - 4i)}{2(4 + 7i)}$$

$$= \frac{3 - 4i}{4 + 7i} \quad \begin{array}{l}\textit{Lowest}\\ \textit{terms}\end{array}$$

$$= \frac{(3 - 4i)(4 - 7i)}{(4 + 7i)(4 - 7i)} \quad \begin{array}{l}\textit{Multiply by}\\ \textit{conjugate of}\\ \textit{denominator}\end{array}$$

$$= \frac{12 - 21i - 16i + 28i^2}{16 - 49i^2}$$

$$= \frac{-16 - 37i}{65}$$

$$= -\frac{16}{65} - \frac{37}{65}i \qquad \textit{Standard form}$$

75. $\dfrac{5 - i}{3 + i} + \dfrac{2 + 7i}{3 - i}$

$= \dfrac{7 + 6i}{3 + i}$

$= \dfrac{(7 + 6i)(3 - i)}{(3 + i)(3 - i)}$

$= \dfrac{21 - 7i + 18i - 6i^2}{9 - i^2}$

$= \dfrac{27 + 11i}{10}$

$= \dfrac{27}{10} + \dfrac{11}{10}i$ *Standard form*

77. $(a + bi)^2 = a^2 + 2abi + b^2i^2$

$\qquad\qquad = (a^2 - b^2) + 2abi$

For the square $(a + bi)^2$ to be real, the coefficient of i would have to be zero, or $2ab = 0$. This means either $a = 0$ or $b = 0$.

79. Show that $\dfrac{\sqrt{2}}{2} + \dfrac{\sqrt{2}}{2}i$ is a square root of i.

We must show that $\left(\dfrac{\sqrt{2}}{2} + \dfrac{\sqrt{2}}{2}i\right)^2 = i$.

$\left(\dfrac{\sqrt{2}}{2} + \dfrac{\sqrt{2}}{2}i\right)^2$

$= \dfrac{2}{4} + 2\left(\dfrac{2}{4}i\right) + \dfrac{2}{4}i^2$ *Square of a binomial*

$= \dfrac{1}{2} + i - \dfrac{1}{2}$

$= i$

81. Evaluate $3z - z^2$ if $z = 3 - 2i$.

$3z - z^2 = 3(3 - 2i) - (3 - 2i)^2$

$\qquad\quad = 9 - 6i - (9 - 12i + 4i^2)$

$\qquad\quad = 9 - 6i - (9 - 12i - 4)$

$\qquad\quad = 9 - 6i - (5 - 12i)$

$\qquad\quad = 9 - 6i - 5 + 12i$

$\qquad\quad = 4 + 6i$

Section 2.4

1. $p^2 = 16$

$\quad p = \sqrt{16}$ or $p = -\sqrt{16}$

$\qquad\qquad\qquad$ *Square root property*

$\quad p = 4$ or $p = -4$

Solution set: $\{\pm 4\}$

3. $x^2 = 27$

$\quad x = \sqrt{27}$ or $x = -\sqrt{27}$

$\qquad\qquad\qquad$ *Square root property*

$\quad x = 3\sqrt{3}$ or $x = -3\sqrt{3}$

Solution set: $\{\pm 3\sqrt{3}\}$

5. $t^2 = -16$

$\quad t = \sqrt{-16}$ or $t = -\sqrt{-16}$

$\qquad\qquad\qquad$ *Square root property*

$\quad t = 4i$ or $t = -4i$

Solution set: $\{\pm 4i\}$

7. $x^2 = -18$

$\quad x = \sqrt{-18}$ or $x = -\sqrt{-18}$

$\quad x = 3i\sqrt{2}$ or $x = -3i\sqrt{2}$

Solution set: $\{\pm 3i\sqrt{2}\}$

9. $(3k - 1)^2 = 12$

$\qquad 3k - 1 = \pm\sqrt{12}$

$\qquad 3k - 1 = \pm 2\sqrt{3}$

$\qquad\quad 3k = 1 \pm 2\sqrt{3}$

$\qquad\qquad k = \dfrac{1 \pm 2\sqrt{3}}{3}$

Solution set: $\left\{\dfrac{1 \pm 2\sqrt{3}}{3}\right\}$

11. $p^2 - 5p + 6 = 0$

$(p - 2)(p - 3) = 0$ *Factor*

$p - 2 = 0$ or $p - 3 = 0$
    *Zero-factor
    property*

$p = 2$ or $p = 3$

Solution set: $\{2, 3\}$

13. $6z^2 - 5z - 50 = 0$

$(2z + 5)(3z - 10) = 0$ *Factor*

$2z + 5 = 0$ or $3z - 10 = 0$
    *Zero-factor
    property*

$2z = -5$ or $3z = 10$

$z = -\dfrac{5}{2}$ or $z = \dfrac{10}{3}$

Solution set: $\left\{-\dfrac{5}{2}, \dfrac{10}{3}\right\}$

15. $(5r - 3)^2 = -3$

$5r - 3 = \pm\sqrt{-3}$

$5r - 3 = \pm i\sqrt{3}$

$5r = 3 \pm i\sqrt{3}$

$r = \dfrac{3 \pm i\sqrt{3}}{5}$

$= \dfrac{3}{5} \pm \dfrac{\sqrt{3}}{5}i$ *Standard form*

Solution set: $\left\{\dfrac{3}{5} \pm \dfrac{\sqrt{3}}{5}i\right\}$

17. $p^2 - 8p + 15 = 0$

$p^2 - 8p = -15$

Half the coefficient of p is −4, and $(-4)^2 = 16$. Add 16 to both sides.

$p^2 - 8p + 16 = -15 + 16$

Factor on the left and combine terms on the right.

$(p - 4)^2 = 1$

Use the square root property to complete the solution.

$p - 4 = \pm 1$

$p = 4 + 1 = 5$ or $p = 4 - 1 = 3$

Solution set: $\{5, 3\}$

19. $x^2 - 2x - 4 = 0$

$x^2 - 2x = 4$

$x^2 - 2x + 1 = 4 + 1$

$(x - 1)^2 = 5$

$x - 1 = \pm\sqrt{5}$

$x = 1 \pm \sqrt{5}$

Solution set: $\{1 \pm \sqrt{5}\}$

21. $2p^2 + 2p + 1 = 0$

$p^2 + p + \dfrac{1}{2} = 0$ *Multiply by 1/2*

$p^2 + p = -\dfrac{1}{2}$ *Subtract 1/2*

Half the coefficient of p is 1/2, and $(1/2)^2 = 1/4$. Add 1/4 to both sides.

$p^2 + p + \dfrac{1}{4} = -\dfrac{1}{2} + \dfrac{1}{4}$

$\left(p + \dfrac{1}{2}\right)^2 = -\dfrac{1}{4}$

$p + \dfrac{1}{2} = \pm\sqrt{-\dfrac{1}{4}} = \dfrac{\pm i}{2}$

$p = -\dfrac{1}{2} \pm \dfrac{1}{2}i$

Solution set: $\left\{-\dfrac{1}{2} \pm \dfrac{1}{2}i\right\}$

25. $y^2 - 3y - 2 = 0$

Here $a = 1$, $b = -3$, and $c = -2$. Substitute these values into the quadratic formula.

$$y = \frac{-b \pm \sqrt{b^2 - 4ac}}{2a}$$

$$= \frac{-(-3) \pm \sqrt{(-3)^2 - 4(1)(-2)}}{2(1)}$$

$$= \frac{3 \pm \sqrt{9 + 8}}{2}$$

$$y = \frac{3 \pm \sqrt{17}}{2}$$

Solution set: $\left\{ \dfrac{3 \pm \sqrt{17}}{2} \right\}$

27. $11p^2 - 7p + 1 = 0$

Here $a = 11$, $b = -7$, and $c = 1$.
Substitute these values into the
quadratic formula.

$$p = \frac{-b \pm \sqrt{b^2 - 4ac}}{2a}$$

$$p = \frac{-(-7) \pm \sqrt{(-7)^2 - 4(11)(1)}}{2(11)}$$

$$= \frac{7 \pm \sqrt{49 - 44}}{22}$$

$$p = \frac{7 \pm \sqrt{5}}{22}$$

Solution set: $\left\{ \dfrac{7 \pm \sqrt{5}}{22} \right\}$

29. $x^2 = 2x - 5$

$x^2 - 2x + 5 = 0$

$a = 1$, $b = -2$, $c = 5$

$$x = \frac{-(-2) \pm \sqrt{(-2)^2 - 4(1)(5)}}{2 \cdot 1}$$

$$= \frac{2 \pm \sqrt{4 - 20}}{2}$$

$$= \frac{2 \pm \sqrt{-16}}{2}$$

$$= \frac{2 \pm 4i}{2}$$

$$= \frac{2(1 \pm 2i)}{2} \quad \textit{Factor numerator}$$

$x = 1 + 2i \qquad \textit{Lowest terms}$

Solution set: $\{1 \pm 2i\}$

31. $\frac{2}{3}x^2 + \frac{1}{4}x = 3$

Multiply both sides of the equation
by the least common denominator, 12.

$$8x^2 + 3x = 36$$

Rewrite this equation in standard
form.

$8x^2 + 3x - 36 = 0$

Here $a = 8$, $b = 3$, and $c = -36$.
Substitute these values into the
quadratic formula.

$$x = \frac{-(3) \pm \sqrt{3^2 - 4(8)(-36)}}{2(8)}$$

$$= \frac{-3 \pm \sqrt{9 + 1152}}{16}$$

$$= \frac{-3 \pm \sqrt{1161}}{16}$$

$$= \frac{-3 \pm \sqrt{9}\sqrt{129}}{16}$$

$$x = \frac{-3 \pm 3\sqrt{129}}{16}$$

Solution set: $\left\{ \dfrac{-3 \pm 3\sqrt{129}}{16} \right\}$

33. $4 + \frac{3}{x} - \frac{2}{x^2} = 0$

Multiply both sides of the equation
by the least common denominator, x^2.

$$4x^2 + 3x - 2 = 0$$

Here $a = 4$, $b = 3$, and $c = -2$.
Substitute these values into the
quadratic formula.

$$x = \frac{-(3) \pm \sqrt{3^2 - 4(4)(-2)}}{2(4)}$$

$$= \frac{-3 \pm \sqrt{9 + 32}}{8}$$

$$x = \frac{-3 \pm \sqrt{41}}{8}$$

Solution set: $\left\{ \dfrac{-3 \pm \sqrt{41}}{8} \right\}$

35. $3 - \dfrac{4}{p} = \dfrac{2}{p^2}$

Multiply both sides of the equation by the least common denominator, p^2.

$$3p^2 - 4p = 2$$

Rewrite this equation in standard form.

$$3p^2 - 4p - 2 = 0$$

Here $a = 3$, $b = 4$, and $c = -2$. Substitute these values into the quadratic formula.

$$p = \frac{-(-4) \pm \sqrt{(-4)^2 - 4(3)(-2)}}{2(3)}$$

$$= \frac{4 \pm \sqrt{16 + 24}}{6} = \frac{4 \pm \sqrt{40}}{6}$$

$$= \frac{4 \pm 2\sqrt{10}}{6} = \frac{2(2 \pm \sqrt{10})}{6}$$

$$p = \frac{2 \pm \sqrt{10}}{3}$$

Solution set: $\left\{ \dfrac{2 \pm \sqrt{10}}{3} \right\}$

37.   (d) $(7x + 4)^2 = 11$

This equation has two real, distinct solutions since the positive number 11 has a positive square root and a negative square root.

39.   (a) $(3x - 4)^2 = -4$

This equation has two imaginary solutions since the negative number -4 has two imaginary square roots.

41.
$$x^3 - 1 = 0$$
$$(x - 1)(x^2 + x + 1) = 0$$
Factor as differ-
ence of two cubes

$x - 1 = 0$ or $x^2 + x + 1 = 0$
Zero-factor
property

$x = 1$ or $x = \dfrac{-1 \pm \sqrt{1 - 4}}{2}$
Quadratic formula

$x = 1$ or $x = -\dfrac{1}{2} \pm \dfrac{\sqrt{3}}{2}i$
Standard form

Solution set: $\left\{ 1, \ -\dfrac{1}{2} \pm \dfrac{\sqrt{3}}{2}i \right\}$

43.
$$x^3 + 27 = 0$$
$$(x + 3)(x^2 - 3x + 9) = 0$$
Factor as sum
of two cubes

$x + 3 = 0$ or $x^2 - 3x + 9 = 0$
Zero-factor
property

$x = -3$ or $x = \dfrac{3 \pm \sqrt{9 - 36}}{2}$
Quadratic
formula

$$x = \frac{3 \pm \sqrt{-27}}{2}$$

$x = -3$ or $x = \dfrac{3}{2} \pm \dfrac{3\sqrt{3}}{2}i$
Standard form

Solution set: $\left\{ -3, \ \dfrac{3}{2} \pm \dfrac{3\sqrt{3}}{2}i \right\}$

45.
$$2 - \frac{5}{k} + \frac{2}{k^2} = 0$$

Multiply both sides by the least common denominator, k^2.

$$2k^2 - 5k + 2 = 0$$
$$(2k - 1)(k - 2) = 0$$
$$2k - 1 = 0 \quad \text{or} \quad k - 2 = 0$$
$$k = \frac{1}{2} \quad \text{or} \quad k = 2$$

Solution set: $\left\{\frac{1}{2}, 2\right\}$

47.
$$t^2 - t = 3$$
$$t^2 - t - 3 = 0$$
$$a = 1, \ b = -1, \ c = -3$$

$$t = \frac{-(-1) \pm \sqrt{(-1)^2 - 4(1)(-3)}}{2(1)}$$

$$t = \frac{1 \pm \sqrt{13}}{2}$$

Solution set: $\left\{\frac{1 \pm \sqrt{13}}{2}\right\}$

49. $64r^3 - 343 = 0$

Factor the left side as the difference of two cubes.

$$(4r)^3 - 7^3 = 0$$
$$(4r - 7)[(4r)^2 + (4r)(7) + 7^2] = 0$$
$$(4r - 7)(16r^2 + 28r + 49) = 0$$

By the zero factor property,

$$4r - 7 = 0 \quad \text{or} \quad 16r^2 + 28r + 49 = 0.$$

Solve $4r - 7 = 0$.

$$r = \frac{7}{4}$$

Solve $16r^2 + 28r + 49 = 0$ by the quadratic formula.

$$a = 16, \ b = 28, \ c = 49$$

$$r = \frac{-28 \pm \sqrt{28^2 - 4(16)(49)}}{2(16)}$$

$$= \frac{-28 \pm \sqrt{784 - 3136}}{32}$$

$$= \frac{-28 \pm \sqrt{-2352}}{32}$$

$$= \frac{-28 \pm i\sqrt{784 \cdot \sqrt{3}}}{32}$$

$$= \frac{-28 \pm 28i\sqrt{3}}{32}$$

$$= \frac{4(-7 \pm 7i\sqrt{3})}{4 \cdot 8}$$

$$= \frac{-7 \pm 7i\sqrt{3}}{8}$$

$$r = -\frac{7}{8} \pm \frac{7\sqrt{3}}{8}i \quad \text{Standard form}$$

Solution set: $\left\{\frac{7}{4}, \ -\frac{7}{8} \pm \frac{7\sqrt{3}}{8}i\right\}$

51. $(3y + 1)^2 = -7$

Use the square root property.

$$3y + 1 = \pm i\sqrt{7}$$
$$3y = -1 \pm i\sqrt{7}$$
$$y = \frac{-1 \pm i\sqrt{7}}{3}$$
$$= -\frac{1}{3} \pm \frac{\sqrt{7}}{3}i \quad \begin{array}{l}\textit{Standard}\\\textit{form}\end{array}$$

Solution set: $\left\{-\frac{1}{3} \pm \frac{\sqrt{7}}{3}i\right\}$

53. $m^2 - \sqrt{2}m - 1 = 0$

Use the quadratic formula with $a = 1, \ b = -\sqrt{2},$ and $c = -1$.

$$m = \frac{-(-2) \pm \sqrt{(-\sqrt{2})^2 - 4(1)(-1)}}{2(1)}$$

$$= \frac{\sqrt{2} \pm \sqrt{2 + 4}}{2}$$

$$= \frac{\sqrt{2} \pm \sqrt{6}}{2}$$

Solution set: $\left\{\frac{\sqrt{2} \pm \sqrt{6}}{2}\right\}$

55. $\sqrt{2}p^2 - 3p + \sqrt{2} = 0$

$$a = \sqrt{2}, \; b = -3, \; c = \sqrt{2}$$

$$p = \frac{-(-3) \pm \sqrt{(-3)^2 - 4(2)(\sqrt{2})}}{2\sqrt{2}}$$

$$= \frac{3 \pm \sqrt{9 - 8}}{2\sqrt{2}}$$

$$= \frac{3 \pm \sqrt{1}}{2\sqrt{2}} = \frac{3 \pm 1}{2\sqrt{2}}$$

This expression gives two solutions:

$$p = \frac{3 + 1}{2\sqrt{2}} = \frac{4}{2\sqrt{2}} = \frac{2}{\sqrt{2}} \cdot \frac{\sqrt{2}}{\sqrt{2}} = \frac{2\sqrt{2}}{2} = \sqrt{2}$$

or $p = \dfrac{3 - 1}{2\sqrt{2}} = \dfrac{2}{2\sqrt{2}} = \dfrac{1}{\sqrt{2}} \cdot \dfrac{\sqrt{2}}{\sqrt{2}} = \dfrac{\sqrt{2}}{2}.$

Solution set: $\left\{ \sqrt{2}, \dfrac{\sqrt{2}}{2} \right\}$

57. $x^2 + \sqrt{5}x + 1 = 0$

$$a = 1, \; b = \sqrt{5}, \; c = 1$$

$$x = \frac{-\sqrt{5} \pm \sqrt{(\sqrt{5})^2 - 4(1)(1)}}{2(1)}$$

$$= \frac{-\sqrt{5} \pm \sqrt{5 - 4}}{2}$$

$$= \frac{-\sqrt{5} \pm 1}{2}$$

Solution set: $\left\{ \dfrac{-\sqrt{5} \pm 1}{2} \right\}$

59. $s = \dfrac{1}{2}gt^2$ for t

First, multiply both sides by 2.

$$2s = gt^2$$

Now divide by g.

$$\frac{2s}{g} = t^2$$

Use the square root property and rationalize the denominator on the right.

$$t = \pm\sqrt{\frac{2s}{g}} = \pm\frac{\sqrt{2s}}{\sqrt{g}} \cdot \frac{\sqrt{g}}{\sqrt{g}}$$

$$= \pm\frac{\sqrt{2sg}}{g}$$

61. $F = \dfrac{kMv^4}{r}$ for r

First, multiply both sides by r.

$$Fr = kMv^4$$

Next, divide both sides by kM.

$$\frac{Fr}{kM} = v^4$$

Now take the fourth root of both sides.

$$v = \pm\sqrt[4]{\frac{Fr}{kM}}$$

Finally, rationalize the denominator on the right.

$$v = \pm\frac{\sqrt[4]{Fr}}{\sqrt[4]{kM}} \cdot \frac{\sqrt[4]{k^3M^3}}{\sqrt[4]{k^3M^3}}$$

$$= \frac{\pm\sqrt[4]{Frk^3M^3}}{kM}$$

63. $P = \dfrac{E^2R}{(r + R)^2}$ for R

$$P(r + R)^2 = E^2R$$

$$P(r^2 + 2rR + R^2) = E^2R$$

$$Pr^2 + 2PrR + PR^2 = E^2R$$

$$PR^2 - E^2R + 2PrR + Pr^2 = 0$$

$$PR^2 + (2Pr - E^2)R + Pr^2 = 0$$

To find R, use the quadratic formula with $a = P$, $b = 2Pr - E^2$, and $c = Pr^2$.

$$R = \frac{-(2Pr - E^2) \pm \sqrt{(2Pr - E^2) - 4(P)(Pr^2)}}{2(P)}$$

$$= \frac{-2Pr + E^2 \pm \sqrt{4P^2r^2 - 4PrE^2 + E^4 - 4P^2r^2}}{2P}$$

$$= \frac{-2Pr + E^2 \pm \sqrt{E^4 - 4PrE^2}}{2P}$$

$$= \frac{-2Pr + E^2 \pm \sqrt{E^2(E^2 - 4Pr)}}{2P}$$

$$R = \frac{E^2 - 2Pr \pm E\sqrt{E^2 - 4Pr}}{2P}$$

65. $4x^2 - 2xy + 3y^2 = 2$

(a) Solve for x in terms of y.

$$4x^2 - 2yx + 3y^2 - 2 = 0$$

$$a = 4, \quad b = -2y, \quad c = 3y^2 - 2$$

$$x = \frac{2y \pm \sqrt{4y^2 - 16(3y^2 - 2)}}{8}$$

$$= \frac{2y \pm \sqrt{4y^2 - 48y^2 + 32}}{8}$$

$$= \frac{2y \pm \sqrt{32 - 44y^2}}{8}$$

$$= \frac{2y \pm \sqrt{4(8 - 11y^2)}}{8}$$

$$= \frac{2y \pm 2\sqrt{8 - 11y^2}}{4}$$

$$x = \frac{y \pm \sqrt{8 - 11y^2}}{4}$$

(b) Solve for y in terms of x.

$$3y^2 - 2xy + 4x^2 - 2 = 0$$

$$a = 3, \quad b = -2x, \quad c = 4x^2 - 2$$

$$y = \frac{2x \pm \sqrt{4x^2 - 12(4x^2 - 2)}}{6}$$

$$= \frac{2x \pm \sqrt{4x^2 - 48x^2 + 24}}{6}$$

$$= \frac{2x \pm \sqrt{24 - 44x^2}}{6}$$

$$= \frac{2x \pm \sqrt{4(6 - 11x^2)}}{6}$$

$$= \frac{2x \pm 2\sqrt{6 - 11x^2}}{6}$$

$$y = \frac{x \pm \sqrt{6 - 11x^2}}{3}$$

67. $x^2 + 8x + 16 = 0$

$$a = 1, \quad b = 8, \quad c = 16$$

$$b^2 - 4ac = 8^2 - 4(1)(16)$$

$$= 64 - 64 = 0$$

The equation has one rational solution since the discriminant is 0.

69. $3m^2 - 5m + 2 = 0$

$$a = 3, \quad b = -5, \quad c = 2$$

$$b^2 - 4ac = (-5)^2 - 4(3)(2)$$

$$= 25 - 24 = 1$$

The equation has two different rational solutions since the discriminant is a positive perfect square.

71. $4p^2 = 6p + 3$

$$a = 4, \quad b = -6, \quad c = -3$$

$$b^2 - 4ac = (-6)^2 - 4(4)(-3)$$

$$= 36 + 48$$

$$= 84$$

The equation has two different irrational solutions since the discriminant is positive but not a perfect square.

73. $9k^2 + 11k + 4 = 0$

$a = 9, \ b = 11, \ c = 4$

$$b^2 - 4ac = 11^2 - 4(9)(4)$$
$$= 121 - 144$$
$$= -23$$

The equation has **two different imaginary** solutions since the discriminant is negative.

75. $8x^2 - 72 = 0$

$a = 8, \ b = 0, \ c = -72$

$$b^2 - 4ac = 0^2 - 4(8)(-72)$$
$$= 0 + 2304$$
$$= 2304 = 48^2$$

The equation has **two different rational** solutions since the discriminant is a positive perfect square.

77. $25m^2 - 10m + k = 0$

This equation will have exactly one solution if the discriminant is zero.

$a = 25, \ b = -10, \ c = k$

$$b^2 - 4ac = (-10)^2 - 4(25)k$$
$$= 100 - 100k$$

Let $b^2 - 4ac = 0$ and solve for k.

$$100 - 100k = 0$$
$$100 = 100k$$
$$k = 1$$

The equation has exactly one solution when $k = 1$.

81. $\dfrac{3 - 2x}{3x^2 - 19x - 14}$

To find the restrictions on the variable, set the denominator equal to zero and solve.

$$3x^2 - 19x - 14 = 0$$
$$(3x + 2)(x - 7) = 0$$
$$3x + 2 = 0 \quad \text{or} \quad x - 7 = 0$$
$$x = -\frac{2}{3} \quad \text{or} \qquad x = 7$$

The restrictions on the variable are $x \neq -\dfrac{2}{3}$ and $x \neq 7$.

83. $\dfrac{-3}{16y^2 + 8y + 1}$

To find the restrictions on the variable, set the denominator equal to zero and solve.

$$16y^2 + 8y + 1 = 0$$
$$(4y + 1)^2 = 0$$
$$4y + 1 = 0$$
$$y = -\frac{1}{4}$$

The restriction on the variable is $x \neq -\dfrac{1}{4}$.

85. $\dfrac{8 + 7x}{x^2 + x + 1}$

To find the restrictions on the variable, set the denominator equal to zero and solve by the quadratic formula with $a = 1$, $b = 1$, and $c = 1$.

$$x^2 + x + 1 = 0$$

$$x = \frac{-1 \pm \sqrt{1 - 4(1)(1)}}{2(1)}$$

$$= \frac{-1 \pm \sqrt{-3}}{2}$$

$$= \frac{-1 \pm i\sqrt{3}}{2}$$

We can also see that the equation has no real solutions by finding the discriminant:

$$b^2 - 4ac = 1 - 4 = -3.$$

Since there are no real solutions to this equation, there are no real numbers that make the denominator equal to zero. Thus, there are no real number restrictions on x.

87. (a) By the quadratic formula the values of r_1 and r_2 are

$$r_1 = \frac{-b + \sqrt{b^2 - 4ac}}{2a} \quad \text{and}$$

$$r_2 = \frac{-b - \sqrt{b^2 - 4ac}}{2a}.$$

It follows that

$$r_1 + r_2$$

$$= \frac{-b + \sqrt{b^2 - 4ac}}{2a} + \frac{-b - \sqrt{b^2 - 4ac}}{2a}$$

$$= \frac{-b + \sqrt{b^2 - 4ac} - b - \sqrt{b^2 - 4ac}}{2a}$$

$$= \frac{-2b}{2a}$$

$$= -\frac{b}{a}.$$

(b) $r_1 \cdot r_2$

$$= \frac{-b + \sqrt{b^2 - 4ac}}{2a} \cdot \frac{-b - \sqrt{b^2 - 4ac}}{2a}$$

$$= \frac{(-b)^2 - (\sqrt{b^2 - 4ac})^2}{4a^2}$$

Difference of two squares

$$= \frac{b^2 - (b^2 - 4ac)}{4a^2}$$

$$= \frac{4ac}{4a^2}$$

$$= \frac{c}{a}$$

Section 2.5

1. Let   x = the width of the parking lot.

Then 2w + 200 = the length.
See the figure in the textbook.
The area of a rectangle is given by the formula A = LW. Here
A = 40,000, L = 2w + 200, W = w.

$$A = LW$$
$$40,000 = (2w + 200)w$$
$$40,000 = 2w^2 + 200w$$
$$0 = 2w^2 + 200w - 40,000$$
$$0 = w^2 + 100w - 20,000$$
$$0 = (w - 100)(w + 200)$$
$$w = 100 \quad \text{or} \quad w = -200$$

A rectangle cannot have a negative width, so reject -200 as a solution.
If w = 100, then 2w + 200 = 400.
The dimensions of the lot are 100 yd by 400 yd.

3. Let   x = outside width of frame.

    Then x + 3 = outside length of
                 frame.

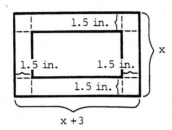

Since 70 sq inches is the area of
the unframed picture, we must
express the dimensions of the
unframed picture in terms of x.
Since the frame extends 1.5 inches
beyond the picture on each side, the
width of the picture is x - 2(1.5) =
x - 3, and the length of the picture
is x + 3 - 2(1.5) = x + 3 - 3 = x.
Apply the formula A = LW to the rectangular picture.

$$A = LW$$
$$70 = x(x - 3)$$
$$70 = x^2 - 3x$$
$$0 = x^2 - 3x - 70$$
$$0 = (x + 7)(x - 10)$$
$$x = -7 \quad or \quad x = 10$$

The width of a rectangle cannot be
negative, so reject -7 as a solution.

If   x = 10, x + 3 = 13.

The outside dimensions of the frame
are 10 inches by 13 inches.

5. Let   x = width of margins.

    Then 23 + 2x = length of page;

        18 + 2x = width of page.

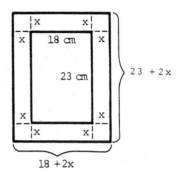

Write an equation for the area of
the page using the formula A = LW.

$$594 = (23 + 2x)(18 + 2x)$$
$$594 = 414 + 36x + 46x + 4x^2$$
$$594 = 414 + 82x + 4x^2$$
$$0 = 4x^2 + 82x - 180$$
$$0 = 2x^2 + 41x - 90$$
$$0 = (2x + 45)(x - 2)$$
$$x = -\frac{45}{2} \quad or \quad x = 2$$

The width of the margin cannot be
negative. The width of the margins
should be 2 cm.

7. Let   x = the number of hours for
                 Felix to clean garage
                 working alone.

    Then x - 9 = the number of hours for
                 Felipe to clean garage
                 working alone.

	Rate	Time	Part of the job accomplished
Felix	$\frac{1}{x}$	20	$20\left(\frac{1}{x}\right) = \frac{20}{x}$
Felipe	$\frac{1}{x - 9}$	20	$20\left(\frac{1}{x - 9}\right) = \frac{20}{x - 9}$

The sum of the two parts of the job accomplished is 1, since one whole job is done.

$$\frac{20}{x} + \frac{20}{x - 9} = 1$$

To clear fractions, multiply both sides by the least common denominator $x(x - 9)$.

$$x(x - 9)\left(\frac{20}{x} + \frac{20}{x - 9}\right) = x(x - 9) \cdot 1$$
$$20(x - 9) + 20x = x^2 - 9x$$
$$20x - 180 + 20x = x^2 - 9x$$
$$40x - 180 = x^2 - 9x$$
$$0 = x^2 - 49x + 180$$
$$0 = (x - 45)(x - 4)$$

$$x = 45 \quad \text{or} \quad x = 4$$
$$x - 9 = 36 \quad \text{or} \quad x - 9 = -5$$

Since the solution $x = 4$ leads to $x - 9 = -5$, this solution must be rejected because the number of hours for Felix working alone cannot be negative.

It would take Felix 45 hr and Felipe 36 hr to clean the garage alone.

9. Let   x = number of hours for
                 new typist to com-
                 plete project.

    Then x − 2 = number of hours for
                 experienced typist
                 to complete project.

	Rate	Time	Part of the job accomplished
New typist	$\frac{1}{x}$	2.4	$2.4\left(\frac{1}{x}\right) = \frac{2.4}{x}$
Experienced typist	$\frac{1}{x-2}$	2.4	$2.4\left(\frac{1}{x-2}\right) = \frac{2.4}{x-2}$

Together, the two typists do one whole job, so

$$\frac{2.4}{x} + \frac{2.4}{x - 2} = 1.$$

To clear fractions, multiply both sides of the equation by the least common denominator, $x(x - 2)$.

$$x(x - 2)\left(\frac{2.4}{x} + \frac{2.4}{x - 2}\right) = x(x - 2) \cdot 1$$
$$2.4(x - 2) + 2.4x = x^2 - 2x$$
$$2.4x - 4.8 + 2.4x = x^2 - 2x$$
$$4.8x - 4.8 = x^2 - 2x$$
$$0 = x^2 - 6.8x + 4.8$$

To clear decimals, multiply both sides of the equation by 10.

$$0 = 10x^2 - 68x + 48$$
$$0 = 5x^2 - 34x + 24$$
$$0 = (5x - 4)(x - 6)$$

$$5x - 4 = 0 \quad \text{or} \quad x - 6 = 0$$
$$x = \frac{4}{5} \quad \text{or} \quad x = 6$$
$$x - 2 = -\frac{6}{5} \qquad x - 2 = 4$$

Since the solution $x = \frac{4}{5}$ leads to $x - 2 = -\frac{6}{5}$, this solution must be rejected because the number of hours for the experienced typist working alone cannot be negative. It would take the experienced typist 4 hr to complete the project alone.

11. Let r = Steve's speed.

	d	r	t
Paula	100	r + 10	$\dfrac{100}{r+10}$
Steve	100	r	$\dfrac{100}{r}$

		d	r	t
1st trip		1000	x	$\dfrac{1000}{x}$
2nd trip		2025	x + 50	$\dfrac{2025}{x+50}$

Steve's time	is	1/3 hour longer than Paula's time.
↓	↓	↓
$\dfrac{100}{r}$	=	$\dfrac{100}{r+10} + \dfrac{1}{3}$

To clear fractions, multiply both sides of the equation by the least common denominator, $3r(r + 10)$.

$$3r(r+10) \cdot \frac{100}{r} = 3r(r+10)\left(\frac{100}{r+10}+\frac{1}{3}\right)$$

$$300(r+10) = 300r + r(r+10)$$

$$300r + 3000 = 300r + r^2 + 10r$$

$$0 = r^2 + 10r - 3000$$

$$0 = (r-50)(r+60)$$

r = 50 or r = –60

Since speed cannot be negative, reject –60 as a solution.

Steve's average speed is 50 mph.

13. Let x = the speed of the plane for the first trip.

Then x + 50 = the speed of the plane for the second trip.

Since d = rt, t = d/r.

Time for 2nd trip	=	Time for 1st trip	+	2 hr
↓	↓	↓	↓	↓
$\dfrac{2025}{x+50}$	=	$\dfrac{1000}{x}$	+	2

To clear fractions, multiply both sides by the least common denominator, $x(x + 50)$.

$$x(x+50)\left(\frac{2025}{x+50}\right) = x(x+50)\left(\frac{1000}{x}\right)+x(x+50)(2)$$

$$2025x = 1000x + 50{,}000 + 2x^2 + 100x$$

$$0 = 2x^2 - 925x + 50{,}000$$

$$0 = (2x-125)(x-400)$$

2x – 125 = 0   or   x – 400 = 0

$$x = \frac{125}{2} \quad \text{or} \quad x = 400$$

The first solution, $x = \dfrac{125}{2}$, means that the speed of the first airplane would be 62.5 mph.

This solution satisfies the equation but is not a realistic airplane speed. The speed of the airplane for the first trip is 400 mph.

To find the time for the first trip, use $t = \dfrac{d}{r} = \dfrac{1000}{x} = \dfrac{1000}{400} = \dfrac{2}{5}$. The time for the first trip was 2.5 hr.

15. (d) The lengths of the legs are x and 2x − 2, and the length of the hypotenuse is x + 4. The Pythagorean theorem leads to
$$x^2 + (2x - 2)^2 = (x + 4)^2.$$

17. Let h = the height of the kite.

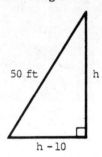

50 ft   h

h − 10

Apply the Pythagorean theorem to the right triangle.

$$a^2 + b^2 = c^2$$
$$h^2 + (h - 10)^2 = 50^2$$
$$h^2 + h^2 - 20h + 100 = 2500$$
$$2h^2 - 20h - 2400 = 0$$
$$h^2 - 10h - 1200 = 0$$
$$(h - 40)(h + 30) = 0$$
$$h = 40 \text{ or } h = -30$$

Reject the negative solution.
The kite is 40 ft above the ground.

19. Let x = the number of hours they can talk to each other on the walkie-talkies.

Use d = rt to determine how far each boy walks in x hours.

Then 2.5x = the number of miles Chris walks north

and   3x = the number of miles Josh walks east.

This forms a right triangle with legs of length 2.5x and 3x, and length of the hypotenuse is the distance between the boys. We want to find x when the length of the hypotenuse is 4 mi.

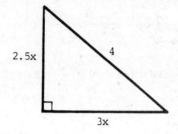

2.5x   4

3x

$$(2.5x)^2 + (3x)^2 = 4^2$$
$$6.25x^2 + 9x^2 = 16$$
$$15.25x^2 = 16$$
$$x^2 = 1.049$$
$$x = \pm 1.02$$

Reject the negative solution.

1.02 hr = 1.02 (60 min)
        ≈ 61 min
They will be able to talk for about 61 min.

21. When $v_0 = 96$,

$s = -16t^2 + v_0 t$ becomes
$s = -16t^2 + 96t.$

(a) When s = 80,

$$80 = -16t^2 + 96t, \text{ or}$$
$$0 = -16t^2 + 96t - 80.$$

Here a = −16, b = 96, and c = −80.

$$t = \frac{-96 \pm \sqrt{96^2 - 4(-16)(-80)}}{2(-16)}$$

$$= \frac{-96 \pm \sqrt{9216 - 5120}}{-32}$$

$$= \frac{-96 \pm \sqrt{4096}}{-32}$$

$$= \frac{-96 \pm 64}{-32} = 1 \text{ or } 5$$

The height will be 80 ft after 1 sec and after 5 sec.

(b) When s = 0 (the height of the ground),

$$0 = -16t^2 + 96t$$

$$0 = -16t(t - 6)$$

$$-16t = 0 \text{ or } t - 6 = 0$$

$$t = 0 \text{ or } \quad t = 6.$$

It will return to the ground after 6 sec.

23. When $v_0 = 32$,

$$s = -16t^2 + v_0t \quad \text{becomes}$$

$$s = -16t^2 + 32t.$$

(a) When s = 80,

$$80 = -16t^2 + 32t$$

$$16t^2 - 32t + 80 = 0$$

$$t^2 - 2t + 5 = 0.$$
$$\textit{Divide by 16}$$

Solve this equation by the quadratic formula,

$$t = \frac{-(-2) \pm \sqrt{(-2)^2 - 4(1)(5)}}{2(1)}$$

$$= \frac{2 \pm \sqrt{-16}}{2}$$

$$= 1 \pm 2i$$

Since the equation has no real solutions, the projectile never reaches a height of 80 ft.

(b) When s = 0,

$$16t^2 - 32t = 0$$

$$t^2 - 2t = 0 \quad \textit{Divide by 16}$$

$$t(t - 2) = 0$$

$$t = 0 \text{ or } t = 2.$$

It will return to the ground after 2 sec.

27. Let x = the number of weeks the manager should wait.

Then 100 + 5x
      = the yield per tree after x weeks;

.40 − .02x = the price per pound after x weeks.

The average revenue per tree is (100 + 5x)(.40 − .02x), so

$$(100 + 5x)(.40 - .02x) = 38.40$$

$$40 - 2x + 2x - .1x^2 = 38.40$$

$$-.1x^2 = -1.60$$

$$x^2 = 16$$

$$x = \pm 4.$$

Reject the negative solution.

The manager should wait 4 weeks.

29. Let x = number of passengers in excess of 75.

Then 225 − 5x
      = the cost per passenger (in dollars);

75 + x = the number of passengers.

Cost per passenger	·	Number of passengers		= revenue
↓	↓	↓	↓	↓
(225 − 5x)	·	(75 + x)		= 16,000

$$16,875 - 150x - 5x^2 = 16,000$$
$$0 = 5x^2 + 150x - 875$$
$$0 = x^2 + 30x - 175$$
$$0 = (x + 35)(x - 5)$$
$$x = -35 \quad \text{or} \quad x = 5$$

Reject the negative solution. Since there are 5 passengers in excess of 75, the total number of passengers is 80.

Section 2.6

1. $m^4 + 2m^2 - 15 = 0$

Let $x = m^2$; then $x^2 = m^4$. With this substitution, the equation becomes

$$x^2 + 2x - 15 = 0.$$

Solve this equation by factoring.

$$(x - 3)(x + 5) = 0$$
$$x = 3 \quad \text{or} \quad x = -5$$

To find m, replace x with m^2.

$$m^2 = 3 \quad \text{or} \quad m^2 = -5$$
$$m = \pm\sqrt{3} \quad \text{or} \quad m = \pm i\sqrt{5}$$

Solution set: $\left\{\pm\sqrt{3}, \pm i\sqrt{5}\right\}$

3. $2r^4 - 7r^2 + 5 = 0$

Let $x = r^2$; then $x^2 = r^4$. With this substitution, the equation becomes

$$2x^2 - 7x + 5 = 0.$$

Solve this equation by factoring.

$$(x - 1)(2x - 5) = 0$$
$$x = 1 \quad \text{or} \quad x = \frac{5}{2}$$

To find r, replace x with r^2.

$$r^2 = 1 \quad \text{or} \quad r^2 = \frac{5}{2}$$
$$r = \pm\sqrt{\frac{5}{2}}$$
$$r = \pm 1 \quad \text{or} \quad r = \pm\frac{\sqrt{10}}{2}$$

Solution set: $\left\{\pm 1, \ \pm\frac{\sqrt{10}}{2}\right\}$

5. $(g - 2)^2 - 6(g - 2) + 8 = 0$

Let $x = g - 2$. Solve the resulting equation by factoring.

$$x^2 - 6x + 8 = 0$$
$$(x - 2)(x - 4) = 0$$
$$x = 2 \quad \text{or} \quad x = 4$$

To find g, replace x with $g - 2$.

$$g - 2 = 2 \quad \text{or} \quad g - 2 = 4$$
$$g = 4 \quad \text{or} \quad g = 6$$

Solution set: $\{4, 6\}$

7. $-(r + 1)^2 - 3(r + 1) + 3 = 0$

Let $x = r + 1$. With this substitution, the equation becomes

$$-x^2 - 3x + 3 = 0$$

or $\quad x^2 + 3x - 3 = 0.$

This equation cannot be solved by factoring, so use the quadratic formula.

$$x = \frac{-3 \pm \sqrt{9 + 12}}{2}$$
$$= \frac{-3 \pm \sqrt{21}}{2}$$

To find r, replace x with r + 1.

$$r + 1 = \frac{-3 \pm \sqrt{21}}{2}$$

$$r = \frac{-3 \pm \sqrt{21}}{2} - \frac{2}{2}$$

$$= \frac{-5 \pm \sqrt{21}}{2}$$

Solution set: $\left\{ \dfrac{-5 \pm \sqrt{21}}{2} \right\}$

9. $6(k + 2)^4 - 11(k + 2)^2 + 4 = 0$

Let $x = (k + 2)^2$.   Solve the resulting equation by factoring.

$$6x^2 - 11x + 4 = 0$$

$$(3x - 4)(2x - 1) = 0$$

$$x = \frac{4}{3} \quad \text{or} \quad x = \frac{1}{2}$$

To find k, replace x with $(k + 2)^2$.

$$(k + 2)^2 = \frac{4}{3}$$

$$k + 2 = \pm\sqrt{\frac{4}{3}}$$

$$= \pm\frac{2\sqrt{3}}{3}$$

$$k = -2 \pm \frac{2\sqrt{3}}{3}$$

$$= -\frac{6}{3} \pm \frac{2\sqrt{3}}{3}$$

$$= \frac{-6 \pm 2\sqrt{3}}{3}$$

or

$$(k + 2)^2 = \frac{1}{2}$$

$$k + 2 = \pm\sqrt{\frac{1}{2}}$$

$$= \pm\frac{\sqrt{2}}{2}$$

$$k = -2 \pm \frac{\sqrt{2}}{2}$$

$$= -\frac{4}{2} \pm \frac{\sqrt{2}}{2}$$

$$= \frac{-4 \pm \sqrt{2}}{2}$$

Solution set: $\left\{ \dfrac{-6 \pm 2\sqrt{3}}{3}, \; \dfrac{-4 \pm \sqrt{2}}{2} \right\}$

11. $7p^{-2} + 19p^{-1} = 6$

Let $x = p^{-1}$.   Solve the resulting equation by factoring.

$$7x^2 + 19x - 6 = 0$$

$$(7x - 2)(x + 3) = 0$$

$$x = \frac{2}{7} \quad \text{or} \quad x = -3$$

To find p, replace x with p^{-1}.

$$p^{-1} = \frac{2}{7} \quad \text{or} \quad p^{-1} = -3$$

$$p = \frac{7}{2} \quad \text{or} \quad p = -\frac{1}{3}$$

Solution set: $\left\{ \dfrac{7}{2}, \; -\dfrac{1}{3} \right\}$

13. $(r - 1)^{2/3} + (r - 1)^{1/3} = 12$

Let $x = (r - 1)^{1/3}$.
Then $x^2 = [(r - 1)^{1/3}]^2 = (r - 1)^{2/3}$.
Solve the resulting equation by factoring.

$$x^2 + x = 12$$

$$x^2 + x - 12 = 0$$

$$(x + 4)(x - 3) = 0$$

$$x = -4 \quad \text{or} \quad x = 3$$

To find r, replace x with $(r - 1)^{1/3}$.

$$(r - 1)^{1/3} = -4 \quad \text{or} \quad (r - 1)^{1/3} = 3$$

Cube both sides in each equation.

$$[(r - 1)^{1/3}]^3 = (-4)^3$$

$$r - 1 = -64$$

$$r = -63$$

or

$$[(r - 1)^{1/3}]^3 = 3^3$$
$$r - 1 = 27$$
$$r = 28$$

Because the original equation contained rational exponents, both solutions must be checked.
Solution set: $\{-63, 28\}$

15. $3 + \dfrac{5}{p^2 + 1} = \dfrac{2}{(p^2 + 1)^2}$

Let $x = p^2 + 1$. With this substitution, the equation becomes

$$3 + \frac{5}{x} = \frac{2}{x^2}.$$

Multiply both sides by the least common denominator, x^2.

$$3x^2 + 5x = 2$$
$$3x^2 + 5x - 2 = 0$$
$$(x + 2)(3x - 1) = 0$$
$$x = -2 \quad \text{or} \quad x = \frac{1}{3}$$

To find p, replace x with $1 + p^2$.

$$1 + p^2 = -2 \quad \text{or} \quad 1 + p^2 = \frac{1}{3}$$
$$p^2 = -3 \qquad\qquad p^2 = -\frac{2}{3}$$
$$p = \pm i\sqrt{\frac{2}{3}}$$
$$p = \pm i\sqrt{3} \text{ or} \qquad p = \pm\frac{i\sqrt{6}}{3}$$

Solution set: $\left\{\pm i\sqrt{3}, \pm\dfrac{i\sqrt{6}}{3}\right\}$

19. $\sqrt{3z + 7} = 3z + 5$

$$(\sqrt{3z + 7})^2 = (3z + 5)^2$$
Square both sides

$$3z + 7 = 9z^2 + 30z + 25$$
Square of a binomial

$$0 = 9z^2 + 27z + 18$$
$$0 = z^2 + 3z + 2$$
Divide by 9

$$0 = (z + 2)(z + 1) \quad \text{Factor}$$
$$z = -2 \quad \text{or} \quad z = -1$$

Check each proposed solution in the original equation.

Let $z = -2$.

$$\sqrt{3z + 7} = 3z + 5$$
$$\sqrt{3(-2) + 7} = 3(-2) + 5 \quad ?$$
$$\sqrt{-6 + 7} = -6 + 5 \qquad ?$$
$$\sqrt{1} = -1 \qquad\qquad ?$$
$$1 = -1 \qquad\qquad\quad False$$

Let $z = -1$.

$$\sqrt{3z + 7} = 3z + 5$$
$$\sqrt{3(-1) + 7} = 3(-1) + 5 \quad ?$$
$$\sqrt{-3 + 7} = -3 + 5 \qquad ?$$
$$\sqrt{4} = 2 \qquad\qquad ?$$
$$2 = 2 \qquad\qquad\quad True$$

These checks show that only -1 is a solution.
Solution set: $\{-1\}$

21. $\sqrt{4k + 5} - 2 = 2k - 7$

$$\sqrt{4k + 5} = 2k - 5$$
$$(\sqrt{4k + 5})^2 = (2k - 5)^2$$
Square both sides

$$4k + 5 = 4k^2 - 20k + 25$$
Square of a binomial

$$0 = 4k^2 - 24k + 20$$
$$0 = k^2 - 6k + 5$$
Divide by 4

$$k = 1 \quad \text{or} \quad k = 5$$

Check each proposed solution in the original equation.

Let k = 1.

$$\sqrt{4k + 5} - 2 = 2k - 7$$
$$\sqrt{4 + 5} - 2 = 2 - 7 \quad ?$$
$$\sqrt{9} - 2 = -5 \quad ?$$
$$3 - 2 = -5 \quad ?$$
$$1 = -5 \qquad \textit{False}$$

Let k = 5.

$$\sqrt{4k + 5} - 2 = 2k - 7$$
$$\sqrt{20 + 5} - 2 = 10 - 7 \quad ?$$
$$\sqrt{25} - 2 = 3 \quad ?$$
$$5 - 2 = 3 \quad ?$$
$$3 = 3 \qquad \textit{True}$$

Solution set: $\{5\}$

23. $\sqrt{4x} - x + 3 = 0$

$$\sqrt{4x} = x - 3$$
$$(\sqrt{4x})^2 = (x - 3)^2$$
$$\qquad\qquad \textit{Square both sides}$$
$$4x = x^2 - 6x + 9$$
$$\qquad\qquad \textit{Square of a binomial}$$
$$0 = x^2 - 10x + 9$$
$$0 = (x - 1)(x - 9)$$
$$x = 1 \quad \text{or} \quad x = 9$$

Check each proposed solution in the original equation.

Let x = 1.

$$\sqrt{4 \cdot 1} - 1 + 3 = 0 \quad ?$$
$$\sqrt{4} - 1 + 3 = 0 \quad ?$$
$$2 - 1 + 3 = 0 \quad ?$$
$$4 = 0 \qquad \textit{False}$$

Let x = 9.

$$\sqrt{4 \cdot 9} - 9 + 3 = 0 \quad ?$$
$$\sqrt{36} - 9 + 3 = 0 \quad ?$$
$$6 - 9 + 3 = 0 \quad ?$$
$$0 = 0 \qquad \textit{True}$$

Solution set: $\{9\}$

25. $\sqrt{y} = \sqrt{y - 5} + 1$

$$(\sqrt{y})^2 = (\sqrt{y - 5} + 1)^2$$
$$\qquad\qquad \textit{Square both sides}$$
$$y = (\sqrt{y - 5})^2 + \sqrt{y - 5} + 1$$
$$\qquad\qquad \textit{Square of a binomial}$$
$$y = y - 5 + 2\sqrt{y - 5} + 1$$
$$0 = -4 + 2\sqrt{y - 5}$$
$$4 = 2\sqrt{y - 5}$$
$$2 = \sqrt{y - 5}$$
$$2^2 = (\sqrt{y - 5})^2 \quad \textit{Square both sides}$$
$$4 = y - 5$$
$$9 = y$$

Check this proposed solution in the original equation.

$$\sqrt{y} = \sqrt{y - 5} + 1$$
$$\sqrt{9} = \sqrt{9 - 5} + 1 \quad ?$$
$$3 = \sqrt{4} + 1 \quad ?$$
$$3 = 3 \qquad \textit{True}$$

Solution set: $\{9\}$

27. $\sqrt{m + 7} + 3 = \sqrt{m - 4}$

$$(\sqrt{m + 7} + 3)^2 = (\sqrt{m - 4})^2$$
$$\qquad\qquad \textit{Square both sides}$$
$$m + 7 + 6\sqrt{m + 7} + 9 = m - 4$$
$$\qquad\qquad \textit{Square of a binomial}$$
$$6\sqrt{m + 7} = -20 \quad \textit{Simplify}$$
$$3\sqrt{m + 7} = -10$$
$$(3\sqrt{m + 7})^2 = (-10)^2$$
$$\qquad\qquad \textit{Square both sides}$$

$$9(m + 7) = 100$$
$$9m + 63 = 100$$
$$9m = 37$$
$$m = \frac{37}{9}$$

Check this proposed solution in the original equation.

$$\sqrt{m + 7} + 3 = \sqrt{m - 4}$$
$$\sqrt{\frac{37}{9} + 7} + 3 = \sqrt{\frac{37}{9} - 4} \quad ?$$
$$\sqrt{\frac{37}{9} + \frac{63}{9}} + 3 = \sqrt{\frac{37}{9} + \frac{36}{9}} \quad ?$$
$$\sqrt{\frac{100}{9}} + 3 = \sqrt{\frac{1}{9}} \quad ?$$
$$\frac{10}{3} + \frac{9}{3} = \frac{1}{3} \quad ?$$
$$\frac{19}{3} = \frac{1}{3} \qquad \textit{False}$$

Since the only proposed solution is not a solution of the original equation, the equation has no solution.
Solution set: Ø

29.
$$\sqrt{2z} = \sqrt{3z + 12} - 2$$
$$(\sqrt{2z})^2 = (\sqrt{3z + 12} - 2)^2$$
$$\qquad\qquad \textit{Square both sides}$$
$$2z = 3z + 12 - 4\sqrt{3z + 12} + 4$$
$$\qquad\qquad \textit{Square of a binomial}$$
$$4\sqrt{3z + 12} = z + 16$$
$$(4\sqrt{3z + 12})^2 = (z + 16)^2$$
$$\qquad\qquad \textit{Square both sides}$$
$$16(3z + 12) = z^2 + 32z + 256$$
$$\qquad\qquad \textit{Square of a binomial}$$
$$48z + 192 = z^2 + 32z + 256$$
$$\qquad\qquad \textit{Distributive property}$$
$$0 = z^2 - 16z + 64$$
$$0 = (z - 8)^2 \quad \textit{Factor}$$
$$z - 8 = 0$$
$$z = 8$$

Check this proposed solution in the original equation.

$$\sqrt{2z} = \sqrt{3z + 12} - 2$$
$$\sqrt{2(8)} = \sqrt{3(8) + 12} - 2 \quad ?$$
$$\sqrt{16} = \sqrt{36} - 2 \qquad ?$$
$$4 = 6 - 2 \qquad\qquad ?$$
$$4 = 4 \qquad\qquad\qquad \textit{True}$$

Solution set: $\{8\}$

31.
$$\sqrt{r + 2} = 1 - \sqrt{3r + 7}$$
$$(\sqrt{r + 2})^2 = (1 - \sqrt{3r + 7})^2$$
$$r + 2 = 1 - 2\sqrt{3r + 7} + 3r + 7$$
$$2\sqrt{3r + 7} = 2r + 6$$
$$2\sqrt{3r + 7} = 2(r + 3)$$
$$\sqrt{3r + 7} = r + 3$$
$$(\sqrt{3r + 7})^2 = (r + 3)^2$$
$$3r + 7 = r^2 + 6r + 9$$
$$0 = r^2 + 3r + 2$$
$$0 = (r + 1)(r + 2)$$
$$r = -1 \text{ or } r = -2$$

Check $r = -1$.

$$\sqrt{-1 + 2} = 1 - \sqrt{-3 + 7} \quad ?$$
$$\sqrt{1} = 1 - \sqrt{4} \qquad ?$$
$$1 = 1 - 2 \qquad\quad ?$$
$$1 = -1 \qquad\qquad \textit{False}$$

Check $r = -2$.

$$\sqrt{-2 + 2} = 1 - \sqrt{-6 + 7} \quad ?$$
$$0 = 1 - \sqrt{1} \qquad ?$$
$$0 = 1 - 1 \qquad\quad ?$$
$$0 = 0 \qquad\qquad \textit{True}$$

Solution set: $\{-2\}$

33.

$$\sqrt{2\sqrt{7x + 2}} = \sqrt{3x + 2}$$

$$\left(\sqrt{2\sqrt{7x + 2}}\right)^2 = (\sqrt{3x + 2})^2$$
Square both sides

$$2\sqrt{7x + 2} = 3x + 2$$

$$(2\sqrt{7x + 2})^2 = (3x + 2)^2$$
Square both sides

$$4(7x + 2) = 9x^2 + 12x + 4$$

$$28x + 8 = 9x^2 + 12x + 4$$

$$0 = 9x^2 - 16x - 4$$

$$0 = (x - 2)(9x + 2)$$

$$x = 2 \quad \text{or} \quad x = -\frac{2}{9}$$

Solution set: $\left\{2, -\frac{2}{9}\right\}$

35.

$$\sqrt[3]{4n + 3} = \sqrt[3]{2n - 1}$$

$$(\sqrt[3]{4n + 3})^3 = (\sqrt[3]{2n - 1})^3$$
Cube both sides

$$4n + 3 = 2n - 1$$

$$2n = -4$$

$$n = -2$$

Check this proposed solution in the original equation.

$$\sqrt[3]{4n + 3} = \sqrt[3]{2n - 1}$$

$$\sqrt[3]{4(-2) + 3} = \sqrt[3]{2(-2) - 1} \quad ?$$

$$\sqrt[3]{-5} = \sqrt[3]{-5} \qquad \text{True}$$

Solution set: $\{-2\}$

37.

$$\sqrt[3]{t^2 + 2t - 1} = \sqrt[3]{t^2 + 3}$$

$$(\sqrt[3]{t^2 + 2t - 1})^3 = (\sqrt[3]{t^2 + 3})^3$$
Cube both sides

$$t^2 + 2t - 1 = t^2 + 3$$

$$2t = 4$$

$$t = 2$$

Check this proposed solution in the original equation.

$$\sqrt[3]{t^2 + 2t - 1} = \sqrt[3]{t^2 = 3}$$

$$\sqrt[3]{2^2 + 2(2) - 1} = \sqrt[3]{2^2 + 3} \quad ?$$

$$\sqrt[3]{4 + 4 - 1} = \sqrt[3]{4 + 3} \quad ?$$

$$\sqrt[3]{7} = \sqrt[3]{7} \qquad \text{True}$$

Solution set: $\{2\}$

39.

$$(2r + 5)^{1/3} = (6r - 1)^{1/3}$$

$$[(2r + 5)^{1/3}]^3 = [(6r - 1)^{1/3}]^3$$
Cube both sides

$$2r + 5 = 6r - 1$$

$$6 = 4r$$

$$r = \frac{3}{2}$$

Check this proposed solution in the original equation.

$$(2r + 5)^{1/3} = (6r - 1)^{1/3}$$

$$\left[2\left(\frac{3}{2}\right) + 5\right]^{1/3} = \left[6\left(\frac{3}{2}\right) - 1\right]^{1/3} \quad ?$$

$$(3 + 5)^{1/3} = (9 - 1)^{1/3} \quad ?$$

$$8^{1/3} = 8^{1/3} \qquad \text{True}$$

Solution set: $\left\{\frac{3}{2}\right\}$

41.

$$\sqrt[4]{q - 15} = 2$$

$$(\sqrt[4]{q - 15})^4 = 2^4 \quad \textit{Raise both sides}$$
to the 4th power

$$q - 15 = 16$$

$$q = 31$$

Check this proposed solution in the original equation.

$$\sqrt[4]{q - 15} = 2$$

$$\sqrt[4]{31 - 15} = 2 \quad ?$$

$$\sqrt[4]{16} = 2 \quad ?$$

$$2 = 2 \qquad \text{True}$$

Solution set: $\{31\}$

43. $\sqrt[4]{y^2 + 2y} = \sqrt[4]{3}$

   $(\sqrt[4]{y + 2y})^4 = (\sqrt[4]{3})^4$
        Raise both sides to 4th power

   $y^2 + 2y = 3$

   $y^2 + 2y - 3 = 0$

   $(y + 3)(y - 1) = 0$

   $y = -3$ or $y = 1$

Checking will show that both of these proposed solutions are solutions of the original equation.

Solution set: $\{-3, 1\}$

45. $(z^2 + 24z)^{1/4} = 3$

   $[(z^2 + 24z)^{1/4}]^4 = 3^4$
        Raise both sides to 4th power

   $z^2 + 24z = 81$

   $z^2 + 24z - 81 = 0$

   $(z + 27)(z - 3) = 0$

   $z = -27$ or $z = 3$

Checking will show that both of these proposed solutions are solutions of the original equation.

Solution set: $\{-27, 3\}$

47. $(2r - 1)^{2/3} = r^{1/3}$

   $[(2r - 1)^{2/3}]^3 = (r^{1/3})^3$
        Cube both sides

   $(2r - 1)^2 = r$

   $4r^2 - 4r + 1 = r$

   $4r^2 - 5r + 1 = 0$

   $(4r - 1)(r - 1) = 0$

   $r = \dfrac{1}{4}$ or $r = 1$

Checking will show that both of these proposed solutions are solutions of the original equation.

Solution set: $\left\{\dfrac{1}{4}, 1\right\}$

51. $d = k\sqrt{h}$ for h

   $\dfrac{d}{k} = \sqrt{h}$

   $\dfrac{d^2}{k^2} = h$

53. $P = 2\sqrt{\dfrac{L}{g}}$ for L

   $p^2 = 4 \cdot \dfrac{L}{g}$ *Square both sides*

   $\dfrac{p^2 g}{4} = L$

55. $x^{2/3} + y^{2/3} = a^{2/3}$ for y

   $y^{2/3} = a^{2/3} - x^{2/3}$

   $(y^{2/3})^3 = (a^{2/3} - x^{2/3})^3$
                Cube both sides

   $y^2 = (a^{2/3} - x^{2/3})^3$

   $y = \sqrt{(a^{2/3} - x^{2/3})^3}$

   $y = (a^{2/3} - x^{2/3})^{3/2}$

Section 2.7

For Exercises 1–7, see the answer graphs in the back of the textbook.

1. $-1 < x < 4$ may be written in interval notation as $(-1, 4)$.

3. $x < 0$ may be written in interval notation as $(-\infty, 0)$.

5. $2 > x \geq 1$

   or $1 \leq x < 2$

   may be written in interval notation as $[1, 2)$.

7. $-9 > x$

or $x < -9$

may be written in interval notation
as $(-\infty, -9)$.

9. The interval $(-4, 3)$ may be written in set-builder notation as
$\{x \mid -4 < x < 3\}$.

11. The interval $(-\infty, -1]$ may be written in set-builder notation as
$\{x \mid x \leq -1\}$.

13. The interval $[-2, 6)$ which is shown
on the graph, may be written in set-
builder notation as $\{x \mid -2 \leq x < 6\}$.

15. The interval $(-\infty, -4]$, which is
shown on the graph, may be written
in interval notation as $\{x \mid x \leq -4\}$.

For Exercises 19–31, see the answer
graphs in the back of the textbook.

19. $-3p - 2 \leq 1$

Add 2 to both sides of the inequality.

$-3p - 2 + 2 \leq 1 + 2$

$-3p \leq 3$

Multiply both sides of the inequality by $-1/3$ and reverse the direction of the inequality symbol.

$$\left(-\frac{1}{3}\right)(-3p) \geq \left(-\frac{1}{3}\right)(3)$$

$$p \geq -1$$

Solution set: $[-1, \infty)$

21. $2(m + 5) - 3m + 1 \geq 5$

$2m + 10 - 3m + 1 \geq 5$
 Distributive property

$-m + 11 \geq 5$ *Combine
terms*

$-m \geq -6$ *Subtract 11*

$-1(-m) \leq -1(-6)$
 *Multiply by -1; re-
verse inequality
symbol*

$m \leq 6$

Solution set: $(-\infty, 6]$

23. $8k - 3k + 2 < 2(k + 7)$

$5k + 2 < 2k + 14$
 Distributive property

$5k + 2 - 2k < 2k + 14 - 2k$
 Subtract 2k

$3k + 2 < 14$

$3k + 2 - 2 < 14 - 2$ *Subtract 2*

$3k < 12$

$\frac{1}{3} \cdot 3k < \frac{1}{3} \cdot 12$ *Multiply
by 1/3*

$k < 4$

Solution set: $(-\infty, 4)$

25. $\dfrac{4x + 7}{-3} \leq 2x + 5$

$(-3)\left(\dfrac{4x + 7}{-3}\right) \geq (-3)(2x + 5)$
 *Multiply by -3; reverse
inequality symbol*

$4x + 7 \geq -6x - 15$

$4x + 7 + 6x \geq -6x - 15 + 6x$
 Add 6x

$10x + 7 \geq -15$

$10x + 7 - 7 \geq -15 - 7$ *Subtract 7*

$10x \geq -22$

$\frac{1}{10}(10x) \geq \frac{1}{10}(-22)$ *Multiply
by 1/10*

$x \geq -\dfrac{11}{5}$

Solution set: $\left[-\dfrac{11}{5}, \infty\right)$

27. $2 \le y + 1 \le 5$

$1 \le y \le 4$ *Subtract 1*

Solution set: $[1, 4]$

29. $-10 > 3r + 2 > -16$

$-10 - 2 > 3r + 2 - 2 > -16 - 2$
                        Subtract 2

$-12 > 3r > -18$

$\frac{1}{3}(-12) > \frac{1}{3}(3r) > \frac{1}{3}(-18)$ *Multiply by 1/3*

$-4 > r > -6$

or $-6 < r < -4$

Solution set: $(-6, -4)$

31. $-3 \le \dfrac{x - 4}{-5} < 4$

$(-5)(-3) \ge (-5)\left(\dfrac{x - 4}{-5}\right) > (-5)(4)$
        Multiply by -5; reverse inequality symbol

$15 \ge x - 4 > -20$

$15 + 4 \ge x - 4 + 4 > -20 + 4$ *Add 4*

$19 \ge x > -16$

or $-16 < x \le 19$

Solution set: $(-16, 19]$

33. $C = 50x + 5000; R = 60x$

The product will at least break even when $R \ge C$.

Set $R \ge C$ and solve for x.

$60x \ge 50x + 5000$

$10x \ge 5000$

$x \ge 500$

The break-even point is at $x = 500$. This product will at least break even if the number of units produced is in the interval $[500, \infty)$.

35. $C = 85x + 900; R = 105x$

$R \ge C$

$105x \ge 85x + 900$

$20x \ge 900$

$x \ge 45$

The break even point is at $x = 45$. This product will at least break even if the number of units produced is in the interval $[45, \infty)$.

For Exercises 37–49, see the answer graphs in the back of the textbook.

37. $x^2 \le 9$

$x^2 - 9 \le 0$

Solve the corresponding quadratic equation by factoring.

$x^2 - 9 = 0$

$(x + 3)(x - 3) = 0$

$x = -3$ or $x = 3$

These two points, -3 and 3, divide a number line into the three regions shown on the following sign graph.

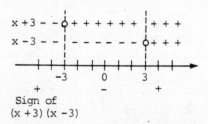

Sign of
$(x + 3)(x - 3)$

The factor $x + 3$ is positive when $x > -3$, and $x - 3$ is positive when $x > 3$.

We want the product to be negative, which happens if the two factors have opposite signs. As the sign graph shows, this happens when x is between -3 and 3. The endpoints of the interval are included in the solution because the product is zero (and therefore "less than or equal to zero") when $x = -3$ or $x = 3$.

Solution set: $[-3, 3]$

39. $r^2 + 4r + 6 \geq 3$

$r^2 + 4r + 3 \geq 0$

Solve the corresponding quadratic equation.

$r^2 + 4r + 3 = 0$

$(r + 3)(r + 1) = 0$

$r = -3$ or $r = -1$

These two points, -3 and -1, divide a number line into the three regions shown on the following sign graph.

Sign of
$(r + 3)(r + 1)$

The factor $r + 3$ is positive when $r > -3$, and $r + 1$ is positive when $r > -1$.

We want the product to be positive, which happens if both factors have the same sign. As the sign graph shows, this happens when $r < -3$ or when $r > -1$.

The inequality is also satisfied when $r = -3$ and when $r = -1$ because of the $\leq$ sign, so square brackets should be used at these endpoints.

Solution set: $(-\infty, -3] \cup [-1, \infty)$

41. $x^2 - x \leq 6$

$x^2 - x - 6 \leq 0$

Solve the corresponding quadratic equation.

$x^2 - x - 6 = 0$

$(x + 2)(x - 3) = 0$

$x + 2 = 0$ or $x - 3 = 0$

$x = -2$ or $x = 3$

These two points divide a number line into three regions.

Sign of
$(x + 2)(x - 3)$

The factor $x - 3$ is positive when $x > 3$ and $x + 2$ is positive when $x > -2$.

The product is negative if the two factors have opposite signs, which happens when x is between -2 and 3. Use square brackets since the original inequality is $\leq$, which means that both -2 and 3 are part of the solution set.

Solution set: $[-2, 3]$

43. $2k^2 - 9k > -4$

$2k^2 - 9k + 4 > 0$

Solve the corresponding quadratic equation.

$2k^2 - 9k + 4 = 0$

$(2k - 1)(k - 4) = 0$

$k = \dfrac{1}{2}$ or $k = 4$

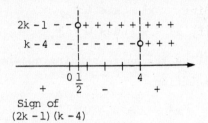

The product is positive when $k < \frac{1}{2}$ or when $k > 4$. No endpoints are included since the inequality symbol is $>$.

Solution set: $\left(-\infty, \frac{1}{2}\right) \cup (4, \infty)$

45. $x^2 > 0$ is true for all values of x except 0, so that the solution is $(-\infty, 0) \cup (0, \infty)$.

47. $x^2 + 5x - 2 < 0$

Solve the corresponding quadratic equation.

$x^2 + 5x - 2 = 0$

The expression $x^2 + 5x - 2$ cannot be factored, so use the quadratic formula.

$$x = \frac{-5 \pm \sqrt{25 + 8}}{2}$$

$$x = \frac{-5 \pm \sqrt{33}}{2}$$

$x \approx .4$ or $x \approx -5.4$

Use these approximations and $x - .4$ and $x + 5.4$ to make a sign graph.

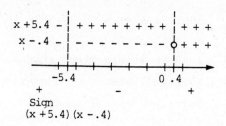

Solution set: $\left(\dfrac{-5 - \sqrt{33}}{2}, \dfrac{-5 + \sqrt{33}}{2}\right)$

49. $m^2 - 2m \le 1$

$m^2 - 2m - 1 \le 0$

Solve the corresponding quadratic equation.

$m^2 - 2m - 1 = 0$

The expression $m^2 - 2m - 1$ cannot be factored, so use the quadratic formula.

$$m = \frac{2 \pm \sqrt{4 + 4}}{2}$$

$$= \frac{2 \pm \sqrt{8}}{2}$$

$$= \frac{2 \pm 2\sqrt{2}}{2}$$

$$m = 1 \pm \sqrt{2}$$

$m \approx 2.4$ or $m \approx -.4$

Use a sign graph.

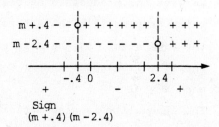

The sign graph shows that the product is negative when x is between $1 - \sqrt{2}$ and $1 + \sqrt{2}$. The endpoints are included because the inequality symbol is $\le$.

Solution set: $[1 - \sqrt{2}, 1 + \sqrt{2}]$

51. $\dfrac{m-3}{m+5} \le 0$

To draw a sign graph, first solve the equations

$$m - 3 = 0 \quad \text{and} \quad m + 5 = 0,$$

getting the solutions

$$m = 3 \text{ and } m = -5.$$

Use the values −5 and 3 to divide the number line into three regions.

The quotient is negative when the factors have different signs, or when m is between −5 and 3. Since the inequality symbol is ≤, we must check each endpoint separately. Here, −5 gives a 0 denominator, but 3 satisfies the inequality.
Solution set: (−5, 3]

53. $\dfrac{k-1}{k+2} > 1$

$\dfrac{k-1}{k+2} - 1 > 0$ *Subtract 1*

$\dfrac{k-1}{k+2} - \dfrac{k+2}{k+2} > 0$ *Common denominator is k + 2*

$\dfrac{k - 1 - (k+2)}{k+2} > 0$

$\dfrac{k - 1 - k - 2}{k+2} > 0$

$\dfrac{-3}{k+2} > 0$ *Combine terms*

Since −3 is negative, this inequality is true when k + 2 is negative, or when k < −2.
Solution set: (−∞, −2)

55. $\dfrac{3}{x-6} \le 2$

$\dfrac{3}{x-6} - 2 \le 0$

$\dfrac{3}{x-6} - \dfrac{2(x-6)}{x-6} \le 0$

$\dfrac{3 - 2(x-6)}{x-6} \le 0$

$\dfrac{3 - 2x + 12}{x-6} \le 0$

$\dfrac{-2x + 15}{x-6} \le 0$

Solve the equations

$$-2x + 15 = 0 \quad \text{and} \quad x - 6 = 0,$$

getting the solutions

$$x = -\dfrac{15}{2} \quad \text{and} \quad x = 6.$$

Draw a sign graph.

The quotient is negative when the factors have different signs, that is, when x < 6 or x > 15/2, Since we have a quotient with "less than or equal to," we must check each endpoint separately. x = 6 doesn't work since it makes a denominator 0. x = 15/2 is acceptable, since it doesn't make any denominator 0.

Thus, the solution set is

$$\left(-\infty,\ 6\right) \cup \left[\frac{15}{2},\ \infty\right).$$

57.

$$\frac{1}{m - 1} < \frac{5}{4}$$

$$\frac{1}{m - 1} - \frac{5}{4} < 0$$

$$\frac{4 - 5(m - 1)}{4(m - 1)} < 0$$

$$\frac{4 - 5m + 5}{4(m - 1)} < 0$$

$$\frac{9 - 5m}{4(m - 1)} < 0$$

Solve the equations

$$9 - 5m = 0 \quad \text{and} \quad 4(m - 1) = 0,$$

getting the solutions

$$m = \frac{9}{5} \quad \text{and} \quad m = 1.$$

Draw a sign graph.

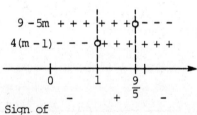

Sign of
$$\frac{9 - 5m}{4(m - 1)}$$

The quotient is negative when $m < 1$ or $m > 9/5$. There is no need to check the endpoints, since the original inequality symbol is $<$. The solution set is

$$\left(-\infty,\ 1\right) \cup \left(\frac{9}{5},\ \infty\right).$$

59.

$$\frac{10}{3 + 2x} \le 5$$

$$\frac{10}{3 + 2x} - 5 \le 0$$

$$\frac{10 - 5(3 + 2x)}{3 + 2x} \le 0$$

$$\frac{10 - 15 - 10x}{3 + 2x} \le 0$$

$$\frac{-5 - 10x}{3 + 2x} \le 0$$

$$\frac{-5(1 + 2x)}{3 + 2x} \le 0$$

Make a sign graph.

Sign of
$$\frac{-5(1 + 2x)}{3 + 2x}$$

The quotient is negative when $x < -3/2$ or $x > -1/2$. Since the inequality symbol is $\le$, we must check each endpoint separately. Here, $-3/2$ gives a 0 denominator but $-1/2$ satisfies the inequality. Thus, the solution set is

$$\left(-\infty,\ -\frac{3}{2}\right) \cup \left[-\frac{1}{2},\ \infty\right).$$

61. $\dfrac{7}{k + 2} \ge \dfrac{1}{k + 2}$

Subtract $\dfrac{1}{k + 2}$ from both sides.

$$\frac{7}{k + 2} - \frac{1}{k + 2} \ge 0$$

$$\frac{6}{k + 2} \ge 0$$

Since 6 is positive, this inequality is true whenever $k + 2$ is positive, or when $k > -2$.

-2 does not satisfy the inequality
because it makes both denominators 0
in the original inequality.
The solution set is $(-2, \infty)$.

63. $\dfrac{3}{2r - 1} > -\dfrac{4}{r}$

Add $\dfrac{4}{r}$ to both sides.

$\dfrac{3}{2r - 1} + \dfrac{4}{r} > 0$ *Add 4/r*

$\dfrac{3r + 4(2r - 1)}{r(2r - 1)} > 0$ *Common denomina-*
tor is r(2r − 1)

$\dfrac{3r + 8r - 4}{r(2r - 1)} > 0$

$\dfrac{11r - 4}{r(2r - 1)} > 0$

Make a sign graph showing the three
factors $11r - 4$, $2r - 1$, and r.

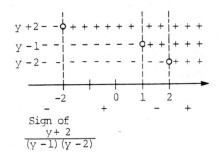

The quotient is positive when
$0 < r < 4/11$ or $r > 1/2$.
Neither endpoint satisfies the inequality since the inequality
symbol is >. The solution set is

$$\left(0, \frac{4}{11}\right) \cup \left(\frac{1}{2}, \infty\right).$$

65. $\dfrac{4}{y - 2} \leq \dfrac{3}{y - 1}$

Subtract $\dfrac{3}{y - 1}$ from both sides.

$\dfrac{4}{y - 2} - \dfrac{3}{y - 1} \leq 0$

$\dfrac{4(y - 1) - 3(y - 2)}{(y - 1)(y - 2)} \leq 0$

    Common denominator
    is (y − 1)(y − 2)

$\dfrac{4y - 4 - 3y + 6}{(y - 1)(y - 2)} \leq 0$

$\dfrac{y + 2}{(y - 1)(y - 2)} \leq 0$

Make a sign graph showing the three
factors $y + 2$, $y - 1$, and $y - 2$.

The quotient is negative when
$y < -2$ or $1 < y < 2$. Checking
the endpoints gives $y = -2$ as
an acceptable solution. However,
$y = 1$ and $y = 2$ are not acceptable since the denominator is
zero at these points. Thus the
solution set is

$$(-\infty, -2] \cup (1, 2).$$

67. $\dfrac{y + 3}{y - 5} \leq 1$

$\dfrac{y + 3}{y - 5} - 1 \leq 0$

$\dfrac{y + 3 - (y - 5)}{y - 5} \leq 0$

$\dfrac{y + 3 - y + 5}{y - 5} \leq 0$

$\dfrac{8}{y - 5} \leq 0$

Since 8 is positive, this inequality is true whenever y - 5 is negative, that is, when y < 5. y = 5 cannot be used since it makes the denominator 0. Thus, the solution set is (-∞, 5).

69. 2 and 5 will both be included in the solution set of (x - 2)(x - 5) ≥ 0, since they make the left side of the inequality equal zero and that is allowed by "≤ 0".

71. 2 and 5 will both be excluded from the solution set of $\frac{x - 2}{x - 5} > 0$ since the endpoints of the solution intervals are never allowed in ">" or "<" inequalities.

73. 2 is included in the solution set of $\frac{x - 2}{x - 5} \geq 0$ since x = 2 makes the fraction $\frac{x - 2}{x - 5}$ have value 0, which is allowed by "≥ 0". 5 is excluded from the solution set since it makes the denominator have value 0, which is never allowed.

75. (2r - 3)(r + 2)(r - 3) ≥ 0

Solve the corresponding equation.

(2r - 3)(r + 2)(r - 3) = 0

2r - 3 = 0   or   r + 2 = 0   or   r - 3 = 0

$r = \frac{3}{2}$ or   r = -2 or   r = 3

The three values divide a number line into four regions.

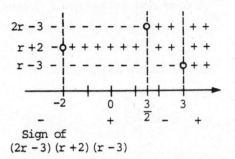

Sign of
(2r - 3) (r + 2) (r - 3)

The product is positive when r is between -2 and 3/2 or when r is greater than 3. Also, the expression is equal to zero at the endpoints. Thus, the solution set is

$$\left[-2, \frac{3}{2}\right] \cup [3, \infty).$$

77. x³ - 4x ≤ 0

Solve the corresponding equation by factoring.

x³ - 4x = 0

x(x² - 4) = 0

x(x + 2)(x - 2) = 0

x = 2   or   x + 2 = 0   or   x - 2 = 0

x = 0   or   x = -2 or   x = 2

The three values divide a number line into three regions.

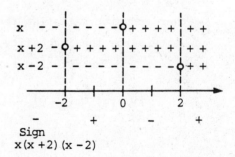

Sign
x(x +2) (x -2)

The product is negative when x is less than -2 or when x is between 0 and 2. Also, the expression is equal to zero at the endpoints.

Thus, the solution set is

$(-\infty, -2] \cup [0, 2]$.

79. $4m^3 + 7m^2 - 2m > 0$

Solve the corresponding equation by factoring.

$m(4m^2 + 7m - 2) = 0$

$m(m + 2)(4m - 1) = 0$

$m = 0$ or $m = -2$ or $m = \frac{1}{4}$

Sign of
m(m+2)(4m-1)

The product is positive when m is between -2 and 0, or when m is greater than 1/4. No endpoints are included since the inequality symbol is $>$. Thus, the solution set is

$$\left(-2,\ 0\right) \cup \left(\frac{1}{4},\ \infty\right).$$

81. (a) and (e)

$(x + 3)^2$ is never negative, so $(x + 3)^2 \geq 0$ has solution set $(-\infty, \infty)$.

$\dfrac{x^2 + 7}{2x^2 + 4}$ is never smaller than $\dfrac{7}{4}$.

This certainly implies that $\dfrac{x^2 + 7}{2x^2 + 4}$

is never negative, so $\dfrac{x^2 + 7}{2x^2 + 4} \geq 0$

has solution set $(-\infty, \infty)$.

85. $P = 4t^2 - 29t + 30$

We want to find the values of t when $P > 0$, so we solve the inequality

$4t^2 - 29t + 30 > 0$.

Solve the corresponding quadratic equation by factoring.

$4t^2 - 29t + 30 = 0$

$(4t - 5)(t - 6) = 0$

$4t - 5 = 0$ or $t - 6 = 0$

$t = \dfrac{5}{4}$ or $t = 6$

Sign of
(4t-5)(t-6)

The expression is positive when $t > \dfrac{5}{4}$ or when $t > 6$. However, since t represents time, we must have $t > 0$. Thus, the investor has been ahead during the time intervals $\left(0,\ \dfrac{5}{4}\right) \cup \left(6,\ \infty\right)$, where t is measured in months.

87. The projectile will be at least 624 ft above the ground whenever

$220t - 16t^2 \geq 624$.

To solve this inequality, first solve the corresponding equation by factoring.

$220t - 16t^2 = 624$

$-16t^2 + 220t - 624 = 0$

$4t^2 - 55t - 156 = 0$ *Divide by -4*

$(t - 4)(4t - 39) = 0$

$t - 4 = 0$ or $4t - 39 = 0$

$t = 4$ or $t = \dfrac{39}{4}$

These two values divide a number line into three regions. Draw a sign graph.

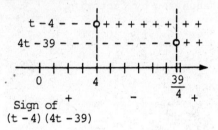

Sign of
$(t - 4)(4t - 39)$

The projectile is at least 624 ft high for values of t in the interval $\left[4, \dfrac{39}{4}\right]$.

89. The apartment complex produces a profit when $P > 0$, so we need to solve the inequality.

$$-x^2 + 250x - 15{,}000 > 0.$$

First, solve the corresponding equation by factoring.

$-x^2 + 250x - 15{,}000 = 0$

$x^2 - 250x + 15{,}000 = 0$
                    Multiply by -1

$(x - 100)(x - 150) = 0$

$x - 100 = 0$ or $x - 150 = 0$

       $x = 100$ or $x = 150$

Draw a sign graph.

```
x -100 - - - - - - - ○+ + + + + +
                      |
x -150 - - - - - - - | - - -| + + +
        ┼     ┼     ┼     ┼
        0    50    100   150
        +           -     +
```

Sign of
$(x - 100)(x - 150)$

The complex produces a profit when the number of units rented is in the interval (100, 150).

Section 2.8

1. $|a - 2| = 1$

$a - 2 = 1$ or $a - 2 = -1$

   $a = 3$ or $a = 1$

Solution set: $\{3, 1\}$

3. $|3m - 1| = 2$

$3m - 1 = 2$ or $3m - 1 = -2$

   $3m = 3$ $3m = -1$

   $m = 1$ or $m = -\dfrac{1}{3}$

Solution set: $\left\{1, -\dfrac{1}{3}\right\}$

5. $|5 - 3x| = 3$

$5 - 3x = 3$ or $5 - 3x = -3$

   $2 = 3x$ $8 = 3x$

   $\dfrac{2}{3} = x$ or $\dfrac{8}{3} = x$

Solution set: $\left\{\dfrac{2}{3}, \dfrac{8}{3}\right\}$

7. $\left|\dfrac{z - 4}{2}\right| = 5$

$\dfrac{z - 4}{2} = 5$ or $\dfrac{z - 4}{2} = -5$

$z - 4 = 10$ $z - 4 = -10$

   $z = 14$ or $z = -6$

Solution set: $\{14, -6\}$

9. $\left|\dfrac{5}{r-3}\right| = 10$

$\dfrac{5}{r-3} = 10$

$5 = 10(r-3)$

$r = 10r - 30$

$35 = 10r$

$\dfrac{7}{2} = r$

or

$\dfrac{5}{r-3} = -10$

$5 = -10(r-3)$

$5 = -10r + 30$

$-25 = -10r$

$\dfrac{5}{2} = r$

Solution set: $\left\{\dfrac{7}{2}, \dfrac{5}{2}\right\}$

11. $|4w + 3| - 2 = 7$

$|4w + 3| = 9$

$4w + 3 = 9$ or $4w + 3 = 9$

$4w = 6$ $\qquad\qquad$ $4w = -12$

$w = \dfrac{3}{2}$ or $\qquad$ $w = -3$

Solution set: $\left\{\dfrac{3}{2}, -3\right\}$

13. $|6x + 9| = 0$

This absolute value equation has only one case.

$6x + 9 = 0$

$6x = -9$

$x = -\dfrac{9}{6} = -\dfrac{3}{2}$

Solution set: $\left\{-\dfrac{3}{2}\right\}$

15. $\left|\dfrac{6y+1}{y-1}\right| = 3$

$\dfrac{6y+1}{y-1} = 3$

$6y + 1 = 3(y - 1)$

$6y + 1 = 3y - 3$

$3y = -4$

$y = -\dfrac{4}{3}$

or

$\dfrac{6y+1}{y-1} = -3$

$6y + 1 = -3(y - 1)$

$6y + 1 = -3y + 3$

$9y = 2$

$y = \dfrac{2}{9}$

Solution set: $\left\{-\dfrac{4}{3}, \dfrac{2}{9}\right\}$

17. $|2k - 3| = |5k + 4|$

$2k - 3 = 5k + 4$

$-7 = 3k$

$-\dfrac{7}{3} = k$

or

$2k - 3 = -(5k + 4)$

$2k - 3 = -5k - 4$

$7k = -1$

$k = -\dfrac{1}{7}$

Solution set: $\left\{-\dfrac{7}{3}, -\dfrac{1}{7}\right\}$

19. $|4 - 3y| = |7 + 2y|$

$4 - 3y = 7 + 2y$

$-3 = 5y$

$-\dfrac{3}{5} = y$

or

$4 - 3y = -(7 + 2y)$

$4 - 3y = -7 - 2y$

$11 = y$

Solution set: $\left\{ -\frac{3}{5}, 11 \right\}$

21. $|8b + 7| = |4b|$

$8b + 7 = 4b$ or $8b + 7 = -4b$

$4b + 7 = 0$ $12b + 7 = 0$

$4b = -7$ $12b = -7$

$b = -\frac{7}{4}$ or $b = -\frac{7}{12}$

Solution set: $\left\{ -\frac{7}{4}, -\frac{7}{12} \right\}$

23. $|x + 2| = |x - 1|$

$x + 2 = x - 1$

$2 = -1$ *False*

or

$x + 2 = -(x - 1)$

$x + 2 = -x + 1$

$2x = -1$

$x = -\frac{1}{2}$

Solution set: $\left\{ -\frac{1}{2} \right\}$

27. $|x| \leq 3$

$-3 \leq x \leq 3$

Solution set: $[-3, 3]$

29. $|m| > 1$

$m < -1$ or $m > 1$

Solution set: $(-\infty, -1) \cup (1, \infty)$

31. $|a| < -2$ has no solution since the absolute value of any number must be positive or zero and therefore cannot be less than -2.

Solution set: Ø

33. $|x| - 3 \leq 7$

$x \leq 10$

$-10 \leq x \leq 10$

Solution set: $[-10, 10]$

35. $|2x + 5| < 3$

$-3 < 2x + 5 < 3$

$-8 < 2x < -2$

$-4 < x < -1$

Solution set: $(-4, -1)$

37. $|3m - 2| > 4$

$3m - 2 < -4$ or $3m - 2 > 4$

$3m < -2$ $3m > 6$

$m < -\frac{2}{3}$ or $m > 2$

Solution set: $\left(-\infty, -\frac{2}{3} \right) \cup (2, \infty)$

39. $|3z + 1| \geq 7$

$3z + 1 \leq -7$ or $3z + 1 \geq 7$

$3z \leq -8$ $3z \geq 6$

$z \leq -\frac{8}{3}$ or $z \geq 2$

Solution set: $\left(-\infty, -\frac{8}{3} \right] \cup [2, \infty)$

41. $\left|\frac{2}{3}t + \frac{1}{2}\right| \leq \frac{1}{6}$

$-\frac{1}{6} \leq \frac{2}{3}t + \frac{1}{2} \leq \frac{1}{6}$

Multiply by the least common denominator, 6, to clear fractions.

$-1 \leq 4t + 3 \leq 1$

$-4 \leq 4t \leq -2$

$-1 \leq t \leq -\frac{1}{2}$

Solution set: $\left[-1, -\frac{1}{2}\right]$

43. $\left|5x + \frac{1}{2}\right| - 2 < 5$

$\left|5x + \frac{1}{2}\right| < 7$ *Add 2*

$-7 < 5x + \frac{1}{2} < 7$

$-\frac{15}{2} < 5x < \frac{13}{2}$ *Subtract 1/2*

$-\frac{3}{2} < x < \frac{13}{10}$ *Divide by 5*

Solution set: $\left(-\frac{3}{2}, \frac{13}{10}\right)$

45. $\left|6x + 3\right| \geq -2$

Since the absolute value of a number is always nonnegative, $\left|6x + 3\right| \geq -2$ is always true.

Solution set: $(-\infty, \infty)$

47. $\left|\frac{1}{2}x + 6\right| > 0$

The absolute value of a number will be positive so long as the number is negative or positive (but not zero).

$\frac{1}{2}x + 6 < 0$ or $\frac{1}{2}x + 6 > 0$

$\frac{1}{2}x < -6$ $\frac{1}{2}x > -6$

$x < -12$ or $x > -12$

Solution set: $(-\infty, -12) \cup (-12, \infty)$

49. $\left|p - q\right| = 5$, which is equivalent to $\left|q - p\right| = 5$, indicates that the distance between p and q is 5 units.

53. "x is within 4 units of 2" means that the distance between x and 2 is less than or equal to 4, or

$$\left|x - 2\right| \leq 4.$$

55. "z is no less than 2 units from 12" means that z is 2 units or more from 12. Thus, the distance between z and 12 is greater than or equal to 2, or

$$\left|z - 12\right| \geq 2.$$

57. "k is 6 units from 1" means that the distance between k and 1 is 6 units, or

$$\left|k - 1\right| = 6.$$

59. "x within .0004 units of 2" means that the distance between x and 2 is less than or equal to .0004, or

$$\left|x - 2\right| \leq .0004.$$

"y within .00001 units of 7" means that the distance between y and 7 is less than or equal to .00001, or

$$\left|y - 7\right| \leq .00001.$$

The statement becomes:

"If $|x - 2| \le .0004$,

then $|y - 7| \le .00001$."

61. $|C + 84| \le 56$

$-56 \le C + 84 \le 56$

$-140 \le C \le -28$

In degrees Celsius, the range
of temperatures is the interval
$[-140, -28]$.

63. 780 is 50 more than 730 and 680 is
50 less than 730, so all of the
temperatures in the acceptable range
are within 50° of 730°. That is

$$|F - 730| \le 50.$$

65. If $|x - 2| < 3$, find m and n so that

$$m < 3x + 5 < n.$$

$|x - 2| < 3$ means

$-3 < x - 2 < 3$

$-1 < x < 5$

$-3 < 3x < 15$ *Multiply by 3*

$2 < 3x + 5 < 20$. *Add 5*

Choose any $m \le 2$ and any $n \ge 20$.

Chapter 2 Review Exercises

1. $2m + 7 = 3m + 1$

$7 = m + 1$ *Subtract 2m*

$6 = m$ *Subtract 1*

Solution set: $\{6\}$

3. $5y - 2(y + 4) = 3(2y + 1)$

$5y - 2y - 8 = 6y + 3$
        Distributive property

$3y - 8 = 6y + 3$
        Combine like terms

$-8 = 3y + 3$ *Subtract 3y*

$-11 = 3y$ *Subtract 3*

$\frac{1}{3}(-11) = \frac{1}{3}(3y)$ *Multiply
                    by 1/3*

$-\frac{11}{3} = y$

Solution set: $\left\{-\frac{11}{3}\right\}$

5. $\dfrac{10}{4z - 4} = \dfrac{1}{1 - z}$

Multiply both sides by the common
denominator $(4z - 4)(1 - z)$, assuming $z = 1$.

$10(1 - z) = 1(4z - 4)$

$10 - 10z = 4z - 4$

$14 = 14z$

$1 = z$

However, substituting 1 for z in the
original equation would result in a
denominator of 0, so 1 is not a
solution.

Solution set: Ø

7. $\dfrac{5}{3r} - 10 = \dfrac{3}{2r}$

Multiply both sides by the least
common denominator, 6r, assuming
$r \ne 0$.

$$6r\left(\frac{5}{3r}\right) - 6r(10) = 6r\left(\frac{3}{2r}\right)$$

$$10 - 60r = 9$$

$$-60r = -1$$

$$r = \frac{1}{60}$$

Solution set: $\left\{\frac{1}{60}\right\}$

9. $3(x + 2b) + a = 2x - 6$ for x

$$3x + 6b + a = 2x - 6$$

$$x + 6b + a = -6$$

$$x = -6b - a - 6$$

11. $\frac{x}{m - 2} = kx - 3$ for x

$$(m - 2)\left(\frac{x}{m - 2}\right) = (m - 2)(kx - 3)$$

$$x = mkx - 3m - 2kx + 6$$

$$x - mkx + 2kx = -3m + 6$$

Put all terms containing x on one side

$$x(1 - mk + 2k) = -3m + 6$$

Factor out x

$$x = \frac{-3m + 6}{1 - mk + 2k}$$

or

$$\frac{6 - 3m}{1 + 2k - km}$$

13. $A = P + Pi$ for P

$$A = P(1 + i) \quad \textit{Factor out } P$$

$$\frac{A}{1 + i} = P \qquad \textit{Divide by } 1 + i$$

15. $A = \frac{24f}{b(p + 1)}$ for f

$$Ab(p + 1) = 24f \quad \textit{Multiply by } b(p+1)$$

$$\frac{Ab(p + 1)}{24} = f \quad \textit{Divide by 24}$$

17. $\frac{xy^2 - 5xy + 4}{3x} = 2p$ for x

$$xy^2 - 5xy + 4 = 6px$$

Multiply by $3x$

$$xy^2 - 5xy - 6px = -4$$

$$x(y^2 - 5y - 6p) = -4$$

Factor out x

$$x = \frac{-4}{y^2 - 5y - 6p}$$

19. Let $\quad x$ = original price.

Then $\quad .15x$ = discount,

so $x - .15x$ = sales price.

$$.85x = 425$$

$$x = 500$$

The original price was $500.

21. Let $\quad x$ = rate biking to library.

Then $x - 8$ = rate biking home.

Set up a chart, using $d = rt$.

	d	r	t
To library	$\frac{1}{3}x$	x	$\frac{1}{3}$
Return	$\frac{1}{2}(x - 8)$	$x - 8$	$\frac{1}{2}$

The times must be changed from minutes to hours, since rates are given in miles per hour.

$$\text{Distance to library} = \text{Return distance}$$

$$\frac{1}{3}x = \frac{1}{2}(x - 8)$$

Multiply both sides by the least common denominator, 6.

$$6\left(\tfrac{1}{3}x\right) = 6\left[\tfrac{1}{2}(x - 8)\right]$$

$$2x = 3(x - 8)$$

$$2x = 3x - 24$$

$$-x = -24$$

$$x = 24$$

To find the distance, substitute

$x = 24$ into $d = \tfrac{1}{3}x$.

$$d = \tfrac{1}{3}(24) = 8$$

Alison lives 8 mi from the library.

23. Let x = amount borrowed at 11.5%.

Then $90,000 - x$

= amount borrowed at 12%.

Interest owed at 11.5%	+	Interest owed at 12%	=	Total interest

$$.115x + .12(90,000 - x) = 10,525$$

$$.115x + 10,800 - .12x = 10,525$$

$$-.005x = -275$$

$$x = 55,000$$

$$90,000 - x = 35,000$$

$55,000 was borrowed at 11.5% and $35,000 was borrowed at 12%.

25. $(6 - i) + (4 - 2i) = 10 - 3i$

27. $15i - (3 + 2i) - 5$

$= 15i - 3 - 2i - 5$

$= -8 + 13i$

29. $(5 - i)(3 + 4i)$

$= 15 + 20i - 3i - 4i^2$

$= 15 + 17i + 4 \quad i^2 = -1$

$= 19 + 17i$

31. $(5 - 11i)(5 + 11i)$

$= 5^2 - (11i)^2$
Difference of two squares

$= 25 - 121i^2$

$= 25 + 121$

$= 146 = 146 + 0i$ *Standard form*

33. $(4 - 3i)^2$

$= 4^2 - 2(4)(3i) + (3i)^2$
Square of a binomial

$= 16 - 24i - 9$

$= 7 - 24i$

35. $\dfrac{6 + i}{1 - i}$

Multiply the numerator and denominator by the conjugate of the denominator.

$$\frac{6 + i}{1 - i} = \frac{(6 + i)(1 + i)}{(1 - i)(1 + i)}$$

$$= \frac{6 + 7i + i^2}{1 - i^2}$$

$$= \frac{6 + 7i - 1}{1 + i}$$

$$= \frac{5 + 7i}{2}$$

$$= \frac{5}{2} + \frac{7}{2}i \quad \textit{Standard form}$$

37. The product of a complex number and its conjugate is always a real number.

$$(a + bi)(a - bi) = a^2 - (bi)^2$$

$$= a^2 - b^2 i^2$$

$$= a^2 - b^2(-1)$$

$$= a^2 + b^2$$

39. $i^7 = i^4 \cdot i^3$

$= 1(-i)$

$= -i$

41. $i^{-35} = (i^{-4})^{-9} \cdot i$

$\qquad = 1^{-9} \cdot i$

$\qquad = 1 \cdot i = i$

43. $(b + 7)^2 = 5$

Use the square root property.

$\qquad b + 7 = \pm\sqrt{5}$

$\qquad\quad b = -7 \pm \sqrt{5}$

Solution set: $\{-7 \pm \sqrt{5}\}$

45. $\qquad 2a^2 + a - 15 = 0$

Solve the equation by factoring.

$(2a - 5)(a + 3) = 0$

$2a - 5 = 0 \quad$ or $\quad a + 3 = 0$

$\qquad\qquad\qquad$ *Zero-factor property*

$\quad a = \dfrac{5}{2} \quad$ or $\qquad a = -3$

Solution set: $\left\{\dfrac{5}{2}, -3\right\}$

47. $\qquad 2q^2 - 11q = 21$

$\quad 2q^2 - 11q - 21 = 0$

$\quad (2q + 3)(q - 7) = 0$

$2q + 3 = 0 \quad$ or $\quad q - 7 = 0$

$\quad q = -\dfrac{3}{2} \quad$ or $\qquad q = 7$

Solution set: $\left\{-\dfrac{3}{2}, 7\right\}$

49. $\qquad 2 - \dfrac{5}{p} = \dfrac{3}{p^2}$

Multiply both sides by the least common denominator, p^2, assuming $p \neq 0$.

$$p^2\left(2 - \dfrac{5}{p}\right) = p^2\left(\dfrac{3}{p^2}\right)$$

$$p^2(2) - p^2\left(\dfrac{5}{p}\right) = p^2\left(\dfrac{3}{p^2}\right)$$

$2p^2 - 5p = 3$

$2p^2 - 5p - 3 = 0$

$(2p + 1)(p - 3) = 0$

$p = -\dfrac{1}{2} \quad$ or $\quad p = 3$

Solution set: $\left\{-\dfrac{1}{2}, 3\right\}$

51. $\sqrt{2}x^2 - 4x + \sqrt{2} = 0$

Use the quadratic formula with $a = \sqrt{2}$, $b = -4$, and $c = \sqrt{2}$.

$$x = \dfrac{4 \pm \sqrt{(-4)^2 - 4 \cdot \sqrt{2} \cdot \sqrt{2}}}{2 \cdot \sqrt{2}}$$

$$= \dfrac{4 \pm \sqrt{16 - 8}}{2\sqrt{2}} = \dfrac{4 \pm \sqrt{8}}{2\sqrt{2}}$$

$$= \dfrac{(4 \pm \sqrt{8})\sqrt{2}}{(2\sqrt{2})\sqrt{2}}$$

$$= \dfrac{4\sqrt{2} \pm \sqrt{16}}{2 \cdot 2} = \dfrac{4\sqrt{2} \pm 4}{4}$$

$$= \sqrt{2} \pm 1 \quad \textit{Lowest terms}$$

Solution set: $\{\sqrt{2} \pm 1\}$

53. $\qquad 8y^2 = 2y - 6$

$8y^2 - 2y + 6 = 0$

$a = 8, \ b = -2, \ c = 6$

$b^2 - 4ac = (-2)^2 - 4(8)(6)$

$\qquad\qquad = 4 - 192$

$\qquad\qquad = -188$

The equation has two different imaginary solutions since the discriminant is negative.

55. $\qquad 16r^2 + 3 = 26r$

$16r^2 - 26r + 3 = 0$

$a = 16, \ b = -26, \ c = 3$

$$b^2 - 4ac = (-26)^2 - 4(16)(3)$$
$$= 676 - 192$$
$$= 484 = 22^2$$

The equation has two different rational solutions since the discriminant is a positive perfect square.

57. $25z^2 - 110z + 121 = 0$

$a = 25, \ b = -110, \ c = 121$

$$b^2 - 4ac = (-110)^2 - 4(25)(121)$$
$$= 12,100 - 12,100$$
$$= 0$$

The equation has one rational solution since the discriminant is 0.

59. Let x = width of border.

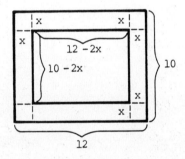

Apply the formula A = LW to both the outside and inside rectangles.

Inside area	=	Outside area	−	Border area
$(12 - 2x)(10 - 2x)$ =		$12 \cdot 10$	−	21

$$120 - 44x + 4x^2 = 120 - 21$$
$$4x^2 - 44x + 120 = 99$$
$$4x^2 - 44x + 21 = 0$$
$$(2x - 21)(2x - 1) = 0$$
$$2x = 21 \ \text{ or } \ 2x - 1 = 0$$
$$x = \frac{21}{2} \ \text{ or } \ \ \ \ \ x = \frac{1}{2}$$

The border width cannot be $\frac{21}{2}$ or $10\frac{1}{2}$ since this exceeds the width of the outside rectangle, so reject this solution. The width of the border is $\frac{1}{2}$ ft.

61. Let x = the number of hours for slower gardener to complete job working alone.

Then x − 1 = the number of hours for faster gardener to complete job working alone.

	Rate	Time	Part of the job accomplished
Slower gardener	$\frac{1}{x}$	3	$3\left(\frac{1}{x}\right) = \frac{3}{x}$
Faster gardener	$\frac{1}{x - 1}$	3	$3\left(\frac{1}{x - 1}\right) = \frac{3}{x - 1}$

Together, the gardeners do one whole job, so

$$\frac{3}{x} + \frac{3}{x - 1} = 1.$$

To clear fractions, multiply both sides by the least common denominator, x(x − 1).

$$x(x - 1)\left(\frac{3}{x} + \frac{3}{x - 1}\right) = x(x - 1) \cdot 1$$
$$3(x - 1) + 3x = x(x - 1)$$
$$3x - 3 + 3x = x^2 - x$$
$$0 = x^2 - 7x + 3$$

This equation cannot be solved by factoring, so use the quadratic formula with a = 1, b = −7, and c = 3.

$$x = \frac{-(-7) \pm \sqrt{(-7)^2 - 4(1)(3)}}{2(1)}$$

$$= \frac{7 \pm \sqrt{49 - 12}}{2}$$

$$= \frac{7 \pm \sqrt{37}}{2}$$

Use a decimal approximation for $\sqrt{37}$ to find the time to the nearest tenth of an hour.

$$\frac{7 + \sqrt{37}}{2} \approx 6.5$$

$$\frac{7 - \sqrt{37}}{2} \approx .5$$

The second answer must be rejected, because the number of hours for the faster gardener $(x - 1)$ would be negative.

It would take the slower gardener 6.5 hr to complete the job alone.

63. Let x = winner's average speed.

Then $x - \frac{4}{3}$ = friend's average speed.

Organize the information in a chart using $t = d/r$.

	d	r	t
Winner	26	x	$\frac{26}{x}$
Friend	26	$x - \frac{4}{3}$	$\frac{46}{x - 4/3}$

Winner's time = Friends time − 2/5 hr

$$\frac{26}{x} = \frac{26}{x - \frac{4}{3}} - \frac{2}{5}$$

Multiply both sides by the least common denominator, $5x\left(x - \frac{4}{3}\right)$.

$$5x\left(x - \frac{4}{3}\right)\left(\frac{26}{x}\right) = 5x\left(x - \frac{4}{3}\right)\left(\frac{26}{x - 4/3}\right)$$
$$- 5x\left(x - \frac{4}{3}\right)\left(\frac{2}{5}\right)$$

$$130\left(x - \frac{4}{3}\right) = 130x - 2x\left(x - \frac{4}{3}\right)$$

$$130x - \frac{520}{3} = 130x - 2x^2 + \frac{8}{3}x$$
Distributive property

$$390x - 520 = 390x - 6x^2 + 8x$$
Multiply by 3

$$6x^2 - 8x - 520 = 0$$

$$3x^2 - 4x - 260 = 0$$
Divide by 2

$$(3x + 26)(x - 10) = 0$$

$$x = -\frac{26}{3} \quad \text{or} \quad x = 10$$

Reject the negative solution. The winner's average speed is 10 mph.

65. $4a^4 + 3a^2 - 1 = 0$

Let $u = a^2$; then $u^2 = a^4$.

With this substitution, the equation becomes

$$4u^2 + 3u - 1 = 0.$$

Solve this equation by factoring.

$$(u + 1)(4u - 1) = 0$$

$$u + 1 = 0 \quad \text{or} \quad 4u - 1 = 0$$

$$u = -1 \quad \text{or} \quad u = \frac{1}{4}$$

To find a, replace u with a^2.

$$a^2 = -1 \quad \text{or} \quad u = \frac{1}{4}$$

$$a = \pm\sqrt{-1} \quad \text{or} \quad u = \pm\sqrt{\frac{1}{4}}$$

$$a = \pm i \quad \text{or} \quad u = \pm\frac{1}{2}$$

Solution set: $\left\{\pm i, \pm\frac{1}{2}\right\}$

67. $(2z + 3)^{2/3} + (2z + 3)^{1/3} = 6$

$(2z + 3)^{2/3} + (2z + 3)^{1/3} - 6 = 0$

Let $x = (2z + 3)^{1/3}$; then

$x^2 = [(2z + 3)^{1/3}]^2 = (2z + 3)^{2/3}$.

With this substitution, the equation becomes

$$x^2 + x - 6 = 0.$$

Solve this equation by factoring.

$$(x + 3)(x - 2) = 0$$
$$x = -3 \quad \text{or} \quad x = 2$$

To find z, replace x with $(2z + 3)^{1/3}$.

If $x = -3$, then

$$(2z + 3)^{1/3} = -3$$
$$2z + 3 = -27 \quad \textit{Cube both sides}$$
$$2z = -30$$
$$z = -15.$$

If $x = 2$, then

$$(2z + 3)^{1/3} = 2$$
$$2z + 3 = 8 \quad \textit{Cube both sides}$$
$$2z = 5$$
$$z = \frac{5}{2}.$$

Since the original equation involves rational exponents, both solutions should be checked. $z = -15$ and $z = \frac{5}{2}$ both satisfy the original equation.

Solution set: $\left\{-15, \frac{5}{2}\right\}$

69. $\sqrt{4y - 2} = \sqrt{3y + 1}$

$(\sqrt{4y - 2})^2 = (\sqrt{3y + 1})^2$

$\qquad\qquad\qquad$ *Square both sides*

$$4y - 2 = 3y + 1$$
$$y = 3$$

Check this proposed solution in the original equation.

$$\sqrt{4y - 2} = \sqrt{3y + 1}$$
$$\sqrt{4(3) - 2} = \sqrt{3(3) + 1} \quad ?$$
$$\sqrt{10} = \sqrt{10} \qquad\qquad \textit{True}$$

Solution set: $\{3\}$

71. $\sqrt{p + 2} = 2 + p$

$(\sqrt{p + 2})^2 = (2 + p)^2$

$\qquad\qquad\qquad$ *Square both sides*

$$p + 2 = p^2 + 4p + 4$$

$\qquad\qquad\qquad$ *Square of a binomial*

$$0 = p^2 + 3p + 2$$
$$o = (p + 2)(p + 1)$$
$$p = -2 \quad \text{or} \quad p = -1$$

Checking these proposed solutions will show that both satisfy the original equation.

Solution set: $\{-2, -1\}$

73. $\sqrt{x + 3} - \sqrt{3x + 10} = 1$

$\sqrt{x + 3} = 1 + \sqrt{3x + 10}$

$(\sqrt{x + 3})^2 = (1 + \sqrt{3x + 10})^2$
Square both sides

$x + 3 = 1 + 2\sqrt{3x + 10}$
$+ 3x + 10$
Square of a binomial

$-2x - 8 = 2\sqrt{3x + 10}$

$2(-x - 4) = 2\sqrt{3x + 10}$

$-x - 4 = \sqrt{3x + 10}$
Divide by 2

$(-x - 4)^2 = (\sqrt{3x + 10})^2$
Square both sides

$x^2 + 8x + 16 = 3x + 10$
Square of a binomial

$x^2 + 5x + 6 = 0$

$(x + 2)(x + 3) = 0$

$x = -2$ or $x = -3$

Check $x = -2$.

$\sqrt{-2 + 3} - \sqrt{-6 + 10} = 1$?

$\sqrt{1} - \sqrt{4} = 1$?

$1 - 2 = 1$?

$-1 = 1$ *False*

Check $x = -3$.

$\sqrt{-3 + 3} - \sqrt{3(-3) + 10} = 1$?

$\sqrt{0} - \sqrt{1} = 1$?

$-1 = 1$ *False*

Since neither of the proposed solutions satisfies the original equation, the equation has no solution.

Solution set: Ø

75. $\sqrt[3]{6y + 2} = \sqrt[3]{4y}$

$(\sqrt[3]{6y + 2})^3 = (\sqrt[3]{4y})^3$ *Cube both sides*

$6y + 2 = 4y$

$2y = -2$

$y = -1$

A check will show that this proposed solution satisfies the original equation.

Solution set: $\{-1\}$

77. $-9x < 4x + 7$

$0 < 13x + 7$

$-7 < 13x$

$-\dfrac{7}{13} < x$ or $x > -\dfrac{7}{13}$

Solution set: $\left(-\dfrac{7}{13}, \infty\right)$

79. $-5z - 4 \geq 3(2z - 5)$

$-5z - 4 \geq 6z - 15$

$-5z - 4 + 5z \geq 6z - 15 + 5z$

$-4 \geq 11z - 15$

$-4 + 15 \geq 11z - 15 + 15$

$11 \geq 11z$

$1 \geq z$ or $z \leq 1$

Solution set: $(-\infty, 1]$

81. $3r - 4 + r > 2(r - 1)$

$4r - 4 > 2r - 2$

$2r - 4 > -2$

$2r > 2$

$r > 1$

Solution set: $(1, \infty)$

83. $5 \leq 2x - 3 \leq 7$

$8 \leq 2x \leq 10$

$4 \leq x \leq 5$

Solution set: $[4, 5]$

85. $x^2 + 3x - 4 \leq 0$

Solve the corresponding quadratic equation by factoring.

$$x^2 + 3x - 4 = 0$$
$$(x + 4)(x - 1) = 0$$
$$x = -4 \quad \text{or} \quad x = 1$$

These two points, -4 and 1, divide a number line into three regions shown on the following sign graph.

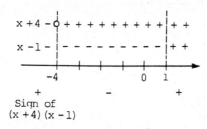

The product is negative when x is between -4 and 1. The endpoints satisfy the inequality because the inequality symbol is $\leq$.

Solution set: $[-4, 1]$

87. $6m^2 - 11m - 10 < 0$

Solve the corresponding quadratic equation by factoring.

$$6m^2 - 11m - 10 = 0$$
$$(3m + 2)(2m - 5) = 0$$
$$3m + 2 = 0 \quad \text{or} \quad 2m - 5 = 0$$
$$m = -\frac{2}{3} \quad \text{or} \quad m = \frac{5}{2}$$

The product is negative when the two factors have opposite signs, which occurs when x is between $-2/3$ and $5/2$. The endpoints are not included because the inequality symbol is $<$.

Solution set: $\left(-\frac{2}{3}, \frac{5}{2}\right)$

89. $x^2 - 6x + 9 \leq 0$

First, solve the corresponding quadratic equation.

$$x^2 - 6x + 9 = 0$$
$$(x - 3)^2 = 0$$
$$x - 3 = 0$$
$$x = 3$$

This point, 3, divides a number line into the region to the left of 3 and the region to the right of 3. Testing any number from either region will result in $(x - 3)^2$ being positive and therefore not less than or equal to zero. The point 3 causes $(x - 3)^2$ to equal 0, so $x = 3$ is the only solution of $x^2 - 6x + 9 \leq 0$.

Solution set: $\{3\}$

91.
$$\frac{3a - 2}{a} > 4$$
$$\frac{3a - 2}{a} - 4 > 0$$
$$\frac{3a - 2}{a} - \frac{4a}{a} > 0$$
$$\frac{3a - 2 - 4a}{a} > 0$$
$$\frac{-a - 2}{a} > 0$$

Make a sign graph.

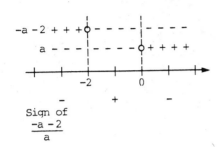

$$\text{Sign of } \frac{-a-2}{a}$$

The quotient is positive when a is between −2 and 0. The endpoints are not included.

Solution set: (−2, 0)

93.
$$\frac{3}{r-1} \le \frac{5}{r+3}$$

$$\frac{3}{r-1} - \frac{5}{r+3} \le 0$$

$$\frac{3(r+3) - 5(r-1)}{(r-1)(r+3)} \le 0$$

Common denominator is (r − 1)(r + 3)

$$\frac{3r + 9 - 5r + 5}{(r-1)(r+3)} \le 0$$

$$\frac{-2r + 14}{(r-1)(r+3)} \le 0$$

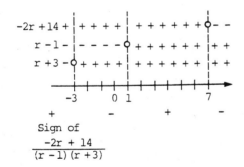

$$\text{Sign of } \frac{-2r+14}{(r-1)(r+3)}$$

The quotient is negative when −3 < r < 1 or r > 7.

Since the inequality symbol is ≤, we must check each endpoint separately. Here, −3 and 1 give 0 denominators, but 7 satisfies the inequality.

Solution set: (−3, 1) ∪ [7, ∞)

95. 0 < a < b divides a number line into the intervals (−∞, a), (a, b), and (b, ∞), as shown on the following sign graph.

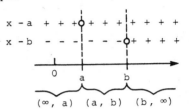

$$\text{Sign of } (x-a)(x-b)$$

x − a is negative on (−∞, a), but positive on (a, b) and (b, ∞), and it is zero at a.

x − b is negative on (−∞, a) and (a, b), but positive on (b, ∞), and it is zero at b.

Therefore, (x − a)(x − b) is positive on the intervals (−∞, a) and (b, ∞), negative on the interval (a, b), and zero at the numbers a and b.

97. C = 3x + 1500, R = 8x

The company will at least break even when R ≥ C.

$$8x \ge 3x + 1500$$
$$5x \ge 1500$$
$$x \ge 300$$

The break−even point is at x = 300. The company will at least break even if the number of units produced is in the interval [300, ∞).

99. |a + 4| = 7
$$a + 4 = 7 \quad \text{or} \quad a + 4 = -7$$
$$a = 3 \quad \text{or} \qquad a = -11$$

Solution set: {3, −11}

101. $\left|\dfrac{7}{2 - 3a}\right| = 9$

$\dfrac{7}{2 - 3a} = 9$

$7 = 9(2 - 3a)$

$7 = 18 - 27a$

$27a = 11$

$a = \dfrac{11}{27}$

or

$\dfrac{7}{2 - 3a} = -9$

$7 = -9(2 - 3a)$

$7 = -18 + 27a$

$25 = 27a$

$a = \dfrac{25}{27}$

Solution set: $\left\{\dfrac{11}{27}, \dfrac{25}{27}\right\}$

103. $|5r - 1| = |2r + 3|$

$5r - 1 = 2r + 3$

$3r = 4$

$r = \dfrac{4}{3}$

or

$5r - 1 = -(2r + 3)$

$5r - 1 = -2r - 3$

$7r = -2$

$r = -\dfrac{2}{7}$

Solution set: $\left\{\dfrac{4}{3}, -\dfrac{2}{7}\right\}$

105. $|m| \le 7$

$-7 \le m \le 7$

Solution set: $[-7, 7]$

107. $|p| > 3$

$p < -3 \ \text{ or } \ p > 3$

Solution set: $(-\infty, -3) \cup (3, \infty)$

109. $|2z + 9| \le 3$

$-3 \le 2z + 9 \le 3$

$-12 \le 2z \le -6$

$-6 \le z \le -3$

Solution set: $[-6, -3]$

111. $|7k - 3| < 5$

$-5 < 7k - 3 < 5$

$-2 < 7k < 8$

$-\dfrac{2}{7} < k < \dfrac{8}{7}$

Solution set: $\left(-\dfrac{2}{7}, \dfrac{8}{7}\right)$

113. $|3r + 7| - 5 > 0$

$|3r + 7| > 5$

$3r + 7 < -5 \ \text{ or } \ 3r + 7 > 5$

$3r < -12 \qquad \qquad 3r > -2$

$r < -4 \ \text{ or } \qquad r > -\dfrac{2}{3}$

Solution set: $(-\infty, -4) \cup \left(-\dfrac{2}{3}, \infty\right)$

Chapter 2   Test

1. $x - (2x + 1) = 7 - 3(x + 1)$

$x - 2x - 1 = 7 - 3x + 3$

$-x - 1 = 4 - 3x$

$2x = 5$

$x = \dfrac{5}{2}$

Solution set: $\left\{\dfrac{5}{2}\right\}$

2. $\dfrac{x}{x - 3} = \dfrac{3}{x - 3} + 4$

Multiply both sides by the least common denominator, x - 3, assuming x ≠ 3.

$$(x - 3)\left(\dfrac{x}{x - 3}\right) = (x - 3)\left(\dfrac{3}{x - 3}\right)$$
$$+ (x - 3)(4)$$
$$x = 3 + 4x - 12$$
$$-3x = -9$$
$$x = 3$$

However, substituting 3 for x in the original equation would result in a denominator of 0, so 3 is not a solution.

Solution set: ∅

3. $\dfrac{1}{c} - \dfrac{1}{a} = \dfrac{1}{b}$ for c

Multiply both sides by the least common denominator, abc.

$$abc\left(\dfrac{1}{c}\right) - abc\left(\dfrac{1}{a}\right) = abc\left(\dfrac{1}{b}\right)$$
$$ab - bc = ac$$
$$ab = ac + bc$$
$$ab = c(a + b) \qquad \textit{Factor out c}$$
$$\dfrac{ab}{a + b} = c$$

4. Let x = number of quarts of 60% alcohol solution.

Strength	Quarts of solution	Quarts of pure alcohol
60%	x	.60x
20%	40	.20(40)
30%	x + 40	.30(x + 40)

$$.60x + .20(40) = .30(x + 40)$$
$$.60x + 8 = .30x + 12$$
$$.30x = 4$$
$$x = \dfrac{4}{.30} = 13\dfrac{1}{3}$$

$13\dfrac{1}{3}$ quarts of 60% alcohol must be added.

5. Let x = time Fred travels.
Then x - 3 = time Wilma travels.

	d	r	t
Fred	30x	30	x
Wilma	50(x - 3)	50	x - 3

They both travel the same distance, so

$$30x = 50(x - 3)$$
$$30x = 50x - 150$$
$$-20x = -150$$
$$x = 7.5$$

and d = 30(7.5) = 225.

They travel 225 mi before meeting.

6. (7 - i) - (6 - 10i)
$$= (7 - 6) + [-1 - (-10)]i$$
$$= 1 + 9i$$

7. (4 + 3i)(-5 + 2i)
$$= 4(-5) + 4(2i) + 3i(-5) + (3i)(2i)$$
$$= -20 - 15i + 8i + 6i^2$$
$$= -20 - 7i - 6 \quad i^2 = -1$$
$$= -26 - 7i$$

8. $\dfrac{5 - 5i}{1 - 3i} = \dfrac{(5 - 5i)(1 + 3i)}{(1 - 3i)(1 + 3i)}$

$= \dfrac{5 + 15i - 5i - 15i^2}{1 - 9i^2}$

$= \dfrac{20 + 10i}{10}$

$= 2 + i$

9. $i^{297} = i^{296} \cdot i$

$= (i^4)^{74} \cdot i$

$= 1^{74} \cdot i$

$= 1 \cdot i = i$

10. $6x(2 - x) = 7$

$12x - 6x^2 = 7$

$0 = 6x^2 - 12x + 7$

Solve this equation by the quadratic formula with $a = 6$, $b = -12$, and $c = 7$.

$x = \dfrac{-(-12) \pm \sqrt{(-12)^2 - 4(6)(7)}}{2(6)}$

$= \dfrac{12 \pm \sqrt{144 - 168}}{12}$

$= \dfrac{12 \pm \sqrt{-24}}{12}$

$= \dfrac{12 \pm 2i\sqrt{6}}{12}$

$= \dfrac{6 \pm i\sqrt{6}}{6}$

or $1 \pm \dfrac{\sqrt{6}}{6}i$

Solution set: $\left\{ 1 \pm \dfrac{\sqrt{6}}{6}i \right\}$

11. $\dfrac{3x - 2}{3x + 2} = \dfrac{2x + 3}{4x - 1}$

$(3x - 2)(4x - 1) = (3x + 2)(2x + 3)$
        Cross products
        are equal

$12x^2 - 11x + 2 = 6x^2 + 13x + 6$

$6x^2 - 24x - 4 = 0$

$3x^2 - 12x - 2 = 0$ *Divide by 2*

Solve this equation by the quadratic formula with $a = 3$, $b = -12$, and $c = -2$.

$x = \dfrac{-(-12) \pm \sqrt{(-12)^2 - 4(3)(-2)}}{2(3)}$

$= \dfrac{12 \pm \sqrt{144 + 24}}{6}$

$= \dfrac{12 \pm \sqrt{168}}{6}$

$= \dfrac{12 \pm 2\sqrt{42}}{6}$

$= \dfrac{6 \pm \sqrt{42}}{3}$

Solution set: $\left\{ \dfrac{6 \pm \sqrt{42}}{3} \right\}$

12. $3x^2 = 5x - 2$

$3x^2 - 5x + 2 = 0$

$a = 3$, $b = -5$, $c = 2$

The discriminant is

$b^2 - 4ac = (-5)^2 - 4(3)(2)$

$= 25 - 24 = 1.$

Because the discriminant is positive and a perfect square, there are two different rational solutions.

13. Let x = width of border.

Then $4 + 2x$ = width of larger
             flower box;

     $6 + 2x$ = length of larger
             flower box.

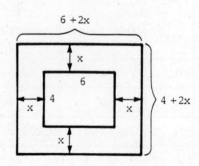

$$\begin{array}{ccc} \text{Area of} & = & \text{Twice area of} \\ \text{larger box} & & \text{smaller box} \end{array}$$

$$(4 + 2x)(6 + 2x) = 2(4 \cdot 6)$$

$$24 + 20x + 4x^2 = 48$$

$$4x^2 + 20x - 24 = 0$$

$$x^2 + 5x - 6 = 0$$

$$(x + 6)(x - 1) = 0$$

$$x = -6 \quad \text{or} \quad x = 1$$

Reject $x = -6$ because width cannot be negative.

$$4 + 2x = 4 + 2(1) = 6$$

$$6 + 2x = 6 + 2(1) = 8$$

The new dimensions should be

6 ft by 8 ft.

14. $\sqrt{5 + 2x} = x + 1$

Square both sides.

$$(\sqrt{5 + 2x})^2 = (x + 1)^2$$

$$5 + 2x = x^2 + 2x + 1$$

$$0 = x^2 - 4$$

$$0 = (x + 2)(x - 2)$$

$$x = -2 \quad \text{or} \quad x = 2$$

Check both proposed solutions in the original equation.

Let $x = -2$.

$$\sqrt{5 + 2x} = x + 1$$

$$\sqrt{5 + 2(-2)} = -2 + 1 \quad ?$$

$$1 = -1 \qquad \textit{False}$$

Let $x = 2$.

$$\sqrt{5 + 2x} = x + 1$$

$$\sqrt{5 + 2(2)} = 2 + 1 \quad ?$$

$$3 = 3 \qquad \textit{True}$$

Solution set: $\{2\}$

15. $\sqrt{2x + 1} - \sqrt{x} = 1$

$$\sqrt{2x + 1} = 1 + \sqrt{x}$$

$$(\sqrt{2x + 1})^2 = (1 + \sqrt{x})^2$$
$$\textit{Square both sides}$$

$$2x + 1 = 1 + 2\sqrt{x} + x$$
$$\textit{Square of a binomial}$$

$$x = 2\sqrt{x}$$

$$x^2 = (2\sqrt{x})^2$$
$$\textit{Square both sides}$$

$$x^2 = 4x$$

$$x^2 - 4x = 0$$

$$x(x - 4) = 0$$

$$x = 0 \quad \text{or} \quad x = 4$$

Checking will show that both proposed solutions satisfy the original equation.

Solution set: $\{0, 4\}$

16. $x^4 - 3x^2 - 10 = 0$

Let $u = x^2$. With this substitution, the equation becomes

$$u^2 - 3u - 10 = 0.$$

Solve this equation by factoring.

$$(u - 5)(u + 2) = 0$$

$$u = 5 \quad \text{or} \quad u = -2$$

To find x, replace u with x^2.

$$x^2 = 5 \quad \text{or} \quad x^2 = -2$$

$$x = \pm\sqrt{5} \quad \text{or} \quad x = \pm i\sqrt{2}$$

Solution set: $\{\pm\sqrt{5}, \pm i\sqrt{2}\}$

17. $2 - \sqrt[3]{2x + x^2} = 0$

$$2 = \sqrt[3]{2x + x^2}$$

$$2^3 = (\sqrt[3]{2x + x^2})^3$$
$$\textit{Cube both sides}$$

$$8 = 2x + x^2$$

$$0 = x^2 + 2x - 8$$

$$0 = (x + 4)(x - 2)$$

x = -4 or x = 2

Checking will show that both proposed solutions satisfy the original equation.

Solution set: $\{-4, 2\}$

18. $-2(x - 1) - 10 \leq 2(2 + x)$

$-2x + 2 - 10 \leq 4 + 2x$

$-2x - 8 \leq 4 + 2x$

$-4x \leq 12$

$\left(-\frac{1}{4}\right)(-4x) \geq \left(-\frac{1}{4}\right)(12)$

$\qquad$ *Multiply by -1/4; reverse inequality symbol*

$x \geq -3$

Solution set: $[-3, \infty)$

19. $4 \geq 3 + \frac{x}{2} \geq -2$

$1 \geq \frac{x}{2} \geq -5$

$2(1) \geq 2\left(\frac{x}{2}\right) \geq 2(-5)$

$2 \geq x \geq -10$

or $-10 \leq x \leq 2$

Solution set: $[-10, 2]$

20. $2x^2 - x - 3 \geq 0$

Solve the corresponding quadratic equation by factoring.

$2x^2 - x - 3 = 0$

$(x + 1)(2x - 3) = 0$

$x = -1$ or $x = \frac{3}{2}$

These two points divide the number line into three regions.

$$\begin{array}{c} x+1 \;\; -\!\!\circ\!+ + + + + + + \;|\; + + + + \\ 2x-3 \;\; -\;|\; \text{-} \text{-} \text{-} \text{-} \text{-} \text{-} \text{-}\!\circ\!+ + + + \\ \hline \hspace{1cm} -1 \hspace{1.3cm} 0 \hspace{1cm} \frac{3}{2} \\ + \hspace{1.5cm} - \hspace{1.5cm} + \\ \text{Sign of} \\ (x+1)(2x-3) \end{array}$$

The sign graphs shows that the product is positive in the intervals $(-\infty, -1)$ and $\left(\frac{3}{2}, \infty\right)$. The endpoints satisfy the inequality since the inequality symbol is $\geq$.

Solution set: $(-\infty, -1] \cup \left[\frac{3}{2}, \infty\right)$

21. $\frac{6}{2x - 5} \leq 2$

$\frac{6}{2x - 5} - 2 \leq 0$

$\frac{6 - 2(2x - 5)}{2x - 5} \leq 0$

$\frac{6 - 4x + 10}{2x - 5} \leq 0$

$\frac{16 - 4x}{2x - 5} \leq 0$

The quotient can change sign when

$16 - 4x = 0$ or $2x - 5 = 0$

$x = 4$ or $\qquad x = \frac{5}{2}$.

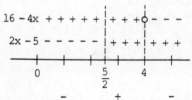

$$\begin{array}{c} 16-4x \;\; + + + + + \;|\; + + +\!\circ\!- - - \\ 2x-5 \;\; \text{-} \text{-} \text{-} \text{-} \text{-} \;|\; + + +\;|\; + + + \\ \hline \hspace{0.6cm} 0 \hspace{1.2cm} \frac{5}{2} \hspace{0.8cm} 4 \\ - \hspace{1.3cm} + \hspace{1.3cm} - \\ \text{Sign of} \\ \frac{16 - 4x}{2x - 5} \end{array}$$

The sign graph shows that $\dfrac{16 - 4x}{2x - 5}$ is negative in the interval $\left(-\infty, \dfrac{5}{2}\right)$ and also in the interval $(4, \infty)$.   The endpoint $x = \dfrac{5}{2}$ cannot be included in the solution because it will make the denominator of $\dfrac{6}{2x - 5}$ equal to zero, but the endpoint $x = 4$ is included.

Solution set: $\left(-\infty, \dfrac{5}{2}\right) \cup [4, \infty)$

23. $\left|-5 - 3x\right| = 4$

$-5 - 3x = 4$ or $-5 - 3x = -4$

$\qquad -3x = 9$ $\qquad\qquad\qquad -3x = 1$

$\qquad\qquad x = -3$ or $\qquad\qquad x = -\dfrac{1}{3}$

Solution set: $\left\{-3, -\dfrac{1}{3}\right\}$

24. $\left|2x - 5\right| < 9$

$\qquad\qquad\qquad -9 < 2x - 5 < 9$

$\qquad\qquad\qquad -4 < 2x < 14$

$\qquad\qquad\qquad -2 < x < 7$

Solution set: $(-2, 7)$

25. $\left|2x + 1\right| - 11 \geq 0$

$\qquad\qquad \left|2x + 1\right| \geq 11$

$2x + 1 \leq -11$ or $2x + 1 \geq 11$

$\qquad 2x \leq -12 \qquad\qquad 2x \geq 10$

$\qquad\quad x \leq -6$ or $\qquad\quad x \geq 5$

Solution set: $(-\infty, -6] \cup [5, \infty)$

CHAPTER 3 RELATIONS AND THEIR GRAPHS

Section 3.1

1. $\{(-3, 5), (-2, 4), (-1, 6), (0, -8), (1, 2)\}$

   $(-3, 5)$, $(-2, 4)$, and $(-1, 6)$ are three of the ordered pairs that belong to this relation.
   The domain is the set of first elements, $\{-3, -2, -1, 0, 1\}$.
   The range is the set of second elements, $\{5, 4, 6, -8, 2\}$.

3. $y = 9x + 2$

   Let $x = 0$.
   Then $y = 9 \cdot 0 + 2 = 2$, giving the ordered pair $(0, 2)$.
   When $x = 1$, $y = 9 \cdot 1 + 2 = 11$.
   When $x = 2$, $y = 9 \cdot 2 + 2 = 20$.

   $(0, 2)$, $(1, 11)$, and $(2, 20)$ are three ordered pairs that belong to this relation.
   x and y can each take any real-number values, so both the domain and range are $(-\infty, \infty)$.

5. $x = y$

   When $y = -1$, $x = -1$.
   When $y = 0$, $x = 0$.
   When $y = 1$, $x = 1$.

   $(-1, -1)$, $(0, 0)$, and $(1, 1)$ are three ordered pairs that belong to this relation.
   x and y can each take any real-number values, so both the domain and range are $(-\infty, \infty)$.

7. $y = \sqrt{x}$

   When $x = 0$, $y = \sqrt{0} = 0$.
   When $x = 1$, $y = \sqrt{1} = 1$.
   When $x = 4$, $y = \sqrt{4} = 2$.

   $(0, 0)$, $(1, 1)$, and $(4, 2)$ are three ordered pairs that belong to this relation.
   Only nonnegative numbers have a square root, so $x \geq 0$ and the domain is $[0, \infty)$.
   Since y equals the principal square root of x, the range is also $[0, \infty)$.

9. $y = x^2$

   When $x = 0$, $y = 0^2 = 0$.
   When $x = 1$, $y = 1^2 = 1$.
   When $x = 2$, $y = 2^2 = 4$.

   $(0, 0)$, $(1, 1)$, and $(2, 4)$ are three ordered pairs that belong to this relation.
   Every real number can be squared, so x can be any real number. The domain is $(-\infty, \infty)$. Squaring a real number always results in a non-negative number, so $y \geq 0$. The range is $[0, \infty)$.

11. $y = \sqrt{x + 3}$

    When $x = -3$, $y = \sqrt{-3 + 3} = \sqrt{0} = 0$.
    When $x = -2$, $y = \sqrt{-2 + 3} = \sqrt{1} = 1$.
    When $x = 1$, $y = \sqrt{1 + 3} = \sqrt{4} = 2$.

    $(-3, 0)$, $(-2, 1)$, $(1, 2)$ are three ordered pairs that belong to this relation.

x + 3 ≥ 0 leads to x ≥ -3, so the domain is [-3, ∞). Since y equals a principal square root, the range is [0, ∞).

15. P(5, 7), Q(13, -1)

To find the distance between P and Q, use the distance formula with $x_1 = 5$, $y_1 = 7$, $x_2 = 13$, and $y_2 = -1$.

$$d(P, Q) = \sqrt{(x_2 - x_1)^2 + (y_2 - y_1)^2}$$
$$= \sqrt{(13 - 5)^2 + (-1 - 7)^2}$$
$$= \sqrt{8^2 + (-8)^2}$$
$$= \sqrt{64 + 64}$$
$$= \sqrt{128} = 8\sqrt{2}$$

To find the midpoint of segment PQ, use the midpoint formula. The midpoint is

$$\left(\frac{x_1 + x_2}{2}, \frac{y_1 + y_2}{2}\right)$$
$$= \left(\frac{5 + 13}{2}, \frac{7 + (-1)}{2}\right)$$
$$= (9, 3).$$

17. P(-8, -2), Q(-3, -5)

$$d(P, Q)$$
$$= \sqrt{[-3 - (-8)]^2 + [-5 - (-2)]^2}$$
$$= \sqrt{5^2 + (-3)^2}$$
$$= \sqrt{25 + 9}$$
$$= \sqrt{34}$$

The midpoint is

$$\left(\frac{-8 + (-3)}{2}, \frac{-2 + (-5)}{2}\right) = \left(-\frac{11}{2}, -\frac{7}{2}\right).$$

19. P(3, -7), Q(-5, 19)

$$d(P, Q) = \sqrt{(-5 - 3)^2 + [19 - (-7)]^2}$$
$$= \sqrt{(-8)^2 + 26^2}$$
$$= \sqrt{64 + 676}$$
$$= \sqrt{740} = 2\sqrt{185}$$

The midpoint is

$$\left(\frac{3 + (-5)}{2}, \frac{-7 + 19}{2}\right) = (-1, 6).$$

21. P(a, b), Q(3a - 4b)

$$d(P, Q)$$
$$= \sqrt{(3a - a)^2 + (-4b - b)^2}$$
$$= \sqrt{(2a)^2 + (-5b)^2}$$
$$= \sqrt{4a^2 + 25b^2}$$

The midpoint is

$$\left(\frac{a + 3a}{2}, \frac{b + (-4b)}{2}\right) = \left(2a, -\frac{3b}{2}\right).$$

23. P($\sqrt{2}$, -$\sqrt{5}$), Q(3$\sqrt{2}$, 4$\sqrt{5}$)

$$d(P, Q)$$
$$= \sqrt{(3\sqrt{2} - \sqrt{2})^2 + [4\sqrt{5} - (-\sqrt{5})]^2}$$
$$= \sqrt{(2\sqrt{2})^2 + (5\sqrt{5})^2}$$
$$= \sqrt{8 + 125}$$
$$= \sqrt{133}$$

The midpoint is

$$\left(\frac{\sqrt{2} + 3\sqrt{2}}{2}, \frac{-\sqrt{5} + 4\sqrt{5}}{2}\right) = \left(\frac{4\sqrt{2}}{2}, \frac{3\sqrt{5}}{2}\right)$$
$$= \left(2\sqrt{2}, \frac{3\sqrt{5}}{2}\right).$$

25. P(2, 8), Q(0, 4), R(4, 7)

$$d(P, Q) = \sqrt{(0 - 2)^2 + (4 - 8)^2} = \sqrt{20}$$
$$d(Q, R) = \sqrt{(4 - 0)^2 + (7 - 4)^2} = \sqrt{25}$$
$$d(P, R) = \sqrt{(4 - 2)^2 + (7 - 8)^2} = \sqrt{5}$$

By these results,

$$[d(Q, R)]^2 = [d(P, Q)]^2 + [d(P, R)]^2,$$

since

$$(\sqrt{25})^2 = (\sqrt{20})^2 + (\sqrt{5})^2,$$

or $25 = 20 + 5$

is a true statement.

This proves that the three points are the vertices of a right triangle.

27. A(-4, 0), B(1, 3), C(-6, -2)

$$d(A, B) = \sqrt{(-4 - 1)^2 + (0 - 3)^2}$$
$$= \sqrt{25 + 9}$$
$$= \sqrt{34}$$

$d(A, C)$
$$= \sqrt{[-4 - (-6)]^2 + [0 - (-2)]^2}$$
$$= \sqrt{4 + 4}$$
$$= \sqrt{8}$$

$$d(B, C) = \sqrt{[1 - (-6)]^2 + [3 - (-2)]^2}$$
$$= \sqrt{49 + 25}$$
$$= \sqrt{74}$$

$[d(A, B)]^2 = 34$

$[d(A, C)]^2 = 8$

$[d(B, C)]^2 = 74$

Since the three sides do not satisfy $a^2 + b^2 = c^2$, the given points are not vertices of a right triangle.

29. A($\sqrt{3}$, $2\sqrt{3} + 3$), B($\sqrt{3} + 4$, $-\sqrt{3} + 3$)
C($2\sqrt{3}$, $2\sqrt{3} + 4$)

$d(A, B)$
$$= \sqrt{(\sqrt{3} + 4) - \sqrt{3}]^2 + [(-\sqrt{3} + 3) - (2\sqrt{3} + 3)]^2}$$
$$= \sqrt{16 + (-3\sqrt{3})^2} = \sqrt{16 + 27} = \sqrt{43}$$

$d(A, C)$
$$= \sqrt{(2\sqrt{3} - \sqrt{3})^2 + [(2\sqrt{3} + 4) - 2\sqrt{3} + 3)]^2}$$
$$= \sqrt{(-\sqrt{3})^2 + 1^2}$$
$$= \sqrt{4} = 2$$

$d(B, C)$
$$= \sqrt{[\sqrt{3} + 4) - 2\sqrt{3}]^2 + [(-\sqrt{3} + 3) - (2\sqrt{3} + 4]^2}$$
$$= \sqrt{(4 - \sqrt{3})^2 + (-3\sqrt{3} - 1)^2}$$
$$= \sqrt{(16 - 8\sqrt{3} + 3) + (27 + 6\sqrt{3} + 1)}$$
$$= \sqrt{47 - 2\sqrt{3}}$$

$[d(A, B)]^2 = 43$

$[d(A, C)]^2 = 4$

$[d(B, C)]^2 = 47 - 2\sqrt{3}$

Since the three sides do not satisfy $a^2 + b^2 = c^2$, the given points are not the vertices of a right triangle.

31. A(0, 7), B(3, -5), C(-2, 15)

$$d(A, B) = \sqrt{3^2 + (-12)^2}$$
$$= \sqrt{9 + 144}$$
$$= \sqrt{153}$$
$$\approx 12.3693$$

$$d(A, C) = \sqrt{4 + 64} = \sqrt{68} \approx 8.2462$$

$$d(B, C) = \sqrt{5^2 + (-20)^2} = \sqrt{425}$$
$$\approx 20.6155$$

$$d(A, B) + d(A, C) = 20.6155$$

Therefore, $d(A, B) + d(A, C) = d(B, C)$, so A, B, and C are collinear.

33. A(0, -9), B(3, 7), C(-2, -19)

$$d(A, B) = \sqrt{3^2 + (-16)^2}$$
$$= \sqrt{265} \approx 16.2788$$

$d(A, C) = \sqrt{(-2)^2 + (-10)^2}$

$= \sqrt{104} \approx 10.1980$

$d(B, C) = \sqrt{(-5)^2 + (-26)^2}$

$= \sqrt{701} \approx 26.4764$

$d(A, B) + d(A, C) = \sqrt{265} + \sqrt{104}$

≈ 26.4768

Therefore, $d(A, B) + d(A, C) \neq$ $d(B, C)$, so A, B, and C are not collinear.

35. $M(2, 7)$, $N(-4, -2)$, $Q(10, 19)$

$d(M, N) = \sqrt{(-4 - 2)^2 + (-2 - 7)^2}$

$= \sqrt{117} = 3\sqrt{13}$

$d(M, Q) = \sqrt{(10 - 2)^2 + (19 - 7)^2}$

$= \sqrt{208} = 4\sqrt{13}$

$d(N, Q) = \sqrt{(-4 - 10)^2 + (-2 - 19)^2}$

$= \sqrt{637} = 7\sqrt{13}$

The points are collinear since $d(M, N) + d(M, Q) = d(N, Q)$, or $3\sqrt{13} + 4\sqrt{13} = 7\sqrt{13}$.

37. Endpoint $(-3, 6)$, midpoint $(5, 8)$ Find the other endpoint, (x_2, y_2).

Let $x = 5$ and $x_1 = -3$.

$5 = \dfrac{-3 + x_2}{2}$

$10 = -3 + x_2$

$13 = x_2$

Let $y = 8$ and $y_1 = 6$.

$8 = \dfrac{6 + y_2}{2}$

$16 = 6 + y_2$

$10 = y_2$

The other endpoint is $(13, 10)$.

39. Endpoint $(5, -4)$, midpoint $(12, 6)$

Let $x = 12$ and $x_1 = 5$.

$12 = \dfrac{5 + x_2}{2}$

$24 = 5 + x_2$

$19 = x_2$

Let $y = 6$ and $y_1 = -4$.

$6 = \dfrac{-4 + y_2}{2}$

$12 = -4 + y_2$

$16 = y_2$

The other endpoint is $(19, 16)$.

41. Endpoint (a, b), midpoint (c, d)

Let $x = c$ and $x_1 = a$.

$c = \dfrac{a + x_2}{2}$

$2c = a + x_2$

$2c - a = x_2$

Let $y = d$ and $y_1 = b$.

$d = \dfrac{b + y_2}{2}$

$2d = b + y_2$

$2d - b = y_2$

The other endpoint is $(2c - a, 2d - b)$.

43. The midpoint of the segment joining $P(x_1, y_1)$ and $Q(x_2, y_2)$ is

$$M\left(\frac{x_1 + x_2}{2}, \frac{y_1 + y_2}{2}\right).$$

Use the distance formula.

$d(P, Q) = \sqrt{(x_2 - x_1)^2 + (y_2 - y_1)^2}$

d(P, M)

$$= \sqrt{\left(\frac{x_1 + x_2}{2} - x_1\right)^2 + \left(\frac{y_1 + y_2}{2} - y_1\right)^2}$$

$$= \sqrt{\left(\frac{x_1 + x_2}{2} - \frac{2x_1}{2}\right)^2 + \left(\frac{y_1 + y_2}{2} - \frac{2y_1}{2}\right)^2}$$

$$= \sqrt{\left(\frac{x_2 - x_1}{2}\right)^2 + \left(\frac{y_2 - y_1}{2}\right)^2}$$

$$= \sqrt{\frac{(x_2 - x_1)^2}{4} + \frac{(y_2 - y_1)^2}{4}}$$

$$= \sqrt{\frac{(x_2 - x_1)^2 + (y_2 - y_1)^2}{4}}$$

$$= \frac{1}{2}\sqrt{(x_2 - x_1)^2 + (y_2 - y_1)^2}$$

d(M, Q)

$$= \sqrt{\left(\frac{x_1 + x_2}{2} - x_2\right)^2 + \left(\frac{y_1 + y_2}{2} - y_2\right)^2}$$

$$= \sqrt{\left(\frac{x_1 - x_2}{2}\right)^2 + \left(\frac{y_1 - y_2}{2}\right)^2}$$

$$= \sqrt{\frac{(x_1 - x_2)^2}{4} + \frac{(y_1 - y_2)^2}{4}}$$

$$= \frac{1}{2}\sqrt{(x_1 - x_2)^2 + (y_2 - y_2)^2}$$

$$= \frac{1}{2}\sqrt{(x_2 - x_1)^2 + (y_2 - y_1)^2}$$

Observe that d(P, M) = d(M, Q) and

d(P, M) + d(M, Q)

$$= \frac{1}{2}\sqrt{(x_2 - x_1)^2 + (y_2 - y_1)^2}$$

$$\quad + \frac{1}{2}\sqrt{(x_2 - x_1)^2 + (y_2 - y_1)^2}$$

$$= \sqrt{(x_2 - x_1)^2 + (y_2 - y_1)^2}$$

$$= d(P, Q).$$

45. The points (x, y) for which x > 0
are located in quadrants I and IV.

47. The points (x, y) for which xy < 0
are located in quadrant II (where
x < 0, y > 0 and quadrant IV (where
x > 0, y < 0).

49. The points (x, y) for which $|x| < 3$
and y < -2 are located in quadrants
III and IV. $|x| < 3$ means that
-3 < x < 3, so x can be positive or
negative, but y must be negative.

51. A(-2, 2), B(13, 10), C(21, -5),
D(6, -13)

$$d(A, B) = \sqrt{(-2 - 13)^2 + (2 - 10)^2}$$
$$= \sqrt{(-15)^2 + (-8)^2}$$
$$= \sqrt{225 + 64}$$
$$= \sqrt{289} = 17$$

$$d(C, D) = \sqrt{(21 - 6)^2 + [-5 - (-13)]^2}$$
$$= \sqrt{15^2 + 8^2}$$
$$= \sqrt{289} = 17$$

$$d(A, D) = \sqrt{(-2 - 6)^2 + [2 - (-13)]^2}$$
$$= \sqrt{(-8)^2 + 15^2}$$
$$= \sqrt{289} = 17$$

$$d(B, C) = \sqrt{(13 - 21)^2 + [10 - (-5)]^2}$$
$$= \sqrt{(-8)^2 + 15^2}$$
$$= \sqrt{289} = 17$$

d(A, B) = d(C, D) = d(A, D) =
d(B, C) = 17

All sides are equal in length, so
the points are the vertices of a
rhombus.

Section 3.2

For Exercises 1–9, see the answer graphs in the back of the textbook.

1. $x - y = 4$

Use the intercepts.

$0 - y = 4$

$\quad y = -4 \quad$ *y-intercept*

$x - 0 = 4$

$\quad x = 4 \quad$ *x-intercept*

Graph the line through $(0, -4)$ and $(4, 0)$. The domain and range are both $(-\infty, \infty)$.

3. $3x - y = 6$

$3 \cdot 0 - y = 6$

$\quad\quad y = -6 \quad$ *y-intercept*

$3x - 0 = 6$

$\quad\quad x = 2 \quad$ *x-intercept*

Graph the line through $(0, -6)$ and $(2, 0)$. The domain and range are both $(-\infty, \infty)$.

5. $2x + 5y = 10$

The y-intercept is 2, and the x-intercept is 5. Graph the line through $(0, 2)$ and $(5, 0)$. The domain and range are both $(-\infty, \infty)$.

7. $x = 2$

The graph of this equation is a vertical line.
The x-intercept is 2.
The line has no y-intercept.

One other point on the line is $(2, 3)$.
The domain is $\{2\}$ and the range is $(-\infty, \infty)$.

9. $y = 3x$

The x- and y-intercepts are both 0. $(1, 3)$ is one other point on the line. Graph the line through $(0, 0)$ and $(1, 3)$. The domain and range are both $(-\infty, \infty)$.

11. Through $(-2, 1)$ and $(3, 2)$

$$m = \frac{y_2 - y_1}{x_2 - x_1}$$

$$= \frac{2 - 1}{3 - (-2)} = \frac{1}{5}$$

13. Through $(8, 4)$ and $(-1, -3)$

$$m = \frac{y_2 - y_1}{x_2 - x_1}$$

$$= \frac{-3 - 4}{-1 - 8} = \frac{-7}{-9} = \frac{7}{9}$$

15. $3x + 4y = 6$

$\left(0, \dfrac{3}{2}\right)$ and $(2, 0)$ are two points on the line.

$$m = \frac{0 - \dfrac{3}{2}}{2 - 0} = \frac{-\dfrac{3}{2}}{2} = -\frac{3}{4}$$

17. $y = 4$

Two points on the line are $(0, 4)$ and $(3, 4)$.

$$m = \frac{4 - 4}{3 - 0} = \frac{0}{3} = 0$$

19. (a) y = 3x + 2

has a positive slope and a
positive y-intercept.
Choice D resembles this.

(b) y = -3x + 2

has a negative slope and a
positive y-intercaept.
Choice B resembles this.

(c) y = 3x - 2

has a positive slope and a
negative y-intercept.
Choice A resembles this.

(d) y = -3x - 2

has a negative slope and a
negative y-intercept.
Choice C resembles this.

For Exercises 23–27, see the answer
graphs in the back of the textbook.

23. Through (-1, 3), m = $\frac{3}{2}$

First locate the point (-1, 3).
Since the slope is $\frac{3}{2}$, a change of 2
units horizontally (2 units to the
right) produces a change of 3 units
vertically (3 units up). This gives
a second point, (1, 6), which can be
used to complete the graph.

25. Through (3, -4), m = $-\frac{1}{3}$

First locate the point (3, -4).
Since the slope is $-\frac{1}{3}$, a change of 3
units horizontally (3 units to the
right) produces a change of -2 units

vertically (2 units down). This
gives a second point, (6, -5), which
can be used to complete the graph.

27. Through (-1, 4), m = 0

The graph is the horizontal line
through (-1, 4).

29. Through (1, 3), m = -2

Write the equation in point-slope
form.

$$y - y_1 = m(x - x_1)$$
$$y - 3 = -2(x - 1)$$

Then, change to standard form.

$$y - 3 = -2x + 2$$
$$2x + y = 5$$

31. Through (-5, 4), m = $-\frac{3}{2}$

Write the equation in point-slope
form.

$$y - 4 = -\frac{3}{2}(x + 5)$$

Change to standard form.

$$2(y - 4) = -3(x + 5)$$
$$2y - 8 = -3x - 15$$
$$3x + 2y = -7$$

33. Through (-8, 1), undefined slope

Since undefined slope indicates a
vertical line, the equation will
have the form x = k. The equation
of the line is x = -8.

35. Through $(-1, 3)$ and $(3, 4)$

First find m.

$$m = \frac{4 - 3}{3 - (-1)} = \frac{1}{4}$$

Use either point and the point-slope form.

$$y - 4 = \frac{1}{4}(x - 3)$$

Change to standard form.

$$4(y - 4) = x - 3$$
$$4y - 16 = x - 3$$
$$-x + 4y = 13$$
$$\text{or } x - 4y = -13$$

37. x-intercept 3, y-intercept -2

The line passes through $(3, 0)$ and $(0, -2)$. Use these points to find m.

$$m = \frac{-2 - 0}{0 - 3} = \frac{2}{3}$$

Then use point-slope form.

$$y - 0 = \frac{2}{3}(x - 3)$$

Change to standard form.

$$3y = 2(x - 3)$$
$$3y = 2x - 6$$
$$6 = 2x - 3y$$
$$\text{or } 2x - 3y = 6$$

39. Vertical, through $(-6, 5)$

The equation of a vertical line has an equation of the form x = k. Since the line passes through $(-6, 5)$, the equation is x = -6.

41. The line x + 2 = 0 has x-intercept -2. It does not have a y-intercept. The slope of this line is undefined. The line 4y = 2 has y-intercept 1/2. It does not have an x-intercept. The slope of this line is zero.

43. The y-axis is a vertical line through $(0, 0)$, so its equation is x = 0.

45. $y = 3x - 1$

This equation is in the slope-intercept form, y = mx + b. The slope is m = 3 and the y-intercept is b = -1.

47. $4x - y = 7$

Solve for y to write the equation in slope-intercept form.

$$-y = -4x + 7$$
$$y = 4x - 7$$

The slope is 4 and the y-intercept is -7.

49. $4y = -3x$

$$y = -\frac{3}{4}x$$

The slope is $-\frac{3}{4}$ and the y-intercept is 0.

51. Through $(-1, 4)$, parallel to x + 3y = 5

First, find the slope of the line x + 3y = 5 by writing this equation in slope-intercept form.

$$x + 3y = 5$$
$$3y = -x + 5$$
$$y = -\frac{1}{3}x + \frac{5}{3}$$

The slope is -1/3. Since the lines are parallel, -1/3 is also the slope of the line whose equation is to be found. Substitute $m = -1/3$, $x_1 = -1$, and $y_1 = 4$ into the point-slope form.

$$y - y_1 = m(x - x_1)$$
$$y - 4 = -\frac{1}{3}[x - (-1)]$$
$$y - 4 = -\frac{1}{3}(x + 1)$$
$$3(y - 4) = -1(x + 1)$$
$$3y - 12 = -x - 1$$
$$x + 3y = 11$$

53. Through (1, 6), perpendicular to $3x + 5y = 1$

First, find the slope of the line $3x + 5y = 1$ by writing this equation in slope-intercept form.

$$3x + 5y = 1$$
$$5y = -3x + 1$$
$$y = -\frac{3}{5}x + \frac{1}{5}$$

This line has a slope of -3/5. Call the line whose equation is to be found L. Since line L is perpendicular to the line $3x + 5y = 1$, the product of their slopes is -1. If line L has slope m, then

$$-\frac{3}{5}m = -1$$
$$m = \frac{5}{3}.$$

To find the equation of the line L, substitute $m = 5/3$, $x_1 = 1$, and $y_1 = 6$ into the point-slope form.

$$y - 6 = \frac{5}{3}(x - 1)$$
$$3(y - 6) = 5(x - 1)$$
$$3y - 18 = 5x - 5$$
$$-13 = 5x - 3y$$
$$\text{or} \quad 5x - 3y = -13$$

55. Through (-5, 7), perpendicular to $y = -2$.

Since $y = -2$ is a horizontal line, any line perpendicular to this line will be vertical and have an equation of the form $x = k$. Since the line passes through (-5, 7), the equation is $x = -5$.

57. M(1, -2), N(3, -18), P(-2, 22)

The slope of line MN is

$$\frac{-18 - (-2)}{3 - 1} = \frac{-16}{2} = -8.$$

The slope of line NP is

$$\frac{22 - (-18)}{-2 - 3} = \frac{40}{-5} = -8.$$

The two slopes are the same, so the points lie on a straight line.

59. X(1, 3), Y(-4, 73), Z(5, -50)

The slope of line XY is

$$\frac{73 - 3}{-4 - 1} = \frac{70}{-5} = -14.$$

The slope of line YZ is

$$\frac{-50 - 73}{5 - (-4)} = \frac{-123}{9} = -\frac{41}{3}.$$

The two slopes are not the same, so the points do not lie on a straight line.

61. (a) Find the slope of the line $3y + 2x = 6$.

$$3y + 2x = 6$$
$$3y = -2x + 6$$
$$y = -\frac{2}{3}x + 2$$
$$m = -\frac{2}{3}$$

A line parallel to $3y + 2x = 6$ also has slope $-\frac{2}{3}$.

$$-\frac{2}{3} = \frac{-1 - 2}{4 - k}$$

Solve for k using the slope formula.

$$-\frac{2}{3} = \frac{-3}{4 - k}$$
$$-9 = -8 + 2k$$
$$-\frac{1}{2} = k$$

(b) Find the slope of the line $2x - 5y = 1$.

$$2y - 5x = 1$$
$$2y = 5x + 1$$
$$y = \frac{5}{2}x + \frac{1}{2}$$
$$m = \frac{5}{2}$$

A line perpendicular to $2y - 5x = 1$ has slope $-\frac{2}{5}$, since $\frac{5}{2}\left(-\frac{2}{5}\right) = -1$. Solve for k using the slope formula.

$$-\frac{2}{5} = \frac{-1 - 2}{4 - k}$$
$$-\frac{2}{5} = \frac{-3}{4 - k}$$
$$-15 = -8 + 2k$$
$$k = -\frac{7}{2}.$$

63. Show that $\overline{MQ}$ has length m_1.

The slope of L_1 is m_1, and points P and M are on L_1. To go from P to M on L_1, $\Delta x = 1$ and $\Delta y = MQ$, so

$$m = \frac{\Delta y}{\Delta x} \text{ becomes}$$

$$m_1 = \frac{MQ}{1} \text{ or}$$

$$MQ = m_1.$$

65. Show that triangles MPQ and PNQ are similar.

Since line PQ is perpendicular to line MN, triangles PQM and PQN are right triangles. Because triangle MPN is a right triangle, the sum of angles N and M is 90°. Also because triangle QPN is a right triangle, the sum of angles N and QPN is 90°. This means that angles QPN and M are equal, since they are complements of the same angle. Therefore, triangles MPQ and PQN are similar, because they have two pairs (including the right angles) of corresponding equal angles.

67. (20, 13,900) and (10, 7500) are two points on the line. Find m.

$$m = \frac{13,900 - 7500}{20 - 10} = \frac{6400}{10} = 640$$

Use $(x_1, y_1) = (10, 7500)$ and
$m = 640$ in the point–slope form.

$$y - y_1 = m(x - x_1)$$
$$y - 7500 = 640(x - 10)$$
$$y - 7500 = 640x - 6400$$
$$y = 640x + 1100$$

69. $(3, 37{,}000)$ and $(12, 28{,}000)$ are two points on the line.   Find m.

$$m = \frac{37{,}000 - 28{,}000}{3 - 12} = \frac{9000}{-9} = -1000$$

Use $(x_1, y_1) = (12, 28{,}000)$ and
$m = -1000$ in the point–slope form.

$$y - 28{,}000 = -1000(x - 12)$$
$$y - 28{,}000 = -1000x + 12{,}000$$
$$1000x + y = 40{,}000$$
$$y = -1000x + 40{,}000$$

71. $(45, 42.5)$ and $(55, 67.5)$ are two points on the line.   Find m.

$$m = \frac{67.5 - 42.5}{55 - 45} = \frac{25}{10} = 2.5$$

Use $(x_1, y_1) = (45, 42.5)$ and
$m = 2.5$ in the point–slope form.

$$y - 42.5 = 2.5(x - 45)$$
$$y - 42.5 = 2.5x - 112.5$$
$$y = 2.5x - 70$$

73. The ordered pairs are $(0, 32)$ and $(100, 212)$.

The slope is $\dfrac{212 - 32}{100 - 0}$

$$m = \frac{180}{100} = \frac{9}{5}.$$

Use $(x_1, y_1) = (0, 32)$ and $m = \dfrac{9}{5}$ in the point–slope form.

$$y - y_1 = m(x - x_1)$$
$$y - 32 = \frac{9}{5}(x - 0)$$
$$y - 32 = \frac{9}{5}x$$
$$y = \frac{9}{5}x + 32$$
$$\text{or}\quad F = \frac{9}{5}C + 32$$

75. In $F = \dfrac{9}{5}C + 32$, let $F = C$.

$$C = \frac{9}{5}C + 32$$
$$C - \frac{9}{5}C = 32$$
$$-\frac{4}{5}C = 32$$
$$C = -\frac{5}{4} \cdot 32 = -40$$

We conclude that

$$-40°C = -40°F.$$

Section 3.3

1. For parts (a), (b), (c), and (d), see the answer graphs in the back of the textbook.

(e) As the coefficient decreases in absolute value, the parabola becomes broader.

3. For parts (a), (b), (c), and (d), see the answer graphs in the back of the textbook.

(e) Each of these graphs is a horizontal translation of the graph of $y = x^2$, so they differ only in the position of their vertices on the x–axis.

The graph of $y = (x - 2)^2$ is the same as that of $y = x^2$, but translated 2 units to the right. The graph of $y = (x + 1)^2$ is the same as that of $y = x^2$, but translated 1 unit to the left. The graph of $y = (x + 3)^2$ is the same as that of $y = x^2$, but translated 3 units to the left. The graph of $y = (x - 4)^2$ is the same as that of $y = x^2$, but translated 4 units to the right.

5. $y = a(x - h)^2 + k$

    (a) If $h < 0$ and $k < 0$, the vertex is in quadrant III.

    (b) If $h < 0$ and $k > 0$, the vertex is in quadrant II.

    (c) If $h > 0$ and $k < 0$, the vertex is in quadrant IV.

    (d) If $h > 0$ and $k > 0$, the vertex is in quadrant I.

For Exercises 7–17, see the answer graphs in the back of the textbook.

7. $y = (x - 2)^2$

This equation is of the form $y = (x - h)^2$, with $h = 2$. The graph opens upward and has the same shape as that of $y = x^2$. It is a horizontal translation of the graph of $y = x^2$ 2 units to the right. The vertex is $(2, 0)$ and the axis is the vertical line $x = 2$. The domain and range can be seen on the graph. The domain is $(-\infty, \infty)$. Since the smallest value of y is 0 and the graph opens upward, the range is $[0, \infty)$.

9. $y = (x + 3)^2 - 4$
$y = [x - (-3)]^2 + (-4)$

This equation is of the form $y = (x - h)^2 + k$, with $h = -3$ and $k = -4$. The vertex is $(-3, -4)$. The graph opens upward and has the same shape as $y = x^2$. It is a translation of $y = x^2$ 3 units to the left and 4 units down. The axis is the vertical line $x = -3$. The domain is $(-\infty, \infty)$. Since the smallest value of y is -4 and the graph opens upward, the range is $[-4, \infty)$.

11. $y = -2(x + 3)^2 + 2$

The vertex is $(-3, 2)$. The graph opens downward and is narrower than $y = x^2$. It is a translation of the graph of $y = -2x^2$ 3 units to the left and 2 units up. The axis is the vertical line $x = -3$. The domain is $(-\infty, \infty)$. Since the largest value of y is 2, the range is $(-\infty, 2]$.

13. $y = -\frac{1}{2}(x + 1)^2 - 3$

The vertex is $(-1, -3)$. The graph opens downward and is wider than $y = x^2$. It is a translation of the graph of $y = -\frac{1}{2}x^2$ 1 unit to the left and 3 units down.

The axis is the vertical line
$x = -1$. The domain is $(-\infty, \infty)$.
Since the largest value of y is -3,
the range is $(-\infty, -3]$.

15. $y = x^2 - 2x + 3$
$\quad = (x^2 - 2x + 1) - 1 + 3$
$\quad = (x - 1)^2 + 2$

The vertex is $(1, 2)$.
The graph opens upward and has the
same shape as $y = x^2$. It is a
translation of the graph of $y = x^2$
1 unit to the right and 2 units up.
The axis is the vertical line $x = 1$.
The domain is $(-\infty, \infty)$. Since the
smallest value of y is 2, the range
is $[2, \infty)$.

17. $y = 2x^2 - 4x + 5$
$\quad = 2(x^2 - 2x \quad) + 5$
$\quad = 2(x^2 - 2x + 1 - 1) + 5$
$\quad = 2(x^2 - 2x + 1) - 2 + 5$
$\quad = 2(x - 1)^2 + 3$

The vertex is $(1, 3)$. The graph
opens upward and has the same shape
as $y = 2x^2$. It is a translation of
the graph of $y = 2x^2$ 1 unit to the
right and 3 units up. The axis is
the vertical line $x = 1$. The domain
is $(-\infty, \infty)$. Since the smallest
value of y is 3, the range is
$[3, \infty)$.

For Exercises 21–29, see the answer
graphs in the back of the textbook.

21. $x = y^2 + 2$
$\quad = (y - 0)^2 + 2$

The vertex is $(2, 0)$.

The graph opens to the right and has
the same shape as $x = y^2$. It is a
translation of the graph of $x = y^2$
2 units to the right. The axis is
the horizontal line $y = 0$ (the
x–axis). Since the smallest value
of x is 2 and the graph opens to the
right, the domain is $[2, \infty)$. The
range is $(-\infty, \infty)$.

x	6	3	2	3	6
y	-2	-1	0	1	2

23. $x = (y + 1)^2$

The vertex is $(0, -1)$.
The graph opens to the right and has
the same shape as $x = y^2$. It is a
translation of the graph of $x = y^2$
1 unit down. The axis is the horizontal line $y = -1$. Since the
smallest value of x is 0, the domain
is $[0, \infty)$.
The range is $(-\infty, \infty)$.

x	4	1	0	1	4
y	-3	-2	-1	0	1

25. $x = (y + 2)^2 - 1$

The vertex is $(-1, -2)$.
The graph opens to the right and has
the same shape as $x = y^2$. It is a
translation of the graph of $x = y^2$
1 unit to the left and 2 units down.
The axis is the horizontal line
$y = -2$. Since the smallest value of
x is -1, the domain is $[-1, \infty)$.

The range is $(-\infty, \infty)$.

x	0	-1	0
y	-3	-2	-1

27. $x = -2(y + 3)^2$

The vertex is $(0, -3)$.
The graph opens to the left and has
the same shape as $x = -2y^2$. It is a
translation of the graph of $x = -2y^2$
3 units down. The axis is the horizontal line $y = -3$. Since the
largest value of x is 0, the domain
is $(-\infty, 0]$.
The range is $(-\infty, \infty)$.

x	-2	0	-2
y	-4	-3	-2

29. $x = y^2 + 2y - 8$

Complete the square on y to find the
vertex and axis.

$x = y^2 + 2y - 8$
$\quad = (y^2 + 2y + 1 - 1) - 8$
$\quad = (y^2 + 2y + 1) - 1 - 8$
$x = (y + 1)^2 - 9$

The vertex is $(-9, -1)$.
The graph opens to the right and has
the same shape as $x = y^2$. It is a
translation of $x = y^2$ 9 units to the
left and 1 unit down. The axis is
the horizontal line $y = -1$. Since
the smallest value of x is -9, the
domain is $[-9, \infty)$.

The range is $(-\infty, \infty)$.

x	-8	-9	-8
y	-2	-1	0

31. $A = x(320 - x)$

Find the vertex.

$A = 320x - 2x^2$
$\quad = -2x^2 + 320x$
$\quad = -2(x^2 - 160x + 6400 - 6400)$
$\quad = -2(x^2 - 160x + 6400) + 12,800$
$\quad = -2(x - 80)^2 + 12,800$

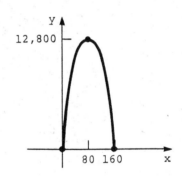

The vertex is $(80, 12,800)$. The
width, x, should be 80 ft in order
to have maximum area of 12,800
sq ft. The length would be 320 -
2(80) = 160 ft.

33. $C = 2x^2 - 1200x + 180,100$

Find the vertex.

$C = 2(x^2 - 600x + 90,000 - 90,000)$
$\qquad + 180,100$
$\quad = 2(x^2 - 600x + 90,000) + 180,100$
$\qquad - 180,000$
$\quad = 2(x - 300)^2 + 100$

The vertex is $(300, 100)$.
In order to minimize the cost, Ed
must sell 300 sandwiches for a cost
of $100.

35. $M = 10x - x^2$

Find the vertex.

$M = -x^2 + 10x$
$= -1(x^2 - 10x + 25 - 25)$
$= -(x^2 - 10x + 25) + 25$
$= -(x - 5)^2 + 25$

The vertex is (5, 25).
A rainfall of 5 inches will produce the maximum number of mosquitoes, 25 million.

37. $h = 32t - 16t^2$
$h = -16t^2 + 32t$
$= -16(t^2 - 2t + 1 - 1)$
$= -16(t^2 - 2t + 1) + 16$
$= -16(t - 1)^2 + 16$

The vertex of the parabola is (1, 16). Since the parabola opens downward, 16 is the maximum value for h. The object reaches its maximum height of 16 ft after 1 sec.

$0 = 32t - 16t^2$
$0 = 16t(2 - t)$

The time it would take to hit the ground is 2 sec. (The height is also 0 at 0 sec, when the object is thrown upward.)

39. $p = 500 - x$

(a) R = demand × price
$= (500 - x) \cdot x$
$= 500x - x^2$

Find the vertex by completing the square.

$R = -x^2 + 500x$
$= -(x^2 - 500x + 62,500)$
$ + 62,500$
$= -(x - 250)^2 + 62,500$

The vertex is (250, 62,500).

(b) Graph $h = -x^2 + 500x$.

x	250	500	0
R	62,500	0	0

See the answer graph in the textbook.

(c) The price that will produce the maximum revenue is $250.

(d) The maximum revenue is $62,500.

41. First sketch the straight line C = 10x + 50 for x = 1 to x = 5. Connect this to the parabola $C = -20(x - 5)^2 + 100$ for x = 5 to x = 7. See the graph in the back of the textbook. The graph shows maximum production in October.

43. Focus (0, 3), directrix y = -3

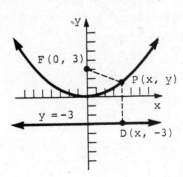

Let F = (0, 3), the focus of the
parabola.

Let P = (x, y) be a point on the
parabola.

Let D = (x, −3), the point on the
directrix with the same x−coordi-
nate as P.

Then

d(P, F) = distance from P to focus,

d(P, D) = distance from P to the
          directrix,

y = −3, since this is the perpendicular distance from P to the line
y = −3.

By the geometric definition of a
parabola,

$$d(P, F) = d(P, D)$$
$$\sqrt{(x - 0)^2 + (y - 3)^2} = |y - (-3)|$$
$$x^2 + (y - 3)^2 = (y + 3)^2$$
$$x^2 + y^2 - 6y + 9 = y^2 + 6y + 9$$
$$x^2 - 12y = 0$$
$$x^2 = 12y.$$

45. Focus (−2, 0), directrix x = 2

Let F = (−2, 0), P = (x, y), and a
point on the parabola, D = (2, y).

$$d(P, F) = d(P, D)$$
$$\sqrt{(x + 2)^2 + (y - 0)^2} = |x - 2|$$
$$(x + 2)^2 + y^2 = (x - 2)^2$$
$$x^2 + 4x + 4 + y^2 = x^2 - 4x + 4$$
$$y^2 = -8x$$

47. Focus (3, 6), vertex (3, 4)

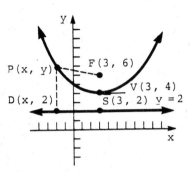

Let F = (3, 6), the focus of the
parabola.

Let V = (3, 4), the vertex of the
parabola.

Since V is a point on the parabola,
the distance from V to F must be
equal to the distance from V to the
directrix. Since d(V, F) = 2,
S = (3, 2), and the equation of the
directrix is y = 2.

Let P = (x, y) be a point on the
parabola.

Let D = (x, 2), the point on the
directrix with the same x−coordi-
nate as P.

By the geometric definition of a
parabola,

d(P, F) = d(P, D)
$$\sqrt{(x - 3)^2 + (y - 6)^2} = |y - 2|$$
$$(x - 3)^2 + (y - 6)^2 = (y - 2)^2$$
$$x^2 - 6x + 9 + y^2 - 12y + 36$$
$$= y^2 - 4y + 4$$
$$x^2 - 6x + 9 = 8y - 32$$
$$(x - 3)^2 = 8(y - 4).$$

49. $y = x^2 - 10x + c$

An x–intercept occurs where $y = 0$, or

$0 = x^2 - 10x + c$.

There will be exactly one x–intercept if this equation has exactly one solution, or if $x^2 - 10x + c$ is a perfect square trinomial.

$0 = x^2 - 10x + 25$

$0 = (x - 5)^2$

$x = 5$

If $c = 25$, the only x–intercept is 5.

51. $a < 0$, $b^2 - 4ac = 0$

The correct choice is E. $a < 0$ indicates that the parabola opens downward, while $b^2 - 4ac = 0$ indicates that the graph has exactly one x–intercept.

53. $a > 0$, $b^2 - 4ac > 0$

The correct choice is C. $a > 0$ indicates that the parabola opens upward, while $b^2 - 4ac > 0$ indicates that the graph has two x–intercepts.

55. $y = 2x^2 + 5x - 3$

$= 2\left(x^2 + \dfrac{5}{2}x + \dfrac{25}{16} - \dfrac{25}{16}\right) - 3$

$= 2\left(x^2 + \dfrac{5}{2}x + \dfrac{25}{16}\right) - \dfrac{25}{8} - 3$

$= 2\left(x + \dfrac{5}{4}\right)^2 - \dfrac{49}{8}$

Vertex: $\left(-\dfrac{5}{4}, -\dfrac{49}{8}\right)$

The graph is narrower than $y = x^2$. See the answer graph in the back of the textbook.

(a) The x–intercepts are -3 and $\dfrac{1}{2}$.

The solution of $2x^2 + 5x - 3 < 0$ is the interval $\left(-3, \dfrac{1}{2}\right)$, since the graph of the parabola is below the x–axis for all x–values between -3 and $\dfrac{1}{2}$. ($y < 0$ for these values.)

(b) The solution of $2x^2 + 5x - 3 > 0$ is

$$\left(-\infty, -3\right) \cup \left(\dfrac{1}{2}, \infty\right),$$

since the graph of the parabola is above the x–axis for all values of x that are less than -3 or greater than $\dfrac{1}{2}$. ($y > 0$ for these values.)

(c) The solutions of $2x^2 + 5x - 3 = 0$ are -3 and $\dfrac{1}{2}$, since the parabola crosses the x–axis at those values ($y = 0$ for these two x–values).

Section 3.4

1. Center (1, 4), radius 3

The circle with center (h, k) and radius r has equation

$$(x - h)^2 + (y - k)^2 = r^2.$$

When $h = 1$, $k = 4$, and $r = 3$,

$$(x - 1)^2 + (y - 4)^2 = 3^2$$

or $(x - 1)^2 + (y - 4)^2 = 9$.

3. Center $(0, 0)$, radius 1

$$(x - 0)^2 + (y - 0)^2 = 1^2$$
$$x^2 + y^2 = 1$$

5. Center $\left(\frac{2}{3}, -\frac{4}{5}\right)$, radius $\frac{3}{7}$

$$\left(x - \frac{2}{3}\right)^2 + \left[y - \left(-\frac{4}{5}\right)\right]^2 = \left(\frac{3}{7}\right)^2$$
$$\left(x - \frac{2}{3}\right)^2 + \left(y + \frac{4}{5}\right)^2 = \frac{9}{49}$$

7. Center $P(-1, 2)$, passing through $Q(2, 6)$

The distance between P and Q is

$$\sqrt{(-1 - 2)^2 + (2 - 6)^2} = \sqrt{(-3)^2 + (-4)^2}$$
$$= \sqrt{9 + 16}$$
$$= \sqrt{25} = 5,$$

so the radius of this circle is 5. Therefore, the equation is

$$[x - (-1)]^2 + (y - 2)^2 = 5^2$$
$$(x + 1)^2 + (y - 2)^2 = 25.$$

9. Center $P(-3, -2)$, tangent to the x-axis

The above two conditions imply that the circle passes through $Q(-3, 0)$. The distance between P and Q is

$$\sqrt{[-3 - (-3)]^2 + (-2 - 0)^2} = \sqrt{0 + 4} = 2,$$

so the radius of this circle is 2. Therefore, the equation is

$$[x - (-3)]^2 + [y - (-2)]^2 = 2^2$$
$$(x + 3)^2 + (y + 2)^2 = 4.$$

For Exercises 11–17, see the answer graphs in the back of the textbook.

11. $x^2 + y^2 = 36$

The graph is a circle with center $(0, 0)$ and radius 6.
From the graph we can see that the domain of the relation is $[-6, 6]$ and the range is also $[-6, 6]$.

13. $(x - 2)^2 + y^2 = 36$

The graph is a circle with center $(2, 0)$ and radius 6.

$$(x - 2)^2 + (y - 0)^2 = 6^2$$

From the graph we can see that the domain of the relation is $[-4, 8]$ and the range is $[-6, 6]$.

15. $(x + 2)^2 + (y - 5)^2 = 16$

The graph is a circle with center $(-2, 5)$ and radius 4. From the graph we can see that the domain of the relation is $[-6, 2]$ and the range is $[1, 9]$.

17. $(x + 3)^2 + (y + 2)^2 = 36$

The graph is a circle with center $(-3, -2)$ and radius 6. From the graph, we can see that the domain of the relation is $[-9, 3]$ and the range is $[-8, 4]$.

19. The graph of the equation

$$(x - 3)^2 + (y - 3)^2 = 0$$

is the single point $(3, 3)$.

21. The graphs of

$$x^2 + y^2 = 4$$

and

$$x^2 + y^2 = 25$$

will not intersect since the smaller circle will lie entirely inside the larger one.

23. $$x^2 + 6x + y^2 + 8y = -9$$

To determine whether this equation represents a circle, complete the square on x and y separately.

$$(x^2 + 6x \quad) + (y^2 + 8y \quad) = -9$$
$$(x^2 + 6x + 9) + (y^2 + 8y + 16) = -9 + 9 + 16$$
$$(x + 3)^2 + (y + 4)^2 = 16$$
$$[x - (-3)]^2 + [y - (-4)]^2 = 4^2$$

The equation represents a circle with center $(-3, -4)$ and radius 4.

25. $$x^2 - 12x + y^2 + 10y + 25 = 0$$
$$(x^2 - 12x \quad) + (y^2 + 10y \quad) = -25$$
$$(x^2 - 12x + 36) + (y^2 + 10y + 25) = -25 + 36 + 25$$
$$(x - 6)^2 + (y + 5)^2 = 36$$
$$(x - 6)^2 + [y - (-5)]^2 = 6^2$$

The equation represents a circle with center $(6, -5)$ and radius 6.

27. $$x^2 + 8x + y^2 - 14y + 64 = 0$$
$$(x^2 + 8x \quad) + (y^2 - 14y \quad) = -64$$
$$(x^2 + 8x + 16) + (y^2 - 14y + 49) = -64 + 16 + 49$$
$$(x + 4)^2 + (x - 7)^2 = 1$$

The equation represents a circle with center $(-4, 7)$ and radius 1.

29. $$x^2 + y^2 = 2y + 48$$
$$x^2 + (y^2 - 2y \quad) = 48$$
$$x^2 + (y^2 - 2y + 1) = 48 + 1$$
$$x^2 + (y - 1)^2 = 49$$
$$(x - 0)^2 + (y - 1)^2 = 7^2$$

The equation represents a circle with center $(0, 1)$ and radius 7.

31. The graph is symmetric with respect to the y-axis, because if the graph were folded along the y-axis, the two halves would coincide.

33. The graph is symmetric with respect to the x-axis, because if the graph were folded along the x-axis, the two halves would coincide.

35. The graph is symmetric with respect to the origin, because a 180° rotation of the graph about the origin would coincide with the original graph.

For Exercises 37 and 39, see the graphs in the answer section of the textbook.

41. If R is symmetric with respect to the x-axis, the point $(s, -t)$ must lie on the graph of R.

43. If R is symmetric with respect to the origin, the point $(-s, -t)$ must lie on the graph of R.

45. $x^2 + y^2 = 5$

(1) To test for symmetry with respect to the x—axis, replace y with —y.

$$x^2 + (-y)^2 = 5$$
$$x^2 + y^2 = 5$$

(2) To test for symmetry with respect to the y—axis, replace x with —x.

$$(-x)^2 + y^2 = 5$$
$$x^2 + y^2 = 5$$

(3) To test for symmetry with respect to the origin, replace x with —x and y with —y.

$$(-x)^2 + (-y)^2 = 5$$
$$x^2 + y^2 = 5$$

Since each time an equation is obtained which is equivalent to the original one, the graph of the relation $x^2 + y^2 = 5$ is symmetric with respect to the x—axis, the y—axis, and the origin.

47. $y = x^2 - 8x$

(1) To test for symmetry with respect to the x—axis, replace y with —y.

$$-y = x^2 - 8x$$
$$y = -x^2 + 8x$$

(2) To test for symmetry with respect to the y—axis, replace x with —x.

$$y = (-x)^2 - 8(-x)$$
$$y = x^2 + 8x$$

(3) To test for symmetry with respect to the origin, replace x with —x and y with —y.

$$-y = (-x)^2 - 8(-x)$$
$$-y = x^2 + 8x$$
$$y = -x^2 - 8x$$

None of the equations obtained by these replacements is equivalent to the original equation. Therefore the graph of $y = x^2 - 8x$ is not symmetric with respect to the x—axis, the y—axis, or the origin.

49. $y = x^3$

(1) Test for symmetry with respect to the x—axis.

$$-y = x^3$$
$$y = -x^3$$

(2) Test for symmetry with respect to the y—axis.

$$y = (-x)^3$$
$$y = -x^3$$

(3) Test for symmetry with respect to the origin.

$$-y = (-x)^3$$
$$-y = -x^3$$
$$y = x^3$$

These test show that the graph of $y = x^3$ is symmetric with respect to the origin but not to either the x—axis or the y—axis.

51. $y = \dfrac{1}{1 + x^2}$

(1) Test for symmetry with respect to the x-axis.

$$-y = \frac{1}{1 + x^2}$$

$$y = -\frac{1}{1 + x^2}$$

(2) Test for symmetry with respect to the y-axis.

$$y = \frac{1}{1 + (-x)^2}$$

$$y = \frac{1}{1 + x^2}$$

(3) Test for symmetry with respect to the origin.

$$-y = \frac{1}{1 + (-x)^2}$$

$$-y = \frac{1}{1 + x^2}$$

$$y = -\frac{1}{1 + x^2}$$

The graph of $y = \dfrac{1}{1 + x^2}$ is symmetric with respect to the y-axis but not with respect to the x-axis or the origin.

For Exercises 53 and 55, see the answer graphs in the back of the textbook.

57. P(-1, 3) and Q(5, -9) are endpoints of a diameter of the circle. The center of the circle is the midpoint of the segment joining P and Q, which is

$$\left(\frac{-1 + 5}{2}, \frac{3 + (-9)}{2}\right) \quad \text{or} \quad (2, -3).$$

The radius of the circle is the distance between the center and one of the original points. Thus,

$$r = \sqrt{[2 - (-1)]^2 + (-3 - 3)^2}$$
$$= \sqrt{3^2 + (-6)^2}$$
$$= \sqrt{45}.$$

The equation of the circle may now be written

$$(x - 2)^2 + [y - (-3)]^2 = (\sqrt{45})^2$$
$$(x - 2)^2 + (y + 3)^2 = 45.$$

59. Find all points (x, y) with x = y that are 4 units from (1, 3). Let R = (1, 3) and let P be a point satisfying the given conditions.

$$d(P, R) = 4$$

Since P is on the line y = x, call P the point (x, x).

$$d(P, R) = \sqrt{(x - 1)^2 + (y - 3)^2}$$

Therefore,

$$\sqrt{(x - 1)^2 + (x - 3)^2} = 4$$
$$x^2 - 2x + 1 + x^2 - 6x + 9 = 16$$
$$2x^2 - 8x + 10 = 16$$
$$2x^2 - 8x - 6 = 0$$
$$x^2 - 4x - 3 = 0.$$

Solve this equation by the quadratic formula.

$$x = \frac{4 \pm \sqrt{16 - (-12)}}{2}$$

$$x = \frac{4 \pm \sqrt{28}}{2}$$

$$x = 2 \pm \sqrt{7}.$$

Two points satisfy the given conditions: $(2 + \sqrt{7}, 2 + \sqrt{7})$ and $(2 - \sqrt{7}, 2 - \sqrt{7})$.

61. If the equation y = F is changed to y = −F, the graph of y = F is reflected about the x-axis.

63. The first circle has center (3, 4) and radius 5. The second circle has center (−1, −3) and radius 4. Using the distance formula, the distance between the centers is

$$\sqrt{[3 - (-1)]^2 + [4 - (-3)]^2}$$
$$= \sqrt{16 + 49}$$
$$= \sqrt{65}.$$

If the 2 circles were tangent to each other, the centers would be 9 units apart since 9 is the sum of the radii. Since the distance between the centers is $\sqrt{65}$, which is less than 9, the circles must intersect.

Section 3.5

For Exercises 1–23, see the answer graphs in the back of the textbook.

1. $\dfrac{x^2}{9} + \dfrac{y^2}{4} = 1$

The graph is an ellipse centered at the origin with x-intercepts 3 and −3 and y-intercepts 2 and −2. From the graph, we can see that the domain of the relation is [−3, 3] and the range is [−2, 2].

3. $\dfrac{x^2}{6} + \dfrac{y^2}{9} = 1$

The graph is an ellipse centered at the origin with x-intercepts $\sqrt{6}$ and −$\sqrt{6}$ and y-intercepts 3 and −3. Use $\sqrt{6} \approx 2.4$ to draw the graph. From the graph, we can see that the domain is [−$\sqrt{6}$, $\sqrt{6}$] and the range is [−3, 3].

5. $\dfrac{x^2}{1/9} + \dfrac{y^2}{1/16} = 1$

The graph is an ellipse centered at the origin with x-intercepts 1/3 and −1/3, and y-intercepts 1/4 and −1/4. The graph shows that the domain is $\left[-\frac{1}{3}, \frac{1}{3}\right]$ and the range is $\left[-\frac{1}{4}, \frac{1}{4}\right]$.

7. $\dfrac{64x^2}{9} + \dfrac{25y^2}{36} = 1$

$\dfrac{x^2}{9/64} + \dfrac{y^2}{36/25} = 1$

The graph is an ellipse centered at the origin with x-intercepts 3/8 and −3/8, and y-intercepts 6/5 and −6/5. The graph shows that the domain is $\left[-\frac{3}{8}, \frac{3}{8}\right]$ and the range is $\left[-\frac{6}{5}, \frac{6}{5}\right]$.

9. $\dfrac{(x - 1)^2}{9} + \dfrac{(y + 3)^2}{25} = 1$

The graph is an ellipse centered at (1, −3). Graph the ellipse using the fact that a = 3 and b = 5. Start at (1, −3) and locate two points 3 units away from (1, −3) on a horizontal line, one to the right

and one to the left. These points
are (-2, -3) and (4, -3). Locate
two other points on a vertical line
through (1, -3), one 5 units up and
one 5 units down. These points are
(1, 2) and (1, -8). The graph shows
that the domain is [-2, 4] and the
range is [-8, 2].

11. $\dfrac{(x - 2)^2}{16} + \dfrac{(y - 1)^2}{9} = 1$

The graph is an ellipse centered at
(2, 1). Graph the ellipse using the
fact that a = 4 and b = 3. Start at
(2, 1) and locate two points 4 units
away from (2, 1) on a horizontal
line, one to the right and one to
the left. These points are (-2, 1),
and (6, 1). Locate two other points
on a vertical line through (2, 1),
one 3 units up and one 3 units down.
These points are (2, 4) and (2, -2).
The graph shows that the domain is
[-2, 6] and the range is [-2, 4].

13. $\dfrac{x^2}{16} - \dfrac{y^2}{9} = 1$

The graph is a hyperbola centered at
(0, 0) with branches opening to the
left and right. The graph has x-
intercepts 4 and -4, but no y-intercepts. Since a = 4 and b = 3, draw
one asymptote through (4, 3) and
(-4, -3), and draw the other through
(-4, 3) and (4, -3). Sketch in the
hyperbola through x-intercepts 4 and

-4 and approaching the two asymptotes. The graph shows that the
domain is (-∞, -4] ∪ [4, ∞) and the
range is (-∞, ∞).

15. $\dfrac{y^2}{36} - \dfrac{x^2}{49} = 1$

The graph is a hyperbola centered at
(0, 0) with branches opening up and
down. The graph has y-intercepts 6
and -6, but no x-intercepts. Since
a = 7 and b = 6, draw one asymptote
through (7, 6) and (-7, -6), and
draw the other through (-7, 6) and
(7, -6). Sketch the hyperbola.
The graph shows that the domain
is (-∞, ∞) and the range is
(-∞, -6] ∪ [6, ∞).

17. $\dfrac{4x^2}{9} - \dfrac{25y^2}{16} = 1$

$\dfrac{x^2}{9/4} - \dfrac{y^2}{16/25} = 1$

The graph is a hyperbola centered at
(0, 0) with branches opening left
and right. The graph has x-intercepts 3/2 and -3/2, but no y-intercepts. Since a = 3/2 and b = 4/5,
draw one asymptote through
(3/2, 4/5) and (-3/2, -4/5), and
draw the other through (-3/2, 4/5)
and (3/2, -4/5). Sketch the
hyperbola. The domain is
$\left(-\infty, -\dfrac{3}{2}\right] \cup \left[\dfrac{3}{2}, \infty\right)$ and the range
is (-∞, ∞).

19. $\dfrac{(x - 1)^2}{9} - \dfrac{(y + 3)^2}{25} = 1$

The graph is a hyperbola centered at (1, -3). This graph may be obtained by translating the graph of $\dfrac{x^2}{9} - \dfrac{y^2}{25} = 1$ one unit to the right and 3 units down. The domain is $(-\infty, -2] \cup [4, \infty)$ and the range is $(-\infty, \infty)$.

21. $\dfrac{(x - 3)^2}{16} - \dfrac{(y + 2)^2}{49} = 1$

The graph is a hyperbola centered at (3, -2), with branches opening left and right. The vertices are (-1, -2) and (7, -2). The asymptotes are the lines passing through (3, -2) with slopes 7/4 and -7/4. The graph shows that the domain of the relation is $(-\infty, -1] \cup [7, \infty)$ and the range is $(-\infty, \infty)$.

23. $\dfrac{(y + 1)^2}{25} - \dfrac{(x - 3)^2}{36} = 1$

The graph is a hyperbola centered at (3, -1) with branches opening up and down. The vertices are (3, 4) and (3, -6). The asymptotes are the lines passing through (3, -1) with slopes 5/6 and -5/6. The graph shows that the domain of the relation is $(-\infty, \infty)$ and the range is $(-\infty, -6] \cup [4, \infty)$.

25. Give the equation of the ellipse with center at the origin, length of major axis = 12, and endpoint of the minor axis at (0, 4).

The equation will have the form

$$\frac{x^2}{a^2} + \frac{y^2}{b^2} = 1.$$

Since an endpoint of the minor axis is (0, 4), b = 4. The length of the major axis, 12, is 2a, so a = 6. The equation is

$$\frac{x^2}{36} + \frac{y^2}{16} = 1.$$

27. Give the equation of the hyperbola with center at the origin, vertex (0, 2), and a = 2b.

Since the vertices of the parabola are the y-intercepts, the equation will have the form

$$\frac{y^2}{b^2} - \frac{x^2}{a^2} = 1.$$

Since (0, 2) is a vertex, b = 2. Since a = 2b, a = 4. The equation is

$$\frac{y^2}{4} - \frac{x^2}{16} = 1.$$

29. Give the equation of the ellipse with center at (2, -2), a = 4, b = 3, and major axis parallel to the x-axis.

The equation will have the form

$$\frac{(x - h)^2}{a^2} + \frac{(y - k)^2}{b^2} = 1$$

where (h, k) is the center. Since (h, k) = (2, -2), a = 4, and b = 3, the equation is

$$\frac{(x - 2)^2}{16} + \frac{(y + 2)^2}{9} = 1.$$

31. Give the equation of the hyperbola with center at (4, 3), vertex at (1, 3), and b = 2.

Since the vertex is to the left of the center, the hyperbola must have branches opening to the left and right, so the equation has the form

$$\frac{(x - h)^2}{a^2} - \frac{(y - k)^2}{b^2} = 1.$$

The distance form the center to the vertex, which is 3, gives the value of a for this hyperbola.

Since (h, k) = (4, 3), a = 3, and b = 2, the equation is

$$\frac{(x - 4)^2}{9} - \frac{(y - 3)^2}{4} = 1.$$

37. An ellipse with major axis 620 ft and minor axis 513 ft has 2a = 620 and 2b = 513, or a = 310 and b = 256.5.

The distance between the center and a focus is c, where

$$c^2 = a^2 - b^2$$
$$c^2 = 310^2 - (256.5)^2$$
$$c^2 = 96,100 - 65,792.25$$
$$c^2 = 30,307.75$$
$$c \approx 174.1.$$

(The negative value of c is rejected.) The distance between the two foci of the ellipse is

$$2c = 2(174.1) = 348.2.$$

There are 348.2 ft between the foci of the Roman Coliseum.

39. Since the distances are proportional to the time a pulse travels, the difference between the distance from P to M and the distance from P to S is a constant on a curve. The curve is a hyperbola.

41. The widest possible distance is the length of the major axis, or 2a. Since a = $\sqrt{5013}$ million ≈ 70.8 million,

$$2a \approx 141.6 \text{ million.}$$

The widest possible distance across the ellipse is about 141.6 million mi.

43. (a) Let R = (-c, 0), S = (c, 0), P = (x, y).

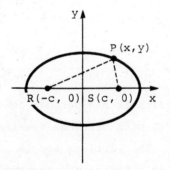

$$\overset{PR}{\sqrt{[x - (-c)]^2 + (y - 0)^2}} + \overset{PS}{\sqrt{(x - c)^2 + (y - 0)^2}}$$
$$= 2a$$
$$\sqrt{(x + c)^2 + y^2} = 2a - \sqrt{(x - c)^2 + y^2}$$

Square both sides.

$$(x + c)^2 + y^2$$
$$= 4a^2 - 2(2a)\sqrt{(x - c)^2 + y^2}$$
$$+ (x - c)^2 + y^2$$
$$x^2 + 2xc + c^2 + y^2$$
$$= 4a^2 - 4a\sqrt{(x - c)^2 + y^2} + x^2$$
$$- 2xc + c^2 + y^2$$
$$4xc = 4a^2 - 4a\sqrt{(x - c)^2 + y^2}$$
$$xc = a^2 - a\sqrt{(x - c)^2 + y^2}$$
$$a\sqrt{(x - c)^2 + y^2} = a^2 - xc$$

Square both sides again.

$$a^2(x^2 - 2xc + c^2 + y^2) = a^4 - 2a^2xc + x^2c^2$$
$$a^2x^2 - 2a^2xc + a^2c^2 + a^2y^2 = a^4 - 2a^2xc + x^2c^2$$
$$a^2x^2 - x^2c^2 + a^2y^2 = a^4 + a^2c^2$$
$$x^2(a^2 - c^2) + a^2y^2 = a^2(a^2 - c^2)$$

Divide both sides of the equation by $a^2(a^2 - c^2)$.

$$\frac{x^2(a^2 - c^2)}{a^2(a^2 - c^2)} + \frac{a^2y^2}{a^2(a^2 - c^2)} = \frac{a^2(a^2 - c^2)}{a^2(a^2 - c^2)}$$

$$\frac{x^2}{a^2} + \frac{y^2}{a^2 - c^2} = 1$$

(b) In part (a), we showed that the equation of the ellipse is

$$\frac{x^2}{a^2} + \frac{y^2}{a^2 - c^2} = 1.$$

Since the x–intercepts are points on the ellipse, they must satisfy this equation, and they are the points for which y = 0.   This gives

$$\frac{x^2}{a^2} + \frac{0^2}{a^2 - c^2} = 1$$

$$\frac{x^2}{a^2} = 1$$

$$x^2 = a^2$$

$$x = \pm a.$$

This shows that a and –a are the x–intercepts.

(c) Since the y–intercepts are points on the ellipse, they must also satisfy the equation of the ellipse.   Since these are the points for which x = 0, this gives

$$\frac{0^2}{a^2} + \frac{y^2}{a^2 - c^2} = 1$$

$$\frac{y^2}{b^2} = 1 \quad Let\ b^2 = a^2 - c^2$$

$$y^2 = b^2$$

$$y = \pm b.$$

This shows that b and –b are the y–intercepts.

45.   A sketch illustrates this exercise.

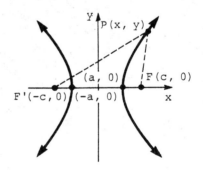

$$d(P,\ F') - d(P,\ F) = 2a$$
$$\sqrt{(x + c)^2 + (y - 0)^2} - \sqrt{(x - c)^2 + (y - 0)^2}$$
$$= 2a$$
$$\sqrt{(x + c)^2 + (y - 0)^2} = 2a + \sqrt{(x + c)^2 + (y - 0)^2}$$
$$(x + c)^2 + y^2 = 4a^2 + 4a\sqrt{(x - c)^2 + y^2}$$
$$+ (x - c)^2 + y^2$$
Square both sides
$$x^2 + 2xc + c^2 + y^2 = 4a^2 + 4a\sqrt{(x - c)^2 + y^2}$$
$$+ x^2 - 2xc + c^2 + y^2$$
Square binomials

$$4xc = 4a^2 + 4a\sqrt{(x - c)^2 + y^2}$$
Simplify

$$xc = a^2 + a\sqrt{(x - c)^2 + y^2}$$
Divide by 4

$$cx - a^2 = a\sqrt{(x - c)^2 + y^2}$$

$$c^2x^2 - 2cxa^2 + a^4 = a^2(x^2 - 2xc + c^2 + y^2)$$
Square both sides again

$$c^2x^2 - 2cxa^2 + a^4 = a^2x^2 - 2a^2xc + a^2c^2 + a^2y^2$$
Distributive property

$$c^2x^2 - a^2x^2 - a^2y^2 = a^2c^2 - a^4 \quad Simplify$$

$$x^2(c^2 - a^2) - a^2y^2 = a^2(c^2 - a^2)$$
Greatest common factors

$$\frac{x^2}{a^2} - \frac{y^2}{c^2 - a^2} = 1 \quad \begin{array}{c}Divide\ by \\ a^2(c^2 - a^2)\end{array}$$

$$\frac{x^2}{a^2} - \frac{y^2}{b^2} = 1 \quad b^2 = c^2 - a^2$$

Section 3.6

1.
$$x^2 + y^2 = 144$$
$$(x - 0)^2 + (y - 0)^2 = 12^2$$

The graph of this equation is a circle.

3. $y = 2x^2 + 3x - 4$

The graph of this equation is a parabola.

5. $x = -3(y - 4)^2 + 1$

The graph of this equation is a parabola.

7.
$$\frac{x^2}{49} + \frac{y^2}{100} = 1$$
$$\frac{x^2}{7^2} + \frac{y^2}{10^2} = 1$$

The graph of this equation is an ellipse.

9.
$$\frac{x^2}{4} - \frac{y^2}{16} = 1$$
$$\frac{x^2}{2^2} - \frac{y^2}{4^2} = 1$$

The graph of this equation is a hyperbola.

11.
$$\frac{x^2}{25} - \frac{y^2}{25} = 1$$
$$\frac{x^2}{5^2} - \frac{y^2}{5^2} = 1$$

The graph of this equation is a hyperbola.

13.
$$\frac{x^2}{4} = 1 - \frac{y^2}{9}$$
$$\frac{x^2}{4} + \frac{y^2}{9} = 1$$

The graph of this equation is an ellipse.

15.
$$\frac{x^2}{4} + \frac{y^2}{4} = 1$$
$$x^2 + y^2 = 4$$

The graph of this equation is a circle.

17. $x^2 + 2x = x^2 + y - 6$

    $2x = y - 6$

    $2x - y = -6$

    This is a linear equation, so its graph is a line.

19. $x^2 = 25 + y^2$

    $x^2 - y^2 = 25$

    $\dfrac{x^2}{25} - \dfrac{y^2}{25} = 1$

    The graph is a hyperbola.

21. $9x^2 + 36y^2 = 36$

    $\dfrac{x^2}{4} + \dfrac{y^2}{1} = 1$

    The graph is an ellipse.

23. $\dfrac{(x + 3)^2}{16} + \dfrac{(y - 2)^2}{16} = 1$

    $(x + 3)^2 + (y - 2)^2 = 16$

    The graph is a circle.

25. $y^2 - 4y = x + 4$

    $y^2 - 4y - 4 = x$

    The graph is a parabola.

27. $(x + 7)^2 + (y - 5)^2 + 4 = 0$

    $(x + y)^2 + (y - 5)^2 = -4$

    This is the form of the equation of a circle, but r^2 cannot be negative, so there is no graph.

29. $3x^2 + 6x + 3y^2 - 12y = 12$

    $x^2 + 2x + y^2 - 4y = 4$

    $(x^2 + 2x + 1) + (y^2 - 4y + 4) = 4 + 1 + 4$

    $(x + 1)^2 + (y - 2)^2 = 9$

    The graph is a circle.

31. $x^2 - 6x + y = 0$

    $y = -x^2 + 6x$

    The graph is a parabola.

33. $4x^2 - 8x - y^2 - 6y = 6$

    $4(x^2 - 2x + 1) - 1(y^2 + 6y + 9) = 6 + 4 - 9$

    $4(x - 1)^2 - 1(y + 3)^2 = 1$

    $\dfrac{(x - 1)^2}{1/4} - \dfrac{(y + 3)^2}{1} = 1$

    The graph is a hyperbola.

35. $4x^2 - 8x + 9y^2 + 54y = -84$

    $4(x^2 - 2x + 1) + 9(y^2 + 6y + 9) = -84 + 4 + 81$

    $4(x - 1)^2 + 9(y + 3)^2 = 1$

    $\dfrac{(x - 1)^2}{1/4} + \dfrac{(y + 3)^2}{1/9} = 1$

    The graph is an ellipse.

37. $6x^2 - 12x + 6y^2 - 18y + 25 = 0$

    $6\left(x^2 - 2x + 1\right) + 6\left(y^2 - 3y + \dfrac{9}{4}\right) = -25 + 6 + \dfrac{27}{2}$

    $6(x - 1)^2 + 6\left(y - \dfrac{3}{2}\right)^2 = -\dfrac{50}{2} + \dfrac{12}{2} + \dfrac{27}{2}$

    $6(x - 1)^2 + 6\left(y - \dfrac{3}{2}\right)^2 = -\dfrac{11}{2}$

    $(x - 1)^2 + \left(y - \dfrac{3}{2}\right)^2 = -\dfrac{11}{12}$

    This is the form of the equation of a circle, but r^2 cannot be negative, so there is no graph.

39. The definition of an ellipse states that "an ellipse is the set of all points in a plane the sum of whose distances from two fixed points is constant."

Therefore, the set of all points in a plane for which the sum of the distances from the points (5, 0) and (-5, 0) is 14 is an ellipse with foci (5, 0) and (-5, 0).

41. (a) Let the plane pass through the cone at the single point where the top and bottom halves of the cone meet. See the sketch in the textbook.

    (b) Let the plane pass through the cone at the point where the top and bottom halves of the cone meet, in such a way so that the entire plane is tangent to the cone along an edge. See the sketch in the textbook.

For Exercises 43–49, see the answer graphs in the back of the textbook.

43. $y = \sqrt{4 - x}$

$y^2 = 4 - x, \; y \geq 0$

$x = 4 - y^2, \; y \geq 0$

The graph is half of a horizontal parabola that opens left and has vertex (4, 0). Only the top half of the parabola is acceptable since the original equation indicated that y had to be nonnegative. The graph

shows that the domain of this relation is $(-\infty, 4)$ and the range is $[0, \infty)$.

45. $y = \sqrt{x^2 - 9}$

$y^2 = x^2 - 9, \; y \geq 0$

$y^2 - x^2 = -9, \; y \geq 0$

$x^2 - y^2 = 9, \; y \geq 0$

The graph is the top half of a hyperbola centered at the origin and whose branches open left and right. The domain is $(-\infty, -3] \cup [3, \infty)$ and the range is $[0, \infty)$.

47. $\dfrac{y}{3} = -\sqrt{1 + \dfrac{x^2}{16}}$

$\dfrac{y^2}{9} = 1 + \dfrac{x^2}{16}, \; y \leq 0$

$\dfrac{y^2}{9} - \dfrac{x^2}{16} = 1, \; y \leq 0$

The graph is the bottom half of a hyperbola centered at the origin. The hyperbola has branches that open up and down, and the graph of $\dfrac{y}{3} = -\sqrt{1 + \dfrac{x^2}{16}}$ is the bottom branch of the hyperbola. The domain is $(-\infty, \infty)$ and the range is $(-\infty, -3]$.

49. $y = \sqrt{1 + \dfrac{x^2}{64}}$

$y^2 = 1 - \dfrac{x^2}{64}, \; y \geq 0$

$\dfrac{x^2}{64} + y^2 = 1$

The graph is the top half of an ellipse centered at the origin. The domain is $[-8, 8]$ and the range is $[0, 1]$.

Section 3.7

For Exercises 1–19, see the answer graphs
in the back of the textbook.

1. x ≤ 3

The boundary is the vertical line
x = 3. Because of the = portion of
≤, the boundary is included in the
graph, so draw a solid line. Select
any test point not on the line, such
as (0, 0). Since 0 ≤ 3 is a true
statement, shade the side of the
line containing (0, 0).

3. x + 2y ≤ 6

The boundary is the line x + 2 = 6,
which can be graphed using the
x–intercept 6 and y–intercept 3.
The boundary is included in the
graph, so draw a solid line.
Use (0, 0) as a test point.
Since 0 + 2(0) ≤ 6 is a true statement, shade the line of the graph
containing (0, 0).

5. 2x + 3y ≥ 4

The boundary is the line 2x + 3y = 4.
The boundary is included in the
graph, so draw a solid line.
Use (0, 0) as a test point.
Since 2(0) + 3(0) ≥ 4 is false,
shade the side of the line that does
not contain (0, 0).

7. 3x − 5y > 6

The boundary is the line 3x − 5y = 6.
Since the inequality symbol is >,
not ≥, the boundary is not included
in the graph, so draw a dashed line.
Use (0, 0) as a test point.
Since 3(0) − 5(0) > 6 is false,
shade the side of the line that does
not include (0, 0).

9. 5x ≤ 4y − 2

The boundary is the line 5x = 4y − 2.
Draw a solid line.
Use (0, 0) as a test point.
Since 5(0) ≤ 4(0) − 2 is false,
shade the side of the line that
does not include (0, 0).

11. y < 3x² + 2

The boundary is the parabola
y = 3x² + 2. Since the inequality
symbol is <, draw a dashed curve.
Use (0, 0) as a test point.
Since 0 < 3(0)² + 2 is true, shade
the region that includes (0, 0).

13. y > (x − 1)² + 2

The boundary is the parabola
y = (x − 1)² + 2, with vertex
(1, 2). Since the inequality
symbol is >, draw a dashed
curve.
Use (0, 0) as a test point.
Since 0 > (0 − 1)² + 2 is false,
shade the region that does not
include (0, 0).

15. $x^2 + (y + 3)^2 \leq 16$

The boundary is a circle with center
(0, -3) and radius 4. Draw a solid
circle to show that the boundary is
included in the graph.
Use (0, 0) as a test point.
Since $0^2 + (0 + 3)^2 \leq 16$ is true,
shade the region that includes
(0, 0), that is, the interior of
the circle.

17. $4x^2 \leq 4 - y^2$

The equation of the boundary is

$$4x^2 = 4 - y^2$$
$$4x^2 + y^2 = 4$$
$$x^2 + \frac{y^2}{4} = 1.$$

This is the equation of an ellipse
centered at the origin with x-intercepts 2 and -2. The boundary is
included, so draw a solid curve.
Use (0, 0) as a test point.
Since $4(0)^2 \leq 4 - 0^2$ is true, shade
the region that contains (0, 0),
that is, the interior of the
ellipse.

19. $9x^2 - 16y^2 > 144$

The equation of the boundary is

$$9x^2 - 16y^2 = 144$$
$$\frac{x^2}{16} - \frac{y^2}{9} = 1.$$

This is the equation of a hyperbola
centered at the origin with x-intercepts 4 and -4. The asymptotes are
$y = \pm\frac{3}{4}x.$

The boundary is not included, so
draw a dashed curve.
Use (0, 0) as a test point.
Since $9(0)^2 - 16(0)^2 > 144$ is false,
the region that does not contain
(0, 0) should be shaded. Additional
test points may be helpful to see
that the regions inside both
branches of the hyperbola must be
shaded.

21. The correct choice is (b).

$(x - 5)^2 + (y - 2)^2 = 4$

would have as its graph the circle
with center (5, 2) and radius 2.

$(x - 5)^2 + (y - 2)^2 < 4$

would have as its graph the interior of the above circle, since the
point (5, 2) is in the interior of
the circle and (5, 2) satisfies the
inequality.

23. The correct choice is (d).

$$\frac{x^2}{16} + \frac{y^2}{81} > 1$$

is the region outside an ellipse
centered at the origin, with
x-intercepts 4 and -4, and y-intercepts 9 and -9.

For Exercises 25–37, see the answer graphs in the back of the textbook.

25. $x - 3y < 4$, where $x \leq 0$

The line $x - 3y = 4$ is the boundary of the inequality $x - 3y < 4$, but the line is dashed because it is not included. The inequality $x \leq 0$ includes all points on or to the left of the y-axis.
The final graph consists of the region above the line $x - 3y = 4$ which is to the left of or on the y-axis.

27. $3x + 2y \geq 6$, where $y \leq 2$

The final graph is the region on and above the line $3x + 2y = 6$ which is on or below the horizontal line $y = 2$.

29. $y^2 \leq x + 3$, where $y \geq 0$

The graph is the region inside and on the horizontal parabola $y^2 = x + 3$ (or $x = y^2 - 3$) which is on or above the x-axis.

31. $y^2 \leq 49 - x^2$, where $y \leq 0$

The graph is the region on and inside the circle $y^2 = 49 - x^2$ (or $x^2 + y^2 = 49$) which is below or on the x-axis.

33. $\dfrac{x^2}{36} < 1 - \dfrac{y^2}{121}$, where $x \geq 0$

The boundary for the first inequality is

$$\frac{x^2}{36} = 1 - \frac{y^2}{121}$$

or $\dfrac{x^2}{36} + \dfrac{y^2}{121} = 1$.

This is an ellipse centered at the origin with x-intercepts 6 and -6 and y-intercepts 11 and -11.
The final graph is the region inside the ellipse which is on or to the right of the y-axis.

35. $x + 2y < 4$, where $3x - y > 5$

The graph is the region which is below the line $x + 2y = 4$ and also below the line $3x - y = 5$.

37. $2x + 3y < 6$, where $x - 5y \geq 10$

The graph is the region which is below the line $2x + 3y = 6$ and also below or on the line $x - 5y = 10$.

Chapter 3 Review Exercises

1. $\{(-3, 6), (-1, 4), (8, 5)\}$

The domain is $\{-3, -1, 8\}$, the set of first elements.
The range is $\{6, 4, 5\}$, the set of second elements.

3. P(3, -1), Q(-4, 5)

$$d(P, Q) = \sqrt{(-4 - 3)^2 + [5 - (-1)]^2}$$
$$= \sqrt{(-7)^2 + 6^2}$$
$$= \sqrt{49 + 36}$$
$$= \sqrt{85}$$

$$\text{midpoint} = \left(\frac{3 + (-4)}{2}, \frac{-1 + 5}{2}\right)$$
$$= \left(-\frac{1}{2}, 2\right)$$

5. A(-6, 3), B(-6, 8)

$$d(A, B) = \sqrt{[-6 - (-6)]^2 + (8 - 3)^2}$$
$$= \sqrt{25} = 5$$

$$\text{midpoint} = \left(\frac{-6 + (-6)}{2}, \frac{3 + 8}{2}\right)$$
$$= \left(-6, \frac{11}{2}\right)$$

7. A(-1, 2), B(-10, 5), C(-4, k)

$$d(A, B) = \sqrt{[-1 - (-10)]^2 + (2 - 5)^2}$$
$$= \sqrt{90}$$

$$d(A, C) = \sqrt{[-4 - (-1)]^2 + (k - 2)^2}$$
$$= \sqrt{9 + (k - 2)^2}$$

$$d(B, C) = \sqrt{[-10 - (-4)]^2 + (5 - k)^2}$$
$$= \sqrt{36 + (k - 5)^2}$$

If segment AB is the hypotenuse,

$$(\sqrt{90})^2$$
$$= \left[\sqrt{9 + (k - 2)^2}\right]^2 + \left[\sqrt{36 + (k - 5)^2}\right]^2$$
$$90 = 9 + k^2 - 4k + 4 + 36 + k^2$$
$$- 10k + 25$$
$$0 = 2k^2 - 14k - 16$$
$$0 = k^2 - 7k - 8$$
$$0 = (k - 8)(k + 1)$$
$$k = 8 \quad \text{or} \quad k = -1.$$

If segment AC is the hypotenuse, the product of the slopes of lines AB and BC is -1 since the product of slopes of perpendicular lines is -1.

$$\left(\frac{5 - 2}{-10 + 1}\right) \cdot \left(\frac{k - 5}{-4 + 10}\right) = -1$$
$$\left(\frac{3}{-9}\right) \cdot \left(\frac{k - 5}{6}\right) = -1$$
$$\frac{k - 5}{-18} = -1$$
$$k - 5 = 18$$
$$k = 23$$

If segment BC is the hypotenuse, the product of the slopes of lines AB and AC is -1.

$$\left(\frac{3}{-9}\right) \cdot \left(\frac{k - 2}{-4 + 1}\right) = -1$$
$$\left(\frac{-1}{3}\right) \cdot \left(\frac{k - 2}{-3}\right) = -1$$
$$\frac{k - 2}{9} = -1$$
$$k - 2 = -9$$
$$k = -7$$

The possible values of k are -7, 23, 8, and -1.

9. Find the slope of the line through (8, 7) and $\left(\frac{1}{2}, -2\right)$.

$$m = \frac{y_2 - y_1}{x_2 - x_1}$$
$$= \frac{-2 - 7}{\frac{1}{2} - 8}$$
$$= \frac{-9}{\frac{-15}{2}}$$
$$= -9\left(-\frac{2}{15}\right)$$
$$= \frac{18}{15} = \frac{6}{5}$$

11. Find the slope of the line through (5, 6) and (5, −2).

$$m = \frac{y_2 - y_1}{x_2 - x_1}$$

$$= \frac{-2 - 6}{5 - 5} = \frac{-8}{0}$$

The slope is undefined.

13. Find the slope of the line

$$9x - 4y = 2.$$

Solve for y to put the equation in slope−intercept form.

$$-4y = -9x + 2$$

$$y = \frac{9}{4}x - \frac{1}{2}$$

$$m = \frac{9}{4}$$

(The slope can also be found by choosing two points on the line and using $m = \frac{y_2 - y_1}{x_2 - x_1}$.)

15. Find the slope of the line

$$x - 5y = 0.$$

Solve for y to put the equation in slope−intercept form.

$$-5y = x$$

$$y = \frac{1}{5}x$$

$$m = \frac{1}{5}$$

17. Find the slope of the line $y + 6 = 0$.

$$y = -6$$

This is a horizontal line, so m = 0. Notice that the slope−intercept form can be used.

$$y = 0x - 6$$

$$m = 0$$

For Exercises 19−23, see the answer graphs in the back of the textbook.

19. $3x + 7y = 14$

$$7y = -3x + 14$$

$$y = -\frac{3}{7}x + 2$$

The graph is the line with slope −3/7 and y−intercept 2. It may also be graphed using the x−intercept 14/3 and y−intercept 2. The domain and range are both (−∞, ∞).

21. $3y = x$

$$y = \frac{1}{3}x$$

The graph is the line with slope 1/3 and y−intercept 0, which means that it passes through the origin. Use another point such as (3, 1) to complete the graph. The domain and range are both (−∞, ∞).

23. x = -5

The graph is the vertical line through (-5, 0).
The domain is {-5} and the range is $(-\infty, \infty)$.

25. Line through (-2, 4) and (1, 3)

First find the slope.

$$m = \frac{3 - 4}{1 - (-2)} = -\frac{1}{3}$$

Now use the point-slope form with

$(x_1, y_1) = (1, 3)$ and $m = -\frac{1}{3}$.

$$y - 3 = -\frac{1}{3}(x - 1)$$
$$3(y - 3) = -1(x - 1)$$
$$3y - 9 = -x + 1$$
$$x + 3y = 10 \quad \textit{Standard form}$$

27. Line through (3, -5) with slope -2

Use the point-slope form.
$$y - (-5) = -2(x - 3)$$
$$y + 5 = -2(x - 3)$$
$$y + 5 = -2x + 6$$
$$2x + y = 1 \quad \textit{Standard form}$$

29. Line through $\left(\frac{1}{5}, \frac{1}{3}\right)$ with slope $-\frac{1}{2}$

Use the point-slope form.

$$y - \frac{1}{3} = -\frac{1}{2}\left(x - \frac{1}{5}\right)$$

To clear fractions, multiply both sides by the least common denominator, 30.

$$30\left(y - \frac{1}{3}\right) = 30\left[-\frac{1}{2}\left(x - \frac{1}{5}\right)\right]$$
$$30y - 10 = -15x + 3$$
$$15x + 30y = 13 \quad \textit{Standard form}$$

31. Line with no x-intercept, y-intercept $\frac{3}{4}$

If a line has no x-intercept, it must be parallel to the x-axis, so it is a horizontal line. Since $b = \frac{3}{4}$, the equation is

$$y = \frac{3}{4}.$$

33. Line through (0, 5), perpendicular to 8x + 5y = 3

First, find the slope of 8x + 5y = 3.

$$8x + 5y = 3$$
$$5y = -8x + 3$$
$$y = -\frac{8}{5}x + \frac{3}{5}$$

Since the slope of 8x + 5y = 3 is -8/5, the slope of a line perpendicular to it is 5/8. Since m = 5/8 and b = 5, the equation is

$$y = \frac{5}{8}x + 5$$
$$8y = 5x + 40$$
$$-40 = -5x - 8y$$
or   5x - 8y = -40.   *Standard form*

35. Line through $(3, -5)$, parallel to $y = 4$

This will be horizontal line through $(3, -5)$. Since y has the same value for all points on the line, $b = -5$. The equation is $y = -5$.

For Exercises 37-51, see the answer graphs in the back of the textbook.

37. Line through $(2, -4)$, $m = \dfrac{3}{4}$

First locate the point $(2, -4)$. Since the slope is 3/4, a change of 4 units horizontally (4 units to the right) produces a change of 3 units vertically (3 units up). This gives a second point, $(6, -1)$, which can be used to complete the graph.

39. Line through $(-4, 1)$, $m = 3$

First locate the point $(-4, 1)$. Since the slope is 3 (or 3/1), a change of 1 unit horizontally (1 unit to the right) produces a change of 3 units vertically (3 units up). This gives a second point, $(-3, 4)$, which can be used to complete the graph.

41. $y = x^2 - 4$

The graph is a parabola with vertex $(0, -4)$ opening upward. The axis is $x = 0$, the y-axis. The domain is $(-\infty, \infty)$. Since the graph opens upward and -4 is the minimum value for y, the range is $[-4, \infty)$, as can be seen from the graph.

43. $y = 3(x + 1)^2 - 5$

The graph is a parabola with vertex $(-1, -5)$ and axis $x = -1$ opening upward. The coefficient 3 produces a narrower curve than the graph of $y = x^2$.
The graph shows that the domain is $(-\infty, \infty)$ and the range is $[-5, \infty)$.

45. $y = x^2 - 4x + 2$

Complete the square to find the vertex.

$$y = x^2 - 4x + 2$$
$$= (x^2 - 4x + 4 - 4) + 2$$
$$= (x - 2)^2 - 2$$

The graph is a parabola with vertex $(2, -2)$ and axis $x = 2$ opening upward. The graph shows that the domain is $(-\infty, \infty)$ and the range is $[-2, \infty)$.

47. $x = y^2 - 2$

The graph is a horizontal parabola which is the same as the graph of $x = y^2$ but translated 2 units to the left. The vertex is $(-2, 0)$ and the axis is $y = 0$ (the x-axis). The parabola opens to the right. The domain is $[-2, \infty)$ and the range is $(-\infty, \infty)$.

49. $x = -(y + 1)^2 + 2$

The graph is a horizontal parabola which is the same as the graph of $x = -y^2$ but translated 2 units to the right and 1 unit down. The vertex is $(2, -1)$ and the axis is $y = -1$. The parabola opens to the left. The domain is $(-\infty, 2]$ and the range is $(-\infty, \infty)$.

51.

$$x = -y^2 + 5y + 1$$

$$\frac{x}{-1} = y^2 - 5y - 1$$

$$\frac{x}{-1} + 1 = y^2 - 5y$$

$$\frac{x}{-1} + 1 + \frac{25}{4} = y^2 - 5y + \frac{25}{4}$$

$$\frac{x}{-1} + \frac{29}{4} = \left(y - \frac{5}{2}\right)^2$$

$$\frac{x}{-1} = \left(y - \frac{5}{2}\right)^2 - \frac{29}{4}$$

$$x = -\left(y - \frac{5}{2}\right)^2 + \frac{29}{4}$$

The graph is a horizontal parabola which is the same as the graph of $x = -y^2$, but translated $\frac{29}{4}$ units to the right and $\frac{5}{2}$ units up. The vertex is $\left(\frac{29}{4}, \frac{5}{2}\right)$ and the axis is $y = \frac{5}{2}$. The parabola opens to the left. The domain is $\left(-\infty, \frac{29}{4}\right]$ and the range is $(-\infty, \infty)$.

53. Let W = width of rectangle.

Since the perimeter of the rectangle is 180 m,

$$2L + 2W = 180$$

$$2L = 180 - 2W$$

$$L = \frac{180 - 2W}{2} = 90 - W.$$

Let A be the area of the rectangle.

$$A = LW$$

$$A = (90 - W)W$$

$$= 90W - W^2$$

$$= -W^2 + 90W$$

$$= -(W^2 - 90W + 45^2) - 45^2$$

$$= -(W - 45)^2 - 45^2$$

The maximum area occurs when $W = 45$, $L = 90 - W = 45$, that is, when the rectangle is a square with side 45 m.

55. The correct choice is (b).

$$y = -3x^2 - 4x + 2$$

has a graph that is a parabola that opens downward and has a vertical axis. The fact that x is the squared variable causes the graph to be a vertical parabola, and the coefficient -3 causes the parabola to open downward.

57. Center $(-2, 3)$, radius 5

$$(x - h)^2 + (y - k)^2 = r^2$$

$$[x - (-2)]^2 + (y - 3)^2 = 5^2$$

$$(x + 2)^2 + (y - 3)^2 = 25$$

59. Center $(-8, -1)$, passing through $(0, 16)$

The radius is the distance from the center to any point on the circle. The distance between $(-8, 1)$ and $(0, 16)$ is

$$r = \sqrt{(-8 - 0)^2 + (1 - 16)^2}$$
$$= \sqrt{8^2 + 15^2}$$
$$= \sqrt{289} = 17.$$

The equation of the circle is

$$[x - (-8)]^2 + (y - 1)^2 = 17^2$$
$$(x + 8)^2 + (y - 1)^2 = 289.$$

61. $x^2 - 4x + y^2 + 6y + 12 = 0$

Complete the square on x and y to put the equation in center–radius form.

$$(x^2 - 4x + \quad) + (y^2 + 6y + \quad) = -12$$
$$(x^2 - 4x + 4) + (y^2 + 6y + 9) = -12 + 4 + 9$$
$$(x - 2)^2 + (y + 3)^2 = 1$$

The circle has center $(2, -3)$ and radius 1.

63.
$$2x^2 + 14x + 2y^2 + 6y + 2 = 0$$
$$x^2 + 7x + y^2 + 3y + 1 = 0$$
$$(x^2 + 7x \quad) + (y^2 + 3y \quad) = -1$$
$$\left(x^2 + 7x + \frac{49}{4}\right) + \left(y^2 + 3y + \frac{9}{4}\right) = -1 + \frac{49}{4} + \frac{9}{4}$$
$$\left(x + \frac{7}{2}\right)^2 + \left(y + \frac{3}{2}\right)^2 = \frac{54}{4}$$

The circle has center $\left(-\frac{7}{2}, -\frac{3}{2}\right)$ and

radius $\sqrt{\frac{54}{4}} = \frac{\sqrt{54}}{\sqrt{4}} = \frac{3\sqrt{6}}{2}$.

65. Find all possible values of x so that the distance between $(x, -9)$ and $(3, -5)$ is 6.

$$\sqrt{(3 - x)^2 + (-5 + 9)^2} = 6$$
$$\sqrt{9 - 6x + x^2 + 16} = 0$$
$$\sqrt{x^2 - 6x + 25} = 0$$
$$x^2 - 6x + 25 = 36$$
$$x^2 - 6x - 11 = 0$$

$$x = \frac{6 \pm \sqrt{36 - 4(1)(-11)}}{2}$$
$$= \frac{6 \pm \sqrt{36 + 44}}{2}$$
$$= \frac{6 \pm \sqrt{80}}{2}$$
$$= \frac{6 \pm 4\sqrt{5}}{2} = \frac{2(3 \pm 2\sqrt{5})}{2}$$
$$x = 3 + 2\sqrt{5} \quad \text{or} \quad x = 3 - 2\sqrt{5}$$

67. Find all points (x, y) with $x + y = 0$ so that (x, y) is 6 units from $(-2, 3)$.

$$6 = \sqrt{(x + 2)^2 + (y - 3)^2}$$
$$6 = \sqrt{(x + 2)^2 + (-x - 3)^2} \quad y = -x$$
$$36 = (x + 2)^2 + (-x - 3)^2$$
$$36 = x^2 + 4x + 4 + x^2 + 6x + 9$$
$$0 = 2x^2 + 10x - 23$$

$$x = \frac{-10 \pm \sqrt{100 - 4(2)(-23)}}{4}$$
$$= \frac{-10 \pm \sqrt{100 + 184}}{4}$$
$$= \frac{-10 \pm \sqrt{284}}{4}$$
$$= \frac{-10 \pm 2\sqrt{71}}{4}$$
$$x = \frac{-5 \pm \sqrt{71}}{2}$$

Since $x + y = 0$ or $y = -x$,

if $x = \dfrac{-5 + \sqrt{71}}{2}$, then $y = \dfrac{5 - \sqrt{71}}{2}$;

if $x = \dfrac{-5 - \sqrt{71}}{2}$, then $y = \dfrac{5 + \sqrt{71}}{2}$.

The points are

$$\left(\frac{-5 + \sqrt{71}}{2}, \frac{5 - \sqrt{71}}{2}\right),$$

$$\left(\frac{-5 - \sqrt{71}}{2}, \frac{5 + \sqrt{71}}{2}\right).$$

69. $3y^2 - 5x^2 = 15$

(1) To test for symmetry with respect to the x-axis, replace y with $-y$.

$$3(-y)^2 - 5x^2 = 15$$
$$3y^2 - 5x^2 = 15$$

(2) To test for symmetry with respect to the y-axis, replace x with $-x$.

$$3y^2 - 5(-x)^2 = 15$$
$$3y^2 - 5x^2 = 15$$

(3) To test for symmetry with respect to the origin, replace x with $-x$ and y with $-y$.

$$3(-y)^2 - 5(-x)^2 = 15$$
$$3y^2 - 5x^2 = 15$$

The graph is symmetric with respect to the x-axis, the y-axis, and the origin.

71. $y^3 = x + 1$

(1) To test for symmetry with respect to the x-axis, replace y with $-y$.

$$(-y)^3 = x + 1$$
$$-y^3 = x + 1$$
$$y^3 = -x - 1$$

(2) To test for symmetry with respect to the y-axis, replace x with $-x$.

$$y^3 = (-x) + 1$$
$$y^3 = -x + 1$$

(3) To test for symmetry with respect to the origin, replace y with $-y$ and x with $-x$.

$$(-y)^3 = (-x) + 1$$
$$-y^3 = -x + 1$$
$$y^3 = x - 1$$

None of the equations obtained is equivalent to $y^3 = x + 1$, so the graph is not symmetric with respect to the x-axis, the y-axis, or the origin.

73. $|y| = -x$

(1) To test for symmetry with respect to the x-axis, replace y with y.

$$|-y| = -x$$
$$|y| = -x \quad Since\ |-y| = |y|$$

(2) To test for symmetry with respect to the y-axis, replace x with $-x$.

$$|y| = -(-x)$$
$$|y| = x$$

(3) To test for symmetry with respect to the origin, replace x with $-x$ and y with $-y$.

$$|-y| = -(-x)$$
$$|y| = x$$

The graph is symmetric with respect to the x-axis, but not with
respect to the y-axis or the origin.

75. $|x| = |y|$

(1) To test for symmetry with
repect to the x-axis, replace
y with -y.

$$|x| = |-y|$$
$$|x| = |y|$$

(2) To test for symmetry with
respect to the y-axis, replace x
with -x.

$$|-x| = |-y|$$
$$|x| = |y|$$

(3) To test for symmetry with respect to the origin, replace x with
the -x and y with -y.

$$|-x| = |-y|$$
$$|x| = |y|$$

The graph is symmetric with respect to the x-axis, the y-axis and
the origin.

For Exercises 77-99, see the answer
graphs in the back of the textbook.

77. $\dfrac{x^2}{25} + \dfrac{y^2}{4} = 1$

The graph is an ellipse centered at
the origin with x-intercepts 5 and
-5 and y-intercepts 2 and -2. The
graph shows that the domain of this
relation is [-5, 5] and the range is
[-2, 2].

79. $\dfrac{x^2}{4} - \dfrac{y^2}{9} = 1$

The graph is a hyperbola centered at
the origin with x-intercepts 2 and
-2. The equations of the asymptotes
are $y = \pm\dfrac{3}{2}x$. The graph shows that
the domain of this relation is
$(-\infty, -2] \cup [2, \infty)$ and the range is
$(-\infty, \infty)$.

81. $\dfrac{(x - 2)^2}{9} + \dfrac{(y + 3)^2}{4} = 1$

The graph is an ellipse centered at
(2, -3).
Since a = 3, the endpoints of the
major axis are (2 - 3, -3) = (-1, -3)
and (2 + 3, -3) = (5, -3).
Since b = 2, the endpoints of the
minor axis are (2, -3 - 2) = (2, -5)
and (2, -3 + 2) = (2, -1).
The domain is [-1, 5] and the range
is [-5, -1].

83. $x^2 = 64 - y^2$
$x^2 + y^2 = 64$

The graph of this equation is a
circle.

85. $\dfrac{(x + 3)^2}{9} + \dfrac{(y - 2)^2}{9} = 1$

$(x + 3)^2 + (y - 2)^2 = 9$

The graph of this equation is a
circle.

87. $y^2 - 3y = x + 4$

$\qquad x = y^2 - 3y - 4$

The graph of this equation is a parabola.

For Exercises 89–99, see the answer graphs in the back of the textbook.

89. $y = \sqrt{100 - x^2}$

$\qquad y^2 = 100 - x^2, \; y \geq 0$

$x^2 + y^2 = 100, \; y \geq 0$

The graph is the top half of the circle centered at the origin with radius 10.   The domain is $[-10, 10]$ and the range is $[0, 10]$.

91. $y = -\sqrt{4 - x}$

$\qquad y^2 = 4 - x, \; y \leq 0$

$\qquad x = 4 - y^2, \; y \leq 0$

The graph is the bottom half of a horizontal parabola.   The vertex is $(4, 0)$ and the parabola opens to the left.   The domain is $(-\infty, 4]$ and the range is $(-\infty, 0]$.

93. $5x - y \leq 20$

The boundary is the line $5x - y = 20$. It can be graphed using the x–intercept 4 and y–intercept -20.   The boundary is included in the graph, so draw a solid line.

Use $(0, 0)$ as a test point.

Since $5(0) - 0 \leq 20$ is a true statement, shade the side of the graph containing $(0, 0)$.

95. $y \leq (x - 4)^2$

The boundary is the parabola $y = (x - 4)^2$, which has vertex $(4, 0)$, axis $x = 4$, and opens upward.

The boundary is included in the graph, so draw a solid curve.

Use $(0, 0)$ as a test point.

Since $0 \leq (0 - 4)^2$ is a true statement, shade the side of the graph containing $(0, 0)$.

97. $25y^2 - 36x^2 > 900$

The boundary has equation

$\qquad 25y^2 - 36x^2 = 900$

or $\dfrac{y^2}{36} - \dfrac{x^2}{25} = 1.$

This is the equation of a hyperbola, centered at the origin with y–intercepts 6 and -6.   The equations of the asymptotes are $y = \pm\dfrac{6}{5}x$.   The boundary is not included in the graph, so draw a dashed curve.

Use $(0, 0)$ as a test point.

Since $25(0)^2 - 36(0)^2 > 900$ is false, shade the region that does not contain $(0, 0)$.   Additional test points may be helpful to see that the regions inside both branches of the hyperbola must be shaded.

99. $2x - y \geq 4$, where $y \geq -2$

The graph is the region that is below the line $2x - y = 4$ and above the horizontal line $y = -2$, with both boundary lines included.

Chapter 3 Test

1. (a) P(-2, 1), Q(3, 4)

$$d(P, Q) = \sqrt{(x_2 - x_1)^2 + (y_2 - y_1)^2}$$
$$= \sqrt{[3 - (-2)]^2 + (4 - 1)^2}$$
$$= \sqrt{5^2 + 3^2}$$
$$= \sqrt{34}$$

(b) The midpoint is

$$\left(\frac{-2 + 3}{2}, \frac{1 + 4}{2}\right) = \left(\frac{1}{2}, \frac{5}{2}\right)$$

2. Slope through (-3, 4) and (5, -6)

$$m = \frac{-6 - 4}{5 - (-3)}$$
$$= \frac{-10}{8}$$
$$= -\frac{5}{4}$$

3. $5x - 4y = 6$

Solve for y.

$$-4y = -5x + 6$$
$$y = \frac{5}{4}x - \frac{3}{2}$$

From this slope-intercept form $m = mx + b$, observe that the slope is $m = \frac{5}{4}$.

5. To find the equation of the line through (-2, 3) and (6, -1), first find the slope.

$$m = \frac{-1 - 3}{6 - (-2)} = \frac{-4}{8} = -\frac{1}{2}$$

Either point can be used for (x_1, y_1) in the point-slope form.

Choosing $x_1 = 2$, $y_1 = 3$ gives

$$y - 3 = -\frac{1}{2}[x - (-2)]$$
$$y - 3 = -\frac{1}{2}(x + 2)$$
$$2(y - 3) = -1(x + 2)$$
$$2y - 6 = -x - 2$$
$$x + 2y = 4.$$

6. To find the equation of the line through (-6, 2) perpendicular to x = 4, begin by realizing that x = 4 is a vertical line and therefore has undefined slope. Any line perpendicular to a vertical line must be a horizontal line and therefore have slope 0. The equation of the horizontal line through (-6, 2) is y = 2.

7. To find the equation of the line through (-6, 2) parallel to 2x - y = 8, first find the slope of the line 2x - y = 8.

$$2x - y = 8$$
$$-y = -2x + 8$$
$$y = 2x - 8$$
$$m = 2$$

Since parallel lines have the same slope, 2 is also the slope of the line whose equation is to be found. To find this equation, substitute $m = 2$, $x_1 = -6$, and $y_1 = 2$ into the point-slope form.

$$y - y_1 = m(x - x_1)$$
$$y - 2 = 2[x - (-6)]$$
$$y - 2 = 2(x + 6)$$
$$y - 2 = 2x + 12$$
$$-14 = 2x - y$$

or $2x - y = -14$

For Problems 8–10, see that answer graphs in the back of the textbook.

8. The line $4x - 5y = 10$ passes through the points $(5, 2)$ and $(0, -2)$, so draw the line through those two points. The domain and range are both $(-\infty, \infty)$.

9. To graph the line through $(2, -3)$ with slope $-\frac{2}{3}$, first locate the point $(2, -3)$. Since the slope is $-\frac{2}{3}$, a change of 3 units horizontally (3 units to the right) produces a change of -2 units vertically (2 units down). This gives a second point, $(5, -5)$, which can be used to complete the graph. See the graph in the back of the textbook.

10. $y = -(x + 3)^2 + 4$

The graph is a parabola with vertex $(-3, 4)$ and axis $x = -3$. The graph opens downward. It is the same as $y = -x^2$ but translated 3 units to the left and 4 units up.
The domain is $(-\infty, \infty)$ and the range is $(-\infty, 4]$.

11. $y = 3x^2 - 12x + 9$
$$\frac{y}{3} = x^2 - 4x + 3$$
$$\frac{y}{3} - 3 = x^2 - 4x$$
$$\frac{y}{3} - 3 + 4 = x^2 - 4x + 4$$
$$\frac{y}{3} + 1 = (x - 2)^2$$
$$\frac{y}{3} = (x - 2)^2 - 1$$
$$y = 3(x - 2)^2 - 3$$

The graph is a parabola with vertex $(2, -3)$ and axis $x = 2$.

See the answer graph in the back of the textbook.

12. $x = -(y + 2)^2 - 4$

The graph is a horizontal parabola with vertex $(-4, -2)$ and axis $y = -2$.
The domain is $(-\infty, -4]$ and the range is $(-\infty, \infty)$.

13. Let x = width of plot.
Then $80 - 2x$ = length of plot.

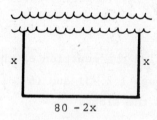

$80 - 2x$

$A = L \cdot W$
$A = (80 - 2x)x$
$\quad = 80x - 2x^2$
$\quad = -2x^2 + 80x$

The maximum area will occur at the vertex, since a = −2 < 0. Complete the square to find the vertex.

$$A = -2x^2 + 80x$$
$$= -2(x^2 - 40x)$$
$$= -2(x^2 - 40x + 400 - 400)$$
$$= -2(x^2 - 40x + 400) + 800$$
$$= -2(x - 20)^2 + 800$$

The vertex is (20, 800). The y-value, 800, gives the maximum area. The maximum area that can be enclosed is 800 sq ft.

14. Center (5, −1), radius 4

$$(x - h)^2 + (y - k)^2 = r^2$$
$$(x - 5)^2 + [y - (-1)]^2 = 4^2$$
$$(x - 5)^2 + (y + 1)^2 = 16$$

The graph is a circle with domain [1, 9] and [−5, 3].

15.
$$x^2 + y^2 + 4x - 6y = -4$$
$$(x^2 + 4x \quad) + (y^2 - 6y \quad) = -4$$
$$(x^2 + 4x + 4) + (y^2 - 6y + 9) = -4 + 4 + 9$$
$$(x + 2)^2 + (y - 3)^2 = 9$$

The circle has center (−2, 3) and radius 3.

16. $y = 2x^3 - x$

(1) To test for symmetry with respect to the x−axis, replace y with −y.

$$-y = 2x^3 - x$$
$$y = -2x^3 + x$$

(2) To test for symmetry with respect to the y−axis, replace x with −x.

$$y = 2(-x)^3 - (-x)$$
$$y = -2x^3 + x$$

(3) To test for symmetry with respect to the origin, replace x with −x and y with −y.

$$-y = 2(-x)^3 - (-x)$$
$$-y = -2x^3 + x$$
$$y = 2x^3 - x$$

These tests show that the graph is symmetric with respect to the origin, but not with respect to either axis.

For Problems 17−20, see the answer graphs in the back of the textbook.

17. $\dfrac{(x + 3)^2}{4} + \dfrac{(y - 2)^2}{9} = 1$

The graph is an ellipse centered at (−3, 2). Use the fact that a = 2 and b = 3 to find points on the ellipse. Start at (−3, 2) and locate two points each 2 units away from (−3, 2) on a horizontal line, one to the left and one to the right. These points are (−5, 2) and (−1, 2). Locate two other points on a vertical line through (−3, 2), one 3 units up and one 3 units down. These points are (−3, 5) and (−3, 1). From the graph we can see that the domain is [−5, −1] and the range is [−1, 5].

18. $9y^2 = 25x^2 + 225$

$9y^2 - 25x^2 = 225$

$\dfrac{y^2}{25} - \dfrac{x^2}{9} = 1$

The graph is a hyperbola centered at the origin. It has y-intercepts 5 and −5 and no x-intercepts. The vertices are (0, 5) and (0, −5). The equations of the asymptotes are $y = \pm\dfrac{5}{3}x$. The graph shows that the domain is (−∞, ∞) and the range is (−∞, −5] ∪ [5, ∞).

19. $\qquad y = \sqrt{25 - x^2}$

$\qquad y^2 = 25 - x^2,\ y \geq 0$

$x^2 + y^2 = 25,\ y \geq 0$

The graph is the top half of the circle with center (0, 0) and radius 5.

The domain is [−5, 5] and the range is [0, 5].

20. $(x - 4)^2 + (y + 3)^2 > 9$

The boundary has equation

$\qquad (x - 4)^2 + (y + 3)^2 = 9.$

The graph of this equation is a circle with center (4, −3) and radius 3. Draw the circle as a dashed curve. To decide which region to shade, choose any point not on the circle, such as (0, 0).

$\qquad (0 - 4)^2 + (0 + 3)^2 > 9$

$\qquad\qquad 16 + 9 > 9$

This is a true statement, which shows that the solution includes the region containing (0, 0). This is the region outside the circle. Shade the region outside the circle to show the solution.

CHAPTER 4 FUNCTIONS

Section 4.1

1. (a) $f(-2) = 0$ since, when $x = -2$, $y = 0$.

    (b) $f(0) = 4$ since, when $x = 0$, $y = 4$.

    (c) $f(1) = 2$ since, when $x = 1$, $y = 2$.

    (d) $f(4) = 4$ since, when $x = 4$, $y = 4$.

3. (a) $f(-2) = -3$ since, when $x = -2$, $y = -3$.

    (b) $f(0) = -2$ since, when $x = -2$, $y = -2$.

    (c) $f(1) = 0$ since, when $x = 1$, $y = 0$.

    (d) $f(4) = 2$ since, when $x = 4$, $y = 2$.

In Exercises 5–17, $f(x) = 3x - 1$ and $g(x) = x^2$.

5. To find $f(-1)$, replace x with -1.
    $$f(x) = 3x - 1$$
    $$f(-1) = 3(-1) - 1$$
    $$= -3 - 1$$
    $$= -4$$

7. To find $f(4)$, replace x with 4.
    $$f(4) = 3(4) - 1$$
    $$= 12 - 1$$
    $$= 11$$

9. To find $f(a)$, replace x with a.
    $$f(a) = 3a - 1$$

11. To find $f(1)$, replace x in $f(x) = 3x - 1$ with 1 to get
    $$f(1) = 3 \cdot 1 - 1 = 2.$$
    To find $g(1)$, replace x in $g(x) = x^2$ with 1 to get
    $$f(1) = 1^2 = 1.$$
    Thus, $f(1) + g(1) = 2 + 1 = 3$.

13. $f(3) \cdot g(3)$
    $$f(3) = 3(3) - 1 = 8$$
    $$g(3) = 3^2 = 9$$
    Thus, $f(3) \cdot g(3) = 8 \cdot 9 = 72$.

15. $f(-2m) = 3(-2m) - 1$
    $$= -6m - 1$$

17. $f(5a - 2)$
    $$= 3(5a - 2) - 1$$
    $$= 15a - 6 - 1$$
    $$= 15a - 7$$

19. If the ordered pair $(2, 5)$ belongs to function g, then $g(2) = 5$.

21. The domain is $[-6, \infty)$, since x takes all values greater than or equal to -6. The range is $[0, \infty)$, since y takes all nonnegative values.

23. The domain and range are both $(-\infty, \infty)$, since both x and y take all real number values.

25. The domain is $[-4, 3]$, since x takes all values from -4 to 3 inclusive. The range is $[-5, 6]$, since x takes all values from -5 to 6 inclusive.

27. The domain and range are both $(-\infty, 0) \cup (0, \infty)$, since both x and y take all real number values except 0.

29. $f(x) = 2x - 1$

Here x can be any real number, and y also takes all real number values. Thus, both the domain and range are the set of all real numbers, which is written in interval notation as $(-\infty, \infty)$.

31. $g(x) = x^4$

Here x can take any value, so the domain is the set of all real numbers, $(-\infty, \infty)$.
Since $x^4 \geq 0$ for all x, the range is the set of all nonnegative real numbers, $[0, \infty)$.

33. $f(x) = \sqrt{8 + x}$

For $\sqrt{8 + x}$ to be a real number, the expression under the radical must be nonnegative, that is,

$$8 + x \geq 0$$
$$x \geq -8.$$

The domain is thus $[-8, \infty)$.
A square root is always nonnegative, so the range is $[0, \infty)$.

35. $h(x) = \sqrt{16 - x^2}$

For $\sqrt{16 - x^2}$ to be a real number, we must have $16 - x^2 \geq 0$. Use a sign graph to verify that

$$16 - x^2 = (4 - x)(4 + x) \geq 0$$

is equivalent to $-4 \leq x \leq 4$. Thus the domain of the function is $[-4, 4]$.
For any x in the domain,

$$0 \leq \sqrt{16 - x^2} \leq 4,$$

so the range is $[0, 4]$.

37. $F(x) = \sqrt[3]{x - 1}$

It is possible to take the cube root of any real number, so x can be any real number.
The domain is $(-\infty, \infty)$. Values of cube roots can be any real number, so y can be any real number. The range is $(-\infty, \infty)$.

39. $f(x) = \dfrac{10}{3 - x}$

The quotient is undefined if the denominator is 0, so x cannot equal 3. The domain is $(-\infty, 3) \cup (3, \infty)$. As x takes all real number values except 3, y will take all real number values except 0. The range is $(-\infty, 0) \cup (0, \infty)$.

41. $g(x) = \dfrac{1}{x^2 - 4}$

The quotient is undefined if the denominator $x^2 - 4 = (x + 2)(x - 2)$ is 0, so x cannot equal -2 or 2.

The domain is

$(-\infty, -2) \cup (-2, 2) \cup (2, \infty)$.

As x takes all real number values
except -2 and 2, y will take all
real number values except 0. The
range is $(-\infty, 0) \cup (0, \infty)$.

43. $f(x) = 6x + 2$

(a) $f(x + h) = 6(x + h) + 2$
$= 6x + 6h + 2$

(b) $f(x + h) - f(x)$
$= (6x + 6h + 2) - (6x + 2)$
$= 6x + 6h + 2 - 6x - 2$
$= 6h$

(c) $\dfrac{f(x + h) - f(x)}{h} = \dfrac{6h}{h}$
$= 6$

45. $f(x) = -2x + 5$

(a) $f(x + h) = -2(x + h) + 5$
$= -2x - 2h + 5$

(b) $f(x + h) - f(x)$
$= (-2x - 2h + 5) - (-2x + 5)$
$= -2x - 2h + 5 + 2x - 5$
$= -2h$

(c) $\dfrac{f(x + h) - f(x)}{h}$
$= \dfrac{-2h}{h}$
$= -2$

47. $f(x) = 1 - x^2$

(a) $f(x + h) = 1 - (x + h)^2$
$= 1 - (x^2 + 2xh + h^2)$
$= 1 - x^2 - 2xh - h^2$

(b) $f(x + h) - f(x)$
$= (1 - x^2 - 2xh - h^2)$
$\quad - (1 - x^2)$
$= 1 - x^2 - 2xh - h^2 - 1 + x^2$
$= -2xh - h^2$

(c) $\dfrac{f(x + h) - f(x)}{h}$
$= \dfrac{-2xh - h^2}{h}$
$= -2x - h$

49. $f(x) = 8 - 3x^2$

(a) $f(x + h)$
$= 8 - 3(x + h)^2$
$= 8 - 3(x^2 + 2xh + h^2)$
$= 8 - 3x^2 - 6xh - 3h^2$

(b) $f(x + h) - f(x)$
$= (8 - 3x^2 - 6xh - 3h^2)$
$\quad - (8 - 3x^2)$
$= 8 - 3x^2 - 6xh - 3h^2$
$\quad - 8 + 3x^2$
$= -6xh - 3h^2$

(c) $\dfrac{f(x + h) - f(x)}{h}$
$= \dfrac{-6xh - 3h^2}{h}$
$= \dfrac{h(-6x - 3h)}{h}$
$= -6x - 3h$

In Exercises 51–57, $f(x) = 2^x$.

51. $f(4) = 2^4 = 16$

53. $f(x) = 2^x$

$f(-4) = 2^{-4}$

$= \dfrac{1}{2^4}$

$= \dfrac{1}{16}$

55. $f(x) = 2^x$

$f(-5r) = 2^{-5r}$ or $\dfrac{1}{2^{5r}}$

57. $f(x) = 2^x$

$f\left(\dfrac{1}{4}\right) = 2^{1/4}$ or $\sqrt[4]{2}$

59.   (a) Since the cost function is linear, it will have the form $C(x) = mx + b$, with $m = 10$ and $b = 500$. That is,

$C(x) = 10x + 500.$

(b) Since each item sells for $35, the revenue function is

$R(x) = px = 35x.$

(c) The profit function is given by

$P(x) = R(x) - C(x)$

$= 35x - (10x + 500)$

$= 35x - 10x - 500$

$= 25x - 500.$

(d) $C(x) = R(x)$

$10x + 500 = 35x$

$500 = 25x$

$20 = x$

The break-even point is 20 units. Do not produce the product, since it is possible to sell only 18 units and no profit is made until after the 20th unit is sold.

61.   (a) $C(x) = mx + b$, with $m = 150$ and $b = 2700$.

$C(x) = 150x + 2700$

(b) $R(x) = 280x$

(c) $P(x) = R(x) - C(x)$

$= 280x - (150x + 2700)$

$= 280x - 150x - 2700$

$= 130x - 2700$

(d) $C(x) = R(x)$

$150x + 2700 = 280x$

$2700 = 130x$

$21 \approx x$

The break-even point is 21 units. Produce the product, since it is possible to sell up to 25 units and a profit will be realized starting with the 21st unit.

63. $C(x) = 200x + 1000$

$R(x) = 240x$

(a) $C(x) = R(x)$

$200x + 1000 = 240x$

$1000 = 40x$

$25 = x$

The break-even point is 25 units.

(b) See the answer graph in the textbook.

(c) $C(25) = 200(25) + 1000$

$= 5000 + 1000$

$= 6000$

$R(25) = 240(25)$

$= 6000$

At the break-even point of 25 units, the cost and the revenue are each $6000.

Section 4.2

1. a varies directly as b.

$$a = kb$$

3. x is inversely proportional to y.

$$x = \frac{k}{y}$$

5. r varies jointly as s and t.

$$r = kst$$

7. w is proportional to x^2 and inversely proportional to y.

$$w = \frac{kx^2}{y}$$

9. $m = kxy$

Substitute m = 10, x = 4, and y = 7 to find k.

$$10 = k \cdot 4 \cdot 7$$
$$10 = 28k$$
$$\frac{5}{14} = k$$

Thus, the relationship between m, x, and y is given by

$$m = \frac{5}{14}xy.$$

Now find m when x = 11 and y = 8.

$$m = \frac{5}{14} \cdot 11 \cdot 8$$
$$= \frac{440}{14}$$
$$= \frac{220}{7}.$$

11. $r = \frac{km^2}{s}$

Substitute r = 12, m = 6, and s = 4 to find k.

$$12 = \frac{k \cdot 6^2}{4}$$
$$12 = 9k$$
$$k = \frac{4}{3}$$

Thus, the relationship between r, m, and s is given by

$$r = \frac{4}{3} \cdot \frac{m^2}{s}.$$

Now find r when m = 4 and s = 10.

$$r = \frac{4}{3} \cdot \frac{4^2}{10}$$
$$= \frac{32}{15}$$

13. $a = \frac{kmn^2}{y^3}$

Substitute a = 9, m = 4, n = 9, and y = 3 to find k.

$$9 = \frac{k \cdot 4 \cdot 9^2}{3^3}$$
$$9 = 12k$$
$$k = \frac{3}{4}$$

Thus, $a = \frac{3}{4} \cdot \frac{mn^2}{y^3}.$

If m = 6, n = 2, and y = 5, then

$$a = \frac{3}{4} \cdot \frac{6 \cdot 2^2}{5^3}$$
$$= \frac{18}{125}.$$

15. For k > 0, if y varies directly as x, when x increases, y increases, and when x decreases, y decreases.

17. $y = \dfrac{4}{x}$

Since the variable x is in the denominator of $\dfrac{4}{x}$, this equation represents inverse variation.

19. $y = 6xz^2$

Since y depends on the product of x and a power of z, this equation represents joint variation (which is a type of combined variation).

21. $y = \dfrac{x^2}{zw^3}$

The variable y depends on the three variables x, z, and w. Whenever one variable depends on more than one other variable, the equation is said to represent combined variation.

23. Let d = distance the spring stretches;
    f = force applied.

$$d = kf$$

Substitute d = 8 and f = 15 to find k.

$$8 = k \cdot 15$$
$$\frac{8}{15} = k$$

Thus, $d = \dfrac{8}{15} \cdot f$.

If f = 30,

$$d = \frac{8}{15} \cdot 30$$
$$d = 16.$$

The spring will stretch 16 in.

25. Let I = illumination;
    d = distance from source.

$$I = \frac{k}{d^2}$$

Substitute I = 70 and d = 5 to find k.

$$70 = \frac{k}{5^2}$$
$$25 \cdot 70 = k$$
$$1750 = k$$

Thus, $I = \dfrac{1750}{d^2}$.

If d = 12,

$$I = \frac{1750}{12^2}$$
$$= \frac{1750}{144}$$
$$= \frac{875}{72}.$$

The illumination is $\dfrac{875}{72}$ candela.

27. Let d = distance a person can see to horizon;
    h = height from the surface of the earth.

$$d = k\sqrt{h}$$

Substitute d = 15 and h = 121 to find k.

$$15 = k\sqrt{121}$$
$$15 = k \cdot 11$$
$$\frac{15}{11} = k$$

Thus, $k = \frac{15}{11}\sqrt{h}$.

If $h = 900$,

$$d = \frac{15}{11}\sqrt{900}$$

$$= \frac{15}{11} \cdot 30$$

$$= \frac{450}{11}.$$

The distance from the hill to the horizon is $\frac{450}{11}$ km.

29. Let V = volume of right circular cylinder;

r = radius of the base;

h = height of the cylinder.

$$V = kr^2h$$

Substitute $V = 300$, $r = 3$, and $h = 10.62$ to find k.

$$300 = k \cdot 3^2 \cdot 10.62$$

$$300 = 95.58k$$

$$3.1387 \approx k$$

Thus, $V = 3.1387r^2h$.

If $h = 15.92$,

$$V = 3.1387 \cdot 4^2 \cdot 15.92$$

$$= 3.1387 \cdot 16 \cdot 15.92$$

$$\approx 799.5.$$

The volume is 799.5 cm³.

31. Let L = load;

w = width;

h = height;

ℓ = length between supports.

$$L = \frac{kwh^2}{\ell}$$

Substitute $L = 400$, $w = 12$, $h = 15$, and $\ell = 8$ to find k.

$$400 = \frac{k \cdot 12 \cdot 15^2}{8}$$

$$400 = \frac{675}{2} \cdot k$$

$$\frac{2}{675} \cdot 400 = k$$

$$\frac{32}{27} = k$$

Thus,

$$L = \frac{\frac{32}{27}wh^2}{\ell}.$$

If $w = 24$, $h = 8$, and $\ell = 16$,

$$L = \frac{\frac{32}{27} \cdot 24 \cdot 8^2}{16}$$

$$= \frac{32 \cdot 24 \cdot 64}{27 \cdot 16}$$

$$= \frac{1024}{9}.$$

The maximum load is $\frac{1024}{9}$ kg.

33. Let F = force to keep car from skidding;

R = radius of curve;

W = weight of car;

S = speed.

$$F = \frac{kWS^2}{R}$$

Substitute $F = 3000$, $W = 2000$, $R = 500$, $S = 30$ to find k.

$$3000 = \frac{k(2000)(30^2)}{500}$$

$$k = \frac{5}{6}$$

Thus $F = \frac{5}{6} \cdot \frac{WS^2}{R}$.

If R = 800, S = 60, and W = 2000, then

$$F = \frac{5}{6} \cdot \frac{(2000)(60^2)}{800}$$

$$= 7500.$$

A force of 7500 lb is needed.

35. $V = \frac{kT}{P}$

Substitute V = 10, T = 280, and P = 6 to find k.

$$10 = \frac{k \cdot 280}{6}$$

$$10 = \frac{140}{3} \cdot k$$

$$\frac{3}{140} \cdot 10 = k$$

$$\frac{3}{14} = k$$

Thus, $V = \frac{\frac{3}{14} \cdot T}{P}$.

If T = 300 and P = 10,

$$V = \frac{\frac{3}{14} \cdot 300}{10}$$

$$= \frac{3}{14} \cdot 30$$

$$= \frac{45}{7}.$$

37.   Let N = number of calls;
       D = distance.

$$N = \frac{k p_1 p_2}{D}$$

Substitute N = 10,000, D = 500, P_1 = 50,000. p_2 = 125,000 to find k.

$$10,000 = \frac{k(50,000)(125,000)}{500}$$

$$k = .0008$$

Thus, $N = \frac{.0008 p_1 p_2}{D}$.

If D = 800, p_1 = 20,000, and p_2 = 80,000, then

$$N = .0008 \cdot \frac{(20,000)(80,000)}{800}$$

$$= 1600.$$

There are 1600 calls between the cities.

39. $R = \frac{k\ell}{r^4}$

Substitute R = 25, ℓ = 12, and r = .2 to find k.

$$25 = \frac{k(12)}{.0016}$$

$$k = \frac{1}{300}$$

Thus, $R = \frac{1}{300} \cdot \frac{\ell}{r^4}$.

If r = .3 and ℓ = 12, then

$$R = \frac{1}{300} \cdot \frac{12}{(.3)^4}$$

$$\approx 4.94.$$

41.   Let D = distance;
       Y = yield.

$$D = k\sqrt[3]{y}$$

Substitute Y = 100 and D = 3 to find k.

$$3 = k\sqrt[3]{100}$$

$$k = \frac{3}{\sqrt[3]{100}}.$$

Thus, $D = \dfrac{3}{\sqrt[3]{100}} \cdot \sqrt[3]{Y}$.

If $Y = 1500$,

$$D = \frac{3}{\sqrt[3]{100}} \cdot \sqrt[3]{1500}$$

$$= 3\sqrt[3]{15}$$

$$\approx 7.4$$

The distance is 7.4 km.

43. Let p = the person's pelidisi;

    w = the person's weight in grams;

    h = the person's sitting height in centimeters.

$$p = \frac{k\sqrt[3]{w}}{h}$$

Substitute $w = 48{,}820$, $h = 78.7$, and $p = 100$ to find k.

$$100 = \frac{k\sqrt[3]{48{,}820}}{78.7}$$

$$7870 = 36.55k$$

$$215.33 \approx k$$

Thus, $p = \dfrac{215.33\sqrt[3]{w}}{h}$.

If $w = 54{,}430$ and $h = 88.9$,

$$p = \frac{215.33\sqrt[3]{54{,}430}}{88.9}$$

$$= \frac{215.33(37.90)}{88.9}$$

$$\approx 92$$

This person's pelidisi is 92. The individual is undernourished since his pelidisi is below 100.

Section 4.3

1. $f(x) = 4x - 1$, $g(x) = 6x + 3$

(a) $(f + g)(x) = f(x) + g(x)$
$$= (4x - 1) + (6x + 3)$$
$$= 10x + 2$$

(b) $(f - g)(x) = f(x) - g(x)$
$$= (4x - 1) - (6x + 3)$$
$$= -2x - 4$$

(c) $(fg)(x) = f(x)g(x)$
$$= (4x - 1)(6x + 3)$$
$$= 24x^2 + 6x - 3$$

(d) $\left(\dfrac{f}{g}\right)(x) = \dfrac{f(x)}{g(x)}$
$$= \frac{4x - 1}{6x + 3}$$

The domains of $f + g$, $f - g$, and fg are the set of all real numbers, or $(-\infty, \infty)$.

The domain of $\dfrac{f}{g}$ is the set of all real numbers except $-\dfrac{1}{2}$, since if $x = -\dfrac{1}{2}$, the denominator is 0. This set is written in interval notation as

$$\left(-\infty, -\frac{1}{2}\right) \cup \left(-\frac{1}{2}, \infty\right).$$

3. $f(x) = 3x^2 - 2x$, $g(x) = x^2 - 2x + 1$

(a) $(f + g)(x)$
$$= f(x) + g(x)$$
$$= (3x^2 - 2x) + (x^2 - 2x + 1)$$
$$= 4x^2 - 4x + 1$$

(b) $(f - g)(x)$
$$= f(x) - g(x)$$
$$= (3x^2 - 2x) - (x^2 - 2x + 1)$$
$$= 2x^2 - 1$$

(c) $(fg)(x)$

$\quad = f(x)g(x)$

$\quad = (3x^2 - 2x)(x^2 - 2x + 1)$

(d) $\left(\dfrac{f}{g}\right)(x)$

$\quad = \dfrac{f(x)}{g(x)}$

$\quad = \dfrac{3x^2 - 2x}{x^2 - 2x + 1}$

The domains of $f + g$, $f - g$, and fg are $(-\infty, \infty)$.

The domain of $\dfrac{f}{g}$ is the set of all real numbers x such that $x^2 - 2x + 1 \neq 0$. Since $x^2 - 2x + 1 = (x - 1)^2$, the only number which gives this denominator a value of 0 is $x = 1$. Therefore, the domain is the set of all real numbers except 1, or $(-\infty, 1) \cup (1, \infty)$.

5. $f(x) = \sqrt{2x + 5}$, $g(x) = \sqrt{4x - 9}$

(a) $(f + g)(x) = \sqrt{2x + 5} + \sqrt{4x - 9}$

The domain of $f + g$ is the set of all real numbers x for which both $2x + 5 \geq 0$ and $4x - 9 \geq 0$, or $x \geq -\dfrac{5}{2}$ and $x \geq \dfrac{9}{4}$. The domain is thus $\left\{x \mid x \geq \dfrac{9}{4}\right\}$ or $\left[\dfrac{9}{4}, \infty\right)$.

(b) $(f - g)(x) = \sqrt{2x + 5} - \sqrt{4x - 9}$

The domain for $f - g$ is the same as for $f + g$.

(c) $(fg)(x) = \sqrt{2x + 5} \cdot \sqrt{4x - 9}$

$\quad\quad\quad = \sqrt{(2x + 5)(4x - 9)}$

The domain for fg is the same as for $f + g$.

(d) $\left(\dfrac{f}{g}\right)(x) = \dfrac{\sqrt{2x + 5}}{\sqrt{4x - 9}}$

The domain for $\dfrac{f}{g}$ is the set of all real numbers for which $2x + 5 \geq 0$ and $4x - 9 > 0$, or $x > \dfrac{9}{4}$. The domain cannot include $\dfrac{9}{4}$ because the denominator cannot be 0. The domain of $\dfrac{f}{g}$ is $\left(\dfrac{9}{4}, \infty\right)$.

In Exercises 7–19, $f(x) = 4x^2 - 2x$, $g(x) = 8x + 1$.

7. $(f + g)(3)$

$\quad = f(3) + g(3)$

$\quad = [4(3)^2 - 2(3)] + [8(3) + 1]$

$\quad = 55$

9. $(fg)(4)$

$\quad = f(4) \cdot g(4)$

$\quad = [4(4)^2 - 2(4)][8(4) + 1]$

$\quad = 56 \cdot 33$

$\quad = 1848$

11. $\left(\dfrac{f}{g}\right)(-1) = \dfrac{f(-1)}{g(-1)}$

$\quad\quad = \dfrac{4(-1)^2 - 2(-1)}{8(-1) + 1}$

$\quad\quad = \dfrac{4 + 2}{-7}$

$\quad\quad = -\dfrac{6}{7}$

13. $(f - g)(m)$

$\quad = f(m) - g(m)$

$\quad = (4m^2 - 2m) - (8m + 1)$

$\quad = 4m^2 - 2m - 8m - 1$

$\quad = 4m^2 - 10m - 1$

15. $(f \circ g)(2) = f[g(2)]$
$= f[8(2) + 1]$
$= f(17)$
$= 4(17)^2 - 2(17)$
$= 1122$

17. $(g \circ f) = g[f(2)]$
$= g[4(2)^2 - 2(2)]$
$= g(16 - 4)$
$= g(12)$
$= 8(12) + 1$
$= 97$

19. $(f \circ g)(k) = f[g(k)]$
$= f(8k + 1)$
$= 4(8k + 1)^2 - 2(8k + 1)$
$= 4(64k^2 + 16k + 1)$
$\quad - 2(8k + 1)$
$= 256k^2 + 48k + 2$

21. $f(x) = 8x + 12$, $g(x) = 3x - 1$
$(f \circ g)(x) = f[g(x)]$
$= f(3x - 1)$
$= 8(3x - 1) + 12$
$= 24x + 4$
$(g \circ f)(x) = g[f(x)]$
$= g(8x + 12)$
$= 3(8x + 12) - 1$
$= 24x + 35$

The domains for both $f \circ g$ and $g \circ f$
are $(-\infty, \infty)$.

23. $f(x) = -x^3 + 2$, $g(x) = 4x$
$(f \circ g)(x) = f[g(x)]$
$= f(4x)$
$= -(4x)^3 + 2$
$= -64x^3 + 2$

$(g \circ f)(x) = g[f(x)]$
$= g(-x^3 + 2)$
$= 4(-x^3 + 2)$
$= -4x^3 + 8$

The domains for both $f \circ g$ and $g \circ f$
are $(-\infty, \infty)$.

25. $f(x) = \frac{1}{x}$, $g(x) = x^2$

$(f \circ g)(x) = f[g(x)]$
$= f(x^2)$
$= \frac{1}{x^2}$

$(g \circ f)(x) = g[f(x)]$
$= g\left(\frac{1}{x}\right)$
$= \left(\frac{1}{x}\right)^2$
$= \frac{1}{x^2}$

The value 0 must be excluded from
the domains of both $f \circ g$ and $g \circ f$.
Therefore, both domains are
$$(-\infty, 0) \cup (0, \infty).$$

27. $f(x) = \sqrt{x + 2}$, $g(x) = 8x - 6$
$(f \circ g)(x) = f[g(x)]$
$= \sqrt{(8x - 6) + 2} = \sqrt{8x - 4}$
$= \sqrt{4(2x - 1)} = 2\sqrt{2x - 1}$

The domain of $f \circ g$ is the set of all
real numbers $2x - 1 \geq 0$, or $x \geq \frac{1}{2}$.
In interval notation, this is
written $\left[\frac{1}{2}, \infty\right)$.

$(g \circ f)(x) = g[f(x)]$
$= g(\sqrt{x + 2})$
$= 8\sqrt{x + 2} - 6$

The domain of g ∘ f is the set of all real numbers for which x + 2 ≥ 0, or x ≥ -2, which may be written [-2, ∞).

29. $f(x) = \dfrac{1}{x - 5}$, $g(x) = \dfrac{2}{x}$

$(f \circ g)(x) = f[g(x)]$

$= f\left(\dfrac{2}{x}\right)$

$= \dfrac{1}{\left(\dfrac{2}{x} - 5\right)} \cdot \dfrac{x}{x}$

$= \dfrac{x}{2 - 5x}$

To find the domain of f ∘ g, focus on the denominators of the fractions that are involved. For $\dfrac{2}{x}$ to be defined, we must have x ≠ 0. For $\dfrac{x}{2 - 5x}$ to be defined, we must have $2 - 5x \neq 0$, or $x \neq \dfrac{5}{2}$.

In interval notation, the domain of f ∘ g is $(-\infty, 0) \cup \left(0, \dfrac{2}{5}\right) \cup \left(\dfrac{2}{5}, \infty\right)$.

$(g \circ f)(x) = g[f(x)]$

$= g\left(\dfrac{1}{x - 5}\right)$

$= \dfrac{2}{\left(\dfrac{1}{x - 5}\right)}$

$= 2(x - 5)$

The domain of g ∘ f is the set of all x such that x ≠ 5, or $(-\infty, 5) \cup (5, \infty)$.

31. $f(x) = \sqrt{x + 1}$, $g(x) = -\dfrac{1}{x}$

The domain of g is the set of all x such that x ≠ 0; the domain of f is the set of all x such that x ≥ -1.

$(f \circ g)(x) = f[g(x)]$

$= f\left(-\dfrac{1}{x}\right)$

$= \sqrt{-\dfrac{1}{x} + 1}$

$= \sqrt{\dfrac{-1 + x}{x}}$

$= \sqrt{\dfrac{x - 1}{x}}$

The domain of f ∘ g is the set of all x ≠ 0 such that $\dfrac{x - 1}{x} \geq 0$. Use a sign graph to solve this inquality. The domain of f ∘ g is $(-\infty, 0) \cup [1, \infty)$.

$(g \circ f)(x) = g[f(x)]$

$= g(\sqrt{x + 1})$

$= -\dfrac{1}{\sqrt{x + 1}}$

The domain of g ∘ f is the set of all x ≥ -1 such that $\sqrt{x + 1} \neq 0$, or x ≠ -1. Thus, the domain of g ∘ f is $(-1, \infty)$.

33. $f(x) = 8x$, $g(x) = \dfrac{1}{8}x$

$(f \circ g)(x) = f[g(x)]$

$= f\left(\dfrac{1}{8}x\right)$

$= 8\left(\dfrac{1}{8}x\right)$

$= x$

$$(g \circ f)(x) = g[f(x)]$$
$$= g(8x)$$
$$= \frac{1}{8}(8x)$$
$$= x$$

35. $f(x) = 8x - 11$, $g(x) = \dfrac{x + 11}{8}$

$$(f \circ g)(x) = f[g(x)]$$
$$= f\left(\frac{x + 11}{8}\right)$$
$$= 8\left(\frac{x + 11}{8}\right) - 11$$
$$= x$$

$$(g \circ f)(x) = g[f(x)]$$
$$= g(8x - 11)$$
$$= \frac{8x - 11 + 11}{8}$$
$$= x$$

37. $f(x) = x^3 + 6$, $g(x) = \sqrt[3]{x - 6}$

$$(f \circ g)(x) = f[g(x)]$$
$$= f(\sqrt[3]{x - 6})$$
$$= (\sqrt[3]{x - 6})^3 + 6$$
$$= x$$

$$(g \circ f)(x) = g[f(x)]$$
$$= g(x^3 + 6)$$
$$= \sqrt{x^3 + 6 - 6}$$
$$= x$$

In Exercises 41–45, examples are given.
Other correct answers are possible.

41. $h(x) = (6x - 2)^2$

One choice is $f(x) = x^2$ and
$g(x) = 6x - 2$; then

$$(f \circ g)(x) = f[g(x)]$$
$$= f(6x - 2)$$
$$= (6x - 2)^2$$
$$= h(x).$$

43. $h(x) = \sqrt{x^2 - 1}$

One choice is $f(x) = \sqrt{x}$ and
$g(x) = x^2 - 1$.
Then

$$(f \circ g)(x) = f[g(x)]$$
$$= f(x^2 - 1)$$
$$= \sqrt{x^2 - 1} = h(x).$$

45. $h(x) = \dfrac{(x - 2)^2 + 1}{5 - (x - 2)^2}$

One choice is $f(x) = \dfrac{x^2 + 1}{5 - x^2}$ and

$g(x) = x - 2$.
Then

$$(f \circ g)(x) = f[g(x)]$$
$$= f(x - 2)$$
$$= \frac{(x - 2)^2 + 1}{5 - (x - 2)^2} = h(x).$$

47. Let x = the number of people less
than 100 people that attend.

(a) x people fewer than 100 attend,
so $100 - x$ people do attend.

$$N(x) = 100 - x$$

(b) The cost per person starts at \$2 and increases by \$.20 for each of the x people that do not attend. The total increase is \$.20x, and the cost per person increases to \$2 + \$.20x.

$$G(x) = 2 + .2x$$

(c) $C(x) = N(x) \cdot G(x)$
$$= (100 - x)(2 + .2x)$$

49. $P(x) = 2x^2 + 1$
$f(a) = 3a + 2$

$(P \circ g)(a) = P[f(a)]$
$$= P(3a + 2)$$
$$= 2(3a + 2)^2 + 1$$
$$= 18a^2 + 24a + 9$$

If the amount of plankton decreases, the fish population will decrease.

51. (a) $A \circ r = A[r(t)]$
$$= A(2t)$$
$$= \pi(2t)^2$$
$$= 4\pi t^2$$

This is the area that is covered by the pollutant at time t.

Section 4.4

For Exercises 1–11, see the answer graphs in the textbook.

1. $f(x) = \frac{3}{4}x - 1$

The graph is the line with slope $\frac{3}{4}$ and y-intercept -1. To graph the line, start at the point (0, -1), which is the y-intercept. Locate the point which is 4 units up from it, which is the point (4, 2). (4, 2) is a second point on the graph. Draw the line through (0, -1) and (4, 2). The domain and range are each $(-\infty, \infty)$.

3. $f(x) = |x + 1|$

The domain of this function is $(-\infty, \infty)$, since x can be any real number.

The value of $f(x) = y$ is always greater than or equal to 0, since $|x + 1| \geq 0$. Thus, the y-value of the "vertex" is 0, and the range is $[0, \infty)$.

The x-value of the "vertex" can be found by substituting 0 for y in the equation.

$$y = |x + 1|$$
$$0 = |x + 1| \quad Let \ y = 0$$
$$0 = x + 1$$
$$-1 = x$$

Plot a few other ordered pairs, like (-2, 1) and (0, 1); then draw the two rays that make up this graph. The graph is the same as the graph of $y = |x|$ except translated 1 unit to the left.

5. $y = |x| + 4$

This graph is the same as the graph of $y = |x|$ except translated 4 units up. The "vertex" is $(0, 4)$. The domain is $(-\infty, \infty)$ and the range is $[4, \infty)$.

7. $y = 3|x - 2| + 1$

$$|x - 2| \geq 0$$
$$3|x - 2| \geq 0 \quad Multiply\ by\ 3$$
$$3|x - 2| + 1 \geq 1 \quad Add\ 1$$
$$y \geq 1 \quad y = 3|x - 2| + 1$$

The domain is $(-\infty, \infty)$ and the range is $[1, \infty)$.
When $y = 1$,

$$1 = 3|x - 2| + 1$$
$$0 = 3|x - 2|$$
$$0 = |x - 2|$$
$$0 = x - 2$$
$$2 = x.$$

The "vertex" is $(2, 1)$.
The graph is the same as $y = 3|x|$ except translated 2 units to the right and 1 unit up. The coefficient 3 indicates that the rays that form this graph are steeper than the rays that form the graph of $y = |x|$.

9. $f(x) = -|x + 1| + 2$

$$|x + 1| \geq 0$$
$$-|x + 1| \leq 0 \quad Multiply\ by\ -1$$
$$-|x + 1| + 2 \leq 2 \quad Add\ 2$$
$$f(x) \leq 2 \quad f(x) = -|x + 1| + 1$$

The domain is $(-\infty, \infty)$ and the range is $(-\infty, 2]$.
When $y = 2$,

$$2 = -|x + 1| + 2$$
$$0 = -|x + 1|$$
$$0 = |x + 1|$$
$$0 = x + 1$$
$$-1 = x.$$

The "vertex" is $(-1, 2)$.
The coefficient -1 indicates that the graph opens downward. The graph is the same as the graph of $y = -|x|$ except translated 1 unit to the right and 2 units up.

11. $f(x) = |x| + x = \begin{cases} x + x & \text{if } x \geq 0 \\ -x + x & \text{if } x < 0 \end{cases}$

$\qquad\qquad$ *By definition of absolute value*

$\qquad = \begin{cases} 2x & \text{of } x \geq 0 \\ 0 & \text{if } x < 0 \end{cases}$

This is a function defined piece-wise.
The domain is $(-\infty, \infty)$ and the range is $[0, \infty)$.

15. $f(x) = \begin{cases} 2x & \text{if } x \leq -1 \\ x - 1 & \text{if } x > -1 \end{cases}$

(a) $f(-5) = 2(-5) = -10$
(b) $f(-1) = 2(-1) = -2$

(c) $f(0) = 0 - 1 = -1$

(d) $f(3) = 3 - 1 = 2$

(e) $f(5) = 5 - 1 = 4$

17. $f(x) = \begin{cases} 3x + 5 \text{ if } x \leq 0 \\ 4 - 2x \text{ if } 0 < x < 2 \\ x \text{ if } x \geq 2 \end{cases}$

   (a) $f(-5) = 3(-5) + 5 = -10$

   (b) $f(-1) = 3(-1) + 5 = 2$

   (c) $f(0) = 3(0) + 5 = 5$

   (d) $f(3) = 3$

   (e) $f(5) = 5$

For Exercises 19–25, see the answer graphs in the textbook.

19. $f(x) = \begin{cases} x - 1 \text{ if } x \leq 3 \\ 2 \text{ if } x > 3 \end{cases}$

   Draw the graph of $y = x - 1$ to the left of $x = 3$, including the endpoint at $x = 3$. Draw the graph of $y = 2$ to the right of $x = 3$, but do not include the endpoint at $x = 3$.

21. $f(x) = \begin{cases} 4 - x \text{ if } x < 2 \\ 1 + 2x \text{ if } x \geq 2 \end{cases}$

   Draw the graph of $y = 4 - x$ to the left of $x = 2$, but do not include the endpoint. Draw the graph of $y = 1 + 2x$ to the right of $x = 2$, including the endpoint.

23. $f(x) = \begin{cases} 2 + x \text{ if } x < -4 \\ -x \text{ if } -4 \leq x < 5 \\ 3x \text{ if } x > 5 \end{cases}$

   Draw the graph of $y = 2 + x$ to the left of -4, but do not include the endpoint at $x = 4$. Draw the graph of $y = -x$ between -4 and 5, including both endpoints.

Draw the graph of $y = 3x$ to the right of 5, but do not include the endpoint at $x = 5$.

25. $f(x) = \begin{cases} |x| \text{ if } x > -2 \\ x \text{ if } x \leq -2 \end{cases}$

Draw the graph of $y = |x|$ to the right of $x = -2$, but do not include the endpoint. Draw the graph of $y = x$ to the left of -2, including the endpoint.

For Exercises 29–33, see the answer graphs in the textbook.

29. $f(x) = [-x]$

   Plot points.

x	-x	$f(x) = [-x]$
-2	2	2
-1.5	1.5	1
-1	1	1
-.5	.5	0
0	0	0
.5	-.5	-1
1	-1	-1
1.5	-1.5	-2
2	-2	2

More generally, to get $y = 0$, we need

$$0 \leq -x < 1$$
$$0 \geq x > -1 \text{ or}$$
$$-1 < x \leq 0.$$

To get $y = 1$, we need

$$-1 \leq -x < 2$$
$$-1 \geq x > -2 \text{ or}$$
$$-2 < x \leq -1.$$

Follow this pattern to graph the step function.

31. $f(x) = [3x + 1]$

To get $y = 0$, we need

$$0 \leq 3x + 1 < 1$$
$$-1 \leq 3x < 0$$
$$-\frac{1}{3} \leq x < 0.$$

To get $y = 1$, we need

$$1 \leq 3x + 1 < 2$$
$$0 \leq 3x < 1$$
$$0 \leq x < \frac{1}{3}.$$

To get $y = 2$, we need

$$2 \leq 3x + 1 < 3$$
$$1 \leq 3x < 2$$
$$\frac{1}{3} \leq x < \frac{2}{3}.$$

Follow this pattern to graph the step function.

33. $f(x) = [3x] + 1$

The graph of this function is the same as that of $f(x) = [3x]$ except translated 1 unit up.
($f(x) = [3x]$ was graphed in Exercise 32.)

35. (a) $L(.75) = 30¢$

(b) $L(1.6) = 30¢ + 27¢ = 57¢$

(c) $L(4) = 30¢ + 3(27¢)$
$$= 30¢ + 81¢ = 111¢$$

(d) See the answer graph in the back of the textbook.

(e) Domain: $(0, \infty)$
Range: $[30, 57, 84, 111, \ldots]$

37. Let x = the number of miles driven;
y = the car rental cost.

For $0 < x \leq 50$, $y = 37$.
For $50 < x \leq 75$, $y = 37 + 10 = 47$.
For $75 < x \leq 100$, $y = 47 + 10 = 57$.
For $100 < x \leq 125$, $y = 57 + 10 = 67$.

Follow this pattern to graph the step function.
See the answer graph in the back of the textbook.

39. (a) $C(1) = \$1.80$

(b) $C(2.3) = C(3) = \$1.80 + 2(\$.20)$
$$= \$2.20$$

(c) $C(8) = \$1.80 + 7(\$.20) = \$3.20$

(d) See the graph in the textbook.

(e) The domain is $(0, \infty)$ (at least in theory) and the range is
$\{1.80, 2.00, 2.20, 2.40, \ldots\}$.

41. (a)

$$f(x) = \begin{cases} 6.5 \text{ if } 0 \leq x \leq 4 \\ -5.5x + 48 \text{ if } 4 < x \leq 6 \\ -30x + 195 \text{ if } 6 \leq x \leq 6.5 \end{cases}$$

Draw the graph of $y = 6.5x$ between 0 and 4, including the endpoints. Draw the graph of $y = -5.5x + 48$ between 4 and 6, including the endpoint at 6 but not the one at 4. Draw the

graph of $y = -30x + 195 + 195$, including the endpoint at 6.5 but not the one at 6. Notice that the endpoints of the three pieces coincide.
See the graph in the textbook.

(b) From the graph, observe that the snow depth y reaches its deepest level (26 in) when $x = 4$. $x = 4$ represents 4 months after the beginning of October, which is the beginning of February.

(c) From the graph, the snow depth y is nonzero when x is between 0 and 6.5. Snow begins at the beginning of October and ends 6.5 months later, in the middle of April.

Section 4.5

1. This is a one-to-one function since every horizontal line intersects the graph in no more than one point.

3. This is a one-to-one function since every horizontal line intersects the graph in no more than one point.

5. This is not a one-to-one function since there is a horizontal line that intersects the graph in more than one point. (Here it intersects the curve at an infinite number of points.)

7. $y = 4x - 5$

If $x_1 \neq x_2$, then

$$4x_1 \neq 4x_2$$
$$4x_1 - 5 \neq 4x_2 - 5$$
$$f(x_1) \neq f(x_2).$$

Thus any two distinct x-values lead to distinct y-values.
The function is thus one-to-one.

9. $y = (x - 2)^2$

If $x = 0$, $y = 4$.
If $x = 4$, $y = 4$.

Thus, there exist two distinct x-values that lead to the same y-value. Thus, the function is not one-to-one.

11. $y = \sqrt{36 - x^2}$

Both $x = 6$ and $x = -6$ lead to the same y-value, 0. Thus, the function is not one-to-one.

13. $y = 2x^3 + 1$

If $x_1 \neq x_2$, then

$$x_1{}^3 \neq x_2{}^3$$
$$2x_1{}^3 \neq 2x_2{}^3$$
$$2x_1{}^3 + 1 \neq 2x_2{}^3 + 1$$
$$f(x_1) \neq f(x_2),$$

so the function is one-to-one.

15. $y = \dfrac{1}{x + 2}$

If $x_1 \neq x_2$, then

$$x_1 + 2 \neq x_2 + 2,$$
$$\dfrac{1}{x_1 + 2} \neq \dfrac{1}{x_2 + 2}$$
$$f(x_1) \neq f(x_2),$$

so the function is one-to-one.

17. $y = 9$

If $x = 1$, $y = 9$.
If $x = 2$, $y = 9$.

Thus, the function is not one-to-one. Here, no matter what x-value is used, the resulting y-value is always the same.

19. The inverse operation of tying your shoelaces would be untying your shoelaces, since untying "undoes" tying.

21. The inverse operation of entering a room would be leaving a room, since leaving "undoes" entering.

23. The inverse operation of taking off in an airplane would be landing in an airplane, since landing "undoes" taking off.

27. These functions are inverses since their graphs are symmetric with respect to the line $y = x$.

29. These functions are not inverses since their graphs are not symmetric with respect to the line $y = x$.

31. These functions are not inverses since their graphs are not symmetric with respect to the line $y = x$.

33. $f(x) = -\dfrac{3}{11}x$, $g(x) = -\dfrac{11}{3}x$

$(f \circ g)(x) = f[g(x)]$
$$= f\left(-\dfrac{11}{3}x\right)$$
$$= -\dfrac{3}{11}\left(-\dfrac{11}{3}x\right) = x$$

$(g \circ f)(x) = g[f(x)]$
$$= g\left(-\dfrac{3}{11}x\right)$$
$$= -\dfrac{11}{3}\left(-\dfrac{3}{11}x\right) = x$$

Since $(f \circ g)(x) = x$ and $(g \circ f)(x) = x$, these functions are inverses.

35. $f(x) = 5x - 5$, $g(x) = \dfrac{1}{5}x + 1$

$(f \circ g)(x) = f[g(x)]$
$$= f\left(\dfrac{1}{5}x + 1\right)$$
$$= 5\left(\dfrac{1}{5}x + 1\right) - 5$$
$$= x + 5 - 5$$
$$= x$$

$(g \circ f)(x) = g[f(x)] = g(5x - 5)$
$$= \dfrac{1}{5}(5x - 5) + 1$$
$$= x - 1 + 1$$
$$= x$$

Since $(f \circ g)(x) = x$ and $(g \circ f)(x) = x$, these functions are inverses.

37. $f(x) = \dfrac{1}{x}$, $g(x) = \dfrac{1}{x}$

$(f \circ g)(x) = f[g(x)] = \dfrac{1}{\dfrac{1}{x}}$

$\qquad\qquad\qquad = x$

$(g \circ f)(x) = g[f(x)] = \dfrac{1}{\dfrac{1}{x}}$

$\qquad\qquad\qquad\qquad = x$

f and g are inverses of each other.

39. $f(x) = \sqrt{x + 8}$, domain $[-8, \infty)$
$g(x) = x^2 - 8$, domain $[0, \infty)$

$(f \circ g)(x) = f[g(x)] = \sqrt{x^2 - 8 + 8}$

$\qquad\qquad\qquad\quad = x^2$

$\qquad\qquad\qquad\quad = x$ for $[0, \infty)$

$(g \circ f)(x) = g[f(x)] = (\sqrt{x + 8})^2 - 8$

$\qquad\qquad\qquad\quad = x + 8 - 8$

$\qquad\qquad\qquad\quad = x$ for $[-8, \infty)$

f and g are inverses of each other.

41. $f(x) = |x - 1|$, domain $[-1, \infty)$,
$g(x) = |x + 1|$, domain $[1, \infty)$

For the interval $[1, \infty)$,

$\qquad |x + 1| = x + 1,$

so the range of g(x) is $[2, \infty)$. The range of g(x) is not the same as the domain of f(x), so these functions are not inverses.

For Exercises 43–53, see the answer graphs in the textbook.

43. To graph the inverse, first draw the line y = x and then draw the mirror image of the graph of the original functions across y = x. The graph of the inverse will be another line that also passes through (0, 0).

45. Draw the mirror image of the original graph across the line y = x. The graph of the inverse will be the top half of a horizontal parabola with vertex (0, 0) and opening to the right.

47. Carefully draw the mirror image of the original graph across the line y = x.

49. $y = 3x - 4$

Solve for x.

$y + 4 = 3x$

$\dfrac{y + 4}{3} = 3x$

Exchange x and y.

$\dfrac{x + 4}{3} = y$

$f^{-1}(x) = \dfrac{x + 4}{3}$

The graph of the original function is a line with slope 3 and y-intercept −4. Since $f^{-1}(x) = \dfrac{x + 4}{3} = \dfrac{1}{3}x + \dfrac{4}{3}$, the graph of the inverse function is a line with slope $\dfrac{1}{3}$ and y-intercept $\dfrac{4}{3}$.

51. $y = \frac{1}{3}x$

$3y = x$

Exchange x and y.

$3x = y$

$f^{-1}(x) = 3x$

The original function is a line with slope $\frac{1}{3}$ and y-intercept 0, and the inverse function is a line with slope 3 and y-intercept 0.

53. $y = x^3 + 1$

$y - 1 = x^3$

$\sqrt[3]{y - 1} = x$

Exchange x and y.

$\sqrt[3]{x - 1} = y$

$f^{-1}(x) = \sqrt[3]{x - 1}$

Plot points to graph these functions.

x	-1	0	1
f(x)	0	1	2

x	0	1	2
$f^{-1}(x)$	-1	0	1

55. $y = x^2$

This is not a one-to-one function since $(2)^2 = 4$ and $(-2)^2 = 4$. Thus, the function has no inverse function.

For Exercise 57-61, see the answer graphs in the textbook.

57. $y = \frac{1}{x}$

$xy = 1$

$x = \frac{1}{x}$

Exchange x and y.

$y = \frac{1}{x}$

$f^{-1}(x) = \frac{1}{x}$

Observe that this function is its own inverse. Plot points to draw the graph, which is a hyperbola.

x	-2	-1	$-\frac{1}{2}$	$\frac{1}{2}$	1	2
y	$-\frac{1}{2}$	-1	-2	2	1	$\frac{1}{2}$

59. $f(x) = 4 - x^2$, domain $(-\infty, 0]$

The restriction of the domain to the interval $(-\infty, 0]$ makes f(x) one-to-one, so it has an inverse function. To find $f^{-1}(x)$, let $y = f(x)$ and solve for x.

$y = 4 - x^2$

$y - 4 = -x^2$

$4 - y = x^2$

$\sqrt{4 - y} = x$ or $-\sqrt{4 - y} = x$

The second choice is the correct one because the domain of f(x) is $(-\infty, 0]$.

Exchange x and y.

$-\sqrt{4 - x} = y$

$f^{-1}(x) = -\sqrt{4 - x}$, domain $(-\infty, 4]$

The domain of $f^{-1}(x)$ is the same as the range of $f(x)$.

The graph of $f(x)$ is the left half of a parabola with vertex $(0, 4)$ that opens downward, and the graph of $f^{-1}(x)$ is its mirror image across the line $y = x$, which is the bottom half of a parabola with vertex $(4, 0)$ that opens left.

61. $f(x) = -\sqrt{x^2 - 16}$, domain $[4, \infty)$

$y = f(x)$ is one-to-one, so it has an inverse function.

Solve for x.

$$y = -\sqrt{x^2 - 16}, \ x \geq 4$$
$$y^2 = x^2 - 16, \ y \leq 0 \text{ and } x \geq 4$$
$$y^2 + 16 = x^2, \ y \leq 0 \text{ and } x \geq 4$$
$$\sqrt{y^2 + 16} = x, \ y \leq 0 \text{ and } x \geq 4$$

Only the positive square root is needed here since it is known that $x \geq 4$.   Exchange x and y.

$$\sqrt{x^2 + 16} = y, \ x \leq 0 \text{ and } y \geq 4$$
$$f^{-1}(x) = \sqrt{x^2 + 16}, \text{ domain } (-\infty, 0]$$

Since the graph of $x^2 - y^2 = 16$ or $\dfrac{x^2}{16} - \dfrac{y^2}{16} = 1$ is a hyperbola with branches opening left and right, the graph of $f(x)$ is a portion of that hyperbola. Specifically, $f(x)$ is the bottom half of the right branch of the hyperbola. The graph of $f^{-1}(x)$ is the mirror image across

$y = x$, which is the left half of the top branch of a hyperbola whose branches open up and down.

Chapter 4   Review Exercises

1. $f(x) = -x$

The domain and range are both $(-\infty, \infty)$.

3. $f(x) = |x| - 4$

Since we can take the absolute value of any real number, the domain is $(-\infty, \infty)$.   For all values of x, $|x| \geq 0$, so $|x| - 4 \geq -4$.   Thus, the range is $[-4, \infty)$.

5. $f(x) = -(x + 3)^2$

Here, x can be any real number, so the domain is $(-\infty, \infty)$.   For every value of x, $(x + 3)^2 \geq 0$, so $-(x + 3)^2 \leq 0$.   Thus, the range is $(-\infty, 0]$.

7. $f(x) = \sqrt{49 - x^2}$

If $\sqrt{49 - x^2}$ is to be a real number, then $49 - x^2 \geq 0$.

Use a sign graph to verify that this inequality is equivalent to $-7 \leq x \leq 7$.

The domain is $[-7, 7]$.

$y = \sqrt{49 - x^2}$ is always nonnegative, but it additionally has a largest value.

$$x^2 \geq 0$$

$$-x^2 \leq 0 \qquad \textit{Multiply by -1}$$

$$49 - x^2 \leq 49 \qquad \textit{Add 49}$$

$$\sqrt{49 - x^2} \leq \sqrt{49} \quad 0 \leq 49 - x^2$$

$$y \leq 7 \qquad y = \sqrt{49 - x^2}$$

The range is $[0, 7]$.

9. $f(x) = \sqrt{\dfrac{1}{x^2 + 9}}$

If $\sqrt{\dfrac{1}{x^2 + 9}}$ is to be a real number,

then

$$\sqrt{\dfrac{1}{x^2 + 9}} > 0.$$

Since $x^2 + 9 > 0$ for all real numbers x, $\dfrac{1}{x^2 + 9} > 0$ for all real numbers x.

The domain is $(-\infty, \infty)$.

$y = f(x)$ is always nonnegative, but it additionally has a largest value.

$$x^2 \geq 0$$

$$x^2 + 9 \geq 9 \qquad \textit{Add 9}$$

$$\dfrac{1}{x^2 + 9} \leq \dfrac{1}{9} \qquad \begin{array}{l}x^2 + 9 \textit{ and 9 are} \\ \textit{both positive}\end{array}$$

$$\sqrt{\dfrac{1}{x^2 + 9}} \leq \sqrt{\dfrac{1}{9}} \qquad \begin{array}{l}1/(x^2 + 9) \textit{ and } 1/9 \\ \textit{are both positive}\end{array}$$

$$y \leq \dfrac{1}{3} \qquad y = \sqrt{\dfrac{1}{x^2 + 9}}$$

The range is $\left[0, \dfrac{1}{3}\right]$.

11. $f(x) = \dfrac{x + 2}{x - 7}$

If $\dfrac{x + 2}{x - 7}$ is to be a real number, we

must have $x - 7 \neq 0$ or $x \neq 7$. Thus, the domain is the set of all real numbers except 7 or

$$(-\infty, 7) \cup (7, \infty).$$

13. $(-2, 3)$ is the ordered pair that corresponds to the notation

$$f(-2) = 3.$$

For Exercises 15–21, $f(x) = -x^2 + 4x + 2$ and $g(x) = 3x + 5$.

15. $f(4) = -(4)^2 + 4(4) + 2$

$$= -16 + 16 + 2$$

$$= 2$$

17. $g(3) = 3(3) + 5$

$$= 9 + 5$$

$$= 14$$

19. $g[f(-2)] = g[-(-2)^2 + 4(-2) + 2]$

$$= g(-4 - 8 + 2)$$

$$= g(-10)$$

$$= 3(-10) + 5$$

$$= -30 + 5$$

$$= -25$$

21. $g(5y - 4) = 3(5y - 4) + 5$

$$= 15y - 12 + 5$$

$$= 15y - 7$$

23.
$$f(x) = -3$$
$$f(x + h) = -3$$
$$\frac{f(x + h) - f(x)}{h} = \frac{-3 - (-3)}{h} = 0$$

25. $f(x) = -x^3 + 2x^2$

$f(x + h)$

$= -(x + h)^3 + 2(x + h)^2$

$= -(x^3 + 3x^2h + 3xh^2 + h^3)$

$\quad + 2(x^2 + 2hx + h^2)$

$= -x^3 - 3x^2h - 3xh^2 - h^3$

$\quad + 2x^2 + 4hx + 2h^3$

$f(x + h) - f(x)$

$= (-x^3 - 3x^2h - 3xh^2 - h^3$

$\quad + 2x^2 + 4hx + 2h^2) - (x^3 + 2x^2)$

$= -x^3 - 3x^2h - 3xh^2 - h^3 + 2x^2$

$\quad + 4hx + 2h^2 + x^3 - 2x^2$

$= -3x^2h - 3xh^2 - h^3 + 4hx + 2h^2$

$\dfrac{f(x + h) - f(x)}{h}$

$= \dfrac{-3x^2h - 3xh^2 - h^3 + 4hx + 2h^2}{h}$

$= -3x^2 - 3xh - h^2 + 4x + 2h$

27. y varies inversely as r and directly as the cube of p.

$$y = \frac{kp^3}{r}$$

29. A varies jointly as the third power of t and the fourth power of s, and inversely as p and the square of h.

$$A = \frac{kt^3s^4}{ph^2}$$

31. $m = \dfrac{knp^2}{q}$

Substitute m = 20, n = 5, p = 6, and q = 18 to find k.

$$20 = \frac{k \cdot 5 \cdot 6^2}{18}$$

$$20 = \frac{k \cdot 5 \cdot 36}{18}$$

$$20 = 10k$$

$$2 = k$$

Thus, $m = \dfrac{2np^2}{q}$.

If n = 7, p = 11, and q = 2,

$$m = \frac{2 \cdot 7 \cdot 11^2}{2}$$

$$= 7 \cdot 121$$

$$= 847.$$

33. Let p = power;

v = wind velocity.

$$p = kv^3$$

Substitute p = 10,000 and v = 10 to find k.

$$10,000 = k \cdot 10^3$$

$$\frac{10,000}{1000} = k$$

$$10 = k$$

Thus, $p = 10v^3$.

If v = 15,

$$p = 10 \cdot 15^3$$

$$= 33,750$$

33,750 units of power are produced.

35. $w = \dfrac{k}{d^2}$

Substitute w = 90 and d = 6400 to find k.

$$90 = \frac{k}{6400^2}$$

$$6400^2 \cdot 90 = k$$

Thus, $w = \dfrac{6400^2 \cdot 90}{d^2}$.

If d = 7200,

$$w = \frac{6400^2 \cdot 90}{7200^2}$$

$$= \frac{6400 \cdot 6400 \cdot 90}{7200 \cdot 7200}$$

$$= \frac{64 \cdot 64 \cdot 90}{72 \cdot 72}$$

$$= \frac{64 \cdot 64 \cdot 10}{8 \cdot 72}$$

$$= \frac{8 \cdot 64 \cdot 10}{72}$$

$$= \frac{8 \cdot 8 \cdot 10}{9} = \frac{640}{9}.$$

The man weighs $\dfrac{640}{9}$ kg.

In Exercises 37–45, $f(x) = 3x^2 - 4$ and $g(x) = x^2 - 3x - 4$.

37. (fg)(x)

$= f(x) \cdot g(x)$

$= (3x^2 - 4)(x^2 - 3x - 4)$

$= 3x^4 - 9x^3 - 12x^2 - 4x^2 + 12x + 16$

$= 3x^4 - 9x^3 - 16x^2 + 12x + 16$

39. (f + g)(-4)

$= f(-4) + g(-4)$

$= [3(-4)^2 - 4] + [(-4)^2 - 3(-4) - 4]$

$= [3(16) - 4] + [16 + 12 - 4]$

$= 44 + 24$

$= 68$

41. (fg)(x)

$= f(x) \cdot g(x)$

$= (3x^2 - 4)(x^2 - 3x - 4)$

$= 3x^4 - 9x^3 - 16x^2 + 12x + 16$

Let x = 1 + r.

(fg)(1 + r)

$= 3(1 + r)^4 - 9(1 + r)^3$

$\quad - 16(1 + r)^2 + 12(1 + r) + 16$

$= 3(1 + 4r + 6r^2 + 4r^3 + r^4)$

$\quad - 9(1 + 3r + 3r^2 + r^3)$

$\quad - 16(1 + 2r + r^2)$

$\quad + 12(1 + r) + 16$

$= 3r^4 + 3r^3 - 25r^2 - 35r + 6$

43. $\left(\dfrac{f}{g}\right)(x) = \dfrac{3x^2 - 4}{x^2 - 3x - 4}$

Since $x^2 - 3x - 4 = 0$ when x = -1 and division by 0 is undefined, $\left(\dfrac{f}{g}\right)(-1)$ is not undefined.

45. $\left(\dfrac{f}{g}\right)(x) = \dfrac{3x^2 - 4}{x^2 - 3x - 4}$

$\qquad\quad = \dfrac{3x^2 - 4}{(x + 1)(x - 4)}$

The expression is not undefined if $(x + 1)(x - 4) = 0$, that is, if x = -1 or x = 4.

Thus, the domain is the set of all real numbers except x = -1 and x = 4, or

$$(-\infty, -1) \cup (-1, 4) \cup (4, \infty).$$

For Exercises 47–51, $f(x) = \sqrt{x - 2}$ and $g(x) = x^2$.

47. $(f \circ g)(x) = f[g(x)]$

$= f(x^2)$

$= \sqrt{x^2 - 2}$

49. $(f \circ g)(-6) = f[g(-6)]$

$= \sqrt{(-6)^2 - 2}$

$= \sqrt{34}$

51. $(g \circ f)(3) = g[f(3)]$

$= g(\sqrt{3 - 2})$

$= g(1)$

$= 1^2$

$= 1$

For Exercises 55–65, see the answer graphs in the textbook.

55. $f(x) = |x| - 3$

The graph is the same as that of $f(x) = |x|$ except translated 3 units downward.
The "vertex" is $(0, -3)$.

57. $f(x) = -|x + 1| + 3$

$= -|x - (-1)| + 3$

The graph opens downward because of the coefficient -1. It is the same as the graph of $f(x) = -|x|$ except translated 1 unit to the left and 3 units up.
The "vertex" is $(-1, 3)$.

59. $f(x) = [x - 3]$

To get $y = 0$, we need

$$0 \le x - 3 < 1$$
$$3 \le x < 4.$$

To get $y = 1$, we need

$$1 \le x - 3 < 2$$
$$4 \le x < 5.$$

Follow this pattern to graph the step function.

61. $f(x) = \begin{cases} -4x + 2 \text{ if } x \le 1 \\ 3x - 5 \text{ if } x > 1 \end{cases}$

Draw the graph of $y = -4x + 2$ to the left of $x = 1$, including the endpoint at $x = 1$. Draw the graph of $y = 3x - 5$ to the right of $x = 1$, but do not include the endpoint at $x = 1$. Observe that the endpoints of the two pieces coincide.

63. $f(x) = \begin{cases} |x| \text{ if } x < 3 \\ 6 - x \text{ if } x \ge 3 \end{cases}$

Draw the graph of $y = |x|$ to the left of $x = 3$, but do not include the endpoint. Draw the graph of $y = 6 - x$ to the right of $x = 3$, including the endpoint. Observe that the endpoints of the two pieces coincide.

65. This is a step function.
The cost is $45 + 1(2) = \$47$ for hauling up to and including one mile;

$$45 + 2(2) = \$49$$

for hauling more than one mile up to and including two miles;

$$45 + 3(2) = \$51$$

for hauling more than two miles up to and including three miles, and so on. The graph is made up of a series of horizontal line segments.
The number of miles can be any positive number, so the domain is $(0, \infty)$.
The range is the set of possible costs, $\{47, 49, 51, \ldots\}$.

67. Since there are horizontal lines that intersect the graph in two points, the function is not one-to-one.

69. $f(x) = \dfrac{8x - 9}{5}$

If $a \neq b$, then $\dfrac{8a - 9}{5} \neq \dfrac{8b - 9}{5}$.
This implies that $f(a) \neq f(b)$, so f is one-to-one.

71. $f(x) = \sqrt{5 - x}$

If $a \neq b$, then $\sqrt{5 - a} \neq \sqrt{5 - b}$.
This implies that $f(a) \neq f(b)$, so f is one-to-one.

73. $f(x) = -\sqrt{1 - \dfrac{x^2}{100}}$, domain $[0, 10]$

Since no two values of x in the domain lead to the same value of $f(x)$, f is one-to-one.

For Exercises 75–79, see the answer graphs in the textbook.

75. $f(x) = 12x + 3$

Since $f(x)$ is one-to-one, it has an inverse function.
To find $f^{-1}(x)$, let $y = f(x)$ and solve for x.

$$y - 3 = 12x$$
$$\frac{y - 3}{12} = x$$

Exchange x and y.

$$\frac{x - 3}{12} = y$$
$$f^{-1}(x) = \frac{x - 3}{12}$$

The graph of the original function is a line with slope 12 and y-intercept 3.
Since $f^{-1}(x) = \dfrac{x - 3}{12} = \dfrac{1}{12}x - \dfrac{1}{4}$, the graph of the inverse function is a line with slope $\dfrac{1}{12}$ and y-intercept $-\dfrac{1}{4}$.

77. $f(x) = x^3 - 3$

$f(x)$ is one-to-one, so it has an inverse function.
Let $y = f(x)$ and solve for x.

$$y + 3 = x^3$$
$$\sqrt[3]{y + 3} = x$$

Exchange x and y.

$$\sqrt[3]{x + 3} = y$$
$$f^{-1}(x) = \sqrt[3]{x + 3}$$

Plot points to graph these functions.

f(x)	-1	0	1
$f^{-1}(x)$	-4	-3	-2

f(x)	-4	-3	-2
$f^{-1}(x)$	-1	0	1

79. $f(x) = \sqrt{25 - x^2}$, domain [0, 5]

The restriction on the domain makes f(x) one-to-one.

Let y = f(x) and solve for x.

$$y = \sqrt{25 - x^2}$$
$$y^2 = 25 - x^2$$
$$x^2 = 25 - y^2$$
$$x = \sqrt{25 - y^2}$$

Exchange x and y.

$$y = \sqrt{25 - x^2}$$
$$f^{-1}(x) = \sqrt{25 - x^2}, \text{ domain } [0, 5]$$

Observe that $f(x) = f^{-1}(x)$, so this function is its own inverse. The largest value of $\sqrt{25 - x^2}$ is 5, so [0, 5] is both the domain and the range of this function.

$y = \sqrt{25 - x^2}$ may be manipulated into the form $x^2 + y^2 = 25$, whose graph would be the circle with center (0, 0) and radius 5. The restricted domain and range of f(x) cause its graph to be the upper right quarter of the above circle.

Chapter 4   Test

1. $f(x) = 3 + |x + 8|$

Since we can take the absolute value of any real number, the domain is $(-\infty, \infty)$. For all values of x, $|x| \geq 0$, so $|x + 8| \geq 0$ and $3 + |x + 8| \geq 3$. Thus, the range is $[3, \infty)$.

2. $f(x) = \sqrt{x^2 + 7x + 12}$

For $\sqrt{x^2 + 7x + 12}$ to be a real number, we must have

$$x^2 + 7x + 12 \geq 0.$$

Use factoring and a sign graph to solve this inequality.
Solve the corresponding quadratic equation.

$$x^2 + 7x + 12 = 0$$
$$(x + 4)(x + 3) = 0$$
$$x = -4 \quad \text{or} \quad x = -3$$

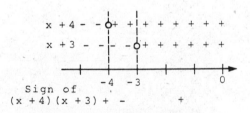

The sign graph shows that $x^2 + 7x + 12 \geq 0$ in the interval $(-\infty, -4]$ and also in the interval $[-3, \infty)$. Thus, the domain is

$$(-\infty, -4] \cup [-3, \infty).$$

The value of y can be any real number, so the range is $[0, \infty)$.

3. $f(x) = \dfrac{x - 1}{2x + 8}$

For $\dfrac{x - 1}{2x + 8}$ to be a real number, we must have $2x + 8 \neq 0$, or $x \neq -4$. The domain is the set of all real numbers except 4, or

$$(-\infty, -4) \cup (-4, \infty).$$

For Problems 4–9, $f(x) = 6 - 3x - x^2$ and $g(x) = 2x + 3$.

4. $f(-2) = 6 - 3(-2) - (-2)^2$

$= 6 + 6 - 4$

$= 8$

5. $(f + g)(2a)$

$= f(2a) + g(2a)$

$= [6 - 3(2a) - (2a)^2] + [2(2a + 3)]$

$= 6 - 6a - 4a^2 + 4a + 3$

$= 9 - 2a - 4a^2$

6. $\left(\dfrac{f}{g}\right)(1) = \dfrac{f(1)}{g(1)}$

$= \dfrac{6 - 3 \cdot 1 - 1^2}{2 \cdot 1 + 3}$

$= \dfrac{6 - 3 - 1}{2 + 3}$

$= \dfrac{2}{5}$

7. $g\left(\frac{1}{2}y - 4\right) = 2\left(\frac{1}{2}y - 4\right) + 3$

$= y - 8 + 3$

$= y - 5$

8. $g[f(2)] = g[6 - 3(2) - 2^2]$

$= g(-4)$

$= 2(-4) + 3$

$= -5$

9. $(g \circ f)(p)$

$= g[f(p)]$

$= g(6 - 3p - p^2)$

$= 2(6 - 3p - p^2) + 3$

$= 12 - 6p - 2p^2 + 3$

$= 15 - 6p - 2p^2$

For Problems 10–12, see the answer graphs in the textbook.

10. $f(x) = |x - 2| - 1$

To obtain the graph of $f(x) = |x - 2| - 1$ translate the graph of $f(x) = |x|$ two units to the right and one unit down.

11. $f(x) = [\![x + 1]\!]$

The graph is a step function. It may be obtained from a table of values or by translating the graph of $f(x) = [\![x]\!]$ one unit to the left.

12. $f(x) = \begin{cases} 3 & \text{if } x < -2 \\ 2 - \frac{1}{2}x & \text{if } x \geq -2 \end{cases}$

Graph the horizontal line $y = 3$ to the left of $x = -2$. Graph the line $y = 2 - \frac{1}{2}x = -\frac{1}{2}x + 2$ (slope $= -\frac{1}{2}$, y-intercept $= 2$) to the right of, and including $x = -2$.

13. There are horizontal lines that intersect the graph in more than one point, so the function is not one-to-one.

14. $f(x) = \sqrt{3x + 2}$

    If $a \neq b$, then

    $$3a \neq 3b$$
    $$3a + 2 \neq 3b + 2$$
    $$\sqrt{3a + 2} \neq \sqrt{3b + 2}$$
    $$f(a) \neq f(b).$$

    Thus, $a \neq b$ implies $f(a) \neq f(b)$, so f is one-to-one.

15. $f(x) = (x - 2)^2$, domain $(2, \infty)$

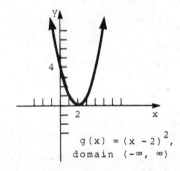

$g(x) = (x - 2)^2$,
domain $(-\infty, \infty)$

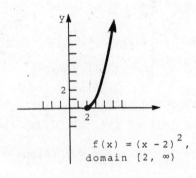

$f(x) = (x - 2)^2$,
domain $[2, \infty)$

These graphs show that if the domain were not restricted, the function would not be one-to-one. However, with the domain restricted to $[2, \infty)$, the graph is just the right half of the parabola, and each horizontal line intersects the graph in no more than one point. Thus, $f(x)$ is one-to-one.

16. $f(x) = 2x + 8$

    $f(x)$ is one-to-one, so it has an inverse function.

    Replace $f(x)$ with y and solve for x.

    $$y = 2x + 8$$
    $$y - 8 = 2x$$
    $$\frac{y - 8}{2} = x$$

    Exchange x and y.

    $$\frac{x - 8}{y} = y$$
    $$f^{-1}(x) = \frac{x - 8}{2}$$

17. $f(x) = 2 + x^2$, domain $(-\infty, 0]$

    The restriction of the domain to the interval $(-\infty, 0]$ makes $f(x)$ one-to-one, so it has an inverse function. To find $f^{-1}(x)$, let $y = f(x)$ and solve for x.

    $$y = 2 + x^2$$
    $$y - 2 = x^2$$
    $$\sqrt{y - 2} = x \quad \text{or} \quad -\sqrt{y - 2} = x$$

    The second choice is the correct one since the domain of $f(x)$ is $(-\infty, 0]$.

Exchange x and y.

$$-\sqrt{x-2} = y$$
$$f^{-1}(x) = -\sqrt{x-2}$$

The domain of $f^{-1}(x)$ is $[2, \infty)$.

18. $f(x) = \dfrac{x}{x+2}$, $g(x) = \dfrac{2x}{x-1}$

The range of $f(x)$ is

$$(-\infty, 0) \cup (0, \infty),$$

but the domain of $g(x)$ is

$$(-\infty, 1) \cup (1, \infty).$$

Since these are not the same, the functions are not inverses.

19. $y = k\sqrt{x}$

(a) To find k, substitute y = 10 and k = 25/9.

$$10 = \sqrt{\dfrac{25}{9}}$$

$$10 = k \cdot \dfrac{5}{3}$$

$$\dfrac{3}{5} \cdot 10 = k$$

$$6 = k$$

Thus, $y = 6\sqrt{x}$.

(b) If x = 144,

$$y = 6\sqrt{144}$$
$$= 6 \cdot 12$$
$$= 72.$$

20. Let F = the force of the wind;
A = the area of the sail;
v = the wind velocity.

$$F = kAv^2$$

To find k, substitute F = 8, v = 15, and A = 3.

$$8 = k(3)(15)^2$$
$$8 = 675k$$
$$\dfrac{8}{675} = k$$

Using this value of k,

$$F = \dfrac{8}{675} Av^2.$$

To find F when A = 6 and v = 22.5, substitute these values into the equation given above.

$$F = \dfrac{8}{675}(6)(22.5)^2$$
$$= \dfrac{8}{675}(6)(506.25)$$
$$= 36$$

The force is 36 lb.

CHAPTER 5 EXPONENTIAL AND LOGARITHMIC FUNCTIONS

Section 5.1

1. $4^x = 2$

    Write both sides as powers of 2.

    $$(2^2)^x = 2^1$$
    $$2^{2x} = 2^1$$
    $$2x = 1 \quad \textit{Property (b)}$$
    $$x = \frac{1}{2}$$

    Solution set: $\left\{\frac{1}{2}\right\}$

3. $\left(\frac{1}{2}\right)^k = 4$

    $$(2^{-1})^k = 2^2 \quad 1/2 = 2^{-1}$$
    $$2^{-k} = 2^2$$
    $$-k = 2 \quad \textit{Property (b)}$$
    $$k = -2$$

    Solution set: $\{-2\}$

5. $2^{3-y} = 8$

    $$2^{3-y} = 2^3$$
    $$3 - y = 3 \quad \textit{Property (b)}$$
    $$-y = 0$$
    $$y = 0$$

    Solution set: $\{0\}$

7. $\frac{1}{27} = b^{-3}$

    $$\left(\frac{1}{27}\right)^{-1/3} = (b^{-3})^{-1/3}$$
    $$\frac{1}{\left(\frac{1}{27}\right)^{1/3}} = b^1$$
    $$\frac{1}{\sqrt[3]{\frac{1}{27}}} = b^1$$
    $$\frac{1}{\frac{1}{3}} = b$$
    $$3 = b$$

    Solution set: $\{3\}$

9. $4 = r^{2/3}$

    Raise both sides of the equation to the 3/2 power, since $(r^{2/3})^{3/2} = r^1 = r$.

    $$4^{3/2} = (r^{2/3})^{3/2}$$
    $$(\pm\sqrt{4})^3 = r^1$$
    $$(\pm 2)^3 = r$$
    $$\pm 8 = r$$

    Since raising both sides of an equation to the same power may result in false "solutions," it is necessary to check all proposed solutions in the original equation.

$4 = 8^{2/3}$	$4 = (-8)^{2/3}$
$= (\sqrt[3]{8})^2$	$= (\sqrt[3]{-8})^2$
$= 2^2$	$= (-2)^2$
$= 4$	$= 4$

    Both proposed solutions check.
    Solution set: $\{-8, 8\}$

11. $27^{4z} = 9^{z+1}$

$(3^3)^{4z} = (3^2)^{z+1}$

$3^{12z} = 3^{2z+2}$

$12z = 2z + 2$

$10z = 2$

$z = \dfrac{1}{5}$

Solution set: $\left\{\dfrac{1}{5}\right\}$

13. $\left(\dfrac{1}{8}\right)^{-2p} = 2^{p+3}$

$(2^{-3})^{-2p} = 2^{p+3}$ $1/8 = 2^{-3}$

$2^{6p} = 2^{p+3}$

$6p = p + 3$

$5p = 3$

$p = \dfrac{3}{5}$

Solution set: $\left\{\dfrac{3}{5}\right\}$

15. $\left(\dfrac{1}{2}\right)^{-x} = \left(\dfrac{1}{4}\right)^{x+1}$

$\left(\dfrac{1}{2}\right)^{-x} = \left[\left(\dfrac{1}{2}\right)^2\right]^{x+1}$

$\left(\dfrac{1}{2}\right)^{-x} = \left(\dfrac{1}{2}\right)^{2x+2}$

$-x = 2x + 2$

$-3x = 2$

$x = -\dfrac{2}{3}$

Solution set: $\left\{-\dfrac{2}{3}\right\}$

For Exercises 17–27, see the answer graphs in the textbook.

17.   (a) $f(x) = 2^x + 1$

This graph is obtained by translating the graph of $f(x) = 2^x$ up 1 unit.

(b) $f(x) = 2^x - 4$

This graph is obtained by translating the graph of $f(x) = 2^x$ down 4 units.

(c) $f(x) = 2^{x+1}$

This graph is obtained by translating the graph of $f(x) = 2^x$ to the left 1 unit.

(d) $f(x) = 2^{x-4}$

This graph is obtained by translating the graph of $f(x) = 2^x$ to the right 4 units.

Sketch the graphs.

21. $f(x) = 4^x$

Make a table of values.

x	-2	-1	0	1	2
y	$\dfrac{1}{16}$	$\dfrac{1}{4}$	1	4	16

Plot these points and draw a smooth curve through them. This is an increasing function. The domain is $(-\infty, \infty)$ and the range is $(0, \infty)$. The x-axis is a horizontal asymptote.

23. $f(x) = e^{-x}$

Use a calculator to find approximate values of y.

x	-2	-1	0	1	2
y	7.39	2.72	1	.368	.135

The graph of $f(x) = e^{-x}$ is the reflection of the graph of $f(x) = e^x$ about the y-axis.

25. $f(x) = e^{x+1}$

Use a calculator to make a table of values.

x	-3	-2	-1	0	1
y	.135	.368	1	2.72	7.39

The graph of $f(x) = e^{x+1}$ can also be drawn by translating the graph of $f(x) = e^x$ to the left one unit.

27. $f(x) = 2^{-|x|}$

Make a table of values.

x	-2	-1	0	1	2
y	$\frac{1}{4}$	$\frac{1}{2}$	1	$\frac{1}{2}$	$\frac{1}{4}$

29. For $a > 1$, the value of $f(x) = a^x$ increases; for $0 < a < 1$, the value of $f(x) = a^x$ decreases.

31. $f(x) = a^x$

$f(0) = a^0 = 1$, so the point $(0, 1)$ will be on the graph.
$f(1) = a^1 = a$, so the point $(1, a)$ will be on the graph.

For Exercises 33–37, see the answer graphs in the textbook.

33. $f(x) = \dfrac{e^x + e^{-x}}{2}$

Use a calculator to make a table of values. All values of y except $y = 1$ are approximate.

x	-2	-1	0	1	2
e^x	.135	.368	1	2.718	7.389
e^{-x}	7.389	2.718	1	.368	.135
y	3.762	1.543	1	1.543	3.762

Sketch the graph.

35. $f(x) = x^2 \cdot 2^{-x}$

x	-2	-1	0	1	2	3	4
x^2	4	1	0	2	4	9	16
2^{-x}	16	2	1	$\frac{1}{2}$	$\frac{1}{4}$	$\frac{1}{8}$	$\frac{1}{16}$
y	64	2	0	$\frac{1}{2}$	1	$\frac{9}{8}$	1

37. $f(x) = xe^x$

Use a calculator to make a table of values. All values of y except $y = 0$ are approximate.

x	-4	-3	-2	-1	0	1	2
e^x	.018	.05	.135	.368	1	2.72	7.39
y	-.073	-.15	-.27	-.368	0	2.72	14.78

39. Use the compound interest formula to find the future amount A if the present amount P = 8906.54, r = .05, m = 2, and t = 9.

$$A = P\left(1 + \frac{r}{m}\right)^{tm}$$

$$= (8906.54)\left(1 + \frac{.05}{2}\right)^{9(2)}$$

$$= (8906.54)(1.025)^{18}$$

$$= (8906.54)(1.5596)$$

$$= 13,891.16$$

The future value is $13,891.16.

41. Use the compound interest formula to find A if P = 45,788, t = 11, m = 365, and r = .06.

$$A = P\left(1 + \frac{r}{m}\right)^{tm}$$

$$= 45,788\left(1 + \frac{.06}{365}\right)^{11(365)}$$

$$= (45,788)(1.0002)^{4015}$$

$$= (45,788)(1.9347)$$

$$= 88,585.47$$

The future value is $88,585.47.

43. Use the compound interest formula to find P if A = 25,000, t = $\frac{11}{4}$, m = 4, and r = .06.

$$A = P\left(1 + \frac{r}{m}\right)^{tm}$$

$$25,000 = P\left(1 + \frac{.06}{4}\right)^{(11/4)(4)}$$

$$25,000 = P(1.015)^{11}$$

$$25,000 = P(1.177948)$$

$$21,223.34 = P$$

The present value is $21,223.34.

45. Use the compound interest formula to find P if A = 123,788, t = $\frac{195}{365}$, m = 365, and r = .087.

$$A = P\left(1 + \frac{r}{m}\right)^{tm}$$

$$123,788 = P\left(1 + \frac{.087}{365}\right)^{(195/365)(365)}$$

$$123,788 = P(1.0002384)^{195}$$

$$118,166.72 = P$$

The present value is $118,166.72.

47. Use the compound interest formula to find r if A = 65,325, P = 65,000, t = $\frac{1}{2}$, and m = 12.

$$65,325 = 65,000\left(1 + \frac{r}{12}\right)^{(1/2)(12)}$$

$$65,325 = 65,000\left(1 + \frac{r}{12}\right)^{6}$$

$$1.005 = \left(1 + \frac{r}{12}\right)^{6}$$

Next, take the sixth root of both sides (or raise both sides to the 1/6 power) using the exponential key on a calculator.

$$(1.005)^{1/6} = 1 + \frac{r}{12}$$

$$1.000832 = 1 + \frac{r}{12}$$

$$.000832 = \frac{r}{12}$$

$$.00998 = r$$

$$.010 \approx r$$

The interest rate, to the nearest tenth, is 1.0%.

49. Use the compound interest formula to find r if A = 19,000, P = 15,000, t = 2, and m = 2.

$$19,000 = 15,000\left(1 + \frac{r}{2}\right)^{2(2)}$$

$$\frac{19}{15} = \left(1 + \frac{r}{2}\right)^{4}$$

$$1.2667 = \left(1 + \frac{r}{2}\right)^{4}$$

Next, take the fourth root of both sides by using the exponential key on a calculator.

$$(1.2667)^{1/4} = 1 + \frac{r}{2}$$

$$1.0609 = 1 + \frac{r}{2}$$

$$.0609 = \frac{r}{2}$$

$$2.1218 = r$$

The interest rate, to the nearest tenth, is 12.2%.

51. To find annual yields, use the compound interet formula

$$A = P\left(1 + \frac{r}{m}\right)^{tm} \text{ with}$$

$P = 1$ and $t = 1$.

(a) 7.00% compound monthly

Use $r = .07$ and $m = 12$.

$$A = P\left(1 + \frac{.07}{12}\right)^{12}$$

$$\approx 1.07229$$

Subtract 1 to get the annual yield, .07229 or 7.229%.

(b) 7.00% compounded daily

Use $r = .07$ and $m = 365$.

$$A = P\left(1 + \frac{.07}{365}\right)^{365}$$

$$\approx 1.07250$$

Subtract 1 to get the annual yield, .07250 or 7.250%.

53. $A(t) = 500e^{-.032t}$

(a) $A(4) = 500e^{-.032(4)}$

$$= 500e^{-.128}$$

$$\approx 500(.8799)$$

$$\approx 440$$

After 4 yr, about 440 g will remain.

(b) $A(8) = 500e^{-.032(8)}$

$$= 500e^{-.256}$$

$$\approx 500(.7741)$$

$$\approx 387$$

After 8 yr, about 387 g will remain.

(c) $A(20) = 500e^{-.032(20)}$

$$= 500e^{-.64}$$

$$\approx 500(.5273)$$

$$\approx 264$$

After 20 yr, about 264 g will remain.

(d) The domain is $[0, \infty)$ and the graph passes through $(4, 440)$, $(8, 387)$, and $(20, 264)$. See the answer graph in the textbook.

55. $S(t) = S_0 e^{-at}$

(a) $a = .10$ and $S_0 = 50{,}000$

Find $S(1)$ and $S(3)$.

$$S(1) = 50{,}000e^{-.10(1)}$$

$$= 50{,}000^{-.10}$$

$$= 50{,}000(.90483)$$

$$= 45{,}241.9 \approx 45{,}200$$

$$S(3) = 50{,}000^{-.10(3)}$$

$$= 50{,}000e^{-.30}$$

$$= 50{,}000(.74081)$$

$$= 37{,}049.9 \approx 37{,}000$$

(b) $S_0 = 80{,}000$ and $a = .05$

Find $S(2)$ and $S(10)$.

$$S(2) = 80{,}000e^{-.05(2)}$$

$$= 80{,}000e^{-.1}$$

$$= 80{,}000(.90483)$$

$$= 72{,}386.4 \approx 72{,}400$$

$$S(10) = 80,000e^{-.05(10)}$$
$$= 80,000^{-.5}$$
$$= 80,000(.60653)$$
$$= 48,522.48 \approx 48,500$$

57. $e^x = 1 + x + \dfrac{x^2}{2 \cdot 1} + \dfrac{x^3}{3 \cdot 2 \cdot 1}$

$$+ \dfrac{x^4}{4 \cdot 3 \cdot 2 \cdot 1}$$

$$+ \dfrac{x^5}{5 \cdot 4 \cdot 3 \cdot 2 \cdot 1} + \cdots$$

$e^1 = 1 + 1 + \dfrac{1}{2} + \dfrac{1}{6} + \dfrac{1}{24} + \dfrac{1}{120} + \cdots$

$$\approx 2 + \dfrac{360}{720} + \dfrac{120}{720} + \dfrac{30}{720} + \dfrac{6}{720}$$

$$= 2\dfrac{516}{720}$$

$$\approx 2.717$$

This is very close to the value of e^1 (2.718, to the nearest thousandth) found with a calculator.

Section 5.2

1. $3^4 = 81$ is equivalent to
$\log_3 81 = 4$.

3. $\left(\dfrac{1}{2}\right)^{-4} = 16$ is equivalent to
$\log_{1/2} 16 = -4$.

5. $10^{-4} = .0001$ is equivalent to
$\log_{10} .0001 = -4$.

7. $\log_6 36 = 2$ is equivalent to
$6^2 = 36$.

9. $\log_{\sqrt{3}} 81 = 8$ is equivalent to
$(\sqrt{3})^8 = 81$.

11. $\log_{10} .0001 = -4$ is equivalent to
$10^{-4} = .0001$.

13. Let $y = \log_5 25$.
Write the equation in exponential form.
$$5^y = 25$$
$$5^y = 5^2$$
$$y = 2$$

Thus, $\log_5 25 = 2$.

15. Let $y = \log_8 8$.
$$8^y = 8^1$$
$$y = 1$$

Thus, $\log_8 8 = 1$.

17. Let $y = \log_{10} .001$.
$$10^y = .001$$
$$10^y = (10)^{-3}$$
$$y = -3$$

Thus, $\log_{10} .001 = -3$.

19. Let $y = \log_4 \dfrac{\sqrt[3]{4}}{2}$.

$$4^y = \dfrac{\sqrt[3]{4}}{2}$$

$$(2^2)^y = \dfrac{\sqrt[3]{2^2}}{2}$$

$$2^{2y} = \dfrac{2^{2/3}}{2}$$

$$2^{2y} = 2^{(2/3)-1}$$

$$2^{2y} = 2^{-1/3}$$

$$2y = -\dfrac{1}{3}$$

$$y = -\dfrac{1}{6}$$

Thus, $\log_4 \dfrac{\sqrt[3]{4}}{2} = -\dfrac{1}{6}$.

21. Let $x = \log_{1/3} \dfrac{9^{-4}}{3}$.

$$\left(\tfrac{1}{3}\right)^x = \dfrac{9^{-4}}{3}$$

Write both sides of the equation as powers of 3.

$$(3^{-1})^x = \dfrac{(3^2)^{-4}}{3}$$

$$3^{-x} = \dfrac{3^{-8}}{3^1}$$

$$3^{-x} = 3^{-9}$$

$$-x = -9$$

$$x = 9$$

Thus, $\log_{1/3} \dfrac{9^{-4}}{3} = 9$.

23. $2^{\log_2 9} = 9$

This is true by the theorem on inverses.

25. $x = \log_2 32$

Write the equation in exponential form.

$$2^x = 32$$

$$2^x = 2^5$$

$$x = 5$$

Solution set: $\{5\}$

27. $\log_x 25 = 2$

$$x^2 = 25$$

$$(x^{-2})^{-1/2} = (25)^{-1/2}$$

$$x = \dfrac{1}{25^{1/2}}$$

$$x = \dfrac{1}{5}$$

Solution set: $\left\{\dfrac{1}{5}\right\}$

For Exercises 31–39, see the answer graphs in the textbook.

31. (a) $f(x) = (\log_2 x) + 3$

This graph is obtained by translating the graph of $f(x) = \log_2 x$ up 3 units.

(b) $f(x) = \log_2 (x + 3)$

This graph is obtained by translating the graph of $f(x) = \log_2 x$ to the left 3 units. The graph has a vertical asymptote at $x = -3$.

(c) $f(x) = \left|\log_2 (x + 3)\right|$

First, translate the graph of $f(x) = \log_2 x$ to the left 3 units to obtain the graph of $\log_2 (x + 3)$. (See part (b).) For the portion of the graph where $f(x) \geq 0$, that is, where $x \geq -2$, use the same graph as in (b). For the portion of the graph in (b) where $f(x) < 0$, that is, where $x < -2$, reflect the graph about the x–axis. In this way, each negative value of $f(x)$ on the graph in (b) is replaced by its opposite, which is positive. The graph has a vertical asymptote at $x = -3$.

33. $f(x) = \log_3 x$

Write $y = \log_3 x$ in exponential form as $x = 3^y$ to find ordered pairs that satisfy the equation. It is easier to choose values for y and find the corresponding values of x.

Make a table of values.

x	$\frac{1}{9}$	$\frac{1}{3}$	1	3	9
y	-2	-1	0	1	2

The graph can also be found by reflecting the graph of $f(x) = 3^x$ about the line $y = x$. The graph has the y-axis as a vertical asymptote. Sketch the graph.

35. $f(x) = \log_{1/2} (1 - x)$

Make a table of values.

x	-7	-3	-1	0	$\frac{1}{2}$	$\frac{3}{4}$	$\frac{7}{8}$
1 - x	8	4	2	1	$\frac{1}{2}$	$\frac{1}{4}$	$\frac{1}{8}$
y	-3	-2	-1	0	1	2	3

The graph has a vertical asymptote at $x = 1$. Sketch the graph.

37. $f(x) = \log_2 x^2$

Make a table of values.

x	$-\sqrt{8} \approx$ -2.8	-2	-1	$-\frac{1}{2}$	$\frac{1}{2}$	1	2	$\sqrt{8} \approx$ 2.8
x^2	8	4	1	$\frac{1}{4}$	$\frac{1}{4}$	1	4	8
y	3	2	0	-2	-2	0	2	3

The graph has the y-axis as a vertical asymptote. Since $f(-x) = f(x)$, the function is even and the graph is symmetric with respect to the y-axis. Sketch the graph.

39. $f(x) = x \log_{10} x$

Make a table of values.

x	.1	1	2	3	5	10
$\log_{10} x$	-1	0	.3010	.4771	.6990	1
y	-.1	0	.6020	.9542	1.398	10

Use a calculator to find decimal approximations for $\log_{10} x$, such as those shown for $x = 2$, $x = 3$, and $x = 5$ on this table. Since $\log_{10} x$ is defined for all $x > 0$, but not at $x = 0$, the graph has an open circle to show that $x = 0$ is not included. The domain of this function is $(0, \infty)$. Sketch the graph.

41. $\log_3 \frac{2}{5} = \log_3 2 - \log_3 5$
Logarithm of a quotient

43. $\log_2 \frac{6x}{y}$

$= \log_2 6x - \log_2 y$
Logarithm of a quotient

$= \log_2 6 + \log_2 x - \log_2 y$
Logarithm of a product

45. $\log_5 \frac{5\sqrt{7}}{3}$

$= \log_5 (5\sqrt{7}) - \log_5 3$
Logarithm of a quotient

$= \log_5 5 + \log_5 \sqrt{7} - \log_5 3$
Logarithm of a product

$= 1 + \log_5 \sqrt{7} - \log_5 3$ $\log_b b = 1$

$= 1 + \log_5 7^{1/2} - \log_5 3$ $\sqrt{7} = 7^{1/2}$

$= 1 + \frac{1}{2} \log_5 7 - \log_5 3$
Logarithm of a power

47. $\log_4 (2x + 5y)$

Since this is a sum, none of the logarithm properties apply, so no change is possible.

49. $\log_k \dfrac{pq^2}{m}$

$= \log_k pq^2 - \log_k m$
Logarithm of a quotient

$= \log_k p + \log_k q^2 - \log_k m$
Logarithm of a product

$= \log_k p + 2 \log_k q - \log_k m$
Logarithm of a power

51. $\log_m \sqrt{\dfrac{5r^3}{z^5}}$

$= \log_m \left(\dfrac{5r^3}{z^2}\right)^{1/2}$

$= \dfrac{1}{2} \log_m \dfrac{5r^3}{z^5}$
Logarithm of a power

$= \dfrac{1}{2}(\log_m 5r^3 - \log_m z^5)$
Logarithm of a quotient

$= \dfrac{1}{2}(\log_m 5 + \log_m r^3 - \log_m z^5)$
Logarithm of a product

$= \dfrac{1}{2}(\log_m 5 + 3 \log_m r - 5 \log_m z)$
Logarithm of a power

53. $\log_a x + \log_a y - \log_a m$

$= \log_a xy - \log_a m$
Logarithm of a product

$= \log_a \dfrac{xy}{m}$ *Logarithm of a quotient*

55. $2 \log_m a - 3 \log_m b^2$

$= \log_m a^2 - \log_m (b^2)^3$
Logarithm of a power

$= \log_m a^2 - \log_m b^6$

$= \log_m \dfrac{a^2}{b^6}$ *Logarithm of a quotient*

57. $2 \log_a (z - 1) + \log_a (3z + 2)$

$= \log_a (z - 1)^2 + \log_a (3z + 2)$
Logarithm of a power

$= \log_a [(z - 1)^2(3z + 2)]$
Logarithm of a product

59. $-\dfrac{2}{3} \log_5 5m^2 + \dfrac{1}{2} \log_5 25m^2$

$= \log_5 (5m^2)^{-2/3} + \log_5 (25m^2)^{1/2}$
Logarithm of a power

$= \log_5 \dfrac{1}{(5m^2)^{2/3}} + \log_5 5m$

$= \log_5 \dfrac{5m}{(5m^2)^{2/3}}$
Logarithm of a product

$= \log_5 \dfrac{5m}{5^{2/3} m^{4/3}}$

$= \log_5 5^{1-2/3} m^{1-4/3}$

$= \log_5 5^{1/3} m^{-1/3}$

$= \log_5 \dfrac{5^{1/3}}{m^{1/3}}$

63. $\log_{10} 12 = \log_{10} (3 \cdot 2^2)$

$= \log_{10} 3 + 2 \cdot \log_{10} 2$

$= .4771 + 2(.3010)$

$= 1.0791$

65. $\log_{10} \dfrac{20}{27}$

$= \log_{10} 20 - \log_{10} 27$

$= \log_{10} 2 \cdot 10 - \log_{10} 3^3$

$= \log_{10} 2 + \log_{10} 10 - 3 \log_{10} 3$

$\approx .3010 + 1 - 3(.4771)$

$= -.1303$

67. $\log_{10} (36)^{1/3}$

$= \dfrac{1}{3} \log_{10} 36$

$= \dfrac{1}{3} \log_{10} (4 \cdot 9)$

$= \dfrac{1}{3}(\log_{10} 4 + \log_{10} 9)$

$= \dfrac{1}{3}(\log_{10} 2^2 + \log_{10} 3^2)$

$= \dfrac{1}{3}(2 \log_{10} 2 + 2 \log_{10} 3)$

$\approx \dfrac{1}{3}[2(.3010) + 2(.4771)]$

$$= \frac{1}{3}(.6020 + .9542)$$

$$= \frac{1}{3}(1.5562)$$

$$\approx .5187$$

69. Prove that $\log_a x^r = r \log_a x$.

Since $x = a^{\log_a x}$,

$\log_a x^r = \log_a (a^{\log_a x})^r$.

Multiply exponents to obtain

$\log_a x^r = \log_a a^{r \log_a x}$.

Use Property (c) of logarithms to obtain

$$\log_a x^r = (r \log_a x) \log_a a.$$

Recall that $\log_a a = 1$ and the desired result follows:

$$\log_a x^r = r \log_a x.$$

Section 5.3

1. $\log 43 = 1.6335$

To find this value, enter 43 and press the log key.

3. $\log 783 = 2.8938$

5. $\log .0069 = -2.1612$

7. $\ln 580 = 6.3630$

To find this value, enter 580 and press the ln key.

9. $\ln .7 = -.3567$

11. $\ln 121,000 = 11.7035$

13. To find $\log_5 10$, use the change of base rule with $a = 5$, $b = e$, and $x = 10$.

$$\log_a x = \frac{\log_b x}{\log_b a}$$

$$\log_5 10 = \frac{\ln 10}{\ln 5}$$

$$\approx \frac{2.3026}{1.6094}$$

$$\approx 1.43$$

This logarithm can also be found by using the change of base rule with $a = 5$, $b = 10$, and $x = 10$.

$$\log_5 10 = \frac{\log_{10} 10}{\log_{10} 5}$$

$$\approx \frac{1}{.6990}$$

$$\approx 1.43$$

15. $\log_{15} 5 = \frac{\ln 5}{\ln 15}$

$$\approx \frac{1.6094}{2.7081}$$

$$\approx .59$$

17. $\log_{100} 83 = \frac{\ln 83}{\ln 100}$

$$\approx \frac{4.4188}{4.6052}$$

$$\approx .96$$

19. $\log_{2.9} 7.5 = \frac{\ln 7.5}{\ln 2.9}$

$$= \frac{2.0149}{1.0647}$$

$$\approx 1.89$$

21. $\log_3 4$ is the logarithm to the base 3 of 4.

($\log_4 3$ would be the logarithm to the base 4 of 3.)

23. Grapefruit, 6.3×10^{-4}

$$\begin{aligned}
pH &= -\log [H_3O^+] \\
&= -\log (6.3 \times 10^{-4}) \\
&= -(\log 6.3 + \log 10^{-4}) \\
&= -(.7793 - 4) \\
&= -.7993 + 4 \\
pH &= 3.2
\end{aligned}$$

The answer is rounded to the nearest tenth because it is customary to round pH values to the nearest tenth.

The pH of grapefruit is 3.2.

25. Limes, 1.6×10^{-2}

$$\begin{aligned}
pH &= -\log [H_3O^+] \\
&= -\log (1.6 \times 10^{-2}) \\
&= -(\log 1.6 + \log 10^{-2}) \\
&= -(.2041 - 2) \\
&= -(-1.7959) \\
pH &= 1.8
\end{aligned}$$

The pH of limes is 1.8.

27. Soda pop, 2.7

$$\begin{aligned}
pH &= -\log [H_3O^+] \\
2.7 &= -\log [H_3O^+] \\
-2.7 &= \log [H_3O^+] \\
[H_3O^+] &= 2.0 \times 10^{-3}
\end{aligned}$$

29. Beer, 4.8

$$\begin{aligned}
pH &= -\log [H_3O^+] \\
4.8 &= -\log [H_3O^+] \\
-4.8 &= \log [H_3O^+] \\
[H_3O^+] &= 10^{-4.8} \\
[H_3O^+] &= 1.6 \times 10^{-5}
\end{aligned}$$

31. Let r = the decibel rating of a sound.

$$r = 10 \cdot \log_{10} \frac{I}{I_0}$$

(a) $$\begin{aligned}
r &= 10 \cdot \log_{10} \frac{100 \cdot I_0}{I_0} \\
&= 10 \cdot \log_{10} 100 \\
&= 10 \cdot 2 \\
&= 20
\end{aligned}$$

(b) $$\begin{aligned}
r &= 10 \cdot \log_{10} \frac{1000 \cdot I_0}{I_0} \\
&= 10 \cdot \log_{10} 1000 \\
&= 10 \cdot 3 \\
&= 30
\end{aligned}$$

$$\begin{aligned}
r &= 10 \cdot \log_{10} \frac{100,000 \cdot I_0}{I_0} \\
&= 10 \cdot 5 \\
&= 50
\end{aligned}$$

(d) $$\begin{aligned}
r &= 10 \cdot \log_{10} \frac{1,000,000 \cdot I_0}{I_0} \\
&= 10 \cdot 6 \\
&= 60
\end{aligned}$$

33. Let r = the Richter scale rating of an earthquake.

$$r = \log_{10} \left(\frac{I}{I_0} \right)$$

(a) $$\begin{aligned}
r &= \log_{10} \frac{1000 \cdot I}{I_0} \\
&= \log_{10} 1000 \\
&= 3
\end{aligned}$$

(b) $$\begin{aligned}
r &= \log_{10} \frac{1,000,000 \cdot I_0}{I_0} \\
&= \log_{10} 1,000,000 \\
&= 6
\end{aligned}$$

(c) $$\begin{aligned}
r &= \log_{10} \frac{100,000,000 \cdot I_0}{I_0} \\
&= \log_{10} 100,000,000 \\
&= 8
\end{aligned}$$

35. (a) $8.3 = \log_{10} \dfrac{I}{I_0}$

$\dfrac{I}{I_0} = 10^{8.3}$

$I = 10^{8.3} \cdot I_0$

$I = 200,000,000 I_0$

(b) $7.1 = \log_{10} \dfrac{I}{I_0}$

$\dfrac{I}{I_0} = 10^{7.1}$

$I = 10^{7.1} \cdot I_0$

$I = 13,000,000 I_0$

(c) $\dfrac{200,000,000 \cdot I_0}{13,000,000 \cdot I_0} \approx 15.38$

The earthquake of 1906 was more than 15 times as intense as the earthquake of 1989.

37. $S_n = a \ln \left(1 + \dfrac{n}{a}\right)$

$= .36 \ln \left(1 + \dfrac{n}{.36}\right)$

(a) $S(100) = .36 \ln \left(1 + \dfrac{100}{.36}\right)$

$= .36 \ln 278.77$

$= (.36)(5.6304)$

$= 2.03$

(b) $S(200) = .36 \ln \left(1 + \dfrac{200}{.36}\right)$

$= .36 \ln (556.555)$

$= .36(6.322)$

$= 2.28$

(c) $S(150) = .36 \ln \left(1 + \dfrac{150}{.36}\right)$

$= .36 \ln 417.666$

$= .36(6.0347)$

$= 2.17$

(d) $S(10) = .36 \ln \left(1 + \dfrac{10}{.36}\right)$

$= .36 \ln 28.777$

$= .36(3.3596)$

$= 1.21$

39. $p = 86.3 \ln h - 680$

(a) For an altitude of 3000 ft, $h = 3000$.

$p = 86.3 \ln 3000 - 680$

$= 86.3(8.0064) - 680$

$= 691 - 680$

$= 11$

At an altitude of 3000 ft, approximately 11% of the moisture falls as snow.

(b) For an altitude of 4000 ft, $h = 4000$.

$p = 86.3 \ln 4000 - 680$

$= 86.3(8.294) - 680$

$= 716 - 680$

≈ 36

At 4000 ft, approximately 36% of the moisture falls as snow.

(c) For an altitude of 7000 ft, $h = 7000$.

$p = 86.3 \ln 7000 - 680$

$= 86.3(8.8537) - 680$

$= 764 - 680$

≈ 84

At 7000 ft, approximately 84% of the moisture falls as snow.

(d) See the answer graph in the textbook.

41. $H = [P_1 \log_2 P_1 + P_2 \log_2 P_2$
$+ P_3 \log_2 P_3 + P_4 \log_2 P_4]$
$= -[.521 \log_2 .521 + .324 \log_2 .324$
$+ .081 \log_2 .081 + .074 \log_2 .074]$

Use the change of base rule to find the required base 2 logarithms.

$$\log_2 .521 = \frac{\ln .521}{\ln 2}$$
$$\approx -.941$$

$$\log_2 .324 = \frac{\ln .324}{\ln 2}$$
$$\approx -1.63$$

$$\log_2 .081 = \frac{\ln .081}{\ln 2}$$
$$\approx -3.63$$

$$\log_2 .074 = \frac{\ln .074}{\ln 2}$$
$$\approx -3.76$$

Therefore,

$H = -[(.521)(-.941) + (.324)(-1.63)$
$+ (.081)(-3.63) + (.074)(-3.76)]$
$= -[(-.490) + (-.528) + (-.294)$
$+ (-2.78)]$
$= -(-1.59)$
$= 1.59.$

43. $f(x) = x \ln x$

Make a table of values, using a calculator to get values for $\ln x$. Notice that since the domain of $f(x) = \ln x$ is $(0, \infty)$, the domain of $f(x) = x \ln x$ is also $(0, \infty)$. The open circle on the graph shows that $x = 0$ is not included.

x	.1	.5	1	2	3	5
ln x	-2.3	-.69	0	.69	1.1	1.6
y	-.23	-.35	0	1.4	3.3	8.0

See the answer graph in the textbook.

45. $g(x) = e^x$

(a) By the theorem on inverses,
$$g(\ln 3) = e^{\ln 3}$$
$$= 3.$$

(b) By the theorem on inverses,
$$g(\ln 5^2) = e^{\ln 5^2}$$
$$= 5^2 = 25$$

(c) By the theorem on inverses,
$$g\left(\ln \frac{1}{e}\right) = e^{\ln (1/e)}$$
$$= \frac{1}{e}$$

47. $f(x) = \ln x$

(a) By the theorem on inverses,
$$f(e^5) = \ln e^5$$
$$= 5.$$

(b) By the theorem on inverses,
$$f(e^{\ln 3}) = \ln e^{\ln 3}$$
$$= \ln 3.$$

(c) By the theorem on inverses,
$$f(e^{2 \ln 3}) = \ln e^{2 \ln 3}$$
$$= 2 \ln 3$$
$$\text{or} \ \ln 9.$$

Section 5.4

1. $3^x = 6$

Take base e (natural) logarithms of both sides.

$\ln 3^x = \ln 6$

$x \ln 3 = \ln 6$ *Logarithm of a power*

$x = \dfrac{\ln 6}{\ln 3}$ *Divide by ln 3*

$= \dfrac{1.7918}{1.0986}$

$= 1.631$

Solution set: $\{1.631\}$

3. $3^{a+2} = 5$

Take natural logarithms of both sides.

$\ln 3^{a+2} = \ln 5$

$(a + 2) \ln 3 = \ln 5$

Logarithm of a power

$a + 2 = \dfrac{\ln 5}{\ln 3}$

Divide by ln 3

$a = \dfrac{\ln 5}{\ln 3} - 2$

Subtract 2

$= \dfrac{1.6094}{1.0986} - 2$

$= 1.465 - 2$

$= -.535$

Solution set: $\{-.535\}$

5. $6^{1-2k} = 8$

Take base 10 (common) logarithms of both sides. (This exercise can also be done using natural logarithms.)

$\log 6^{1-2k} = \log 8$

$(1 - 2k) \log 6 = \log 8$

$1 - 2k = \dfrac{\log 8}{\log 6}$

$2k = 1 - \dfrac{\log 8}{\log 6}$

$k = \dfrac{1}{2}\left(1 - \dfrac{\log 8}{\log 6}\right)$

$= \dfrac{1}{2}\left(1 - \dfrac{.9031}{.7782}\right)$

$= \dfrac{1}{2}(1 - 1.1606)$

$= \dfrac{1}{2}(-.1606)$

$= -.080$

Solution set: $\{-.080\}$

7. $e^{k-1} = 4$

$\ln e^{k-1} = \ln 4$

$k - 1 = \ln 4$ *Theorem on inverses*

$k = \ln 4 + 1$

$= 1.3863 + 1$

$= 2.386$

Solution set: $\{2.386\}$

9. $2e^{5a+2} = 8$

$e^{5a+2} = 4$

$\ln e^{5a+2} = \ln 4$

$5a + 2 = \ln 4$ *Theorem on inverses*

$5a = \ln 4 - 2$

$a = \dfrac{1}{5}(\ln 4 - 2)$

$= \dfrac{1}{5}(1.3863 - 2)$

$= \dfrac{1}{5}(-.6137)$

$= -.123$

Solution set: $\{-.123\}$

11. $2^x = -3$ has no solution since 2 raised to any power is positive.

Solution set: $\emptyset$

13. $e^{2x} \cdot e^{5x} = e^{14}$

$\qquad e^{7x} = e^{14}$ *Product rule for exponents*

$\qquad 7x = 14$ *Property 1*

$\qquad x = 2$

Solution set: $\{2\}$

15. $e^{-\ln x} = 2$

$e^{\ln x^{-1}} = 2$ *Logarithm of a power*

$\qquad x^{-1} = 2$ *Theorem on inverses*

$\qquad \dfrac{1}{x} = 2$

$\qquad x = \dfrac{1}{2} = .5$

Solution set: $\{.5\}$

17. $e^{\ln x + \ln (x-2)} = 8$

$\qquad e^{\ln x(x-2)} = 8$ *Logarithm of a product*

$\qquad x(x - 2) = 8$ *Theorem on inverses*

$\qquad x^2 - 2x - 8 = 0$

$(x - 4)(x + 2) = 0$ *Factor*

$x = 4$ or $x = -2$

We reject -2 since it is not in the domain of $\ln x$.

Solution set: $\{4\}$

19. $100(1 + .02)^{3+n} = 150$

$\qquad (1.02)^{3+n} = 1.5$

$(3 + n) \log 1.02 = \log 1.5$

$\qquad\qquad$ *Logarithm of a power*

$\qquad 3 + n = \dfrac{\log 1.5}{\log 1.02}$

$\qquad 3 + n = \dfrac{.1761}{.0086}$

$\qquad 3 + n = 20.475$

$\qquad n = 17.475$

Solution set: $\{17.475\}$

21. $\log (t - 1) = 1$

$\qquad t - 1 = 10^1$

$\qquad\qquad$ *Exponential form*

$\qquad t - 1 = 10$

$\qquad t = 11$

Solution set: $\{11\}$

23. $\log (x - 3) = 1 - \log x$

$\log (x - 3) = \log 10 - \log x$

$\qquad\qquad$ *log 10 = 1*

$\log (x - 3) = \log \left(\dfrac{10}{x}\right)$

$\qquad\qquad$ *Logarithm of a quotient*

$\qquad x - 3 = \dfrac{10}{x}$

$\qquad\qquad$ *Property 2*

$\qquad x^2 - 3x = 10$

$\qquad\qquad$ *Multiply by LCD, x*

$\qquad x^2 - 3x - 10 = 0$

$(x - 5)(x + 2) = 0$ *Factor*

$x - 5 = 0$ or $x + 2 = 0$

$x = 5$ or $\qquad x = -2$

We reject -2 since it is not in the domain of $\log x$.

Solution set: $\{5\}$

This equation can also be solved by changing the equation to exponential form.

$$\log (x - 3) = 1 - \log x$$
$$\log (x - 3) + \log x = 1$$
$$\log x(x - 3) = 1$$

Logarithm of a product

$$x(x - 3) = 10^1$$

Exponential form

$$x^2 - 3x = 10$$
$$x^2 - 3x - 10 = 0$$
$$(x - 5)(x + 2) = 0$$
$$x = 5 \quad \text{or} \quad x = -2$$

-2 is not in the domain of log x.

Solution set: $\{5\}$

25. $\ln (y + 2) = \ln (y - 7) + \ln 4$
$$\ln (y + 2) = \ln [(y - 7) \cdot 4]$$

Logarithm of a product

$$y + 2 = 4(y - 7) \quad \text{Property 2}$$
$$y + 2 = 4y - 28$$
$$-3y = -30$$
$$y = 10$$

Solution set: $\{10\}$

27. $\ln (5 + 4y) - \ln (3 + y) = \ln 3$
$$\ln \frac{5 + 4y}{3 + y} = \ln 3$$

Logarithm of a quotient

$$\frac{5 + 4y}{3 + y} = 3$$

Property 2

$$5 + 4y = 3(3 + y)$$
$$5 + 4y = 9 + 3y$$
$$y = 4$$

Solution set: $\{4\}$

29. $\ln x + 1 = \ln (x - 4)$
$$1 = \ln (x - 4) - \ln x$$
$$1 = \ln \left(\frac{x - 4}{x}\right)$$

Logarithm of a quotient

$$e^1 = \frac{x - 4}{x} \quad \text{Exponential form}$$
$$ex = x - 4 \quad \text{Multiply by } x$$
$$ex - x = -4$$
$$x(e - 1) = -4 \quad \text{Factor out } x$$
$$x = \frac{-4}{e - 1}$$
$$= \frac{-4}{2.718 - 1}$$
$$= \frac{-4}{1.718}$$
$$= -2.328$$

This potential solution cannot be used since the domain of ln x is $(0, \infty)$.

Solution set: $\emptyset$

31. $2 \ln (x - 3) = \ln (x + 5) + \ln 4$
$$\ln (x - 3)^2 = \ln [4(x + 5)]$$

Logarithm of a power; logarithm of a product

$$(x - 3)^2 = 4(x + 5) \quad \text{Property 2}$$
$$x^2 - 6x + 9 = 4x + 20$$
$$x^2 - 10x - 11 = 0$$
$$(x - 11)(x + 1) = 0$$
$$x - 11 = 0 \quad \text{or} \quad x + 1 = 0$$
$$x = 11 \quad \text{or} \qquad x = -1$$

If x = -1, x - 3 = -4, so -1 is not in the domain of ln (x - 3) and cannot be used.

Solution set: $\{11\}$

33. $\log_5 (r + 2) + \log_5 (r - 2) = 1$

$\log_5 [(r + 2)(r - 2)] = \log_5 5$
Logarithm of a product

$(r + 2)(r - 2) = 5$
Property 2

$r^2 - 4 = 5$

$r^2 - 9 = 0$

$(r + 3)(r - 3) = 0$

$r = -3 \quad \text{or} \quad r = 3$

−3 is not in the domain of
log (r + 2) or log (r − 2)
and therefore cannot be used.

Solution set: $\{3\}$

35. $\log_3 (a - 3) = 1 + \log_3 (a + 1)$

$\log_3 (a - 3) = \log_3 3 + \log_3 (a + 1)$
$\log_3 3 = 1$

$\log_3 (a - 3) = \log_3 [3(a + 1)]$
Logarithm of a product

$a - 3 = 3(a + 1)$
Property 2

$a - 3 = 3a + 3$

$-2a = 6$

$a = -3$

−3 is not in the domain of
$\log_3 (a - 3)$ or $\log_3 (a + 1)$
and therefore cannot be used.
Solution set: $\emptyset$

37. $\ln e^x - \ln e^3 = \ln e^5$

$x - 3 = 5$ *Theorem on inverses*

$x = 8$

Solution set: $\{8\}$

39. $\log_2 \sqrt{2y^2} - 1 = \dfrac{1}{2}$

$\log_2 \sqrt{2y^2} = \dfrac{3}{2}$
Add 1

$2^{3/2} = \sqrt{2y^2}$
Change to exponential form

$(2^{3/2})^2 = (\sqrt{2y^2})^2$
Square both sides

$2^3 = 2y^2$

$8 = 2y^2$

$4 = y^2$

$\pm 2 = y$

Since the solution involves squaring both sides, both proposed solutions must be checked in the original equation.   Both answers check.

Solution set: $\{-2, 2\}$

41. $\log z = \sqrt{\log z}$

$(\log z)^2 = (\sqrt{\log z})^2$
Square both sides

$(\log z)^2 = \log z$

$(\log z)^2 - \log z = 0$

$\log z(\log z - 1) = 0$
Factor out log z

$\log z = 0 \quad \text{or} \quad \log z - 1 = 0$

$\log z = 1$

$10^0 = z \qquad\qquad 10^1 = z$

$1 = z \quad \text{or} \qquad 10 = z$

Since the work involves squaring both sides, both proposed solutions must be checked in the original equation.   Both answers check.

Solution set: $\{1, 10\}$

45. $P = P_0 e^{kt/1000}$ for t

$$\frac{P}{P_0} = e^{kt/1000}$$

$$\ln\left(\frac{P}{P_0}\right) = \ln\left(e^{kt/1000}\right)$$

$$\ln\frac{P}{P_0} = \frac{kt}{1000}$$

$$1000 \ln\frac{P}{P_0} = kt$$

$$\frac{1000}{k} \ln\frac{P}{P_0} = t$$

47. $r = p - k \ln t$ for t

$$r - p = -k \ln t$$

$$p - r = k \ln t$$

$$\frac{(p - r)}{k} = \ln t$$

$$e^{(p-r)/k} = t$$

$$t = e^{(p-r)/k}$$

49. $T = T_0 + (T_1 - T_0)10^{-kt}$ for t

$$T - T_0 = (T_1 - T_0)10^{-kt}$$

$$\frac{T - T_0}{T_1 - T_0} = 10^{-kt}$$

$$\log\left[\frac{T - T_0}{T_1 - T_0}\right] = \log 10^{-kt}$$

$$\log\left[\frac{T - T_0}{T_1 - T_0}\right] = -kt \cdot \log 10$$

$$\log\left[\frac{T - T_0}{T_1 - T_0}\right] = -kt$$

$$-\frac{1}{k} \log\left[\frac{T - T_0}{T_1 - T_0}\right] = t$$

51. $d = 10 \log \frac{I}{I_0}$

$$\frac{d}{10} = \log \frac{I}{I_0}$$

$$10^{d/10} = \frac{I}{I_0} \quad \textit{Definition of logarithm}$$

$$I = I_0 \cdot 10^{d/10}$$

53. $A = P\left(1 + \frac{r}{m}\right)^{tm}$

To solve for n, substitute
$A = 10,000$, $P = 7500$, $r = .06$,
and $m = 2$.

$$10,000 = 7500\left(1 + \frac{.06}{2}\right)^{(t)(2)}$$

$$1.3333 = (1.03)^{2t}$$

$$\ln 1.3333 = 2t \ln 1.03$$

$$\frac{\ln 1.3333}{2 \ln 1.03} = t$$

$$\frac{.2877}{.0591} = t$$

$$4.9 \approx t$$

To the nearest tenth, $t = 4.9$ yr.

55. $A = P\left(1 + \frac{r}{m}\right)^{tm}$

To solve for t, substitute
$A = 30,000$, $P = 27,000$,
$r = .06$, and $m = 4$.

$$30,000 = 27,000\left(1 + \frac{.06}{4}\right)^{(t)(4)}$$

$$1.1111 \approx (1.015)^{4t}$$

$$\ln 1.1111 = \ln 1.015^{4t}$$

$$\ln 1.1111 = 4t \ln 1.015$$

$$\frac{\ln 1.1111}{4 \ln 1.015} = t$$

$$1.8 \approx t$$

To the nearest tenth of a year,
George will be ready to buy the
car in 1.8 yr.

57. $A = P\left(1 + \frac{r}{m}\right)^{tm}$

To find t, substitute $A = 2500$,
$P = 2000$, $r = .06$, and $m = 2$.

$$2500 = 2000\left(1 + \frac{.06}{2}\right)^{(t)(2)}$$

$$1.25 = (1.03)^{2t}$$

$$\ln 1.25 = 2t \ln 1.03$$

$$\frac{\ln 1.25}{2 \ln 1.03} = t$$

$$\frac{.2231}{.0591} = t$$

$$3.8 \approx t$$

It will take about 3.8 yr for $2000 to amount to $2500 at 6% compounded semiannually.

59. $f(t) = 500e^{.1t}$

(a) For 3,000,000,000 bacteria, or 3000 million bacteria, $f(t) = 3000$.
Solve $f(t) = 3000$ for t.

$$3000 = 500e^{.1t}$$

$$6 = e^{.1t}$$

$$\ln 6 = .1t$$

$$1.7918 \approx .1t$$

$$18 \approx t$$

The product will be unsafe after about 18 days.

(b) The date that should be placed on the product is 18 days after January 1, or January 19.

61. $F = 500 \log (2t + 3)$

$$\frac{F}{500} = \log (2t + 3)$$

$$10^{F/500} = 2t + 3$$

$$10^{F/500} - 3 = 2t$$

$$\frac{10^{F/500} - 3}{2} = t$$

(a) When F = 600,

$$t = \frac{10^{600/500} - 3}{2}$$

$$= \frac{10^{1.2} - 3}{2}$$

$$= \frac{15.8 - 3}{2}$$

$$= \frac{12.8}{2}$$

$$= 6.4.$$

F = 600 when t = 6.4 mo.

(b) When F = 1000,

$$t = \frac{10^{1000/500} - 3}{2}$$

$$= \frac{10^2 - 3}{2}$$

$$= \frac{97}{2}$$

$$= 48.5$$

F = 1000 when t = 48.5 mo.

Section 5.5

1. $A = Pe^{rt}$

P = 10, t = 3

(a) When r = .03,

$$A = 10e^{.03(3)}$$

$$= 10e^{.09}$$

$$\approx 10.94.$$

The item will cost about $10.94.

(b) When r = .04,

$$A = 10e^{.04(3)}$$

$$= 10e^{.12}$$

$$\approx 11.27.$$

The item will cost about $11.27.

(c) When r = .05,

$$A = 10e^{.05(3)}$$
$$= 10e^{.15}$$
$$\approx 11.62.$$

The item will cost about $11.62.

3. P = 60,000 and t = 5

Substitute r = .07 and m = 4 into the compound interest formula.

$$A = 60,000\left(1 + \frac{.07}{4}\right)^{5(4)}$$
$$= 60,000(1.0175)^{20}$$
$$\approx 60,000(1.4148)$$
$$\approx 84,886.69$$

The interest from this investment would be $84,886.69 − $60,000 = $24,886.69.

Substitute r = .0675 into the formula for continuous compounding.

$$A = 60,000e^{.0675(5)}$$
$$= 60,000e^{.3375}$$
$$\approx 60,000(1.4014)$$
$$\approx 84,086.38$$

The interest from this investment would be $84,086.38 − $60,000 = $24,086.38.

Note that $24,886.69 − $24,086.38 = $800.31.

The investment that offers 7% compounded quarterly will earn $800.31 more in interest.

5. Use $A = Pe^{rt}$ with P = 1, A = 2 · 1 = 2, and r = .06.

$$2 = 1 \cdot e^{.06t}$$
$$2 = e^{.06t}$$

Take natural logarithms of both sides.

$$\ln 2 = .06t$$
$$.6931 \approx .06t$$
$$11.6 \approx t$$

The doubling time is about 11.6 yr.

7. Use $A = Pe^{rt}$ with A = 7250, P = 5000, and t = 4.

$$7250 = 5000e^{(r)(4)}$$
$$1.45 = e^{4r}$$
$$\ln 1.45 = 4r$$
$$.3716 \approx 4r$$
$$.093 \approx r$$

The interest rate is about .093 or 9.3%.

9. $L = 9 + 2e^{.15t}$

(a) Use t = 1982 − 1982 = 0.

$$L = 9 + 2e^{.15(0)}$$
$$= 9 + 2e^{0}$$
$$= 9 + 2$$
$$= 11$$

(b) Use t = 1986 − 1982 = 4.

$$L = 9 + 2e^{.15(4)}$$
$$= 9 + 2e^{.6}$$
$$\approx 9 + 2(1.8221)$$
$$= 9 + 3.6442$$
$$\approx 12.6$$

(c) Use t = 1992 − 1982 = 10.

$$L = 9 + 2e^{.15(10)}$$
$$= 9 + 2e^{1.5}$$
$$\approx 9 + 2(4.4817)$$
$$= 9 + 8.9634$$
$$\approx 18$$

(d) The domain is $[0, \infty)$. Draw a smooth curve through the points $(0, 11)$, $(4, 12.6)$, and $(10, 18)$. See the answer graph in the textbook.

(e) According to this equation, the living standards in the U.S. are growing exponentially, but at a slow rate.

11. $A(t) = A_0 e^{-.053t}$

Solve $A(t) = \frac{1}{2}A_0$ for t.

$$\frac{1}{2}A = A_0 e^{-.053t}$$

$$.5 = e^{-.053t}$$

$$\ln .5 = -.053t$$

$$-.6931 \approx -.053t$$

$$13.1 \approx t$$

The half-life of plutonium 241 is about 13.1 yr.

13. $A(t) = A_0 e^{-.087t}$

Solve $A(t) = \frac{1}{2}A_0$ for t.

$$\frac{1}{2}A_0 = A_0 e^{-.087t}$$

$$.5 = e^{-.087t}$$

$$\ln .5 = -.087t$$

$$-.6931 \approx -.087t$$

$$7.97 \approx t$$

The half-life of iodine 131 is about 7.97 yr.

15. $P = P_0 e^{-.04t}$

(a) $P = 1,000,000e^{-.04(1)}$

$P = 10^6 \cdot .961$

$P = 961,000$

The population at time $t = 1$ is about 961,000.

(b) $750,000 = 1,000,000e^{-.04t}$

$$.75 = e^{-.04t}$$

$$\ln .75 = \ln e^{-.04t}$$

$$\ln .75 = -.04t \ln e$$

$$\frac{\ln .75}{-.04} = t$$

$$7.192 = t$$

It will take about 7.2 yr for the population to be reduced to 750,000.

(c) $500,000 = 1,000,000e^{-.04t}$

$$.5 = e^{-.04t}$$

$$\ln .5 = \ln e^{-.04t}$$

$$\ln .5 = -.04t$$

$$\frac{\ln .5}{-.04} = t$$

$$17.329 = t$$

It will take about 17.3 yr for the population to be cut in half.

17. $y = y_0 e^{-(\ln 2)(1/5700)t}$

or

$y \approx y_0 e^{-.0001216t}$

Find t when $y = .60y_0$.

$$.60 \; y_0 = y_0 e^{-.0001216t}$$

$$.60 = e^{-.0001216t}$$

$$\ln .60 = -.0001216t$$

$$\frac{\ln .60}{-.0001216} = t$$

$$4200 \approx t$$

The sample was about 4200 yr old.

19.

$$A = A_0 e^{kt} \quad \text{for } t$$

$$\frac{A}{A_0} = e^{kt}$$

$$\ln \left(\frac{A}{A_0}\right) = kt$$

$$\frac{\ln (A/A_0)}{k} = t$$

$$t = \frac{\ln (A/A_0)}{k}$$

21.

$$y = y_0 e^{-(\ln 2)(1/5700)t}$$

or

$$y \approx y_0 e^{-.0001216t}$$

Find t when $y = .20 \; y_0$.

$$.20 \; y_0 = y_0 e^{-.0001216t}$$

$$.20 = e^{-.0001216t}$$

$$\ln .20 = -.0001216t$$

$$-1.6094 \approx -.0001216t$$

$$13,000 \approx t$$

The specimen is about 13,000 yr old.

23.

$$P(t) = 100e^{-.1t}$$

$$1 = 100e^{-.1t}$$

$$\ln 1 = \ln 10^2 + \ln e^{-.1t}$$

$$\ln 1 = 2 \ln 10 - .1t \ln e$$

$$\ln 1 - 2 \ln 10 = -.1t$$

$$\frac{-\ln 1 + 2 \ln 10}{.1} = t$$

$$\frac{0 + 4.605}{.1} = t$$

$$46.05 = t$$

It would take 46 days.

25.

$$A(t) = T_0 + Ce^{-kt}$$

Substitute $T_0 = 20$, $C = 100$, and $k = .1$ into the formula, and solve $A(t) = 25$ for t.

$$25 = 20 + 100e^{-.1t}$$

$$5 = 100e^{-.1t}$$

$$.05 = e^{-.1t}$$

$$\ln .05 = -.1t$$

$$-2.9957 \approx -.1t$$

$$30 \approx t$$

It will take about 30 min.

Chapter 5 Review Exercises

1.

$$8^p = 32$$

$$(2^3)^p = 2^5$$

$$2^{3p} = 2^5$$

$$3p = 5$$

$$p = \frac{5}{3}$$

Solution set: $\left\{\dfrac{5}{3}\right\}$

3.

$$\frac{8}{27} = b^{-3}$$

$$\left(\frac{8}{27}\right)^{-1/3} = (b^{-3})^{-1/3}$$

$$\frac{1}{\sqrt[3]{\dfrac{8}{27}}} = b^1$$

$$\frac{1}{\dfrac{2}{3}} = b$$

$$\frac{3}{2} = b$$

Solution set: $\left\{\dfrac{3}{2}\right\}$

5. $A(t) = 800e^{-.04t}$

    $A(5) = 800e^{-.04(5)}$

    $\quad\quad = 800 \cdot e^{-.20}$

    $\quad\quad = 800(.81873)$

    $A(5) \approx 655$

    The amount present after 5 days is 655 g.

7. $y = e^x$

    The point $(0, 1)$ is on the graph since $e^0 = 1$, so the correct choice must be either (a) or (d). Since the base is e and $e > 1$, $y = e^x$ is an increasing function, and so the correct choice must be (a).

9. $y = (.3)^x$

    The point $(0, 1)$ is on the graph since $(.3)^0 = 1$, so the correct choice must be either (a) or (d). Since the base is .3 and $0 < .3 < 1$, $y = (.3)^x$ is a decreasing function, and so the correct choice must be (d).

For Exercises 11 and 13, see the answer graphs in the textbook.

11. $y = 2^{-x} + 1$

    Make a table of values.

x	-3	-2	-1	0	1	2	3
y	9	5	2	1	$\frac{3}{2}$	$\frac{5}{4}$	$\frac{9}{8}$

Draw a smooth curve through these points. This graph may also be found by translating the graph of $y = 2^{-x}$ up 1 unit.

13. $y = \log_{1/2} (x - 1)$

    The domain is the set of all real numbers for which $x - 1 > 0$ or $x > 1$, that is, the interval $(1, \infty)$.

    Make a table of values.

x	1.125	1.25	1.5	2	3	5	9
y	3	2	1	0	-1	-2	-3

    Draw a smooth curve through these points. This graph may also be found by translating the graph of $y = \log_{1/2} x$ to the right 1 unit.

15. $f(x) = \log_{2/3} x$ defines a decreasing function since the base, $2/3$, is between 0 and 1.

17. $2^5 = 32$ is written in logarithmic form as

    $$\log_2 32 = 5.$$

19. $\left(\frac{1}{16}\right)^{1/4} = \frac{1}{2}$ is written in logarithmic form as

    $$\log_{1/16} \frac{1}{2} = \frac{1}{4}.$$

21. $10^{.4771} = 3$ is written in logarithmic form as

    $$\log_{10} 3 = .4771.$$

23. $\log_{10} .001 = -3$ is written in exponential form as

$$10^{-3} = .001.$$

25. $\log 3.45 = .537819$ is written in exponential form as

$$10^{.537819} = 3.45$$

27. $\log_3 \dfrac{mn}{5r}$

$= \log_3 mn - \log_3 5r$
Logarithm of a quotient

$= \log_3 m + \log_3 n - \log_3 5 - \log_3 r$
Logarithm of a product

29. $\log_5 x^2 y^4 \sqrt[5]{m^3 p}$

$= \log_5 x^2 y^4 (m^3 p)^{1/5}$

$= \log_5 x^2 + \log_5 y^4 + \log_5 (m^3 p)^{1/5}$
Logarithm of a product

$= 2 \log_5 x + 4 \log_5 y$

$\quad + \dfrac{1}{5}(\log_5 m^3 p)$
Logarithm of a power

$= 2 \log_5 x + 4 \log_5 y$

$\quad + \dfrac{1}{5}(\log_5 m^3 + \log_5 p)$
Logarithm of a product

$= 2 \log_5 x + 4 \log_5 y$

$\quad + \dfrac{1}{5}(3 \log_5 m + \log_5 p)$
Logarithm of a power

31. $\log 45.6 = 1.659$

33. $\log .00056 = -3.252$

35. $\ln 35 = 3.555$

37. $\ln 144,000 = 11.878$

39. To find $\log_{3.4} 15.8$, use the change of base rule.

$$\log_{3.4} 15.8 = \frac{\ln 15.8}{\ln 3.4}$$

$$\approx \frac{2.7600}{1.2238}$$

$$\approx 2.255$$

41. $\log_3 769 = \dfrac{\ln 769}{\ln 3}$

$\approx \dfrac{6.6451}{1.0986}$

≈ 6.049

43. $h = .5 + \log t$

$= .5 + \log 2$

$\approx .5 + .3010$

$= .8010$

The height of a 2 year-old tribe member is about .8 m.

45. $h = .5 + \log t$

$= .5 + \log 10$

$= .5 + 1$

$= 1.5$

The height of a 10 year-old tribe member is about 1.5 m.

47. $A = P\left(1 + \dfrac{r}{m}\right)^{tm}$

To find A, substitute $P = 1000$, $r = .08$, $t = 9$, and $m = 1$.

$A = 1000\left(1 + \dfrac{.08}{1}\right)^{9(1)}$

$= 1000(1.08)^9$

$= 1000(1.999)$

$= 1999.00$

The amount on deposit will be $1999.

49. $A = P\left(1 + \dfrac{r}{m}\right)^{tm}$

To find P, substitute A = 2000,
r = .04, t = 11, and m = 2.

$2000 = P\left(1 + \dfrac{.04}{2}\right)^{11(2)}$

$2000 = P(1.02)^{22}$

$1293.68 \approx P$

The present value is $1293.68.

51. $A = P\left(1 + \dfrac{r}{m}\right)^{tm}$

First, substitute P = 10,000,
r = .12, t = 12, and m = 1
into the formula.

$A = 10,000\left(1 + \dfrac{.12}{1}\right)^{12(1)}$

$= 10,000(1.12)^{12}$

$\approx 10,000(3.8960)$

$\approx 38,959.76$

After the first 12 yr, there would
be about $38,959.76 in the account.
To finish off the 21-year period,
substitute P = 38,959.76, r = .10,
t = 9, and m = 2 into the original
formula.

$A = 38,959.76\left(1 + \dfrac{.10}{2}\right)^{9(2)}$

$= 38,959.76(1.05)^{18}$

$\approx 38,959.76(2.4066)$

$\approx 93,761.31$

At the end of the 21-year period,
about $93,761.31 would be in the
account.

53. $5^r = 11$

Take natural logarithms of both
sides.

$\ln 5^r = \ln 11$

$r \ln 5 = \ln 11$

$r = \dfrac{\ln 11}{\ln 5}$

$= \dfrac{2.3979}{1.6094}$

$= 1.490$

Solution set: $\{1.490\}$

55. $e^{p+1} = 10$

$\ln e^{p+1} = \ln 10$

$(p + 1) \ln e = \ln 10$

$p + 1 = \ln 10$

$p = \ln 10 - 1$

$= 2.3026 - 1$

$= 1.303$

Solution set: $\{1.303\}$

57. $\log_{64} y = \dfrac{1}{3}$

$64^{1/3} = y$ *Change to exponential
form*

$4 = y$

Solution set: $\{4\}$

59. $\log_{16} \sqrt{x + 1} = \dfrac{1}{4}$

$16^{1/4} = \sqrt{x + 1}$
     *Change to exponential
form*

$(16^{1/4})^2 = (\sqrt{x + 1})^2$
     Square both sides

$16^{1/2} = x + 1$

$4 = x + 1$

$3 = x$

The solution must be checked in the original equation since both sides were squared.

Solution set: $\{3\}$

61. $\ln (\ln e^{-x}) = \ln 3$

$\quad \ln (-x) = \ln 3$ *ln $3^{-x} = -x$ by theorem on inverses*

$\quad\quad -x = 3$

$\quad\quad\quad x = 3$

Solution set: $\{-3\}$

63. $A = 27,208.60$, $P = 12,700$, $t = 10$, $m = 2$

$$A = P\left(1 + \frac{r}{m}\right)^{tm}$$

$$27,208.60 = 12,700\left(1 + \frac{r}{2}\right)^{10(2)}$$

$$2.1424 \approx \left(1 + \frac{r}{2}\right)^{20}$$

$$(2.1424)^{1/20} = 1 + \frac{r}{2}$$

$$1.0388 \approx 1 + \frac{r}{2}$$

$$.0388 = \frac{r}{2}$$

$$.078 \approx r$$

The annual interest rate, to the nearest tenth, is 7.8%.

65. Substitute $P = 48,000$, $A = 58,344$, $r = .05$, and $m = 2$ into the formula.

$$A = P\left(1 + \frac{r}{m}\right)^{tm}$$

$$58,344 = 48,000\left(1 + \frac{.05}{2}\right)^{(t)(2)}$$

$$58,344 = 48,000(1.025)^{2t}$$

$$1.2155 = (1.025)^{2t}$$

$$\ln 1.2155 = \ln (1.025)^{2t}$$

$$\ln 1.2155 = 2t \ln 1.025$$

$$\frac{\ln 1.2155}{2 \ln 1.025} = t$$

$$4.0 \approx t$$

\$48,000 will increase to \$58,344 in about 4.0 yr.

67. $\quad A = Pe^{rt}$

$\quad\quad 2 = 1 \cdot e^{.04t}$

$\quad\quad 2 = e^{.04t}$

$\quad \ln 2 = .04t$

$\quad .6931 \approx .04t$

$\quad 17.3 \approx t$

It would take about 17.3 yr.

69. $A = Pe^{rt}$

$A = 1200e^{10(.10)}$

$A = 1200e^1$

$A = 1200(2.71828)$

$A = 3261.94$

After 10 yr, \$1200 would amount to \$3261.94.

71. $A = Pe^{rt}$

Let $A = 2P$ and $r = .02$.

$\quad 2P = Pe^{.02t}$

$\quad\quad 2 = e^{.02t}$

Take natural logarithms of both sides.

$\quad \ln 2 = .02t$

$\quad \dfrac{\ln 2}{.02} = t$

$\quad 34.657 = t$

It would take about 35 yr.

73. $t = T \cdot \dfrac{\ln\,[1 + 8.33(A/K)]}{\ln 2}$

When $T = 1.26 \times 10^9$ and $A/K = .103$.

$t = (1.26 \times 10^9)\dfrac{\ln\,[1 + 8.33(.103)]}{\ln 2}$

$\quad = (1.26 \times 10^9)\dfrac{\ln 1.85799}{\ln 2}$

$\quad \approx (1.26 \times 10^9)\dfrac{.6195}{.6931}$

$\quad \approx 1.126 \times 10^9$

The rock sample is about 1.126 billion yr old.

Chapter 5   Test

1. $25^{2x-1} = 125^{x+1}$

$(5^2)^{2x-1} = (5^3)^{x+1}$

$5^{4x-2} = 5^{3x+3}$

$4x - 2 = 3x + 3$

$x = 5$

Solution set: $\{5\}$

2. $\left(\dfrac{a}{2}\right)^{-1/3} = 4$

Raise both sides of the equation to the −3 power.

$\left[\left(\dfrac{a}{2}\right)^{-1/3}\right]^{-3} = 4^{-3}$

$\left(\dfrac{a}{2}\right)^{1} = \dfrac{1}{4^3}$

$\dfrac{a}{2} = \dfrac{1}{64}$

$a = \dfrac{1}{32}$

Solution set: $\left\{\dfrac{1}{32}\right\}$

3. $a^2 = b$ is written in logarithmic form as

$\log_a b = 2.$

4. $e^c = 4.82$

is written in logarithmic form as

$\ln 4.82 = c.$

5. $\log_3 \sqrt{27} = \dfrac{3}{2}$ is written in exponential form as

$3^{3/2} = \sqrt{27}.$

6. $\ln 5 = a$ is written in exponential form as

$e^a = 5.$

7. $y = \log_a x$

$\log_a a = 1$, so $(a, 1)$ is a point on the graph.

$\log_a 1 = 0$, so $(1, 0)$ is a point on the graph.

For Problems 8 and 9, see the answer graphs in the textbook.

8. $y = (1.5)^{x+2}$

x	−4	−3	−2	−1	0	1	2
x + 2	−2	−1	0	1	2	3	4
y	.444	.667	1	1.5	2.25	3.375	5.0625

9. $y = \log_{1/2} x$

x	$\frac{1}{16}$	$\frac{1}{8}$	$\frac{1}{4}$	$\frac{1}{2}$	1	2	4	8	16
y	4	3	2	1	0	−1	−2	−3	−4

10. $\log_7 \dfrac{x^2 \sqrt[4]{y}}{z^3}$

  $= \log_7 x^2 + \log_7 \sqrt[4]{y} - \log_7 z^3$

  $= 2 \log_7 x + \dfrac{1}{4} \log_7 y - 3 \log_7 z$

11. $\ln 2300 = 7.741$

12. $\log_{2.7} 94.6 = \dfrac{\ln 94.6}{\ln 2.7}$

  $= \dfrac{4.5497}{.99325}$

  $= 4.581$

13. $\log_a (2x - 3) = -1$

  $\log_a t$ is defined when t is in the interval $(0, \infty)$, but it is undefined when t is in the interval $(-\infty, 0]$. $\log_a (2x - 3)$ will be undefined when $2x - 3$ is in the interval $(-\infty, 0]$, which corresponds to

  $$2x - 3 \leq 0$$
  $$2x \leq 3$$
  $$x \leq \dfrac{3}{2}.$$

  The numbers in the interval $\left(-\infty, \dfrac{3}{2}\right]$ cannot be solutions of the given equation.

14. $8^{2w-4} = 100$

  Take natural logarithms of both sides.

  $$(2w - 4) \ln 8 = \ln 100$$
  $$2w - 4 = \dfrac{\ln 100}{\ln 8}$$
  $$= \dfrac{4.6052}{2.0794}$$

  $$2w - 4 = 2.2147$$
  $$2w = 6.2147$$
  $$w = 3.107$$

  Solution set: $\{3.107\}$

15. $\log_3 (m + 2) = 2$

  $$3^2 = m + 2$$
  $$9 = m + 2$$
  $$7 = m$$

  Solution set: $\{7\}$

16. $\ln x - 4 \ln 3 = \ln \dfrac{5}{x}$

  $$\ln \dfrac{x}{3^4} = \ln \dfrac{5}{x}$$
  $$\dfrac{x}{81} = \dfrac{5}{x}$$
  $$x^2 = 405$$
  $$x = 20.125$$

  (Since the domain of $\ln x$ is $(0, \infty)$, only the positive square root of 405 can be a solution.)

  Solution set: $\{20.125\}$

17. $A(t) = 600e^{-.05t}$

  (a) $A(12) = 600e^{-.05(12)}$

  $= 600e^{-.6}$

  $\approx 600(.5488)$

  ≈ 329.3

  The amount of radioactive material present after 12 days is about 329.3 g.

  (b) Solve $A(t) = \dfrac{1}{2} A_0$ for t.

  Note that $\dfrac{1}{2}A_0 = \dfrac{1}{2}A(0) = \dfrac{1}{2}(600) = 300$.

$$300 = 600e^{-.05t}$$

$$.5 = e^{-.05t}$$

$$\ln 5 = -.05t$$

$$-.691 \approx -.05t$$

$$13.9 \approx t$$

The half-life of the material is about 13.9 days.

18. $A = Pe^{rt}$

To find the amount on deposit after 12 yr, substitute $P = 2500$, $t = 12$, and $r = .08$.

$$A = 2500e^{.08(12)}$$

$$= 2500e^{.96}$$

$$= 2500(2.611696)$$

$$= 6529.24$$

The amount on deposit after 12 yr will be $6529.24.

19.   To find the number of years it will take the population to double, substitute $A = 2P$ and $r = .04$ into the continuous compounding formula.

$$A = Pe^{rt}$$

$$2P = Pe^{.04t}$$

$$2 = e^{.04t}$$

Take natural logarithms of both sides.

$$\ln 2 = .04t$$

$$\frac{\ln 2}{.04} = t$$

$$\frac{.69315}{.04} = t$$

$$17.3 \approx t$$

It will take about 17.3 yr for the population to double.

20. $A = P\left(1 + \dfrac{r}{m}\right)^{tm}$

To find t, substitute $P = 5000$, $A = 18,000$, $r = .12$, and $m = 12$.

$$18,000 = 5000\left(1 + \frac{.12}{12}\right)^{t(12)}$$

$$3.6 = (1.01)^{12t}$$

Take natural logarithms of both sides.

$$\ln 3.6 = 12t(\ln 1.01)$$

$$\frac{\ln 3.6}{12 \ln 1.01} = t$$

$$\frac{1.2809}{.1194} = t$$

$$10.7 = t$$

It will take 10.7 yr for $5000 to increase to $18,000 at 12% compounded monthly.

CHAPTER 6   POLYNOMIAL AND RATIONAL
            FUNCTIONS

Section 6.1

For graphs in this section, see the
answer graphs in the textbook.

1. $P(x) = \frac{1}{4}x^6$

The graph is broader than that of
$P(x) = x^6$ but has the same general
shape. It includes the points
$(-2, 16)$, $\left(-1, \frac{1}{4}\right)$, $(0, 0)$, $\left(1, \frac{1}{4}\right)$,
and $(2, 16)$. Connect these points
with a smooth curve.

3. $P(x) = -\frac{5}{4}x^5$

The graph is narrower than that of
$P(x) = x^5$. The negative sign causes
the graph to be the reflection of
the graph of $P(x) = \frac{5}{4}x^5$ about the
x-axis. The graph includes the
points $(-2, 40)$, $\left(-1, \frac{5}{4}\right)$, $(0, 0)$,
$\left(1, -\frac{5}{4}\right)$ and $(2, -40)$.

5. $P(x) = \frac{1}{2}x^3 + 1$

The graph of $P(x) = \frac{1}{2}x^3 + 1$ looks
like $y = x^3$ but is broader and is
translated 1 unit up. The graph
includes the points $(-2, -3)$,
$\left(-1, \frac{1}{2}\right)$, $(0, 1)$, $\left(1, \frac{3}{2}\right)$, and $(2, 5)$.

7. $P(x) = -(x + 1)^3$

The graph can be obtained by reflecting the graph of $P(x) = x^3$
about the x-axis and then translating it 1 unit to the left.

9. $P(x) = (x - 1)^4 + 2$

This graph has the same shape as
$y = x^4$, but is translated 1 unit
to the right and 2 units up.

11. $P(x) = 2x(x - 3)(x + 2)$

First set each of the three factors
equal to 0 and solve the resulting
equations to find the zeros of the
function.

$2x = 0$ or $x - 3 = 0$ or $x + 2 = 0$
   $x = 0$ or $x = 3$ or $x = -2$

The three zeros, 0, 3, and -2,
divide the x-axis into four intervals:
$(-\infty, -2)$, $(-2, 0)$, $(0, 3)$ and
$(3, \infty)$.
These intervals can be shown on a
number line.

To find the sign of $P(x)$ in each
interval, select a value of x in the
interval and determine by substitution whether the function values
are positive or negative in that
interval.

Interval	Test point	Value of $P(x)$	Location relative to x-axis
$(-\infty, -2)$	-3	-36	Below
$(-2, 0)$	-1	8	Above
$(0, 3)$	1	-12	Below
$(3, \infty)$	4	48	Above

Sketch the graph.

13. $P(x) = x^2(x - 2)(x + 3)^2$

Set each factor equal to zero and solve the resulting equations to get the zeros 0, 2, and -3. The zeros divide the x-axis into four intervals:
$(-\infty, -3)$, $(-3, 0)$, $(0, 2)$, and $(2, \infty)$.
Test a point into each interval to find the sign of $P(x)$ in that interval.

Interval	Test point	Value of $P(x)$	Location relative to x-axis
$(-\infty, -3)$	-4	-96	Below
$(-3, 0)$	-1	-12	Below
$(0, 2)$	1	-16	Below
$(2, \infty)$	3	324	Above

Sketch the graph.

15. $P(x) = x^3 - x^2 - 2x$
      $\qquad = x(x^2 - x - 2)$
      $\qquad = x(x - 2)(x + 1)$

Set each factor equal to zero and solve the resulting equations to get the zeros 0, 2, and -1. The zeros divide the x-axis into four intervals:

$(-\infty, -1)$, $(-1, 0)$, $(0, 2)$, and $(2, \infty)$.

Test a point in each interval to find the sign of $P(x)$ in that interval.

Interval	Test point	Value of $P(x)$	Location relative to x-axis
$(-\infty, -1)$	-2	-8	Below
$(-1, 0)$	$-\dfrac{1}{2}$	$\dfrac{5}{8}$	Above
$(0, 2)$	1	-2	Below
$(2, \infty)$	3	12	Above

Sketch the graph.

17. $P(x) = (x + 2)(x - 1)(x + 1)$

Set each factor equal to zero and solve the resulting equations to get the zeros -2, 1, and -1. The zeros divide the x-axis into four intervals:
$(-\infty, -2)$, $(-2, -1)$, $(-1, 1)$ and $(1, \infty)$.
Test a point in each interval to find the sign of $P(x)$ in that interval.

Interval	Test point	Value of $P(x)$	Location relative to x-axis
$(-\infty, -2)$	-3	-8	Below
$(-2, -1)$	$-\dfrac{3}{2}$	$\dfrac{5}{8}$	Above
$(-1, 1)$	0	-2	Below
$(1, \infty)$	2	12	Above

Sketch the graph.

19.   P(x) = (3x − 1)(x + 2)²

Set each factor equal to zero and solve the resulting equations to get the zeros $\frac{1}{3}$ and −2. The zeros divide the x-axis into three intervals:

$$(-\infty,\ -2),\ \left(-2,\ \tfrac{1}{3}\right),\ \text{and}\ \left(\tfrac{1}{3},\ \infty\right).$$

Test a point in each interval to find the sign of P(x) in that interval.

Interval	Test point	Value of P(x)	Location relative to x-axis
$(-\infty,\ -2)$	−3	−10	Below
$\left(-2,\ \tfrac{1}{3}\right)$	0	−4	Below
$\left(\tfrac{1}{3},\ \infty\right)$	1	18	Above

Sketch the graph.

21.   P(x) = x³ + 5x² − x − 5

= x²(x + 5) − 1(x + 5)

= (x + 5)(x² − 1)

= (x + 5)(x + 1)(x − 1)

Set each factor equal to zero, and solve the resulting equations to get the zeros −5, −1, and 1. The zeros divide the x-axis into four intervals:

(−∞, −5), (−5, −1), (−1, 1), and (1, ∞).

Test a point in each interval to find the sign of P(x) in that interval.

Interval	Test point	Value of P(x)	Location relative to x-axis
$(-\infty,\ -5)$	−6	−35	Below
$(-5,\ -1)$	−2	9	Above
$(-1,\ 1)$	0	−5	Below
$(1,\ \infty)$	2	21	Above

Sketch the graph.

23.   (c) $P(x) = \dfrac{1}{x}$

does not define a polynomial function since polynomials never have variables in a denominator.

25.   D(x) = −.125x⁵ + 3.125x⁴ + 4000

(a) Make a table.

x	D(x)	
0	4000	Stable from 1905 to 1910
2	4046	
4	4672	
6	7078	Increasing from 1910 to 1925
8	12,704	
10	22,750	
12	37,696	
14	56,822	
16	77,728	
18	95,854	
20	104,000	
22	91,846	Decreasing from 1925 to 1930
24	45,472	
26	−53,122	

Use these values to draw the graph.

(b) The table and the graph both show that the population was increasing from 1910 to 1925, relatively stable from 1905 to 1910, and decreasing from 1925 to 1930.

29. The graph of any even function will be symmetric with respect to the y-axis. One example is $P(x) = x^2$.

31. $P(x) = x^3 + 4x^2 - 8x - 8$, $[-3.8, -3]$

Find the values of the function at the appropriate x-values.

x	P(x)
-3.8	25.288
-3.7	25.707
-3.6	25.984
-3.5	26.125
-3.4	26.136
-3.3	26.023
-3.2	25.792
-3.1	25.449
-3	25

The maximum value on the interval is 26.136 when $x = -3.4$. The minimum value on the interval is 25 when $x = -3$.

33. $P(x) = 2x^3 - 5x^2 - x + 1$, $[-1, 0]$

Find the values of the function at the appropriate x-values.

x	P(x)
-1	-5
-.9	-3.608
-.8	-2.424
-.7	-1.436
-.6	-.632
-.5	0
-.4	.472
-.3	.796
-.2	.984
-.1	1.048
0	1

The maximum value on the interval is 1.048 when $x = -.1$. The minimum value on the interval is -5 when $x = -1$.

35. $P(x) = x^4 - 7x^3 + 13x^2 + 6x - 28$, $[-2, -1]$

Find the values of the function at the appropriate x-values.

x	P(x)
-2	84
-1.9	68.575
-1.8	54.642
-1.7	42.113
-1.6	30.906
-1.5	20.938
-1.4	12.130
-1.3	4.405
-1.2	2.310
-1.1	-8.089
-1	-13

The maximum value on the interval is 84 when $x = -2$. The minimum value on the interval is -13 when $x = -1$.

37. As the exponent increases, the part of the graph in the interval $[-1, 1]$ gets flatter, while the parts in the intervals $(-\infty, -1)$ and $(1, \infty)$ get steeper. Thus, the graph of $P(x) = x^{11}$ would be flatter than the others on the interval $[-1, 1]$.

See the graph in the textbook.

Section 6.2

1. $\dfrac{4m^3 - 8m^2 + 16m}{2m}$

$= \dfrac{4m^3}{2m} - \dfrac{8m^2}{2m} + \dfrac{16m}{2m}$

$= 2m^2 - 4m + 8$

3. $\dfrac{25x^2y^4 - 15x^3y^3 + 40x^4y^2}{5x^2y^2}$

$= \dfrac{25x^2y^4}{5x^2y^2} - \dfrac{15x^3y^3}{5x^2y^2} + \dfrac{40x^4y^2}{5x^2y^2}$

$= 5y^2 - 3xy + 8x^2$

5. $\dfrac{6y^2 + y - 2}{2y - 1}$

$$
\begin{array}{r}
3y + 2 \\
2y - 1 \overline{\smash{)}6y^2 + y - 2} \\
\underline{6y^2 - 3y} \\
4y - 2 \\
\underline{4y - 2} \\
0
\end{array}
$$

Thus,

$\dfrac{6y^2 + y - 2}{2y - 1} = 3y + 2.$

7. $\dfrac{8z^2 + 14z - 20}{2z + 5}$

$$
\begin{array}{r}
4z - 3 \\
2z + 5 \overline{\smash{)}8z^2 + 14z - 20} \\
\underline{8z^2 + 20z} \\
-6z - 20 \\
\underline{-6z - 15} \\
-5
\end{array}
$$

Thus,

$\dfrac{8z^2 + 14z - 20}{2z + 5} = 4z - 3 + \dfrac{-5}{2z + 5}.$

9. $\dfrac{2x^3 - 11x^2 + 19x - 10}{2x - 5}$

$$
\begin{array}{r}
x^2 - 3x + 2 \\
2x - 5 \overline{\smash{)}2x^3 - 11x^2 + 19x - 10} \\
\underline{2x^3 - 5x^2} \\
-6x^2 + 19x \\
\underline{-6x^2 + 15x} \\
4x - 10 \\
\underline{4x - 10} \\
0
\end{array}
$$

Thus,

$\dfrac{2x^3 - 11x^2 + 19x - 10}{2x - 5}$

$= x^2 - 3x + 2.$

11. $\dfrac{15x^3 + 11x^2 + 20}{3x + 4}$

$$
\begin{array}{r}
5x^2 - 3x + 4 \\
3x + 4 \overline{\smash{)}15x^3 + 11x^2 + 0x + 20} \\
\underline{15x^3 + 20x^2} \\
-9x^2 \\
\underline{-9x^2 - 12x} \\
12x + 20 \\
\underline{12x + 16} \\
4
\end{array}
$$

Thus,

$\dfrac{15x^3 + 11x^2 + 20}{3x + 4}$

$= 5x^2 - 3x + 4 + \dfrac{4}{3x + 4}.$

13. $\dfrac{x^4 + 2x^3 + 2x^2 - 2x - 3}{x^2 - 1}$

$$
\begin{array}{r}
x^2 + 2x + 3 \\
x^2 - 1 \overline{\smash{)}x^4 + 2x^3 + 2x^2 - 2x - 3} \\
\underline{x^4 \qquad - x^2} \\
2x^3 + 3x^2 - 2x \\
\underline{2x^3 \qquad - 2x} \\
3x^2 \qquad - 3 \\
\underline{3x^2 \qquad - 3} \\
0
\end{array}
$$

Thus,

$\dfrac{x^4 + 2x^3 + 2x^2 - 2x - 3}{x^2 - 1}$

$= x^2 + 2x + 3.$

15. $\dfrac{4z^5 - 4z^2 - 5z + 3}{2z^2 + z + 1}$

$$
\begin{array}{r}
2z^3 - z^2 - \tfrac{1}{2}z \qquad\quad - \tfrac{5}{4} \\[4pt]
\hline
2z^2 + z + 1\,\big)\,4z^5 + 0z^4 + 0z^3 - 4z^2 - 5z + 3 \\[2pt]
\underline{4z^5 + 2z^4 + 2z^3} \\[2pt]
-2z^4 - 2z^3 - 4z^2 \\[2pt]
\underline{-2z^4 - z^3 - z^2} \\[2pt]
-z^3 - 3z^2 - 5z \\[2pt]
\underline{-z^3 - \tfrac{1}{2}z^2 - \tfrac{1}{2}z} \\[2pt]
-\tfrac{5}{2}z^2 - \tfrac{9}{2}z + 3 \\[2pt]
\underline{-\tfrac{5}{2}z^2 - \tfrac{5}{4}z - \tfrac{5}{4}} \\[2pt]
-\tfrac{13}{4}z + \tfrac{17}{4}
\end{array}
$$

Thus,

$\dfrac{4z^5 - 4z^2 - 5z + 3}{2x^2 + z + 1}$

$= 2z^3 - z^2 - \dfrac{1}{2}z - \dfrac{5}{4} + \dfrac{-\tfrac{13}{4}z + \tfrac{17}{4}}{2z^2 + z + 1}.$

19. $\dfrac{x^3 + 2x^2 - 17x - 10}{x + 5}$

Use −5 since x + 5 = x − (−5).

$$
\begin{array}{r}
-5\,\big)\,1 \quad 2 \quad -17 \quad -10 \\
\underline{\quad\;\; -5 \quad 15 \quad\;\; 10} \\
1 \quad -3 \quad -2 \quad\;\; 0
\end{array}
$$

Bring down the 1. Multiply 1 by
−5. Add 2 to −5, resulting in −3.
Next (−5)(−3) = 15; −17 + 15 = −2.
Now (−5)(−2) = 10; −10 + 10 = 0.
Thus,

$\dfrac{x^3 + 2x^2 - 17x - 10}{x + 5} = x^2 - 3x - 2.$

21. $\dfrac{m^4 - 3m^3 - 4m^2 + 12m}{m - 2}$

Use 2 since m − 2 = m − (2).

$$
\begin{array}{r}
2\,\big)\,1 \quad -3 \quad -4 \quad\;\; 12 \\
\underline{\quad\;\; 2 \quad -2 \quad -12} \\
1 \quad -1 \quad -6 \quad\;\; 0
\end{array}
$$

Thus,

$\dfrac{m^4 - 3m^3 - 4m^2 + 12m}{m - 2} = m^3 - m^2 - 6m.$

23. $\dfrac{3x^3 - 11x^2 - 20x + 3}{x - 5}$

Use 5 since x − 5 = x − (5).

$$
\begin{array}{r}
5\,\big)\,3 \quad -11 \quad -20 \quad\;\; 3 \\
\underline{\quad\;\; 15 \quad\;\; 20 \quad\;\; 0} \\
3 \quad\;\; 4 \quad\;\;\; 0 \quad\;\; 3
\end{array}
$$

Thus,

$\dfrac{3x^3 - 11x^2 - 20x + 3}{x - 5}$

$= 3x^2 + 4x + \dfrac{3}{x - 5}.$

25. $\dfrac{x^5 + 3x^4 + 2x^3 + 2x^2 + 3x + 1}{x + 2}$

Use −2 since x + 2 = x − (−2).

$$
\begin{array}{r}
-2\,\big)\,1 \quad 3 \quad 2 \quad\;\; 2 \quad\;\; 3 \quad\;\; 1 \\
\underline{\quad\;\; -2 \quad -2 \quad\;\; 0 \quad -4 \quad\;\; 2} \\
1 \quad 1 \quad 0 \quad\;\; 2 \quad -1 \quad\;\; 3
\end{array}
$$

Thus,

$\dfrac{x^5 + 3x^4 + 2x^3 + 2x^2 + 3x + 1}{x + 2}$

$= x^4 + x^3 + 2x - 1 + \dfrac{3}{x + 2}.$

27. $\dfrac{\tfrac{1}{3}x^3 - \tfrac{2}{9}x^2 + \tfrac{1}{27}x + 1}{x - \tfrac{1}{3}}$

Use $\dfrac{1}{3}$ since $x - \dfrac{1}{3} = x - \left(\dfrac{1}{3}\right)$.

$$
\begin{array}{r}
\tfrac{1}{3}\,\big)\,\tfrac{1}{3} \quad -\tfrac{2}{9} \quad\;\; \tfrac{1}{27} \quad\;\; 1 \\
\underline{\qquad\;\; \tfrac{1}{9} \quad -\tfrac{1}{27} \quad\;\; 0} \\
\tfrac{1}{3} \quad -\tfrac{1}{9} \quad\;\; 0 \quad\;\; 1
\end{array}
$$

Thus,

$$\frac{\frac{1}{3}x^3 - \frac{2}{9}x^2 + \frac{1}{27}x + 1}{x - \frac{1}{3}}$$

$$= \frac{1}{3}x^2 - \frac{1}{9}x + \frac{1}{x - \frac{1}{3}}.$$

29. $\dfrac{y^3 - 1}{y - 1}$

$y^3 - 1 = y^3 + 0y^2 + 0y - 1$

Use 1 since $y - 1 = y - (-1)$.

$$\begin{array}{r} 1\overline{)\,1 \quad 0 \quad 0 \quad -1} \\ 1 \quad 1 \quad 1 \\ \hline 1 \quad 1 \quad 1 \quad 0 \end{array}$$

Thus,

$$\frac{y^3 - 1}{y - 1} = y^2 + y + 1.$$

31. $\dfrac{x^4 - 1}{x - 1}$

$x^4 - 1 = x^4 + 0x^3 + 0x^2 + 0x - 1$

Use 1 since $x - 1 = x - (1)$.

$$\begin{array}{r} 1\overline{)\,1 \quad 0 \quad 0 \quad 0 \quad -1} \\ 1 \quad 1 \quad 1 \quad 1 \\ \hline 1 \quad 1 \quad 1 \quad 1 \quad 0 \end{array}$$

Thus,

$$\frac{x^4 - 1}{x - 1} = x^3 + x^2 + x + 1.$$

33. $P(x) = x^3 + x^2 + x - 8;\ k = 1$

Use synthetic division.

$$\begin{array}{r} 1\overline{)\,1 \quad 1 \quad 1 \quad -8} \\ 1 \quad 2 \quad 3 \\ \hline 1 \quad 2 \quad 3 \quad -5 \end{array}$$

Thus,

$$\frac{x^3 + x^2 + x - 8}{x - 1}$$

$$= x^2 + 2x + 3 - \frac{5}{x - 1}.$$

$P(x) = (x - 1)(x^2 + 2x + 3) - 5$

35. $P(x) = -x^3 + 2x^2 + 4;\ k = -2$

Use synthetic division.

$$\begin{array}{r} -2\overline{)\,-1 \quad 2 \quad 0 \quad 4} \\ 2 \quad -8 \quad 16 \\ \hline -1 \quad 4 \quad -8 \quad 20 \end{array}$$

Thus,

$$\frac{-x^3 + 2x^2 + 4}{x + 2}$$

$$= -x^2 + 4x - 8 + \frac{20}{x + 2}.$$

$P(x) = (x + 2)(-x^2 + 4x - 8) + 20$

37. $P(x) = x^4 - 3x^3 + 2x^2 - x + 5;$
$k = 3$

Use synthetic division.

$$\begin{array}{r} 3\overline{)\,1 \quad -3 \quad 2 \quad -1 \quad 5} \\ 3 \quad 0 \quad 6 \quad 15 \\ \hline 1 \quad 0 \quad 2 \quad 5 \quad 20 \end{array}$$

Thus,

$$\frac{x^4 - 3x^3 + 2x^2 - x + 5}{x - 3}$$

$$= x^3 + 2x + 5 + \frac{20}{x - 3}.$$

$P(x) = (x - 3)(x^3 + 2x + 5) + 20$

39. $k = 3;\ P(x) = x^2 - 4x + 5$

Divide $P(x)$ by $x - 3$. The remainder gives $P(3)$.

$$\begin{array}{r} 3\overline{)\,1 \quad -4 \quad 5} \\ 3 \quad -3 \\ \hline 1 \quad -1 \quad 2 \end{array}$$

$P(3) = 2$

41. $k = -2$; $P(x) = 5x^3 + 2x^2 - x$

Divide $P(x)$ by $x + 2$. The remainder gives $P(-2)$.

$$
\begin{array}{r|rrrr}
-2) & 5 & 2 & -1 & 0 \\
   &   & -10 & 16 & -30 \\
\hline
   & 5 & -8 & 15 & -30
\end{array}
$$

$P(-2) = -30$

43. $k = 2 + i$; $P(x) = x^2 - 5x + 1$

Divide $P(x)$ by $x - (2 + i)$; the
remainder gives $P(2 + i)$.

$$
\begin{array}{r|rrr}
2 + i) & 1 & -5 & 1 \\
       &   & 2 + i & -7 - i \\
\hline
       & 1 & -3 + i & -6 - i
\end{array}
$$

$P(2 + i) = -6 - i$

45. $k = 1 - i$; $P(x) = x^3 + x^2 - x + 1$

Divide $P(x) = x - (1 - i)$; the
remainder gives $P(1 - i)$.

$$
\begin{array}{r|rrrr}
1 - i) & 1 & 1 & -1 & 1 \\
       &   & 1 - i & 1 - 3i & -3 - 3i \\
\hline
       & 1 & 2 - i & -3i & -2 - 2i
\end{array}
$$

$P(1 - i) = -2 - 3i$

47. $P(x) = x^2 + 2x - 8$

Is 2 a zero of $P(x) = x^2 + 2x - 8$?
We want to see if $P(2) = 0$.

$$
\begin{array}{r|rrr}
2) & 1 & 2 & -8 \\
   &   & 2 & 8 \\
\hline
   & 1 & 4 & 0
\end{array}
$$

The remainder is 0, so 2 is zero.

49. 2; $P(g) = g^3 - 3g^2 + 4g - 4$

We want to see if $P(2) = 0$.

$$
\begin{array}{r|rrrr}
2) & 1 & -3 & 4 & -4 \\
   &   & 2 & -2 & 4 \\
\hline
   & 1 & -1 & 2 & 0
\end{array}
$$

The remainder is 0, so 2 is a zero.

51. 4; $P(r) = 2r^3 - 6r^2 - 9r + 6$

We want to see if $P(4) = 0$.

$$
\begin{array}{r|rrrr}
4) & 2 & -6 & -9 & 6 \\
   &   & 8 & 8 & -4 \\
\hline
   & 2 & 2 & -1 & 2
\end{array}
$$

The remainder is not 0, so 4 is not
a zero.

53. $2 + i$; $P(k) = k^2 + 3k + 4$

We want to see if $P(2 + i) = 0$.

$$
\begin{array}{r|rrr}
2 + i) & 1 & 3 & 4 \\
       &   & 2 + i & 9 + 7i \\
\hline
       & 1 & 5 + i & 13 + 7i
\end{array}
$$

Here the remainder is not zero, so
$2 + i$ is not a zero.

55. i; $P(x) = x^3 + 2ix^2 + 2x + i$

We want to see if $P(i) = 0$.

$$
\begin{array}{r|rrrr}
i) & 1 & 2i & 2 & i \\
   &   & i & -3 & -i \\
\hline
   & 1 & 3i & -1 & 0
\end{array}
$$

The remainder is 0, so i is a zero
of $P(x)$.

Section 6.3

1. Let $P(x) = 4x^2 + 2x + 42$.

    By the factor theorem, $x - 3$ will be a factor of $P(x)$ only if $P(3) = 0$.

    $$3 \overline{)\begin{array}{ccc} 4 & 2 & 42 \\ & 12 & 42 \\ \hline 4 & 14 & 84 \end{array}}$$

    Since the remainder is 84, $P(3) = 84$, not 0, so $x - 3$ is not a factor of $4x^2 + 2x + 42$.

3. Let $P(x) = x^3 + 2x^2 - 3$.

    $x - 1$ is a factor of $P(x)$ only if $P(1) = 0$.

    $$1 \overline{)\begin{array}{cccc} 1 & 2 & 0 & -3 \\ & 1 & 3 & 3 \\ \hline 1 & 3 & 3 & 0 \end{array}}$$

    Since the remainder is 0, $P(1) = 0$, so $x - 1$ is a factor of $P(x)$.

5. Let $P(x) = 3x^3 - 12x^2 - 11x - 20$.

    $x - 5$ is a factor of $P(x)$ only if $P(5) = 0$.

    $$5 \overline{)\begin{array}{cccc} 3 & -12 & -11 & -20 \\ & 15 & 15 & 20 \\ \hline 3 & 3 & 4 & 0 \end{array}}$$

    Since the remainder is 0, $P(5) = 0$, so $x - 5$ is a factor of $P(x)$.

7. Let $P(x) = 2x^4 + 5x^3 - 2x^2 + 5x + 3$.

    $x + 3$ is a factor of $P(x)$ only if $P(-3) = 0$.

    $$-3 \overline{)\begin{array}{ccccc} 2 & 5 & -2 & 5 & 3 \\ & -6 & 3 & -3 & -6 \\ \hline 2 & -1 & 1 & 2 & -3 \end{array}}$$

    The remainder is -3. $P(-3) = -3$, not 0, so $x + 3$ is not a factor of $P(x)$.

9. $P(x) = x^4(x + 3)^5(x - 8)^2$

    The factor $x^4 = (x - 0)^4$ indicates that 0 is a zero of multiplicity 4. The factor $(x + 3)^5 = [x - (-3)]^5$ indicates that -3 is a zero of multiplicity 5. The factor $(x - 8)^2$ indicates that 8 is a zero of multiplicity 2.

11. $P(x) = (4x - 7)^3(x - 5)$

    The factor $(4x - 7)^3$ indicates that $\frac{7}{4}$ is a zero of multiplicity 3.

    The factor $x - 5$ indicates that 5 is a zero of multiplicity 1.

13. $P(x) = (x - i)^4(x + i)^4$

    The factor $(x - i)^4$ indicates that i is a zero of multiplicity 4. The factor $(x + i)^4 = [x - (-i)]^4$ indicates that $-i$ is a zero of multiplicity 4.

15. Zeros of -3, -1, and 4; $P(2) = 5$

    These three zeros give $x - (-3) = x + 3$, $x - (-1) = x + 1$, and $x - 4$ as factors of $P(x)$. Since $P(x)$ has degree 3, these are the only possible factors, so $P(x)$ has the form

    $$P(x) = a(x + 3)(x + 1)(x - 4)$$

    for some nonzero real number a.

To find a, use the fact that P(2) = 5.

$$P(2) = a(2 + 3)(2 + 1)(2 - 4) = 5$$
$$\text{Let } x = 2, \ P(2) = 5$$
$$a(5)(3)(-2) = 5$$
$$-30a = 5$$
$$a = -\frac{1}{6}$$

Thus,

$$P(x) = -\frac{1}{6}(x + 3)(x + 1)(x - 4)$$
$$= -\frac{1}{6}(x + 3)(x^2 - 3x - 4)$$
$$= -\frac{1}{6}(x^3 - 13x - 12)$$
$$= -\frac{1}{6}x^3 + \frac{13}{6}x + 2.$$

17. Zeros of −2, 1, and 0; P(−1) = −1

The polynomial P(x) has the form

$$P(x) = a(x + 2)(x - 1)(x - 0).$$

Use the condition P(−1) = −1 to find a.

$$P(-1) = a(-1 + 2)(-1 - 1)(-1 - 0) = -1$$
$$\text{Let } x = -1; \ P(-1) = -1$$
$$a(1)(-2)(-1) = -1$$
$$2a = -1$$
$$a = -\frac{1}{2}$$

$$P(x) = -\frac{1}{2}(x + 2)(x - 1)x$$
$$= -\frac{1}{2}(x + 2)(x^2 - x)$$
$$= -\frac{1}{2}(x^3 + x^2 - 2x)$$
$$= -\frac{1}{2}x^3 - \frac{1}{2}x^2 + x$$

19. Zeros of 3, i, and −i; P(2) = 50

The polynomial P(x) has the form

$$P(x) = a(x - 3)(x - i)(x + i).$$

Use the condition P(2) = 50 to find a.

$$P(2) = a(2 - 3)(2 - i)(2 + i) = 50$$
$$\text{Let } x = 2; \ P(2) = 50$$
$$a(-1)(5) = 50$$
$$(2 + i)(2 - i) = 2^2 - i^2$$
$$= 4 - (-1) = 5$$
$$-5a = 50$$
$$a = -10$$

$$P(x) = -10(x - 3)(x - i)(x + i)$$
$$= -10(x - 3)(x^2 + 1)$$
$$= -10(x^3 - 3x^2 + x - 3)$$
$$= -10x^3 + 30x^2 - 10x + 30$$

For Exercises 21–35, find a function P defined by a polynomial of lowest degree with real coefficients having the given zeros.

21. 5 + i and 5 − i

$$P(x) = [x - (5 + i)][x - (5 - i)]$$
$$= [(x - 5) - i][(x - 5) + i]$$
$$= (x - 5)^2 - i^2$$
$$\quad (a - b)(a + b) = a^2 - b^2$$
$$= (x - 5)^2 + 1 \qquad i^2 = -1$$
$$= x^2 - 10x + 25 + 1$$
$$= x^2 - 10x + 26$$

23. 2, 1 − i, and 1 + i

$$P(x) = (x - 2)[x - (1 - i)][x - (1 + i)]$$
$$= (x - 2)[(x - 1) + i][(x - 1) - i]$$
$$= (x - 2)[(x - 1)^2 - i^2]$$
$$\quad (a + b)(a - b) = a^2 - b^2$$
$$= (x - 2)[(x - 1)^2 + 1]$$
$$= (x - 2)(x^2 - 2x + 1 + 1)$$
$$= (x - 2)(x^2 - 2x + 2)$$
$$= x^3 - 4x^2 + 6x - 4$$

25. $1 + \sqrt{2}$, $1 - \sqrt{2}$, and 1

$P(x)$

$= (x - 1)[x - (1 + \sqrt{2})][x - (1 - \sqrt{2})]$

$= (x - 1)[(x - 1) - \sqrt{2}][(x - 1) + \sqrt{2}]$

$= (x - 1)[(x - 1)^2 - (\sqrt{2})^2]$

$= (x - 1)[(x - 1)^2 - 2]$

$= (x - 1)(x^2 - 2x + 1 - 2)$

$= (x - 1)(x^2 - 2x - 1)$

$= x^3 - 3x^2 + x + 1$

27. $2 + i$, $2 - i$, 3, and -1

$P(x) = [x - (2 + i)][x - (2 - i)]$

$\quad \cdot (x - 3)(x + 1)$

$= [(x - 2) - i][(x - 2) + i]$

$\quad \cdot (x - 3)(x + 1)$

$= [(x - 2)^2 - i^2](x^2 - 2x - 3)$

$= [(x^2 - 4x + 4) + 1](x^2 - 2x - 3)$

$= (x^2 - 4x + 5)(x^2 - 2x - 3)$

$= x^4 - 6x^3 + 10x^2 + 2x - 15$

29. 2 and $3 + i$

Since the polynomial is required to have real coefficients, by the conjugate zeros theorem, the conjugate of $3 + i$, which is $3 - i$, must also be a zero.

Thus, the polynomial is

$P(x) = (x - 2)[x - (3 + i)][x - (3 - i)]$

$= (x - 2)[(x - 3) - i][(x - 3) + i]$

$= (x - 2)[(x - 3)^2 - i^2]$

$= (x - 2)(x^2 - 6x + 9 + 1)$

$= (x - 2)(x^2 - 6x + 10)$

$= x^3 - 8x^2 + 22x - 20.$

31. $2 - i$ and $3 + 2i$

Since the polynomial is required to have real coefficients, the conjugates of $2 - i$ and $3 + 2i$, which are $2 + i$ and $3 - 2i$, must also be zeros.

$P(x) = [x - (2 - i)][x - (2 + i)]$

$\quad \cdot [x - (3 + 2i)][x - (3 - 2i)]$

$= [(x - 2) + i][(x - 2) - i]$

$\quad \cdot [(x - 3) - 2i][(x - 3) + 2i]$

$= [(x - 2)^2 - i^2][(x - 3)^2 - 4i^2]$

$= (x^2 - 4x + 4 + 1)(x^2 - 6x + 9 + 4)$

$= (x^2 - 4x + 5)(x^2 - 6x + 13)$

$= x^4 - 10x^3 + 42x^2 - 82x + 65$

33. 4, $1 - 2i$, and $3 + 4i$

Since the polynomial is required to have real coefficients, the conjugates of $1 - 2i$ and $3 + 4i$, which are $1 + 2i$ and $3 - 4i$, must also be zeros.

$P(x) = (x - 4)[x - (1 - 2i)][x - (1 + 2i)]$

$\quad \cdot [x - (3 + 4i)][x - (3 - 4i)]$

$= (x - 4)[(x - 1) + 2i][(x - 1) - 2i]$

$\quad \cdot [(x - 3) - 4i][(x - 3) + 4i]$

$= (x - 4)[(x - 1)^2 - 4i^2]$

$\quad \cdot [(x - 3)^2 - 16i^2]$

$= (x - 4)(x^2 - 2x + 1 + 4)$

$\quad \cdot (x^2 - 6x + 9 + 16)$

$= (x - 4)(x^2 - 2x + 5)(x^2 - 6x + 25)$

$= (x - 4)$

$\quad \cdot (x^4 - 8x^3 + 42x^2 - 80x + 125)$

$= x^5 - 12x^4 + 74x^3 - 248x^2$

$\quad + 445x - 500$

35. 1 + 2i and 2 (multiplicity 2)

Since the polynomial is required to have real coefficients, the conjugate of 1 + 2i, which is 1 − 2i, must also be a zero.
Since 2 is a zero of multiplicity 2, two factors of (x − 2) are needed.

$$P(x) = [x - (1 + 2i)][x - (1 - 2i)]$$
$$\cdot (x - 2)(x - 2)$$
$$= [(x - 1) - 2i][(x - 1) + 2i]$$
$$\cdot (x - 2)^2$$
$$= [(x - 1)^2 - 4i^2](x^2 - 4x + 4)$$
$$= (x^2 - 2x + 1 + 4)(x^2 - 4x + 4)$$
$$= (x^2 - 2x + 5)(x^2 - 4x + 4)$$
$$= x^4 - 6x^3 + 17x^2 - 28x + 20$$

37. $P(x) = x^3 - x^2 - 4x - 6$; 3 is a zero.

Divide $P(x)$ by x − 3.

$$\begin{array}{r|rrrr} 3) & 1 & -1 & -4 & -6 \\ & & 3 & 6 & 6 \\ \hline & 1 & 2 & 2 & 0 \end{array}$$

Thus,

$$P(x) = (x - 3)(x^2 + 2x + 2).$$

The other zeros of P(x) will be the zeros of the quadratic polynomial $x^2 + 2x + 2$. To find them, solve the quadratic equation

$$x^2 + 2x + 2 = 0.$$

Since $x^2 + 2x + 2$ cannot be factored, use the quadratic formula with a = 1, b = 2, and c = 2.

$$x = \frac{-b \pm \sqrt{b^2 - 4ac}}{2a}$$
$$= \frac{-2 \pm \sqrt{4 - 4(1)(2)}}{2}$$

$$= \frac{-2 \pm \sqrt{-4}}{2}$$
$$= \frac{-2 \pm 2i}{2}$$
$$= \frac{2(-1 \pm i)}{2}$$
$$= -1 \pm i$$

The other zeros are −1 + i and −1 − i.

39. $P(x) = 2x^3 - 2x^2 - x - 6$; 2 is a zero.

Divide $P(x)$ by x − 2.

$$\begin{array}{r|rrrr} 2) & 2 & -2 & -1 & -6 \\ & & 4 & 4 & 6 \\ \hline & 2 & 2 & 3 & 0 \end{array}$$

Thus,

$$P(x) = (x - 2)(2x^2 + 2x + 3).$$

To find the other zeros, solve the quadratic equation

$$2x^2 + 2x + 3 = 0.$$

Since $2x^2 + 2x + 3$ cannot be factored, use the quadratic formula.

$$x = \frac{-2 \pm \sqrt{4 - 4(2)(3)}}{4}$$
$$= \frac{-2 \pm \sqrt{4 - 24}}{4}$$
$$= \frac{-2 \pm \sqrt{-20}}{4}$$
$$= \frac{-2 \pm 2i\sqrt{5}}{4}$$
$$= \frac{2(-1 \pm i\sqrt{5})}{2 \cdot 2}$$
$$= \frac{-1 \pm i\sqrt{5}}{2}$$
$$= -\frac{1}{2} \pm \frac{\sqrt{5}}{2}i \quad a + bi \text{ form}$$

The other zeros are

$$-\frac{1}{2} + \frac{\sqrt{5}}{2}i \text{ and } -\frac{1}{2} - \frac{\sqrt{5}}{2}i.$$

41. $P(x) = x^4 + 5x^2 + 4$; $-i$ is a zero.

First divide $P(x)$ by $x - (-i) =$ $x + i$.

```
-i)1   0    5    0    4
      -i   -1  -4i  -4
   1  -i    4  -4i    0
```

Now use the quotient polynomial for the next division.
Since $P(x)$ has real coefficients and $-i$ is a zero, its conjugate, i, is also a zero.

```
i)1  -i    4   -4i
      i    0    4i
   1   0    4    0
```

To find the remaining zeros, find the zeros of the quotient $x^2 + 4$.

$$x^2 + 4 = 0$$
$$x^2 = -4$$
$$x = \pm 2i$$

The other zeros are i, $2i$, and $-2i$.

43. $P(x) = x^4 - 3x^3 + 6x^2 + 2x - 60$; $1 + 3i$ is a zero.

Since $P(x)$ has real coefficients and $1 + 3i$ is a zero, $1 - 3i$ is also a zero.
First divide the original polynomial by $x - (1 + 3i)$.

```
1+3i)1   -3         6        2        -60
          1+3i   -11-3i    4-18i       60
      1  -2+3i    -5-3i    6-18i        0
```

Now divide the quotient from the first division by $x - (1 - 3i)$.

```
1-3i)1   -2+3i    -5-3i     6-18i
          1-3i     1+3i    -6+18i
      1    1        -6        0
```

To find the zeros of the quadratic polynomial $x^2 - x - 6$, solve the equation $x^2 - x - 6 = 0$ by factoring.

$$x^2 - x - 6 = 0$$
$$(x - 3)(x + 2) = 0$$
$$x = 3 \quad \text{or} \quad x = -2$$

The other zeros are 3, -2, and $1 - 3i$.

45. $P(x) = 2x^3 - 3x^2 - 17x + 30$; $k = 2$

Since 2 is a zero of $P(x)$, $x - 2$ is a factor of $P(x)$.
Divide $P(x)$ by $x - 2$.

```
2)2   -3   -17    30
       4    2    -30
   2    1   -15    0
```

The quotient is $2x^2 + x - 15$, so

$$P(x) = (x - 2)(2x^2 + x - 15).$$

Factor $2x^2 + x - 15$ as $(2x - 5)(x + 3)$ to get

$$P(x) = (x - 2)(2x - 5)(x + 3).$$

47. $P(x) = 6x^3 + 25x^2 + 3x - 4$; $k = -4$

Since -4 is a zero of $P(x)$, $x + 4$ is a factor.
Divide $P(x)$ by $x + 4$.

```
-4)6   25    3   -4
      -24   -4    4
   6    1   -1    0
```

Factor the quotient.

$$6x^2 + x - 1 = (3x - 1)(2x + 1)$$

Thus,

$$P(x) = (x + 4)(3x - 1)(2x + 1).$$

49. $P(x) = x^3 + (7 - 3i)x^2 + (12 - 21i)x - 36i$;

$k = 3i$

Since $3i$ is a zero of $P(x)$, $x - 3i$ is a factor.

Divide $P(x)$ by $x - 3i$.

$$3i\overline{)1 \quad 7 - 3i \quad 12 - 21i \quad -36i}$$
$$3i \quad 21i \quad 36i$$
$$\overline{1 \quad 7 \quad 12 \quad 0}$$

Factor the quotient.

$x^2 + 7x + 12 = (x + 4)(x + 3)$

Thus,

$P(x) = (x - 3i)(x + 4)(x + 3)$.

51. $P(x) = 2x^3 + (3 - 2i)x^2 + (-8 - 5i)x + 3 + 3i$;

$k = 1 + i$

Since $1 + i$ is a zero of $P(x)$,

$x - (1 + i)$ is a factor.

Divide $P(x)$ by $x - (1 + i)$.

$$1 + i\overline{)2 \quad 3 - 2i \quad -8 - 5i \quad 3 + 3i}$$
$$2 + 2i \quad 5 + 5i \quad -3 + 3i$$
$$\overline{2 \quad 5 \quad -3 \quad 0}$$

Factor the quotient.

$2x^2 + 5x - 3 = (2x - 1)(x + 3)$

Therefore,

$P(x) = [x - (1 + i)](2x - 1)(x + 3)$

$ = (x - 1 - i)(2x - 1)(x + 3)$.

53. $P(x) = x^4 + 2x^3 - 7x^2 - 20x - 12$

For -2 to be a zero of multiplicity 2, $P(x)$ must have two factors of $x + 2$.

Divide $P(x)$ by $x + 2$.

$$-2\overline{)1 \quad 2 \quad -7 \quad -20 \quad -12}$$
$$-2 \quad 0 \quad 14 \quad 12$$
$$\overline{1 \quad 0 \quad -7 \quad -6 \quad 0}$$

Now, divide the quotient by $x + 2$.

$$-2\overline{)1 \quad 0 \quad -7 \quad -6}$$
$$-2 \quad 4 \quad 6$$
$$\overline{1 \quad -2 \quad -3 \quad 0}$$

We have shown that -2 is a zero of multiplicity 2.

Factor $x^2 - 2x - 3$ to obtain the other zeros.

$x^2 - 2x - 3 = (x + 1)(x - 3)$

The other zeros are 3 and -1.

Therefore,

$P(x) = (x + 2)^2(x + 1)(x - 3)$.

55. If $a + bi$ is a zero of a polynomial function with real coefficients, then by the conjugate zeros theorem $a - bi$ must also be a zero. That is, complex zeros must occur in pairs for such polynomials. There may be 0, 2, or 4 or any even number of complex zeros, and the remaining zeros would have to be real. This polynomial function with real coefficients of degree five could possibly have 0, 2, or 4 complex zeros (but not 6 or more, since that would exceed the degree of the polynomial).

If there are 0 complex zeros, then all 5 must be real. If there are 2 complex zeros, then the remaining 3 must be real. If there are 4 complex zeros, then the remaining 1 must be real.

Therefore, the possible numbers of real zeros in this case are 1, 3, or 5.

59. $s(t) = t^3 - 2t^2 - 5t + 6$

The displacement is 0 after 1 sec has elapsed which means $s(t) = 0$ at $t = 1$, or equivalently $t - 1$, is a factor of $s(t)$.

Divide $s(t)$ by $t - 1$.

$$\begin{array}{r} 1)\overline{1 \quad -2 \quad -5 \quad 6} \\ \underline{1 \quad -1 \quad -6} \\ 1 \quad -1 \quad -6 \quad 0 \end{array}$$

To find the other zeros, factor the quotient.

$$t^2 - t - 6 = 0$$
$$(t - 3)(t + 2) = 0$$
$$t = 3 \quad \text{or} \quad t = -2$$

Discard -2 since the time must be positive. The other time that the displacement is 0 is at $t = 3$ sec.

61. Let $c = a + bi$, $d = m + ni$.

$$\overline{c + d} = \overline{(a + bi) + (m + ni)}$$
$$= \overline{(a + m) + (b + n)i}$$
$$= (a + m) - (b + n)i$$

$$\overline{c} + \overline{d} = \overline{a + bi} + \overline{m + ni}$$
$$= (a - bi) + (m - ni)$$
$$= (a + m) - (b + n)i$$

Therefore, $\overline{c + d} = \overline{c} + \overline{d}$.

63. If a is a real number, then

$$a = a + 0i$$
$$\overline{a} = \overline{a + 0i}$$
$$= a - 0i$$
$$= a.$$

65. Prove that a polynomial of degree n has at most n distinct zeros. Let $P(x)$ be a polynomial of degree n. Assume that the statement to be proved is false. If $P(x)$ has more than n distinct zeros, it must have at least $n + 1$ distinct zeros. Let x_1, x_2, ..., x_n, x_{n+1} be the zeros of $P(x)$. Then $P(x)$ can be written in factored form as

$$P(x) = (x - x_1)(x - x_2)$$
$$\ldots (x - x_n)(x - x_{n+1}),$$

so that $P(x)$ is the product of $n + 1$ distinct linear factors. If these factors are multiplied together, the leading coefficient will be x^{n+1}, which means that $P(x)$ is of degree $n + 1$.

This contradicts the requirement that $P(x)$ be of degree n. Since the assumption that $P(x)$ has more than n distinct zeros leads to a contradiction, this assumption must be false. Thus, we have shown by indirect proof that a polynomial of degree n is at most n distinct zeros.

Section 6.4

1. $P(x) = 6x^3 + 17x^2 - 31x - 1$

By the rational zeros theorem, if p/q is a rational zero of $P(x)$, p must be a factor of the constant term $a_0 = -1$ and q must be a factor of the leading coefficient $a_3 = 6$.

The possible values of p are ±1.

The possible values of q are ±1, ±2, ±3, ±6.

The possible rational zeros are all possible quotients of the form p/q:

$$\pm 1, \ \pm\frac{1}{2}, \ \pm\frac{1}{3}, \ \pm\frac{1}{6}.$$

3. $P(x) = 12x^3 + 20x^2 - x - 2$

The possible values of p are the factors of $a_0 = -2$: ±1, ±2.

The possible values of q are the factors of $a_3 = 12$: ±1, ±2, ±3, ±4, ±6, ±12.

The possible rational zeros, p/q, are

$$\pm 1, \ \pm\frac{1}{2}, \ \pm\frac{1}{3}, \ \pm\frac{1}{4}, \ \pm\frac{1}{6}, \ \pm\frac{1}{12}, \ \pm 2, \ \pm\frac{2}{3}.$$

5. $P(x) = 2x^3 + 7x^2 + 12x - 8$

The possible values of p are the factors of $a_0 = -8$: ±1, ±2, ±4, ±8.

The possible values of q are the factors of $a_3 = 2$: ±1, ±2.

The possible rational zeros, p/q, are

$$\pm 1, \ \pm\frac{1}{2}, \ \pm 2, \ \pm 4, \ \pm 8.$$

7. $P(x) = x^4 + 4x^3 + 3x^2 - 10x + 50$

The possible values of p are the factors of $a_0 = 50$: ±1, ±2, ±5, ±10, ±25, ±50.

The possible values of q are the factors $a_4 = 1$: ±1.

The possible rational zeros, p/q, are

±1, ±2, ±5, ±10, ±25, ±50.

11. $P(x) = x^3 - 2x^2 - 13x - 10$

The possible rational zeros are ±1, ±2, ±5, ±10.

$$
\begin{array}{r|rrr}
-1) & 1 & -2 & -13 & -10 \\
   &   & -1 &   3 & 10 \\
\hline
   & 1 & -3 & -10 &   0
\end{array}
$$

-1 is a zero, so P(x) may be factored as

$$P(x) = (x + 2)(x^2 - 3x - 10).$$

To find the remaining zeros, factor the quotient, $x^2 - 3x - 10$.

$$
\begin{aligned}
x^2 - 3x - 10 &= 0 \\
(x + 2)(x - 5) &= 0 \\
x = -2 \ \ \text{or} \ \ x &= 5
\end{aligned}
$$

The rational zeros are -1, -2, and 5.

13. $P(x) = x^3 + 6x^2 - x - 30$

The possible rational zeros are ±1, ±2, ±3, ±5, ±6, ±10, ±15, ±30.

$$
\begin{array}{r|rrr}
2) & 1 & 6 & -1 & 30 \\
   &   & 2 & 16 & 30 \\
\hline
   & 1 & 8 & 15 & 0
\end{array}
$$

2 is a zero.

The remaining zeros are the zeros of the quotient, $x^2 + 8x + 15$. These zeros may be found by factoring.

$$
\begin{aligned}
x^2 + 8x + 15 &= 0 \\
(x + 3)(x + 5) &= 0 \\
x = -3 \ \ \text{or} \ \ x &= -5
\end{aligned}
$$

The rational zeros are 2, -3, and -5.

15. $P(x) = x^3 + 9x^2 - 14x - 24$

The possible rational zeros are

$\pm 1, \pm 2, \pm 3, \pm 4, \pm 6, \pm 8, \pm 12, \pm 24$.

Synthetic division will show that no number on this list is a zero.
Thus, there are no rational zeros.

17. $P(x) = x^4 + 9x^3 + 21x^2 - x - 30$

The possible rational zeros are

$\pm 1, \pm 2, \pm 3, \pm 5, \pm 6, \pm 10, \pm 15, \pm 30$.

$$\begin{array}{r|rrrrr} 1) & 1 & 9 & 21 & -1 & -30 \\ & & 1 & 10 & 31 & 30 \\ \hline & 1 & 10 & 31 & 30 & 0 \end{array}$$

1 is a zero.
Now find zeros of the quotient polynomial

$$x^3 + 10x^2 + 31x + 30.$$

Possible zeros are

$\pm 1, \pm 2, \pm 3, \pm 5, \pm 6, \pm 10, \pm 15, \pm 30$.

$$\begin{array}{r|rrrr} -2) & 1 & 10 & 31 & 30 \\ & & -2 & -16 & -30 \\ \hline & 1 & 8 & 15 & 0 \end{array}$$

-2 is a zero.
Finally, find the zeros of $x^2 + 8x + 15$ by factoring.

$$x^2 + 8x + 15 = 0$$
$$(x + 3)(x + 5) = 0$$
$$x = -3 \quad \text{or} \quad x = -5$$

The rational zeros are 1, -2, -3, and -5.

19. $P(x) = 6x^3 + 17x^2 - 31x - 12$

The possible rational zeros are

$\pm 1, \pm 2, \pm 3, \pm 4, \pm 6, \pm 12,$

$\pm \frac{1}{2}, \pm \frac{3}{2}, \pm \frac{1}{3}, \pm \frac{2}{3}, \pm \frac{4}{3}, \pm \frac{1}{6}$.

Synthetic division shows that -4 is a zero.

$$\begin{array}{r|rrrr} -4) & 6 & 17 & -31 & -12 \\ & & -24 & 28 & 12 \\ \hline & 6 & -7 & -3 & 0 \end{array}$$

The remaining zeros can be found by factoring the quotient polynomial, $6x^2 - 7x - 3$.

$$6x^2 - 7x - 3 = 0$$
$$(2x - 3)(3x + 1) = 0$$
$$2x - 3 = 0 \quad \text{or} \quad 3x + 1 = 0$$
$$x = \frac{3}{2} \quad \text{or} \quad x = -\frac{1}{3}$$

The rational zeros are -4, $\frac{3}{2}$, and $-\frac{1}{3}$.

Since the three zeros of $P(x)$ are -4, $\frac{3}{2}$, and $-\frac{1}{3}$,

$$P(x) = a_3 [x - (-4)]\left(x - \frac{3}{2}\right)\left[x - \left(-\frac{1}{3}\right)\right].$$

Since $a_3 = 6 = 2 \cdot 3$, fractions can be eliminated in the last two factors by writing the product as

$$P(x) = 6(x + 4)\left(x - \frac{3}{2}\right)\left(x + \frac{1}{3}\right)$$
$$= (x + 4) \cdot 2\left(x - \frac{3}{2}\right) \cdot 3\left(x + \frac{1}{3}\right)$$
$$= (x + 4)(2x - 3)(3x + 1).$$

21. $P(x) = 12x^3 + 20x^2 - x - 6$

The possible rational zeros are

$\pm 1, \pm 2, \pm 3, \pm 6, \pm \frac{1}{2}, \pm \frac{3}{2},$

$\pm \frac{1}{3}, \pm \frac{2}{3}, \pm \frac{1}{4}, \pm \frac{3}{4}, \pm \frac{1}{6}, \pm \frac{1}{12}$.

$$\frac{1}{2}\overline{)\begin{array}{rrrr} 12 & 20 & -1 & -6 \\ & 6 & 13 & 6 \\ \hline 12 & 26 & 12 & 0 \end{array}}$$

$\frac{1}{2}$ is a zero.

Now find the zeros of $12x^2 + 26x + 12$ by factoring.

$$2(6x^2 + 13x + 6) = 0$$
$$2(3x + 2)(2x + 3) = 0$$
$$x = -\frac{2}{3} \quad \text{or} \quad x = -\frac{3}{2}$$

Thus, the rational zeros are $\frac{1}{2}$, $-\frac{2}{3}$, and $-\frac{3}{2}$.

$$P(x) = a_3\left(x - \frac{1}{2}\right)\left(x + \frac{2}{3}\right)\left(x + \frac{3}{2}\right)$$

Since $a_3 = 12 = 2 \cdot 3 \cdot 2$, fractions can be cleared by writing the product as

$$P(x) = 12\left(x - \frac{1}{2}\right)\left(x + \frac{2}{3}\right)\left(x + \frac{3}{2}\right)$$
$$= 2\left(x - \frac{1}{2}\right) \cdot 3\left(x + \frac{2}{3}\right) \cdot 2\left(x + \frac{3}{2}\right)$$
$$= (2x - 1)(3x + 2)(2x + 3).$$

23. $P(x) = 2x^3 + 7x^2 + 12x - 8$

The possible rational zeros are

$$\pm 1, \ \pm 2, \ \pm 4, \ \pm 8, \ \pm\frac{1}{2}.$$

$$\frac{1}{2}\overline{)\begin{array}{rrrr} 2 & 7 & 12 & -8 \\ & 1 & 4 & 8 \\ \hline 2 & 8 & 16 & 0 \end{array}}$$

$\frac{1}{2}$ is a zero.

Now find the zeros of
$2x^2 + 8x + 16 = 2(x^2 + 4x + 8)$
by the quadratic formula.

$$x = \frac{-4 \pm \sqrt{16 - 32}}{2}$$
$$= \frac{-4 \pm 4i}{2}$$
$$= -2 \pm 2i$$

Thus,

$x = \frac{1}{2}$ is the only rational zero.

$$P(x) = 2\left(x - \frac{1}{2}\right)(x^2 + 4x + 8)$$
$$= (2x - 1)(x^2 + 4x + 8)$$

25. $P(x) = 2x^4 + 3x^3 - 4x^2 - 3x + 2$

The possible rational zeros are

$$\pm 1, \ \pm\frac{1}{2}, \ \pm 2.$$

$$1\overline{)\begin{array}{rrrrr} 2 & 3 & -4 & -3 & -2 \\ & 2 & 5 & 1 & -2 \\ \hline 2 & 5 & 1 & -2 & 0 \end{array}}$$

$$-1\overline{)\begin{array}{rrrr} 2 & 5 & 1 & -2 \\ & -2 & -3 & 2 \\ \hline 2 & 3 & -2 & 0 \end{array}}$$

$$2x^2 + 3x - 2 = 0$$
$$(2x - 1)(x + 2) = 0$$
$$x = \frac{1}{2} \quad \text{or} \quad x = -2$$

The rational zeros are 1, -1, $\frac{1}{2}$, and -2.

$$P(x) = 2(x - 1)(x + 1)\left(x - \frac{1}{2}\right)(x + 2)$$
$$= (x - 1)(x + 1)(2x - 1)(x + 2)$$

27. $P(x) = 3x^4 + 5x^3 - 10x^2 - 20x - 8$

The possible rational zeros are

$$\pm 1, \ \pm\frac{1}{3}, \ \pm 2, \ \pm\frac{2}{3}, \ \pm 4, \ \pm\frac{4}{3}, \ \pm 8, \ \pm\frac{8}{3}.$$

$$\begin{array}{r|rrrr} -1) & 3 & 5 & -10 & -20 & -8 \\ & & -3 & -2 & 12 & 8 \\ \hline & 3 & 2 & -12 & -8 & 0 \end{array}$$

$$\begin{array}{r|rrrr} -2) & 3 & 2 & -12 & -8 \\ & & -6 & 8 & 8 \\ \hline & 3 & -4 & -4 & 0 \end{array}$$

$$3x^2 - 4x - 4 = 0$$

$$(3x + 2)(x - 2) = 0$$

$$x = -\frac{2}{3} \quad \text{or} \quad x = 2$$

The rational zeros are -1, -2, $-\frac{2}{3}$, and 2.

$$P(x) = 3(x + 1)(x + 2)\left(x + \frac{2}{3}\right)(x - 2)$$

$$= (x + 1)(x + 2)(3x + 2)(x - 2)$$

29. $P(x) = x^3 - \frac{4}{3}x^2 - \frac{13}{3}x - 2$

Multiply by 3 to clear fractions.
$3P(x) = 3x^3 - 4x^2 - 13x - 6$
has the same zeros as $P(x)$.
Since we now have a polynomial with integer coefficients, the rational zeros theorem can be used.
The possible rational zeros are

$$\pm 1, \ \pm 2, \ \pm 3, \ \pm 6, \ \pm\frac{1}{3}, \ \pm\frac{2}{3}.$$

Synthetic division shows that 3 is a zero.

$$\begin{array}{r|rrrr} 3) & 3 & -4 & -13 & -6 \\ & & 9 & 15 & 6 \\ \hline & 3 & 5 & 2 & 0 \end{array}$$

The remaining zeros can be found by factoring.

$$3x^2 + 5x + 2 = 0$$

$$(3x + 2)(x + 1) = 0$$

$$x = -\frac{2}{3} \quad \text{or} \quad x = -1$$

The rational zeros are 3, $-\frac{2}{3}$, and -1.

31. $P(x) = x^4 + \frac{1}{4}x^3 + \frac{11}{4}x^2 + x - 5$

Multiply by 4 to clear fractions.
$4P(x) = 4x^4 + x^3 + 11x^2 + 4x - 20$
has the same zeros as $P(x)$.
The possible rational zeros are

$$\pm 1, \ \pm 2, \ \pm 4, \ \pm 5, \ \pm 10,$$

$$\pm 20, \ \pm\frac{1}{2}, \ \pm\frac{5}{2}, \ \pm\frac{1}{4}, \ \pm\frac{5}{4}.$$

1 is a zero.

$$\begin{array}{r|rrrr} 1) & 4 & 1 & 11 & 4 & -20 \\ & & 4 & 5 & 16 & 20 \\ \hline & 4 & 5 & 16 & 20 & 0 \end{array}$$

Now find the zeros of the quotient $4x^3 + 5x^2 + 16x + 20$, which has the same set of possible rational zeros.

$-\frac{5}{4}$ is a zero.

$$\begin{array}{r|rrr} -\frac{5}{4}) & 4 & 5 & 16 & 20 \\ & & -5 & 0 & -20 \\ \hline & 4 & 0 & 16 & 0 \end{array}$$

Now find the zeros of $4x^2 + 16$.

$$4x^2 + 16 = 0$$

$$4(x^2 + 4) = 0$$

$$x^2 + 4 = 0$$

$$x^2 = -4$$

$$x = \pm 2i$$

These are not rational.
Thus, the rational zeros are 1 and $-\frac{5}{4}$.

33. $P(x) = \frac{1}{3}x^5 + x^4 - \frac{5}{3}x^3 - \frac{11}{3}x^2 + 4$

Multiply by 3 to clear fractions.
$3P(x) = x^5 + 3x^4 - 5x^3 - 11x^2 + 12$
has the same zeros as $P(x)$.

The possible rational roots are

$$\pm 1, \ \pm 2, \ \pm 3, \ \pm 6, \ \pm 12.$$

1 is a zero.

```
1)1   3   -5   -11    0    12
      1    4    -1  -12   -12
  ───────────────────────────
   1   4   -1   -12  -12    0
```

Now try to find the zeros of the quotient, $x^4 + 4x^3 - x^2 - 12x - 12x$, which has the same set of possible rational zeros.

Synthetic division will show that no number on the list of possibilities is a zero.

Thus, the only rational zero is 1.

For Exercises 35–43, see the answer graphs in the textbook.

35. $P(x) = 2x^3 - 5x^2 - x + 6$

The possible rational zeros are

$$\pm 1, \ \pm 2, \ \pm 3, \ \pm 6, \ \pm\frac{1}{2}, \ \pm\frac{3}{2}.$$

```
-1)2   -5   -1    6
       -2    7   -6
   ─────────────────
    2   -7    6    0
```

−1 is a zero.

Find the other zeros by factoring.

$$2x^2 - 7x + 6 = 0$$

$$(2x - 3)(x - 2) = 0$$

$$x = \frac{3}{2} \quad \text{or} \quad x = 2$$

Thus, the rational zeros are −1, $\frac{3}{2}$, and 2.

P(x) may be written in factored form as

$$P(x) = (x + 1)(2x - 3)(x - 2).$$

The three zeros divide the x-axis into four intervals:

$$(-\infty, \ -1), \ \left(1, \ \frac{3}{2}\right), \ \left(\frac{3}{2}, \ 2\right), \ \text{and}$$

$$(2, \ \infty).$$

Test a point in each interval to find the sign of P(x) (and thus the location of the graph relative to the x-axis) in that interval.

Interval	Test point	Value of P(x)	Location relative to x-axis
$(-\infty, \ -1)$	−2	−28	Below
$\left(-1, \ \frac{3}{2}\right)$	0	6	Above
$\left(\frac{3}{2}, \ 2\right)$	$\frac{7}{4}$	$-\frac{11}{32}$	Below
$(2, \ \infty)$	3	12	Above

Sketch the graph.

37. $P(x) = x^3 + x^2 - 8x - 12$

The possible rational zeros are

$$\pm 1, \ \pm 2, \ \pm 3, \ \pm 4, \ \pm 6, \ \pm 12.$$

```
-2)1    1   -8   -12
       -2    2    12
   ──────────────────
    1   -1   -6    0
```

$$x^2 - x - 6 = 0$$

$$(x - 3)(x + 2) = 0$$

$$x = 3 \quad \text{or} \quad x = -2$$

Thus, the rational zeros are −2 (multiplicity 2) and 3. P(x) may be written in factored form as

$$P(x) = (x + 2)^2 (x - 3).$$

These zeros divide the x-axis into three intervals. Test a point in each interval to find the sign of P(x) in that interval.

Interval	Test point	Value of $P(x)$	Location relative to x—axis
$(-\infty, -2)$	-3	-6	Below
$(-2, 3)$	0	-12	Below
$(3, \infty)$	4	36	Above

Sketch the graph.

39. $P(x) = -x^3 - x^2 + 8x + 12$

The possible rational zeros are
$\pm 1, \pm 2, \pm 3, \pm 4, \pm 6, \pm 12$.

$$-2\overline{)\begin{array}{rrrr} -1 & -1 & 8 & 12 \\ & 2 & -2 & -12 \\ \hline -1 & 1 & 6 & 0 \end{array}}$$

-2 is a zero.

Find the other zeros by factoring.

$$-x^2 + x + 6 = 0$$
$$-1(x^2 - x - 6) = 0$$
$$-1(x - 3)(x + 2) = 0$$
$$x = 3 \quad \text{or} \quad x = -2$$

Thus, the rational zeros are -2
(multiplicity 2) and 3. $P(x)$ may
be written in factored form as

$P(x) = -(x - 3)(x + 2)^2$ or
$P(x) = (-x + 3)(x + 2)^2$.

The two zeros divide the x—axis into
three intervals.
Test a point in each interval to
find the sign of $P(x)$ in that
interval.

Interval	Test point	Value of $P(x)$	Location relative to x—axis
$(-\infty, -2)$	-3	6	Above
$(-2, 3)$	0	-12	Below
$(3, \infty)$	4	36	Below

Sketch the graph.

41. $P(x) = x^4 - 18x^2 + 81$

The possible rational zeros are
$\pm 1, \pm 3, \pm 9, \pm 27, \pm 81$.

$$3\overline{)\begin{array}{rrrrr} 1 & 0 & -18 & 0 & 81 \\ & 3 & 9 & -27 & -81 \\ \hline 1 & 3 & -9 & -27 & 0 \end{array}}$$

3 is a zero.

$$-3\overline{)\begin{array}{rrrr} 1 & 3 & -9 & -27 \\ & -3 & 0 & 27 \\ \hline 1 & 0 & -9 & 0 \end{array}}$$

-3 is a zero.

Find the other zeros by factoring.

$$x^2 - 9 = 0$$
$$(x + 3)(x - 3) = 0$$
$$x = -3 \quad \text{or} \quad x = 3$$

Thus, the rational zeros are -3
(multiplicity 2) and 3 (multiplicity 2).

$P(x)$ may be written in factored form
as

$$P(x) = (x + 3)^2(x - 3)^2.$$

The two zeros divide the x—axis into
three intervals.
Test a point in each interval to
find the sign of $P(x)$ in that
interval.

Interval	Test point	Value of $P(x)$	Location relative to x—axis
$(-\infty, -3)$	-4	49	Above
$(-3, 3)$	0	81	Above
$(3, \infty)$	4	49	Above

Sketch the graph.

43. $P(x) = 2x^4 + x^3 - 6x^2 - 7x - 2$

The possible rational zeros are ± 1, ± 2, $\pm\frac{1}{2}$.

$$
\begin{array}{r|rrrr}
2) & 2 & 1 & -6 & -7 & -2 \\
   &   & 4 & 10 & 8 & 2 \\
\hline
   & 2 & 5 & 4 & 1 & 0
\end{array}
$$

2 is a zero.

$$
\begin{array}{r|rrrr}
-1) & 2 & 5 & 4 & 1 \\
    &   & -2 & -3 & -1 \\
\hline
    & 2 & 3 & 1 & 0
\end{array}
$$

−1 is a zero.

Find the other zeros by factoring.

$$2x^2 + 3x + 1 = 0$$

$$(2x + 1)(x + 1) = 0$$

$$2x + 1 = 0 \quad \text{or} \quad x + 1 = 0$$

$$x = -\frac{1}{2} \quad \text{or} \qquad x = -1$$

Thus, the rational zeros are −1 (multiplicity 2), 2, and $-\frac{1}{2}$.

P(x) may be written in factored form as

$$P(x) = (2x + 1)(x - 2)(x + 1)^2.$$

The three zeros divide the x-axis into four intervals.
Test a point in each interval to find the sign of P(x) in that interval.

Interval	Test point	Value of P(x)	Location relative to x-axis
$(-\infty, -1)$	−2	12	Above
$(-1, -\frac{1}{2})$	−.9	.0232	Above
$(-\frac{1}{2}, 2)$	0	−2	Below
$(2, \infty)$	3	112	Above

Sketch the graph.

45. Show that $P(x) = x^2 - 2$ has no rational zeros, so that $\sqrt{2}$ must be irrational.

$$P(x) = x^2 - 2$$

By the rational zeros theorem, the possible rational zeros are ± 1, ± 2.

Try 1.
$$
\begin{array}{r|rrr}
1) & 1 & 0 & -2 \\
   &   & 1 & 1 \\
\hline
   & 1 & 1 & -1
\end{array}
$$

Try 2.
$$
\begin{array}{r|rrr}
2) & 1 & 0 & -2 \\
   &   & 2 & 4 \\
\hline
   & 1 & 2 & 2
\end{array}
$$

Try −1.
$$
\begin{array}{r|rrr}
-1) & 1 & 0 & -2 \\
    &   & -1 & 1 \\
\hline
    & 1 & -1 & 1
\end{array}
$$

Try −2.
$$
\begin{array}{r|rrr}
-2) & 1 & 0 & -2 \\
    &   & -2 & 4 \\
\hline
    & 1 & -2 & 2
\end{array}
$$

Therefore, there are no rational zeros of $x^2 - 2$. However, $\sqrt{2}$ is a zero of P(x) since

$$P(\sqrt{2}) = (\sqrt{2})^2 - 2 = 2 - 2 = 0.$$

Since $\sqrt{2}$ is a real zero that is not rational, it must be irrational.

47. Show that $P(x) = x^4 + 5x^2 + 4$ has no rational zeros.

The possible rational zeros are

$$\pm 1, \ \pm 2, \ \pm 4.$$

Each of these can be tested by synthetic division. Two of these synthetic divisions are shown here.

$$
\begin{array}{r|rrrrr}
1) & 1 & 0 & 5 & 0 & 4 \\
   &   & 1 & 1 & 6 & 6 \\
\hline
   & 1 & 1 & 6 & 6 & 10
\end{array}
$$

$$
\begin{array}{r|rrrrr}
-1) & 1 & 0 & 5 & 0 & 4 \\
    &   & -1 & 1 & -6 & 6 \\
\hline
    & 1 & -1 & 6 & -6 & 10
\end{array}
$$

When all six of these divisions have been tried, it will be shown that $P(x)$ has no rational zeros.

This exercise can also be solved by factoring, rather than synthetic division.

$$x^4 + 5x^2 + 4 = 0$$
$$(x^2 + 4)(x^2 + 1) = 0$$
$$x^2 + 4 = 0 \quad \text{or} \quad x^2 + 1 = 0$$
$$x^2 = -4 \quad \text{or} \quad x^2 = -1$$
$$x = \pm 2i \quad \text{or} \quad x = \pm i$$

This shows that all of the zeros are imaginary and, therefore, there are no rational zeros.

49. Let $P(x) = a_n x^n + a_{n-1} x^{n-1}$
$$+ \ldots + a_1 x + a_0.$$

By the rational zeros theorem, for any rational zero p/q, p is a factor of a_0 and q is a factor of a_n. Any integer k can be written as the rational number $k/1$. Since 1 is a factor of any value of a_n, the integer k must be a factor of a_0.

Section 6.5

1. $P(x) = 2x^3 - 4x^2 + 2x + 7$

P(x) has 2 variations in sign:
$$P(x) = +2x^3 - 4x^2 + 2x + 7.$$
$$\underbrace{}_{1 \qquad 2}$$

Thus, P has either 2 or $2 - 2 = 0$ positive real zeros.

$P(-x)$ has 1 variation in sign:
$$P(-x) = 2(-x)^3 - 4(-x)^2 + 2(-x) + 7$$
$$= -2x^3 - 4x^2 - 2x + 7.$$
$$\underbrace{}_{1}$$

Thus, P has 1 negative real zero.

3. $P(x) = 5x^4 + 3x^2 + 2x - 9$

P(x) has 1 variation in sign:
$$P(x) = +5x^4 + 3x^2 + 2x - 9.$$
$$\underbrace{}_{1}$$

Thus, P has 1 positive real zero.

$P(-x)$ has 1 variation in sign:
$$P(-x) = 5(-x)^4 + 3(-x)^2 + 2(-x) - 9$$
$$= 5x^4 + 3x^2 - 2x - 9.$$
$$\underbrace{}_{1}$$

Thus, P has 1 negative real zero.

5. $P(x) = x^5 + 3x^4 - x^3 + 2x + 3$

P(x) has 2 variations in sign:
$$P(x) = x^5 + 3x^4 - x^3 + 2x + 3.$$
$$\underbrace{}_{1 \qquad 2}$$

Thus, P has 2 or $2 - 2 = 0$ positive real zeros.

$P(-x)$ has 3 variations in sign:
$$P(-x) = (-x)^5 + 3(-x)^4 - (-x)^3$$
$$+ 2(-x) + 3$$
$$= -x^5 + 3x^4 + x^3 - 2x + 3.$$
$$\underbrace{}_{1} \quad \underbrace{}_{2 \qquad 3}$$

Thus, P has 3 or $3 - 2 = 1$ negative real zeros.

7. $P(x) = 3x^2 - 2x - 6$; 1 and 2

Find P(1).

$$
\begin{array}{r}
1\overline{)\;3 \quad -2 \quad -6} \\
\underline{\quad 3 \quad\;\; 1} \\
3 \quad\;\; 1 \quad -5
\end{array}
$$

$P(1) = -5 < 0$

Find P(2).

$$
\begin{array}{r}
2\overline{)\;3 \quad -2 \quad -6} \\
\underline{\quad 6 \quad\;\; 8} \\
3 \quad\;\; 4 \quad\;\; 2
\end{array}
$$

$P(2) = 2 > 0$

Since P(1) and P(2) are of opposite sign, P has a zero between 1 and 2.

9. $P(x) = 2x^3 - 8x^2 + x + 16$; 2 and 2.5

Find P(2).

$$
\begin{array}{r}
2\overline{)\;2 \quad -8 \quad\;\; 1 \quad\;\; 16} \\
\underline{\quad 4 \quad -8 \quad -14} \\
2 \quad -4 \quad -7 \quad\;\; 2
\end{array}
$$

$P(2) = 2 > 0$

Find P(2.5).

$$
\begin{array}{r}
2.5\overline{)\;2 \quad -8 \quad\;\; 1 \quad\;\;\;\; 16} \\
\underline{\quad\quad\;\; 7.5 \quad -16.25} \\
2 \quad -3 \quad -6.5 \quad\;\; -.25
\end{array}
$$

$P(2.5) = -.25 < 0$

Since P(2) and P(2.5) are of opposite sign, P has a zero between 2 and 2.5.

11. $P(x) = 2x^4 - 4x^2 + 3x - 6$; 2 and 1.5

Find P(2).

$$
\begin{array}{r}
2\overline{)\;2 \quad\;\; 0 \quad -4 \quad\;\; 3 \quad -6} \\
\underline{\quad 4 \quad\;\; 8 \quad\;\; 8 \quad\;\; 22} \\
2 \quad\;\; 4 \quad\;\; 4 \quad\;\; 11 \quad\;\; 16
\end{array}
$$

$P(2) = 16 > 0$

Find P(1.5).

$$
\begin{array}{r}
1.5\overline{)\;2 \quad\;\; 0 \quad -4 \quad\;\; 3 \quad\;\;\;\; -6} \\
\underline{\quad 3 \quad\;\; 4.5 \quad .75 \quad 5.625} \\
2 \quad\;\; 3 \quad\;\; .5 \quad\;\; 3.75 \quad -.375
\end{array}
$$

$P(1.5) = -.375 < 0$

Since P(1.5) and P(2) are of opposite sign, P has a zero between 1.5 and 2.

13. $P(2) = -4$, $P(2.5) = 2$

If P(x) is a polynomial function with P(2) having a negative value and P(2.5) having a positive value, then the intermediate value theorem guarantees that P has a zero between 2 and 2.5.

15. $P(x) = x^4 - x^3 + 3x^2 - 8x + 8$; no real zero greater than 2

$$
\begin{array}{r}
2\overline{)\;1 \quad -1 \quad\;\; 3 \quad -8 \quad\;\; 8} \\
\underline{\quad 2 \quad\;\; 2 \quad\;\; 10 \quad\;\; 4} \\
1 \quad\;\; 1 \quad\;\; 5 \quad\;\; 2 \quad\;\; 12
\end{array}
$$

By the boundedness theorem, since 2 > 0 and all numbers in the bottom row of the synthetic division are nonnegative, P has no real zero greater than 2.

17. $P(x) = x^4 + x^3 - x^2 + 3$; no real zero less than −2

$$
\begin{array}{r}
-2\overline{)\;1 \quad\;\; 1 \quad -1 \quad\;\; 0 \quad\;\; 3} \\
\underline{\quad -2 \quad\;\; 2 \quad -2 \quad\;\; 4} \\
1 \quad -1 \quad\;\; 1 \quad -2 \quad\;\; 7
\end{array}
$$

By the boundedness theorem, since −2 < 0 and since the numbers in the bottom row of the synthetic division alternate in sign, P has no real zero less than −2.

19. $P(x) = 3x^4 + 2x^3 - 4x^2 + x - 1$; no real zero greater than 1

Try 1.

```
1)3  2  -4   1  -1
      3   5   1   2
   3  5   1   2   1
```

By the boundedness theorem, since $1 > 0$ and all numbers in the bottom row of the synthetic division are nonnegative, P has no real zero greater than 1.

21. $P(x) = x^5 - 3x^3 + x + 2$; no real zero greater than 2

```
2)1  0  -3   0   1   2
     2   4   2   4  10
   1  2   1   2   5  12
```

By the boundedness theorem, since $2 > 0$ and all numbers in the bottom row of the synthetic division are nonnegative, P has no real zero greater than 2.

23. $P(x) = x^3 + 3x^2 - 2x - 6$

(a) There is one change of sign in $P(x)$, so P has 1 positive real zero.

$$P(-x) = (-x)^3 + 3(-x)^2 - 2(-x) - 6$$
$$= -x^3 + 3x^2 + 2x - 6$$

There are two changes of sign in $P(-x)$, so P has either 2 or 0 negative zeros.

(b) Now, search for the positive zero.

```
1)1   3   -2   -6
      1    4    2
   1  4    2   -4
```

```
2)1   3   -2   -6
      2   10   16
   1  5    8   10
```

Since $P(1) < 0$ and $P(2) > 0$, the positive real zero is between 1 and 2.

Try 1.5.

```
1.5)1   3     -2     -6
        1.5   6.75   7.125
    1   4.5   4.75   1.125
```

$P(1) < 0$ and $P(1.5) > 0$, so the positive zero is between 1 and 1.5.

Try 1.4.

```
1.4)1   3     -2     -6
        1.4   6.16   5.824
    1   4.4   4.16   -.176
```

The zero is between 1.4 and 1.5, but closer to 1.4, so, to the nearest tenth, the positive zero is 1.4.

Now search for any negative zeros.

```
-3)1   3   -2   -6
      -3    0    6
   1   0   -2    0
```

-3 is a zero.

Now find the zeros of $x^2 - 2$.

$$x^2 - 2 = 0$$
$$x^2 = 2$$
$$x = \pm\sqrt{2} \approx \pm 1.4.$$

Therefore, to the nearest tenth, the zeros are -3, -1.4, and 1.4.

25. $P(x) = x^3 - 4x^2 - 5x + 14$

(a) There are two changes of sign in $P(x)$, so P has either 2 or 0 positive zeros.

$$P(-x) = (-x)^3 - 4(-x)^2 - 5(-x) + 14$$
$$= -x^3 - 4x^2 + 5x + 14$$

There is one change of sign in $P(-x)$, so P has 1 negative zero.

(b) Look for the negative zero first (since there is definitely one of them).

$$
\begin{array}{r|rrrr}
-2) & 1 & -4 & -5 & 14 \\
   &   & -2 & 12 & -14 \\
\hline
   & 1 & -6 & 7 & 0
\end{array}
$$

The only negative zero of $P(x)$ is -2. Now look for the positive zeros of $x^2 - 6x + 7$. Solve $x^2 - 6 + 7 = 0$ by the quadratic formula to get

$$x = 3 \pm \sqrt{2} \approx 3 \pm 1.41$$
$$x \approx 3 + 1.4 = 4.4$$
or $\quad x \approx 3 - 1.4 = 1.6.$

To the nearest tenth, the zeros are -2, 4.4, and 1.6.

27. $P(x) = x^3 + 6x - 13$

(a) There is one change of sign in $P(x)$, so $P(x)$ has 1 positive real zero.

$$P(x) = -x^3 - 6x - 13$$

There is no change of sign in $P(-x)$, so P has 0 negative real zeros. The other zeros are imaginary numbers.

(b) Since $P(1) = -6$ and $P(2) = 7$, there is a positive real zero between 1 and 2.

Try 1.5.

$$
\begin{array}{r|rrrr}
1.5) & 1 & 0 & 6 & -13 \\
    &   & 1.5 & 2.25 & 12.375 \\
\hline
    & 1 & 1.5 & 8.25 & -.625
\end{array}
$$

Try 1.6.

$$
\begin{array}{r|rrrr}
1.6) & 1 & 0 & 6 & -13 \\
    &   & 1.6 & 2.56 & 13.696 \\
\hline
    & 1 & 1.6 & 8.56 & .696
\end{array}
$$

The zero is between 1.5 and 1.6. Since $-.625$ is closer to 0 than $.696$, to the nearest tenth, the positive zero is 1.5. Since P has 1 positive real zero and no negative real zero, the only real zero of P is 1.5.

29. $P(x) = 4x^4 - 8x^3 + 17x^2 - 2x - 14$

(a) There are three changes of sign in $P(x)$, so $P(x)$ has 3 or 1 positive real zeros.

$$P(-x) = 4x^4 + 8x^3 + 17x^2 + 2x - 14$$

There is one change of sign in $P(-x)$, so $P(x)$ has 1 negative real zero.

(b) Look for the negative zero (since there is definitely one of them).

$$P(-1) = 17 > 0 \text{ and}$$
$$P\left(-\frac{1}{2}\right) = -18\frac{1}{2} < 0.$$

Thus, there is a real zero between $x = -1$ and $x = -\frac{1}{2}$.

Try -.75.

```
-.75)4   -8    17       -2        -14
         -3   8.25  -18.9375   15.703125
      4 -11  25.25  -20.9375    1.703125
```

Try -.7.

```
-.7)4   -8     17        -2       -14
        -2.8  7.56   -17.192   13.4344
     4 -10.8 24.56  -19.192     .5656
```

Try -.8.

```
-.8)4   -8     17        -2       -14
        -3.2  8.96   -20.768   18.2144
     4 -11.2 25.96  -22.768    4.2144
```

The only negative zero of P(x) (to the nearest tenth) is -.7. $P(1) = -5$ and $P(2) = 50$. Thus, there is a real positive zero between 1 and 2.

Try 1.1.

```
1.1)4   -8      17       -2       -14
        4.4   -3.96   14.344   13.5784
     4  4.4   13.04   12.344     .4216
```

To the nearest tenth, there is a zero at 1.1.

Find P(2).

```
2)4   -8   17   -2   -14
       8    0   34    64
   4   0   17   32    50
```

By the boundedness theorem, since 2 > 0 for all the numbers in the bottom row of the synthetic division are positive, P has no zero greater than 2. To the nearest tenth, the real zeros are -.7 and 1.1.

31. $P(x) = -x^4 + 2x^3 + 3x^2 + 6$

(a) There is one change in sign, so P(x) has 1 positive real zero.

$P(-x) = -(-x)^4 + 2(-x)^3 + 3(-x)^2 + 6$

$ = -x^4 - 2x^3 + 3x^2 + 6$

There is one change in sign in P(-x), so P has 1 negative real zero.

(b) First look for the positive zero.

```
Try 3.   3)-1  2   3   0   6
            -3  -3   0   0
          -1  -1   0   0   6
```

Since P(6) < 0 and P(3) > 0, there is a zero between 3 and 6.

```
Try 4.   4)-1   2    3     0     6
            -4   -8   -20   -80
          -1   -2   -5   -20   -74
```

P(4) < 0 and P(3) > 0, so the zero is between 3 and 4.

Try 3.2.

```
3.2)-1    2      3       0        6
       -3.2   -3.84   -2.688   -8.6016
   -1  -1.2   -.84   -2.688   -2.6016
```

P(3.2) < 0 and P(3) > 0, so try 3.1.

```
3.1)-1    2      3       0        6
       -3.1   -3.41   -1.271   -3.9401
   -1  -1.1   -.41   -1.271    2.0599
```

The positive zero, to the nearest tenth, is 3.1. Now try to find the negative zero.

```
Try -1.   -1)-1   2    3   0   6
               1   -3   0   0
             -1   3    0   0   6
```

```
Try -2.   -2)-1   2    3    0     6
               2   -8   10   -20
             -1   4   -5   10   -14
```

The signs alternate and -2 < 0, so the negative zero is between -2 and -1. Since P(-2) < 0 and

$P(-1) > 0$, the negative zero is between -2 and -1.

Try -1.5.

```
-1.5)-1   2      3      0       6
          1.5  -5.25  3.375  -5.0625
      -1   3.5  -2.25  3.375    .9375
```

The zero is between -2 and -1.5.

Try -1.6.

```
-1.6)-1   2      3      0       6
          1.6  -5.76  4.416  -7.07
      -1   3.6  -2.76  4.416  -1.07
```

The negative zero is between -1.5 and -1.6. To the nearest tenth, this zero is -1.5. Thus, to the nearest tenth, the real zeros are 3.1 and -1.5.

For Exercises 33–39, see the answer graphs in the textbook.

33. $P(x) = x^3 - 7x - 6$

P(x) has 1 positive zero and either 2 or 0 negative zeros.

x	1	0	-7	P(x) -6
0	1	0	-7	-6
1	1	1	-6	-12
2	1	2	-3	-12
3	1	3	2	0
4	1	4	9	30
-1	1	-1	-6	0
-2	1	-2	-3	0
-3	1	-3	2	-12
-1.5	1	-1.5	-4.75	1.25
1.5	1	1.5	-4.75	-13.125

The zeros are 3, -1, and -2.
Sketch the graph.

35. $P(x) = x^4 - 5x^2 + 6$
$= (x^2 - 3)(x^2 - 2)$

The zeros are $\pm\sqrt{3}$ and $\pm\sqrt{2}$, so approximate points for plotting are $(\pm1.73, 0)$ and $(\pm1.41, 0)$.

x	1	0	-5	0	P(x) -6
-3	1	-3	4	-12	42
-2	1	-2	-1	2	2
-1.5	1	-1.5	-2.75	4.125	-.1875
1	1	-1	-4	4	2
0	1	0	-5	0	6
1	1	1	-4	-4	2
1.5	1	1.5	-2.75	-4.125	-.1875
2	1	2	-1	-2	2
3	1	3	4	12	42

Sketch the graph.

37. $P(x) = 6x^3 + 11x^2 - x - 6$

x	6	11	-1	P(x) -6
-3	6	-7	20	-66
-2	6	-1	1	-8
-1.5	6	2	-4	0
-1	6	5	-6	0
0	6	11	-1	-6 } a zero
1	6	17	16	10 } between 0 and 1

-1.5 and -1 are zeros, and there is a zero between 0 and 1.
Sketch the graph.

39. $P(x) = -x^3 + 6x^2 - x - 14$

x				P(x)	
	-1	6	-1	-14	
-2	-1	8	-17	20	a zero
-1	-1	7	-8	-6	between
					-2 and -1
0	-1	6	-1	-14	
1	-1	5	4	-10	
2	-1	4	7	0	
3	-1	3	8	10	
5	-1	1	4	-6	a zero
6	-1	0	-1	-20	between
					5 and 6

2 is a zero. There is also a zero between -2 and -1 and a zero between 5 and 6.

Sketch the graph.

41. $P(x) = x^4 + x^3 - 6x^2 - 20x - 16$

Make a table. First find the values at the endpoints of each interval in order to know which way to go to select refinements.

x					P(x)
	1	1	-6	-20	-16
-1.4	1	-.4	-5.44	-12.384	1.3376
-1.1	1	-.1	-5.89	-13.521	-1.1269
3.2	1	4.2	7.44	3.808	-3.8144
3.3	1	4.3	8.19	7.027	7.1891

The interval $[-1.4, -1.1]$:

x					P(x)
	1	1	-6	-20	-16
-1.4	1	-.4	-5.44	-12.384	1.3376
-1.1	1	-.1	-5.89	-13.521	-1.1269
-1.3	1	-.3	-5.61	-12.707	.5194
-1.2	1	-.2	-5.76	-13.088	-.2944
-1.25	1	-.25	-5.6875	-12.890625	.1132812
-1.24	1	-.24	-5.7024	-12.9290	.03199
-1.23	1	-.23	-5.7171	-12.9680	-.0494

To the nearest hundredth, the zero in $[-1.4, -1.1]$ is -1.24.

The interval $[3.2, 3.3]$:

x				P(x)
	1	1	-6	-16
3.2	1	4.2	7.44	-3.8144
3.3	1	4.3	8.19	7.1891
3.25	1	4.25	7.8125	1.5195
3.24	1	4.24	7.7376	.42623
3.23	1	4.23	7.6629	-.6537

To the nearest hundredth, the zero is $[3.2, 3.3]$ is 3.24.

43. $P(x) = x^4 - 4x^3 - 20x^2 + 32x + 12$

Make a table. First find the values at the endpoints of each interval so that we can tell which way to go to select refinements.

x					P(x)
	1	-4	-20	32	12
-4	1	-8	12	-16	76
-3	1	-7	1	29	-75
-1	1	-5	-15	47	-35
0	1	-4	-20	32	12
1	1	-3	-23	9	21
2	1	-2	-24	-16	-20
6	1	2	-8	-16	-84
7	1	3	1	39	285

The interval [-4, -3]:

x					P(x)
	1	-4	-20	32	12
-4	1	-8	12	-16	76
-3	1	-7	1	29	-75
-3.5	1	-7.5	6.25	-10.125	-23.4375
-3.75	1	-7.75	9.0625	-1.984375	19.44141
-3.7	1	-7.7	8.49	.587	9.8281
-3.65	1	-7.65	7.9225	3.082875	.74751
-3.64	1	-7.64	7.8096	3.573056	-1.005924

The zero in [-4, -3] to the nearest hundredth is -3.65.

The interval [-1, 0]:

x					P(x)
	1	-4	-20	32	12
0	1	-4	-20	32	12
-1	1	-5	-15	47	-35
-.5	1	-4.5	-17.75	40.875	-8.8164063
-.25	1	-4.25	-18.9375	36.734375	2.8164063
-.37	1	-4.37	-18.3831	38.801747	-2.35665
-.35	1	-4.35	-18.4775	38.467125	-1.463493
-.30	1	-4.30	-18.71	37.613	.7161
-.32	1	-4.32	-18.6176	37.957632	-.146442
-.31	1	-4.31	-18.6639	37.785809	.2864

To the nearest hundredth, the zero in [-1, 0] is -.32.

The interval [1, 2]:

x					P(x)
	1	-4	-20	32	12
1	1	-3	-23	9	3
2	1	-2	-24	-16	-20
1.5	1	-2.5	-23.75	-3.625	6.5625
1.75	1	-2.25	-23.9375	-9.890625	-5.308593
1.62	1	-2.38	-23.8556	-6.646072	1.233364
1.65	1	-2.35	-23.8775	-7.397875	-.206493
1.64	1	-2.36	-23.8704	-7.147456	.278173

To the nearest hundredth, the zero in [1, 2], is 1.65.

The interval [6, 7]:

x					P(x)
	1	-4	-20	32	12
6	1	2	-8	-16	-84
7	1	3	1	39	285
6.25	1	2.25	-5.9375	-5.109375	-19.933593
6.40	1	2.40	-4.64	2.304	26.7456
6.30	1	2.3	-5.51	-2.713	-5.0919
6.35	1	2.35	-5.0775	-2.242125	10.4625063
6.32	1	2.32	-5.3376	-1.733632	1.043446
6.31	1	2.31	-5.4239	-2.224809	-2.038544

To the nearest hundredth, the zero in [6, 7] is 6.32.

To the nearest hundredth, the zeros are -3.65, -.32, 1.65, and 6.32.

45. $g(x) = -.006x^4 + .140x^3 - .053x^2$
$+ 1.79x$

(a) Find g(20).

$$20 \overline{)\begin{array}{ccccc} -.006 & .140 & -.053 & 1.79 & 0 \\ & -.12 & .4 & 6.94 & 174.6 \\ \hline -.006 & .02 & .347 & 8.73 & 174.6 \end{array}}$$

Therefore, $g(20) = 174.6 \approx 175$.

(b) Graph g.

Note that negative numbers will
not be included in the domain,
since negative numbers of seconds will not be considered.
There are three changes of sign
in g(x), so g has either 3 or
1 positive real zeros.

Make a table of rounded values.

x	g(x)
0	0
5	21
10	93
15	184
20	175
25	-145
24	-43
23	38

Note that there is a zero
between 23 and 24. Sketch the
graph. (See the answer graph in
the textbook.)

47. $P(x) = (x + 3)(x - 2)$
has -3 and 2 as zeros, but it
is negative between -3 and 2.

$P(x) = (x + 3)^2(x - 2)^2$ has -3 and
2 as zeros and it is always nonnegative; it may be rewritten as

$P(x) = x^4 + 2x^3 - 11x^2 - 12x + 36$.

(Other answers are possible.)

Section 6.6

For Exercises 1-7, see the answer graphs
in the textbook.

1. $f(x) = \dfrac{2}{x}$

Since $\dfrac{2}{x} = 2 \cdot \dfrac{1}{x}$, the graph of $f(x) = \dfrac{2}{x}$ will be similar to the graph

of $f(x) = \dfrac{1}{x}$, except that each point

will be twice as far from the
x-axis. Just as with the graph of

$f(x) = \dfrac{1}{x}$, y = 0 is the horizontal

asymptote and x = 0 is the vertical
asymptote. Sketch the graph.

3. $f(x) = \dfrac{1}{x + 2}$

Since $\dfrac{1}{x + 2} = \dfrac{1}{x - (-2)}$, the graph of

$f(x) = \dfrac{1}{x + 2}$ will be similar to the

graph of $f(x) = \dfrac{1}{x}$, except that each

point will be translated 2 units to

the left. Just as with $f(x) = \dfrac{1}{x}$,

y = 0 is the horizontal asymptote, but this graph has x = -2 as its vertical asymptote. Sketch the graph.

5. $f(x) = \frac{1}{x} + 1$

The graph of this function will be similar to the graph of $f(x) = \frac{1}{x}$, except that each point will be translated 1 unit upward. Just as with $f(x) = \frac{1}{x}$, x = 0 is the vertical asymptote, but this graph has y = 1 as its horizontal asymptote. Sketch the graph.

7. (a) The graph of

$$f(x) = \frac{1}{(x - 3)^2}$$

is the same as the graph of

$$f(x) = \frac{1}{x^2},$$

but it is translated 3 units to the right.

(b) The graph of

$$f(x) = -\frac{2}{x^2}$$

is the reflection of

$$f(x) = \frac{1}{x^2}$$

about the x-axis and with each point twice as far from the x-axis.

(c) The graph of

$$f(x) = \frac{-2}{(x - 3)^2}$$

is the graph in part (b) translated 3 units to the right.

9. $f(x) = \frac{2}{x - 5}$

To find any vertical asymptotes, set the denominator equal to 0 and solve for x.

$$x - 5 = 0$$
$$x = 5$$

Thus, x = 5 is a vertical asymptote. Since the numerator has lower degree than the denominator, y = 0 is the horizontal asymptote.

11. $f(x) = \frac{-8}{3x - 7}$

To find any vertical asymptotes, solve 3x - 7 = 0. Thus, $x = \frac{7}{3}$ is a vertical asymptote.
The horizontal asymptote is y = 0, since the numerator has lower degree than the denominator.

13. $f(x) = \frac{2 - x}{x + 2}$

Setting the denominator equal to 0 shows that the vertical asymptote is x = -2.
Since the numerator and the denominator have the same degree, to find the horizontal asymptote, divide the numerator and denominator by x to get

$$f(x) = \frac{\frac{2}{x} - \frac{x}{x}}{\frac{x}{x} + 2} = \frac{\frac{2}{x} - 1}{1 - \frac{2}{x}}.$$

As $|x|$ gets larger, $\frac{2}{x}$ approaches 0, and the value of $f(x)$ approaches

$$\frac{0 - 1}{1 - 0} = -1.$$

The line $y = -1$ is therefore a horizontal asymptote.

15. $f(x) = \frac{3x - 5}{2x + 9}$

$f(x)$ has vertical asymptote

$x = -\frac{9}{2}$ since the denominator is 0

if $x = -\frac{9}{2}$. Dividing the numerator

and denominator by x, we find that

the horizontal asymptote is $y = \frac{3}{2}$.

17. $f(x) = \frac{2}{x^2 - 4x + 3}$

To find any vertical asymptotes, set the denominator equal to 0 and solve for x.

$$x^2 - 4x + 3 = 0$$
$$(x - 3)(x - 1) = 0$$
$$x - 3 = 0 \quad \text{or} \quad x - 1 = 0$$
$$x = 3 \quad \text{or} \quad x = 1$$

The vertical asymptotes are $x = 3$ and $x = 1$.

Since the numerator has lower degree than the denominator, $y = 0$ is a horizontal asymptote.

19. $f(x) = \frac{x^2 - 1}{x + 3}$

The vertical asymptote is $x = -3$, found by solving $x + 3 = 0$. Since the numerator is of degree exactly one more than the denominator, there is no horizontal asymptote, but there may be an oblique asymptote. To find it, divide the numerator by the denominator and disregard any remainder.

$$\begin{array}{r} -3 \overline{)\begin{array}{rrr} 1 & 0 & -1 \\ & -3 & 9 \\ \hline 1 & -3 & 8 \end{array}} \end{array}$$

Thus,

$$f(x) = \frac{x^2 - 1}{x + 3} = x - 3 + \frac{8}{x + 3}.$$

The oblique asymptote is the line $y = x - 3$.

21. $f(x) = \frac{(x - 3)(x + 1)}{(x + 2)(2x - 5)}$

The vertical asymptotes are $x = -2$ and $x = \frac{5}{2}$, since these values make the denominator equal to 0. Multiply the factors in the numerator and denominator to get

$$f(x) = \frac{x^2 - 2x - 3}{2x^2 - x - 10}.$$

Thus, the horizontal asymptote, found by dividing the numerator and denominator by x^2, is $y = \frac{1}{2}$.

23.   (a) $f(x) = \dfrac{1}{x^2 + 2}$

has a graph that does not have a vertical asymptote. This is because there is no real number that can make the denominator $x^2 + 2$ become 0.

For Exercises 25–49, see the answer graphs in the textbook.

25. $f(x) = \dfrac{4}{5 + 3x}$

The line $x = -\dfrac{5}{3}$ is a vertical asymptote.
Since the numerator has lower degree than the denominator, the line $y = 0$ is a horizontal asymptote.
To find the y-intercept, evaluate $f(0)$.

$$f(0) = \frac{4}{5 + 3 \cdot 0} = \frac{4}{5},$$

so the y-intercept is $\dfrac{4}{5}$.

There are no x-intercepts since the numerator is never equal to 0.
Make a table of values.

x	-3	-2	-1	0	1
y	-1	-4	2	$\frac{4}{5}$	$\frac{1}{2}$

Use the asymptotes and these points to sketch the graph.

27. $f(x) = \dfrac{3}{(x + 4)^2}$

The vertical asymptote is $x = -4$.
The horizontal asymptote is $y = 0$.

$$f(0) = \frac{3}{(0 + 4)^2} = \frac{3}{16},$$

so the y-intercept is $\dfrac{3}{16}$.

There are no x-intercepts. Notice that the value of $y = f(x)$ will always be positive.

x	-8	-5	-3	0
y	$\frac{3}{16}$	3	3	$\frac{3}{16}$

Sketch the graph.

29. $f(x) = \dfrac{2x + 1}{(x + 2)(x + 4)}$

The vertical asymptotes are $x = -2$ and $x = -4$.
Since the numerator has lower degree than the denominator, $y = 0$ is a horizontal asymptote.

$$f(0) = \frac{2(0) + 1}{(0 + 2)(0 + 4)} = \frac{1}{8},$$

so the y-intercept is $\dfrac{1}{8}$.

Any x-intercepts can be found by solving $f(x) = 0$.

$$\frac{2x + 1}{(x + 2)(x + 4)} = 0$$
$$2x + 1 = 0$$
$$x = -\frac{1}{2}$$

The x-intercept is $-\dfrac{1}{2}$.

x	-6	-5	-3	-1	$-\frac{1}{2}$	0	1
y	$-\frac{11}{8}$	-3	5	$-\frac{1}{3}$	0	$\frac{1}{8}$	$\frac{1}{5}$

Sketch the graph.

31. $f(x) = \dfrac{-x}{x^2 - 4}$

The vertical asymptotes are $x = 2$ and $x = -2$.

Since the numerator has lower degree than the denominator, $y = 0$ is the horizontal asymptote.

$$f(0) = \frac{-0}{0^2 - 4} = 0,$$

so the y-intercept is 0.
Any x-intercepts will be the zeros of the numerator.

$$-x = 0$$
$$x = 0$$

Thus, 0 is the only x-intercept.

x	-3	-1	0	1	3
y	$-\frac{3}{5}$	$-\frac{1}{3}$	0	$\frac{1}{3}$	$-\frac{3}{5}$

Sketch the graph.

33. $f(x) = \dfrac{4x}{1 - 3x}$

The vertical asymptote is $x = \frac{1}{3}$.

Since the numerator and denominator have the same degree, divide numerator and denominator by x to find the horizontal asymptote, $y = -\frac{4}{3}$.

$$f(0) = \frac{4 \cdot 0}{1 - 3 \cdot 0} = 0,$$

so the y-intercept is 0. Since 0 is the only zero of the numerator, 0 is the only x-intercept.

x	-1	0	$\frac{1}{4}$	1	2
y	-1	0	4	-2	$-\frac{8}{5}$

Sketch the graph.

35. $f(x) = \dfrac{x - 5}{x + 3}$

The vertical asymptote is $x = -3$.

$$\frac{x - 5}{x + 3} = \frac{1 - \dfrac{5}{x}}{1 + \dfrac{3}{x}}, \text{ so}$$

as $|x| \to \infty$, $f(x) \to 1$.
Thus, the horizontal asymptote is $y = 1$.

$$f(0) = \frac{0 - 5}{0 + 3} = -\frac{5}{3},$$

so the y-intercept is $-\frac{5}{3}$.
The only zero of the numerator is 5, so 5 is the only x-intercept.

x	-6	-5	-4	-2	-1	0	1	3	5
y	$\frac{11}{3}$	5	9	-7	-3	$-\frac{5}{3}$	-1	$-\frac{1}{3}$	0

Sketch the graph.

37. $f(x) = \dfrac{3x}{x^2 - 16}$

The vertical asymptotes are $x = 4$ and $x = -4$. Since the numerator has lower degree than the denominator, the horizontal asymptote is $y = 0$.

x	−5	−3	−1	0	1	3	5
y	$-\dfrac{5}{3}$	$\dfrac{9}{7}$	$\dfrac{1}{5}$	0	$-\dfrac{1}{5}$	$-\dfrac{9}{7}$	$\dfrac{5}{3}$

Sketch the graph.

39. $f(x) = \dfrac{x^2 + 1}{x + 3}$

The vertical asymptote is $x = -3$. Since the numerator is of degree exactly one more than the denominator, there is no horizontal asymptote, but there may be an oblique asymptote.

Use synthetic division.

$$-3\overline{)\begin{array}{ccc} 1 & 0 & 1 \\ & -3 & 9 \end{array}}$$
$$\begin{array}{ccc} 1 & -3 & 10 \end{array}$$

$$f(x) = \frac{x^2 + 1}{x + 3} = x - 3 + \frac{10}{x + 3}$$

The oblique asymptote is $y = x - 3$.

$$f(0) = \frac{0^2 + 1}{0 + 3} = \frac{1}{3},$$

so the y-intercept is $\dfrac{1}{3}$.

Note that $f(x)$ can never be zero, so the graph has no x-intercepts.

x	−4	−5	−2	−1	0	1	3
y	−17	−13	5	1	$\dfrac{1}{3}$	$\dfrac{1}{2}$	$\dfrac{5}{3}$

Sketch the graph.

41. $f(x) = \dfrac{x^2 - x}{x + 2}$

The vertical asymptote is $x = -2$. Since the numerator is of degree exactly one more than the denominator, there may be an oblique asymptote.

$$-2\overline{)\begin{array}{ccc} 1 & -1 & 0 \\ & -2 & 6 \end{array}}$$
$$\begin{array}{ccc} 1 & -3 & 6 \end{array}$$

$$f(x) = \frac{x^2 - x}{x + 2} = x - 3 + \frac{6}{x + 2}$$

The oblique asymptote is $y = x - 3$.

$$f(0) = \frac{0^2 - 0}{0 + 2} = \frac{0}{2} = 0,$$

so the y-intercept is 0.
To find any x-intercepts, find the zeros of the numerator.

$$x^2 - x = 0$$
$$x(x - 1) = 0$$
$$x = 0 \quad \text{or} \quad x = 1$$

The x-intercepts are 0 and 1.

x	−4	−3	−1	0	1	2	3	4
y	−10	−12	2	0	0	.5	1.2	2

Sketch the graph.

43. $f(x) = \dfrac{x(x - 2)}{(x + 3)^2}$

The vertical asymptote is $x = -3$. The numerator and denominator have the same degree, so divide by x^2 to find the horizontal asymptote.

$$\frac{x(x - 2)}{(x + 3)^2} = \frac{x^2 - 2x}{x^2 + 6x + 9}$$

$$= \frac{1 - \frac{2}{x}}{1 + \frac{6}{x} + \frac{9}{x^2}}$$

As $|x| \to \infty$, $f(x) \to 1$, so $y = 1$ is a horizontal asymptote.

$$f(0) = \frac{0(0 - 2)}{(0 + 3)^2} = \frac{0}{9} = 0,$$

so the y-intercept is 0.

Any x-intercepts will be the zeros of the numerator.

$$x(x - 2) = 0$$
$$x = 0 \quad \text{and} \quad x = 2$$

The graph has two x-intercepts, 0 and 2.

x	-4	-5	-6	-2	-1	0	1	2	3
y	24	$\frac{35}{4}$	$\frac{48}{9}$	8	$\frac{3}{4}$	0	$-\frac{1}{16}$	0	$\frac{1}{12}$

Sketch the graph.

45. $f(x) = \dfrac{1}{x^2 + 1}$

Since $x^2 + 1$ has no real zeros, there are no vertical asymptotes. Since $f(x) = f(-x)$, the graph is symmetric with respect to the y-axis. Note that $f(x) > 0$ for all values of x, so the graph is entirely above the x-axis. Since the numerator has lower degree than the denominator, $y = 0$ is the horizontal asymptote.

$$f(0) = \frac{1}{0^2 + 1} = 1,$$

so the y-intercept is 1.

There are no x-intercepts.

x	-2	-1	0	1	2
y	$\frac{1}{5}$	$\frac{1}{2}$	1	$\frac{1}{2}$	$\frac{1}{5}$

Sketch the graph.

47. $f(x) = \dfrac{x^2 - 16}{x + 4}$

$$= \frac{(x + 4)(x - 4)}{x + 4}$$

$$= x - 4 \quad (x \neq -4)$$

Therefore, the graph of this function will be the same as the graph of $y = x - 4$ (a straight line), with the exception of the point with x-value -4. A "hole" appears in the graph at $(-4, -8)$. Sketch the graph.

49. $C(x) = \dfrac{10x}{49(101 - x)}$

The graph has a vertical asymptote at $x = 101$. The x- and y-intercepts are both 0. In this application, x cannot be negative, since the number of points awarded is never negative. Therefore, the function is graphed only for $x \geq 0$. Sketch the graph.

51. $C(x) = \dfrac{6.7x}{100 - x}$

(a) The graph has a vertical asymptote at $x = 100$. The x- and y-intercepts are both $(0, 0)$. In this application x cannot be negative, so the function is graphed only for $x \geq 0$.

See the answer graph in the the textbook.

(b) $x = 100$ is the vertical asymptote of the graph. This means that x will never be equal to 100. Since x represents the percent of the pollutant removed, this shows that it is not possible to remove all (100%) of the pollutant.

53. $R(x) = \dfrac{80x - 8000}{x - 110}$

(a) If $x = 55$,

$R(x) = \dfrac{80(55) - 8000}{55 - 110}$

≈ 65.5 tens of millions of dollars

or $655,000,000.

(b) If $x = 60$,

$R(x) = \dfrac{80(60) - 8000}{60 - 110}$

$= 64$ tens of millions of dollars

or $640,000,000.

(c) If $x = 70$,

$R(x) = \dfrac{80(70) - 8000}{70 - 110}$

$= 60$ tens of millions of dollars

or $600,000,000.

(d) If $x = 90$,

$R(x) = \dfrac{80(90) - 8000}{90 - 110}$

$= 40$ tens of millions of dollars

or $400,000,000.

(e) If $x = 100$,

$R(x) = \dfrac{80(100) - 8000}{90 - 110}$

$= 0$ tens of millions of dollars

or $0.

(f) See the graph in the textbook.

Chapter 6   Review Exercises

For Exercises 1–5, see the answer graphs in the textbook.

1. $P(x) = x^3 + 5$

The graph will be the same as that of $P(x) = x^3$, but translated 5 units up.

Sketch the graph.

3. $P(x) = x^2(2x + 1)(x - 2)$

The zeros are 0, $-\dfrac{1}{2}$, and 2.

The zeros divide the x-axis into four intervals:

$(-\infty, -\dfrac{1}{2})$, $(-\dfrac{1}{2}, 0)$, $(0, 2)$, and $(2, \infty)$.

Test a point in each interval to find the sign of $P(x)$ in that interval.

Interval	Test point	Value of $P(x)$	Location relative to x-axis
$(-\infty, -\frac{1}{2})$	-1	3	Above
$(-\frac{1}{2}, 0)$	$-\frac{1}{4}$	$-\frac{9}{128}$	Below
$(0, 2)$	1	-3	Below
$(2, \infty)$	3	63	Above

Sketch the graph.

5. $P(x) = 2x^3 + 13x^2 + 15x$

$\qquad = x(2x^2 + 13x + 15)$

$\qquad = x(2x + 3)(x + 5)$

The zeros are 0, $-\frac{3}{2}$, and -5.

Interval	Test point	Value of $P(x)$	Location relative to x-axis
$(-\infty, -5)$	-6	-54	Below
$(-5, -\frac{3}{2})$	-2	6	Above
$(-\frac{3}{2}, 0)$	-1	-4	Below
$(0, \infty)$	1	30	Above

Sketch the graph.

7. For the polynomial function defined by $P(x) = x^3 - 3x^2 - 7x + 12$, $P(4) = 0$. Therefore, we can say that 4 is a zero of the function, 4 is a solution of the equation $x^3 - 3x^2 - 7x + 12 = 0$, and that 4 is an x-intercept of the graph of the function.

9. $\dfrac{15m^4n^5 - 10m^3n^7 + 20m^7n^3}{10mn^4}$

$= \dfrac{15m^4n^5}{10mn^4} - \dfrac{10m^3n^7}{10mn^4} + \dfrac{20m^7n^3}{10mn^4}$

$= \dfrac{3m^3n}{2} - m^2n^3 + \dfrac{2m^6}{n}$

11. $\dfrac{6p^2 - 5p - 56}{2p - 7}$

$$2p - 7 \overline{)\begin{array}{l} 3p\ +\ \ 8 \\ 6p^2 -\ \ 5p\ -\ 56 \\ \underline{6p^2 -\ 21p} \\ \qquad 16p - 56 \\ \qquad \underline{16p - 56} \\ \qquad\qquad\quad 0 \end{array}}$$

Therefore,

$\dfrac{6p^2 - 5p - 56}{2p - 7} = 3p + 8.$

13. $\dfrac{5m^3 - 7m^2 - 10m + 14}{m^2 - 2}$

$$m^2 - 2 \overline{)\begin{array}{l} 5m\ \ -\ 7 \\ 5m^3 -\ 7m^2 - 10m + 14 \\ \underline{5m^3 \qquad\quad - 10m} \\ \quad -7m^2 \qquad\qquad + 14 \\ \quad \underline{-7m^2 \qquad\qquad + 14} \\ \qquad\qquad\qquad\qquad\quad 0 \end{array}}$$

Therefore,

$\dfrac{5m^3 - 7m^2 - 10m + 14}{m^2 - 2} = 5m - 7.$

15. $(2x^3 - x^2 - x - 6) \div (x + 1)$

$$\begin{array}{r} -1\overline{)\begin{array}{rrrr} 2 & -1 & -1 & -6 \\ & -2 & 3 & -2 \\ \hline 2 & -3 & 2 & -8 \end{array}} \end{array}$$

Thus,

$\dfrac{2x^3 - x^2 - x - 6}{x + 1}$

$= 2x^2 - 3x + 2 + \dfrac{-8}{x + 1}.$

17. $(4m^3 + m^2 - 12) \div (m - 2)$

$$
\begin{array}{r|rrrr}
2) & 4 & 1 & 0 & -12 \\
   &   & 8 & 18 & 36 \\
\hline
   & 4 & 9 & 18 & 24
\end{array}
$$

Thus,

$$\frac{4m^3 + m^2 - 12}{m - 2}$$

$$= 4m^2 + 9m + 18 + \frac{24}{m - 2}.$$

19. $P(x) = x^3 + 3x^2 - 5x + 1$

Find $P(2)$.

$$
\begin{array}{r|rrrr}
2) & 1 & 3 & -5 & 1 \\
   &   & 2 & 10 & 10 \\
\hline
   & 1 & 5 & 5 & 11
\end{array}
$$

$P(2) = 11$

21. $P(x) = 5x^4 - 12x^2 + 2x - 8$

Find $P(2)$.

$$
\begin{array}{r|rrrrr}
2) & 5 & 0 & -12 & 2 & -8 \\
   &   & 10 & 20 & 16 & 36 \\
\hline
   & 5 & 10 & 8 & 18 & 28
\end{array}
$$

$P(2) = 28$

In Exercises 23 and 25, we give only one such polynomial. There are others.

23. Zeros are -1, 4, and 7.

$P(x) = (x + 1)(x - 4)(x - 7)$

$\quad = (x + 1)(x^2 - 11x + 28)$

$\quad = x^3 - 10x^2 + 17x + 28$

25. Zeros are $-\sqrt{7}$, $\sqrt{7}$, 2, and -1.

$P(x) = (x + \sqrt{7})(x - \sqrt{7})$

$\qquad \cdot (x - 2)(x + 1)$

$\quad = (x^2 - 7)(x^2 - x - 2)$

$\quad = x^4 - x^3 - 9x^2 + 7x + 14$

27. $P(x) = 2x^4 + x^3 - 4x^2 + 3x + 1$

Use synthetic division to determine whether -1 is a zero of P.

$$
\begin{array}{r|rrrrr}
-1) & 2 & 1 & -4 & 3 & 1 \\
    &   & -2 & 1 & 3 & -6 \\
\hline
    & 2 & -1 & -3 & 6 & -5
\end{array}
$$

Since the remainder is not 0, -1 is not a zero of P.

29. $P(x) = x^3 + 2x^2 + 3x - 1$

$x + 1$ is a factor of $P(x)$ if $P(-1) = 0$.

$$
\begin{array}{r|rrrr}
-1) & 1 & 2 & 3 & -1 \\
    &   & -1 & -1 & -2 \\
\hline
    & 1 & 1 & 2 & -3
\end{array}
$$

$P(-1) = -3 \neq 0$

Therefore, $x + 1$ is not a factor of P.

31. Zeros are -2, 1, and 4; $P(2) = 16$.

$P(x) = a(x + 2)(x - 1)(x - 4)$

Use $P(2) = 16$ to determine the value of a.

$P(2) = a(2 + 2)(2 - 1)(2 - 4) = 16$

$\qquad$ *Let x = 2; P(2) = 16*

$\qquad\qquad a(4)(1)(-2) = 16$

$\qquad\qquad\qquad -8a = 16$

$\qquad\qquad\qquad\qquad a = -2$

$P(x) = -2(x + 2)(x - 1)(x - 4)$

$\quad = -2(x + 2)(x^2 - 5x + 4)$

$\quad = -2(x^3 - 3x^2 - 6x + 8)$

$\quad = -2x^3 + 6x^2 + 12x - 16$

In Exercises 33 and 35, we give only one such polynomial. Others are possible.

33. If the polynomial has zeros of 2, −2, and −i and is required to have real coefficients, then the conjugate of −i, which is i, must also be a zero.

$$P(x) = (x + 2)(x - 2)(x + i)(x - i)$$
$$= (x^2 - 4)(x^2 + 1)$$
$$= x^4 - 3x^2 - 4$$

35. If −3 and 1 − i are zeros of a polynomial with real coefficients, then the conjugate of 1 − i, which is 1 + i, must also be a zero.

$$P(x) = (x + 3)[x - (1 - i)]$$
$$\cdot [x - (1 + i)]$$
$$= (x + 3)[(x - 1) + i)]$$
$$\cdot [(x - 1) - i)]$$
$$= (x + 3)([(x - 1)^2 + 1]$$
$$= (x + 3)(x^2 - 2x + 1 + 1)$$
$$= (x + 3)(x^2 - 2x + 2)$$
$$= x^3 + x^2 - 4x + 6$$

37. $P(x) = x^4 - 6x^3 + 14x^2 - 24x + 40$

Since $P(x)$ has real coefficients and $3 + i$ is a zero of P, $3 - i$ is also a zero.

First divide $P(x)$ by $x - (3 + i)$.

```
3 + i)1  −6        14      −24      −40
        3 + i   −10      12 + 4i   −40
     1  −3 + i    4    −12 + 4i     0
```

Now divide the quotient from the first division by $x - (3 - i)$.

```
3 − i)1  −3 + i   4   −12 + 4i
        3 − i    0    12 − 4i
     1   0       4      0
```

The remaining zeros are the zeros of the quotient, $x^2 + 4$.

$$x^2 + 4 = 0$$
$$x^2 = -4$$
$$x = \pm 2i$$

The zeros of $P(x)$ are $3 + i$, $3 - i$, $2i$, and $-2i$.
Thus,

$$P(x) = [x - (3 + i)][x - (3 - i)]$$
$$\cdot (x - 2i)(x + 2i)$$
$$= (x - 3 - i)(x - 3 + i)$$
$$\cdot (x - 2i)(x + 2i).$$

39. $P(x) = 2x^3 - 9x^2 - 6x + 5$

The possible rational zeros are

$$\pm 1, \ \pm 5, \ \pm\frac{1}{2}, \ \pm\frac{5}{2}.$$

```
−1)2  −9   −6    5
      −2   11   −5
   2  −11   5    0
```

−1 is a zero.
The remaining zeros can be found by factoring the quotient.

$$2x^2 - 11x + 5 = 0$$
$$(2x - 1)(x - 5) = 0$$
$$x = \frac{1}{2} \ \text{ or } \ x = 5$$

The rational zeros are -1, $\frac{1}{2}$, and 5.

41. $P(x) = x^3 - \dfrac{17}{6}x^2 - \dfrac{13}{3}x - \dfrac{4}{3}$

Multiply by 6 to clear fractions.

$$6P(x) = 6x^3 - 17x^2 - 26x - 8$$

has the same zeros as $P(x)$.

Possible rational zeros are

$$\pm 1,\ \pm 2,\ \pm 4,\ \pm 8,\ \pm\frac{1}{2},$$

$$\pm\frac{1}{3},\ \pm\frac{2}{3},\ \pm\frac{4}{3},\ \pm\frac{8}{3},\ \pm\frac{1}{6}.$$

Synthetic division shows that 4 is a zero.

$$\begin{array}{r}
4\,\overline{)\,6 \quad -17 \quad -26 \quad -8} \\
\ 24 \quad\ \ 28 \quad\ \ 8 \\
\hline
6 \qquad 7 \qquad\ 2 \qquad\ 0
\end{array}$$

Now find the zeros of the quotient $6x^2 + 7x + 2$ by factoring.

$$6x^2 + 7x + 2 = 0$$

$$(2x + 1)(3x + 2) = 0$$

$$x = -\frac{1}{2} \quad\text{or}\quad x = -\frac{2}{3}$$

The rational zeros are 4, $-\frac{1}{2}$, and $-\frac{2}{3}$.

43. $P(x) = x^3 - 5x^2 - 13x - 7$

The possible rational zeros are $\pm 1,\ \pm 7$.

$$\begin{array}{r}
-1\,\overline{)\,1 \quad -5 \quad -13 \quad -7} \\
\ -1 \qquad\ 6 \qquad\ 7 \\
\hline
1 \quad -6 \quad\ -7 \qquad 0
\end{array}$$

-1 is a zero.

Find the remaining zeros by factoring the quotient.

$$x^2 - 6x - 7 = 0$$

$$(x - 7)(x + 1) = 0$$

$$x = 7 \quad\text{or}\quad x = -1$$

The rational zeros are -1 (multiplicity 2) and 7.

In factored form,

$$P(x) = [x - (-1)][x - (-1)](x - 7)$$

$$= (x + 1)^2(x - 7).$$

Interval	Test point	Value of $P(x)$	Location relative to x-axis
$(-\infty, -1)$	-2	-9	Below
$(-1, 7)$	0	-7	Below
$(7, \infty)$	8	81	Above

In order to get an accurate graph, a few additional points should be plotted for values of x between 0 and 7. Three such points are $(1, -24)$, $(3, -64)$, and $(5, -72)$. Sketch the graph. See the answer graph in the textbook.

45. Let $P(x) = x^2 - 11$.

By Descartes' rule of signs, this polynomial has one positive and one negative real zero. The only possible rational zeros are ± 1 and ± 11.

x			$P(x)$
	1	0	-11
1	1	1	-10
-1	1	-1	-10
11	1	11	110
-11	1	-11	110

The synthetic divisions show that $P(x)$ has no rational zeros, but $\sqrt{11}$ is a zero since

$$P(\sqrt{11}) = (\sqrt{11})^2 - 11 = 11 - 11 = 0.$$

Since $\sqrt{11}$ is a real zero that is not rational, it must be irrational.

47. Show $P(x) = 3x^3 - 8x^2 + x + 2$ has real zeros in $[-1, 0]$ and $[2, 3]$.

x				P(x)
	3	−8	1	2
−1	3	−11	12	−10
0	3	8	1	2
2	3	−2	−3	−4
3	3	1	4	14

$P(-1) = -10$ and $P(0) = 2$. By the intermediate value theorem, since $P(-1) < 0$ and $P(0) > 0$, there is a zero between -1 and 0.

$P(2) = -4 < 0$ and $P(3) = 14 > 0$. Since $P(2) < 0$ and $P(3) > 0$, there is a zero between 2 and 3.

49. Show $P(x) = x^3 + 2x^2 - 22x - 8$ has real zeros in $[-1, 0]$ and $[-6, -5]$.

x				P(x)
	1	2	−22	−8
−6	1	−4	2	−20
−5	1	−3	−7	27
−1	1	1	−23	15
0	1	2	−22	−8

$P(-1) = 15 > 0$ and $P(0) = -8 < 0$. Thus, there is a zero between -1 and 0.

$P(-6) = -20 < 0$ and $P(-5) = 27 > 0$. Thus, there is a zero between -6 and -5.

51. Show $P(x) = 6x^4 + 13x^3 - 11x^2 - 3x + 5$ has no real zero greater than 1 or less than -3.

$$
\begin{array}{r|rrrrr}
1) & 6 & 13 & -11 & -3 & 5 \\
& & 6 & 19 & 8 & 5 \\
\hline
& 6 & 19 & 8 & 5 & 10
\end{array}
$$

All numbers in the bottom row are nonnegative and $1 > 0$, so by the boundedness theorem, P has no real zero greater than 1.

$$
\begin{array}{r|rrrrr}
-3) & 6 & 13 & -11 & -3 & 5 \\
& & -18 & 15 & -12 & 45 \\
\hline
& 6 & -5 & 4 & -15 & 50
\end{array}
$$

The signs alternate and $-3 < 0$, so, by the boundedness theorem, P has no real zero less than -3.

53. $P(x) = x^4 - 4x^3 - 5x^2 + 14x - 15$

There are three variations of sign in $P(x)$, so P has 3 or 1 positive real zeros.

$P(-x) = x^4 + 4x^3 - 5x^2 - 14x - 15$

There is one variation of sign in $P(-x)$, so P has one negative real zero.

Set up a table of values to find the zeros.

x					P(x)	
	1	-4	-5	14	-15	
-3	1	-7	16	-34	87	} a zero
-2	1	-6	7	0	-15	between
-1	1	-5	0	14	-29	-3 and 2
0	1	-4	-5	14	-15	
1	1	-3	-8	6	-9	
2	1	-2	-9	-4	-23	
4	1	0	-5	-6	-39	} a zero
5	1	1	0	14	55	between 4 and 5

No rational zero is greater than 5, since all the numbers in the bottom row are nonnegative.

First locate the zero in the interval [-3, -2].

x					P(x)
	1	-4	-5	14	-15
-2	1	-6	7	0	-15
-2.3	1	-6.3	9.49	-7.827	3.002
-2.2	1	-6.2	8.64	-5.008	-3.9824

To the nearest tenth, the negative zero is -2.3.
Now locate the zero in the interval [4, 5].

x					P(x)
	1	-4	-5	14	-15
4	1	0	-5	-6	-39
4.4	1	.4	-3.24	-.256	-16.1264
4.5	1	.5	-2.75	1.625	-7.6875
4.6	1	.6	-2.24	3.696	2.0016

To the nearest tenth, the positive zero is 4.6.
Therefore, to the nearest tenth, the real zeros are -2.3 and 4.6.

For Exercises 55–67, see the answer graphs in the textbook.

55. $P(x) = 2x^3 - 11x^4 - 2x + 2$

By Descartes' rule of signs, P has either 2 or 0 positive real zeros and 1 negative real zero.
Try to locate the zeros by looking for changes in the sign of the value of P(x).

x				P(x)
	2	-11	-2	2
-3	2	-17	49	-145
-2	2	-15	28	-54
-1	2	-13	11	-9
0	2	-11	-2	2
1	2	-9	-11	-9
2	2	-7	-16	-30
3	2	-5	-17	-49
4	2	-3	-14	-54
5	2	-1	-7	-33
6	2	1	4	26

There is one zero in [-1, 0], one zero in [0, 1], and one zero in [5, 6]. Sketch the graph.

57. $P(x) = x^3 + 3x^2 - 4x - 2$

There is 1 positive real zero and either 2 or 0 negative real zeros.

x				P(x)
	1	3	−4	−2
−4	1	−1	0	−2
−3	1	0	−4	10
−2	1	1	−6	10
−1	1	2	−6	4
0	1	3	−4	−2
1	1	4	0	−2
2	1	5	6	10
3	1	6	14	40

There is one zero in $[-1, 0]$, one zero in $[1, 2]$, and one zero in $[-4, -3]$.

Sketch the graph.

59. $f(x) = \dfrac{8}{x}$

$\qquad = 8\left(\dfrac{1}{x}\right)$

The graph is similar to that of $f(x) = \dfrac{1}{x}$, but each value is multiplied by 8. The x- and y-axes are the asymptotes. Sketch the graph.

61. $f(x) = \dfrac{4x - 2}{3x + 1}$

Find the vertical asymptote by solving $3x + 1 = 0$. There is one vertical asymptote,

$$x = -\frac{1}{3}.$$

Since the numerator and denominator have the same degree, find the horizontal asymptote by dividing numerator and denominator by x and letting $|x| \to \infty$.

The horizontal asymptote is $y = \dfrac{4}{3}$.

$$f(0) = \frac{4 \cdot 0 - 2}{3 \cdot 0 + 1} = -2,$$

so the y-intercept is −2.

Any x-intercepts will be the zeros of the numerator.

$$4x - 2 = 0$$
$$4x = 2$$
$$x = \frac{1}{2}$$

There is one x-intercept, $\dfrac{1}{2}$.

Sketch the graph.

63. $f(x) = \dfrac{2x}{x^2 - 1}$

$\qquad f(x) = \dfrac{2x}{(x + 1)(x - 1)}$

The vertical asymptotes are $x = 1$ and $x = -1$, since 1 and −1 are the zeros of the denominator. Since the numerator has lower degree than the denominator, the horizontal asymptote is $y = 0$.

Notice that

$$f(-x) = \frac{2(-x)}{(-x)^2 - 1}$$

$$= \frac{-2x}{x^2 - 1}$$

$$= -f(x),$$

so the function is odd, and symmetric with respect to the origin. The x- and y-intercepts are both 0. Sketch the graph.

65. $f(x) = \dfrac{x^2 - 1}{x}$

There is one vertical asymptote, $x = 0$.

Since the numerator is of degree exactly one more than the denominator, there is no horizontal asymptote, but there may be an oblique asymptote.

$$f(x) = \frac{x^2 - 1}{x} = x - \frac{1}{x},$$

so as $|x| \to \infty$, $f(x) \to x$.

Thus, the line $y = x$ is an oblique asymptote.

There is no y-intercept because $f(0)$ is undefined.

The zeros of the numerator are -1 and 1, so the graph has two x-intercepts, -1 and 1.

Use the intercepts and asymptotes to sketch the graph.

67. $f(x) = \dfrac{4x^2 - 9}{2x + 3}$

$$= \frac{(2x + 3)(2x - 3)}{2x + 3}$$

$$= 2x - 3 \ \left(x \neq -\frac{3}{2}\right)$$

The graph is the same as that of $f(x) = 2x - 3$, the line with slope 2 and y-intercept -3, except that the point with x-value $-3/2$ is missing. This is shown by graphing the line $y = 2x - 3$ with an open circle at $\left(-\frac{3}{2}, -6\right)$ to indicate the missing point.

Chapter 6   Test

1. $P(x) = (1 - x)^4$

$$= [-(x - 1)]^4$$

$$= (x - 1)^4$$

Translate the graph of $y = x^4$ one unit to the right. See the answer graph in the textbook.

2. $P(x) = (x + 1)(x - 2)x$

The zeros of P are -1, 2 and 0. Find a test point in each interval and make a table.

Interval	Test point	Value of P(x)	Location relative to x-axis
$(-\infty, -1)$	-2	-8	Below
$(-1, 0)$	$-\dfrac{1}{2}$	$\dfrac{5}{8}$	Above
$(0, 2)$	1	-2	Below
$(2, \infty)$	2	12	Above

Use the zeros and these points to sketch the graph. See the answer graph in the textbook.

3. $(3x^3 + 4x^2 - 9x + 6) \div (x + 2)$

$$\begin{array}{r}
-2)\overline{\begin{array}{cccc} 3 & 4 & -9 & 6 \end{array}} \\
\begin{array}{cccc} & -6 & 4 & 10 \end{array} \\
\hline
\begin{array}{cccc} 3 & -2 & -5 & 16 \end{array}
\end{array}$$

$$\frac{3x^3 + 4x^2 - 9x + 6}{x + 2}$$

$$= 3x^2 - 2x - 5 + \frac{16}{x + 2}$$

4. $(2x^3 - 11x^2 + 28) \div (x - 5)$

$$5\overline{)\begin{array}{cccc} 2 & -11 & 0 & 28 \\ & 10 & -5 & -25 \\ \hline 2 & -1 & -5 & 3 \end{array}}$$

$$\frac{2x^3 - 11x^2 + 28}{x - 5}$$

$$= 2x^2 - x - 5 + \frac{3}{x - 5}$$

5. $P(x) = 2x^3 - 9x^2 + 4x + 8$; find $P(3)$.

$$3\overline{)\begin{array}{cccc} 2 & -9 & 4 & 8 \\ & 6 & -9 & -15 \\ \hline 3 & -3 & -5 & -7 \end{array}}$$

Since the remainder is -7, $P(3) = -7$.

6. If -1, -2, and 4 are zeros of $P(x)$, then $x - (-1) = x + 1$, $x - (-2) = x + 2$, and $x - 4$ will be factors of $P(x)$.

$$\begin{aligned} P(x) &= (x + 1)(x + 2)(x - 4) \\ &= (x + 1)(x^2 - 2x - 8) \\ &= x^3 - x^2 - 10x - 8 \end{aligned}$$

This is one polynomial function that satisfies the given conditions. Others are possible.

7. $P(x) = 6x^4 - 11x^3 - 35x^2 + 34x + 24$

$$3\overline{)\begin{array}{ccccc} 6 & -11 & -35 & 34 & 24 \\ & 18 & 21 & -42 & -24 \\ \hline 6 & 7 & -14 & -8 & 0 \end{array}}$$

When $P(x)$ is divided by $x - 3$, the remainder is 0. Therefore, by the remainder theorem, 3 is a zero of P.

8. $P(x) = 6x^4 - 11x^3 - 35x^2 + 34x + 24$

$$-2\overline{)\begin{array}{ccccc} 6 & -11 & -35 & 34 & 24 \\ & -12 & 46 & -22 & -24 \\ \hline 6 & -23 & 11 & 12 & 0 \end{array}}$$

When $P(x)$ is divided by $x + 2$, the remainder is 0, so by the remainder theorem, $P(-2) = 0$. Therefore, by the factor theorem, $x + 2$ is a factor of $P(x)$.

9. A polynomial having zeros 2, -1, and $-i$ must also have i (the conjugate of $-i$) as a zero if its coefficients are real.

The polynomial has the form

$$P(x) = a(x - 2)(x + 1)(x + i)(x - i).$$

Use the condition $P(3) = 80$ to find the value of a.

$$\begin{aligned} P(3) &= a(3 - 2)(3 + 1)(3 + i)(3 - i) \\ &= 80 \\ a(1)(4)(10) &= 80 \\ 40a &= 80 \\ a &= 2 \\ P(x) &= 2(x - 2)(x + 1)(x - i)(x + 1) \\ &= 2(x^2 - x - 2)(x^2 + 1) \\ &= 2(x^4 - x^3 - x^2 - x - 2) \\ &= 2x^4 - 2x^3 - 2x^2 - 2x - 4 \end{aligned}$$

10. $P(x) = 2x^3 - x^2 - 13x - 6$; -2 is a zero.

Since -2 is a zero of $P(x)$, $x + 2$ is a factor.

Divide by x + 2.

$$-2\overline{)\begin{array}{ccc} 2 & -1 & -13 & -6 \\ & -4 & 10 & 6 \end{array}}$$
$$\begin{array}{cccc} 2 & -5 & -3 & 0 \end{array}$$

The quotient is $2x^2 - 5x - 3$, so

$P(x) = (x + 2)(2x^2 - 5x - 3)$

$\quad\quad = (x + 2)(2x + 1)(x - 3).$

11. $P(x) = 6x^3 - 25x^2 + 12x + 7$

(a) The possible rational zeros are

$\pm 1,\ \pm 7,\ \pm\dfrac{1}{2},\ \pm\dfrac{1}{3},\ \pm\dfrac{1}{6},\ \pm\dfrac{7}{2},\ \pm\dfrac{7}{3},\ \pm\dfrac{7}{6}.$

(b) $1\overline{)\begin{array}{cccc} 6 & -25 & 12 & 7 \\ & 6 & -19 & -7 \end{array}}$
$\quad\quad\overline{\begin{array}{cccc} 6 & -19 & -7 & 0 \end{array}}$

1 is a zero.
The other zeros may be found
by factoring the quotient.

$\quad\quad 6x^2 - 19x - 7 = 0$

$\quad\quad (2x - 7)(3x + 1) = 0$

$\quad\quad x = \dfrac{7}{2}\ \text{ or }\ x = -\dfrac{1}{3}$

The rational zeros are

$1,\ \dfrac{7}{2},\ \text{ and } -\dfrac{1}{3}.$

13. $P(x) = 2x^4 - 3x^3 + 4x^2 - 5x - 1$

$2\overline{)\begin{array}{ccccc} 2 & -3 & 4 & -5 & -1 \\ & 4 & 2 & 12 & 14 \end{array}}$
$\quad\overline{\begin{array}{ccccc} 2 & 1 & 6 & 7 & 13 \end{array}}$

Since 2 > 0 and all numbers in the
bottom row are nonnegative, the
boundedness theorem guarantees that
P has no real zeros greater than 2.

$-1\overline{)\begin{array}{ccccc} 2 & -3 & 4 & -5 & -1 \\ & -2 & 5 & -9 & 14 \end{array}}$
$\quad\overline{\begin{array}{ccccc} 2 & -5 & 9 & -14 & 13 \end{array}}$

Since $-1 < 0$ and the numbers in the
bottom row alternate in sign, the
boundedness theorem guarantees that
P has no real zeros less than -1.

14. $P(x) = x^4 + 3x^3 - x^2 - 2x + 1$

P(x) has 2 variations in sign:

$P(x) = +x^4 \underbrace{+ 3x^3 - }_{1} x^2 \underbrace{- 2x +}_{2} 1.$

Thus, P(x) has either 2 or 2 − 2 = 0
positive real zeros.

P(−x) has 2 variations in sign:

$P(-x) = (-x)^4 + 3(-x)^3 - (-x)^2 - 2(-x) + 1$

$\quad\quad = \underbrace{+x^4 -}_{1} 3x^3 \underbrace{- x^2 +}_{2} 2x + 1.$

Thus, P(x) has either 2 or 2 − 2 = 0
negative real zeros.

For Problems 15–18, see the answer graphs
in the textbook.

15. $P(x) = x^3 - 5x^2 + 8x - 4$

The possible rational zeros are

$\pm 1,\ \pm 2,\ \pm 4.$

$2\overline{)\begin{array}{cccc} 1 & -5 & 8 & -4 \\ & 2 & -6 & -4 \end{array}}$
$\quad\overline{\begin{array}{cccc} 1 & -3 & 2 & 0 \end{array}}$

2 is a zero. The other zeros
are the zeros of the quotient,
$x^2 - 3x + 2.$

$\quad\quad x^2 - 3x + 2 = 0$

$\quad\quad (x - 2)(x - 1) = 0$

$\quad\quad x = 2\ \text{ or }\ x = 1$

The rational zeros are 2 (multiplicity 2) and 1.

Interval	Test point	Value of $P(x)$	Location relative to x-axis
$(-\infty, 1)$	0	-4	Below
$(1, 2)$	$\dfrac{3}{2}$	$\dfrac{1}{8}$	Above
$(2, \infty)$	3	2	Above

Use the zeros and these points to sketch the graph.

16. $f(x) = \dfrac{-2}{x + 3}$

$\qquad = (-2) \cdot \dfrac{1}{x - (-3)}$

Compared to $y = \dfrac{1}{x}$, the graph will be reflected about the x-axis (because of the negative sign) and each point will be twice as far from the axis and translated 3 units to the left. There is a vertical asymptote at $x = -3$ and a horizontal asymptote at $y = 0$.

Sketch the graph.

17. $f(x) = \dfrac{3x - 1}{x - 2}$

To find any vertical asymptotes, solve the equation $x - 2 = 0$. There is one vertical asymptote, $x = 2$.

$$f(x) = \dfrac{3x - 1}{x - 2} = \dfrac{3 - \dfrac{1}{x}}{1 - \dfrac{5}{x}},$$

so, as $|x| \to \infty$, $f(x) \to 3$.

The horizontal asymptote is $y = 3$.

$$f(0) = \dfrac{3 \cdot 0 - 1}{0 - 2} = \dfrac{1}{2},$$

so the y-intercept is $\dfrac{1}{2}$.

The only zero of the numerator is $\dfrac{1}{3}$, so $\dfrac{1}{3}$ is the only x-intercept.

Sketch the graph.

18. $f(x) = \dfrac{x^2 - 1}{x^2 - 9}$

$\qquad = \dfrac{(x + 1)(x - 1)}{(x + 3)(x - 3)}$

There are vertical asymptotes at $x = -3$ and $x = 3$.

Since the numerator and denominator have the same degree, divide by x^2 to find the horizontal asymptote.

$$f(x) = \dfrac{1 - \dfrac{1}{x^2}}{1 - \dfrac{9}{x^2}} \to \dfrac{1}{1} \text{ as } |x| \to \infty,$$

so $y = 1$ is a horizontal asymptote.

The y-intercept is $\dfrac{1}{9}$.

The x-intercepts are -1 and 1.

Sketch the graph.

19. $f(x) = \dfrac{2x^2 + x - 6}{x - 1}$

To find the oblique asymptote, find the quotient determined by $(2x + x - 6) \div (x - 1)$ and disregard the remainder. Divide synthetically.

$$\begin{array}{r|rrr} 1) & 2 & 1 & -6 \\ & & 2 & 3 \\ \hline & 2 & 3 & -3 \end{array}$$

$(2x^2 + x - 6) \div (x - 1)$

$= 2x + 3 + \dfrac{-3}{x - 1}$

The oblique asymptote is $y = 2x + 3$.

20. (d) $f(x) = \dfrac{1}{x^2 + 4}$

has no x-intercepts since the function never attains the value 0. Since the numerator is never equal to 0, the expression $\dfrac{1}{x^2 + 4}$ can never be equal to 0.

**CHAPTER 7   SYSTEMS OF EQUATIONS
AND INEQUALITIES**

Section 7.1

1. One line has positive slope and one line has negative slope.

   The line with positive slope will slant upward from left to right and the line with negative slope will slant downward from left to right. Two such lines will intersect in a single point, so this system has a single solution.

3. Both lines have slope 0 and have the same y–intercept.

   Since they have slope 0, both lines are horizontal. Since they have the same y–intercept, the two horizontal lines are in fact the same line. The equations are dependent, and the system has infinitely many solutions.

5. $y = 3x + 6$
   $y = 3x + 9$

   These two lines have the same slope but different y–intercepts, so they are parallel. Parallel lines do not intersect, so the system has no solution.

7. $x + y = 4$
   $kx + ky = 4k$ $(k \neq 0)$

   $\frac{k}{1} = k, \frac{k}{1} = k,$ and $\frac{4k}{4} = k.$

   The second equation is a multiple of the first, so the graphs of the equations are the same line. The system has infinitely many solutions.

9. $y = 2x + 3$ (1)
   $3x + 4y = 78$ (2)

   Substitute $2x + 3$ for y into equation (2) and solve for x.

   $$3x + 4(2x + 3) = 78$$
   $$3x + 8x + 12 = 78$$
   $$11x = 66$$
   $$x = 6$$

   Substitute $x = 6$ back into equation (1) to find y.

   $$y = 2(6) + 3$$
   $$= 15$$

   Check by substituting 6 for x and 15 for y in each of the equations of the system.

   Solution set: $\{(6, 15)\}$

11. $3x - 2y = 12$ (1)
    $5x = 4 - 2y$ (2)

    First, solve equation (2) for x.

    $$x = \frac{4 - 2y}{5}   (3)$$

Now substitute this result into equation (1) and solve for y.

$$3\left(\frac{4 - 2y}{5}\right) - 2y = 12$$

$$5 \cdot 3\left(\frac{4 - 2y}{5}\right) - 5 \cdot 2y = 5(12)$$

$$3(4 - 2y) - 10y = 60$$

$$12 - 6y - 10y = 60$$

$$12 - 16y = 60$$

$$-16y = 48$$

$$y = -3$$

Substitute y = -3 back into equation (3) to find x.

$$x = \frac{4 - 2(-3)}{5}$$

$$= \frac{4 + 6}{5}$$

$$= 2$$

Check by substituting 2 for x and -3 for y in equations (1) and (2).

Solution set: $\{(2, -3)\}$

13. $\frac{x}{2} = \frac{7}{6} - \frac{y}{3}$ (1)

$\frac{2x}{3} = \frac{3y}{2} - \frac{7}{3}$ (2)

To eliminate fractions, multiply both sides of each equation by 6.

$$6\left(\frac{x}{2}\right) = 6\left(\frac{7}{6}\right) - 6\left(\frac{y}{3}\right)$$

$$6\left(\frac{2x}{3}\right) = 6\left(\frac{3y}{2}\right) - 6\left(\frac{7}{3}\right)$$

This gives a new system, which is equivalent to the original one.

$$3x = 7 - 2y   \quad (3)$$

$$4x = 9y - 14   \quad (4)$$

Solve equation (3) for x.

$$x = \frac{7 - 2y}{3}   \quad (5)$$

Now substitute this result into equation (4) and find y.

$$4\left(\frac{7 - 2y}{3}\right) = 9y - 14$$

$$3 \cdot 4\left(\frac{7 - 2y}{3}\right) = 3(9y - 14)$$

$$4(7 - 2y) = 27y - 42$$

$$28 - 8y = 27y - 42$$

$$70 = 35y$$

$$2 = y$$

Substitute y = 2 back into equation (5) to find x.

$$x = \frac{7 - 2(2)}{3}$$

$$= \frac{7 - 4}{3} = 1$$

Solution set: $\{(1, 2)\}$

15. $3x - 4y = 1$ (4)

$2x + 3y = 12$ (5)

To eliminate y, multiply both sides of equation (4) by 3 and both sides of equation (5) by 4 to get equations (6) and (7).

$$9x - 12y = 3 \quad (6)$$

$$\underline{8x + 12y = 48} \quad (7)$$

$$17x \qquad = 51$$

Thus, y has been eliminated.

17. $5x + 3y = 7 \quad (1)$

$\underline{7x - 3y = -19} \quad (2)$

$12x \qquad = -12$

$x = -1$

Substitute x = -1 in equation (1) to find y.

$$5(-1) + 3y = 7$$
$$3y = 12$$
$$y = 4$$

Check by substituting -1 for x and 4 for y in equation (2).

Solution set: $\{(-1, 4)\}$

19.   3x + 2y = 5   (*1*)
6x + 4y = 8   (*2*)

Multiply equation (1) by -2 and then add the result to equation (2).

$$\begin{array}{rcr} -6x - 4y &=& -10 \\ 6x + 4y &=& 8 \\ \hline 0 &=& -2 \end{array}$$

Both variables were eliminated, leaving the false statement 0 = -2. This shows that the two equations have no common solution and the system is inconsistent.

Solution set: Ø

21.   2x - 3y = -7   (*1*)
5x + 4y = 17   (*2*)

Multiply equation (1) by 4 and equation (2) by 3, and then add the resulting equations.

$$\begin{array}{rcr} 8x - 12y &=& -28 \\ 15x + 12y &=& 51 \\ \hline 23x &=& 23 \\ x &=& 1 \end{array}$$

To find y, substitute x = 1 in equation (1).

$$2(1) - 3y = -7$$
$$-3y = -9$$
$$y = 3$$

Check the solution by substituting 1 for x and 3 for y in equation (2).

Solution set: $\{(1, 3)\}$

23.   5x + 7y = 6   (*1*)
10x - 3y = 46   (*2*)

Multiply equation (1) by -2 and then add to equation (2).

$$\begin{array}{rcr} -10x - 14y &=& -12 \\ 10x - 3y &=& 46 \\ \hline -17y &=& 34 \\ y &=& -2 \end{array}$$

To find x, substitute y = -2 in equation (1).

$$5x + 7(-2) = 6$$
$$x = 4$$

Solution set: $\{(4, -2)\}$

25.   4x - y = 9   (*1*)
-8x + 2y = -18   (*2*)

Multiply equation (1) by 2 and then add to equation (2).

$$\begin{array}{rcr} 8x - 2y &=& 18 \\ -8x + 2y &=& -18 \\ \hline 0 &=& 0 \end{array}$$

These equations are dependent, that is, they represent the same line, and the solution set is the set of all points on the line. We will write the solution as an ordered

pair by expressing x in terms of y.
Solve equation (1) for x.

$$4x - y = 9$$
$$4x = y + 9$$
$$x = \frac{y + 9}{4}$$

We write the solution set as

$$\left\{ \left(\frac{y + 9}{4}, \ y \right) \right\}.$$

27. $\dfrac{x}{2} + \dfrac{y}{3} = 8$ (1)

$\dfrac{2x}{3} + \dfrac{3y}{2} = 17$ (2)

First, eliminate fractions by multiplying both sides of each equation
by 6.

$$6\left(\frac{x}{2}\right) + 6\left(\frac{y}{3}\right) = 6 \cdot 8$$
$$6\left(\frac{2x}{3}\right) + 6\left(\frac{3y}{2}\right) = 6 \cdot 17$$

This gives a new system, which is
equivalent to the original one.

$$3x + 2y = 48 \quad (3)$$
$$4x + 9y = 102 \quad (4)$$

Multiply equation (3) by 4 and
equation (4) by -3 and then add
them.

$$\begin{aligned} 12x + 8y &= 192 \\ -12x - 27y &= -306 \\ \hline -19y &= -114 \\ y &= 6 \end{aligned}$$

Substitute y = 6 in equation (1) to
find x.

$$\frac{x}{2} + \frac{6}{3} = 8$$
$$\frac{x}{2} + 2 = 8$$
$$\frac{x}{2} = 6$$
$$x = 12$$

Check by substituting 12 for x and 6
for y in equation (2).

Solution set: $\{(12, \ 6)\}$

29. $\dfrac{2}{x} + \dfrac{1}{y} = \dfrac{3}{2}$ (1)

$\dfrac{3}{x} - \dfrac{1}{y} = 1$ (2)

Let $\dfrac{1}{x} = t$, $\dfrac{1}{y} = u$.

Making these substitutions, we get
the following system.

$$2t + u = \frac{3}{2} \quad (3)$$
$$3t - u = 1 \quad (4)$$

Multiply equation (3) by 2 to eliminate fractions, multiply equation
(4) by 2, and then add the resulting
equations.

$$\begin{aligned} 4t + 2u &= 3 \quad (5) \\ 6t - 2u &= 2 \quad (6) \\ \hline 10t &= 5 \\ t &= \frac{1}{2} \end{aligned}$$

Substitute $t = \frac{1}{2}$ in equation (5) to find u.

$$4\left(\frac{1}{2}\right) + 2u = 3$$
$$2 + 2u = 3$$
$$2u = 1$$
$$u = \frac{1}{2}$$

Now find x and y.

$$\frac{1}{x} = t \qquad \frac{1}{y} = u$$
$$\frac{1}{x} = \frac{1}{2} \qquad \frac{1}{y} = \frac{1}{2}$$
$$x = 2 \qquad y = 2$$

Check by substituting x = 2 and y = 2 in the original system, equations (1) and (2).

Solution set: $\{(2, 2)\}$

31. $\frac{2}{x} + \frac{1}{y} = 11$ $\qquad$ (1)

$\frac{3}{x} - \frac{5}{y} = 10$ $\qquad$ (2)

Let $\frac{1}{x} = t$, $\frac{1}{y} = u$.

Making these substitutions, we get the following system.

$$2t + \ u = 11 \quad (3)$$
$$3t - 5u = 10 \quad (4)$$

Multiply equation (3) by 5 and then add the result to equation (4).

$$10t + 5u = 55 \quad (5)$$
$$\underline{3t - 5u = 10} \quad (6)$$
$$13t \qquad = 65$$
$$t = 5$$

To find u, substitute t = 5 in equation (3). (Any of the equations from any of the systems may be used.)

$$2(5) + u = 11$$
$$10 + u = 11$$
$$u = 1$$

Now find x and y.

$$\frac{1}{x} = t \qquad \frac{1}{y} = u$$
$$\frac{1}{x} = 5 \qquad \frac{1}{y} = 1$$
$$1 = 5x \qquad y = 1$$
$$x = \frac{1}{5}$$

Check by substituting $x = \frac{1}{5}$ and y = 1 in the original system, equations (1) and (2).

Solution set: $\left\{\left(\frac{1}{5}, 1\right)\right\}$

33. $\qquad\qquad ax + by = 5$

Since (-2, 1) is on the line,

$$-2a + b = 5.$$

Since (-1, -2) is on the line,

$$-a - 2b = 5.$$

This gives us the system

$$-2a + b = 5 \quad (1)$$
$$-a - 2b = 5. \quad (2)$$

Multiply equation (1) by 2 and add to equation (2).

$$-4a + 2b = 10$$
$$\underline{-a - 2b = 5}$$
$$-5a \qquad = 15$$
$$a = -3$$

Substitute a = −3 in equation (1) to find b.

$$(-2)(-3) + b = 5$$
$$6 + b = 5$$
$$b = -1$$

The line with equation ax + by = 5 passes through (−2, 1) and (−1, −2) if a = −3 and b = −1.

35.
$$y = mx + b$$

Since (2, 5) is on the line,

$$5 = 2m + b.$$

Since (−1, 4) is on the line,

$$4 = -m + b.$$

This gives us the system

$$2m + b = 5 \quad (1)$$
$$-m + b = 4. \quad (2)$$

Multiply equation (2) by −1 and then add to equation (1).

$$2m + b = 5 \quad (1)$$
$$\underline{m - b = -4} \quad (3)$$
$$3m \quad = 1$$
$$m = \frac{1}{3}$$

To find b, substitute m = 1/3 in equation (3).

$$\frac{1}{3} - b = -4$$
$$-b = -\frac{12}{3} - \frac{3}{3}$$
$$-b = -\frac{13}{3}$$
$$b = \frac{13}{3}$$

The line y = mx + b passes through (2, 5) and (−1, 4) if m = $\frac{1}{3}$ and b = $\frac{13}{3}$.

37. Let x = the amount borrowed at 7%; y = the amount borrowed at 8%.

$$x = 3y \quad (1)$$
$$.07x + .08y = 1160 \quad (2)$$

Substitute 3y for x in (2) and solve for y.

$$.07(3y) + .08y = 1160$$
$$.21y + .08y = 1160$$
$$.29y = 1160$$
$$y = 4000$$

Substitute y = 4000 in equation (1).

$$x = 3(4000)$$
$$= 12,000$$

She borrowed $4000 at 8% and $12,000 at 7%.

39. Let x = the number of kilograms of chemical x; y = the number of kilograms of chemical y.

Since the ratio of x to y must be 3 to 2,

$$\frac{x}{y} = \frac{3}{2} \quad \text{or} \quad x = \frac{3}{2}y.$$

$$x = \frac{3}{2}y \quad (1)$$
$$x + y = 800 \quad (2)$$

Substitute $\frac{3}{2}y$ for x in (2) and solve for y.

$$\frac{3}{2}y + y = 800$$

$$\frac{5}{2}y = 800$$

$$y = 320$$

Substitute y = 320 in equation (1).

$$x = \frac{3}{2}(320)$$

$$= 480$$

480 kg of chemical x and 320 kg of chemical y should be used.

41.   Let x = number of gallons of 98-octane gasoline;

y = number of gallons of 92-octane gasoline.

There are 40 gal in the mixture, so

$$x + y = 40.$$

Amount of isooctane in 98-octane gas		amount of isooctane in 92-octane gas		amount of isooctane in 94-octane gas
.98x	+	.92x	=	.94(40)

These two equations give us the system

$$x + y = 40 \qquad (1)$$
$$.98x + .92y = .94(40). \quad (2)$$

The decimals in equation (2) may be eliminated by multiplying both sides by 100.

$$98x + 92y = 94(40)$$

This gives us the following system, which can be solved by the addition method.

$$x + y = 40 \qquad (3)$$
$$98x + 92y = 3760 \quad (4)$$
$$-98x - 98y = -3920$$
$$98x + 92y = 3760$$
$$-6y = -160$$
$$y = \frac{160}{6} = \frac{80}{3}$$

Substitute y = 80/3 in equation (3) to find x.

$$x + \frac{80}{3} = 40$$

$$x = 40 - \frac{80}{3}$$

$$= \frac{120}{3} - \frac{80}{3}$$

$$= \frac{40}{3}$$

The 94-octane mixture will contain $\frac{40}{3}$ gal of 98-octane gasoline and $\frac{80}{3}$ gal of 92-octane gasoline.

43.   Let t = the cost of one turkey;

c = the cost of one chicken.

$$20t + \ 8c = 74 \quad (1)$$
$$15t + 24c = 87 \quad (2)$$

Multiply equation (1) by -3.

$$-60t - 24c = -222$$
$$\underline{15t + 24c = \quad 87}$$
$$-45t \qquad = -135$$
$$t = 3$$

Substitute t = 3 in equation (1).

$$20(3) + 8c = 74$$
$$60 + 8c = 74$$
$$8c = 14$$
$$c = 1.75$$

Each turkey costs $3 and each chicken costs $1.75.

45. Let d = the number of days Hector worked;

w = the daily wage Hector earned.

It follows from the given information that Ann worked d + 8 days for a daily wage of w + 4 dollars.

$$d + (d + 8) = 72 \quad (1)$$
$$dw + (d + 8)(w + 4) = 6496 \quad (2)$$

Solve equation (1).

$$d + (d + 8) = 72$$
$$2d + 8 = 72$$
$$2d = 64$$
$$d = 32$$

Substitute d = 32 in equation (2).

$$32w + (32 + 8)(w + 4) = 6496$$
$$32w + 40(w + 4) = 6496$$
$$32w + 40w + 160 = 6496$$
$$72w = 6336$$
$$w = 88$$

Hector worked 32 days at $88 per day.

47. Let x = the amount invested in the mutual fund;

y = the amount invested in bonds.

$$x + \quad y = 30,000 \quad (1)$$
$$.045x + .05y = 1410 \quad (2)$$

Multiply equation (1) by -.045.

$$-.045x - .045y = -1350 \quad (3)$$
$$\underline{.045x + .045y = \quad 1410} \quad (2)$$
$$.005y = \quad 60$$
$$y = 12,000$$

Substitute y = 12,000 in equation (1).

$$x + 12,000 = 30,000$$
$$x = 18,000$$

$18,000 is invested in the mutual fund at 4.5% and $12,000 is invested in the bonds at 5%.

49.
$$C = 1.5x + 252$$
$$R = 5.5x$$
$$C = R$$
$$1.5x + 252 = 5.5x$$
$$252 = 4x$$
$$63 = x$$

The break-even point is 63 items. Thus, at the break-even point,

$$C = R = 1.5x + 252$$
$$= 1.5(63) + 252$$
$$= 346.5.$$

51.
$$C = 20x + 10,000$$
$$R = 30x - 11,000$$
$$C = R$$
$$20x + 10,000 = 30x - 11,000$$
$$21,000 = 10x$$
$$2100 = x$$

The break-even point is 2100 items.

$$C = R = 20x + 10{,}000$$
$$= 20(2100) + 10{,}000$$
$$= 20(2100) + 10{,}000$$
$$= 52{,}000$$

The equilibrium price is $\frac{28}{3}$.

The equilibrium supply/demand is 28.

53. $p = 80 - \frac{3}{5}q$ (1)

$p = \frac{2}{5}q$ (2)

Solve this system by the substitution method.

$$\frac{2}{5}q = 80 - \frac{3}{5}q$$

$$5\left(\frac{2}{5}q\right) = 5(80) - 5\left(\frac{3}{5}q\right)$$

$$2q = 400 - 3q$$

$$5q = 400$$

$$q = 80$$

$$p = \frac{2}{5}(80) = 32$$

The equilibrium price is 32.
The equilibrium supply/demand is 80.

55. $3p = 84 - 2q$ (1)
$3p - q = 0$ (2)

Solve equation (2) for x and substitute the result into equation (1).

$$3p = q$$
$$3p = 84 - 2(3p)$$
$$3p = 84 - 6p$$
$$9p = 84$$
$$p = \frac{84}{9} = \frac{28}{3}$$
$$q = 3\left(\frac{28}{3}\right) = 28$$

57. $p = 16 - \frac{5}{4}q$

(a) $p = 16 - \frac{5}{4} \cdot 0$

   $= 16$

(b) $p = 16 - \frac{5}{4} \cdot 4$

   $= 16 - 5$

   $= 11$

(c) $p = 16 - \frac{5}{4} \cdot 8$

   $= 16 - 10$

   $= 6$

(d) $6 = 16 - \frac{5}{4}q$

   $-10 = -\frac{5}{4}q$

   $-40 = -5q$

   $8 = q$

(e) $11 = 16 - \frac{5}{4}q$

   $-5 = -\frac{5}{4}q$

   $-20 = -5q$

   $4 = q$

(f) $16 = 16 - \frac{5}{4}q$

   $0 = -\frac{5}{4}q$

   $0 = q$

(g) See the graph of $p = 16 - \frac{5}{4}q$

   in the textbook.

(h) $p = \frac{3}{4}q$

$0 = \frac{3}{4}q$

$0 = q$

(i) $10 = \frac{3}{4}q$

$\frac{4}{3}(10) = q$

$\frac{40}{3} = q$

(j) $20 = \frac{3}{4}q$

$\frac{4}{3}(20) = q$

$\frac{80}{3} = q$

(k) See the graph of $p = \frac{3}{4}q$ in the textbook.

(1) To find the equilibrium supply, solve the system

$$p = 16 - \frac{5}{4}q \quad (1)$$

$$p = \frac{3}{4}q. \quad (2)$$

The value of q will give the equilibrium supply.

$$\frac{3}{4}q = 16 - \frac{5}{4}q$$

$$4\left(\frac{3}{4}q\right) = 4(16) - 4\left(\frac{5}{4}q\right)$$

$$3x = 64 - 5q$$

$$8q = 64$$

$$q = 8$$

The equilibrium supply is 8.

(m) To find p, substitute q = 8 into equation (2).

$$p = \frac{3}{4}(8)$$

$$= 6$$

The equilibrium price is 6.

59. supply: $p = \frac{2}{5}q$

demand: $p = 100 - \frac{2}{5}q$

(a) See the graph in the textbook.

(b) The equilibrium demand may be found from reading the graph. The lines intersect at (125, 50), so the equilibrium demand is the x-coordinate of this point, 125. To solve algebraically, solve the system of the two given equations by the substitution method.

$$p = \frac{2}{5}q \quad (1)$$

$$p = 100 - \frac{2}{q} \quad (2)$$

$$\frac{2}{5}q = 100 - \frac{2}{q}$$

$$5\left(\frac{2}{5}q\right) = 5(100) - 5\left(\frac{2}{5}q\right)$$

$$2q = 500 - 2q$$

$$4q = 500$$

$$q = 125$$

The equilibrium demand is 125.

(c) The equilibrium may be found by reading the graph. This price is the y-coordinate of the intersection point of the two lines, 25.

To find p algebraically, substitute q = 125 into equation (1).

$$p = \frac{2}{5}(125)$$

$$= 50$$

The equilibrium price is 50.

Section 7.2

1. $x + y + z = 2$ (1)
    $2x + y - z = 5$ (2)
    $x - y + z = -2$ (3)

Eliminate z first.
Add equations (1) and (2) to get

$$3x + 2y = 7. \quad (4)$$

Add equations (2) and (3) to get

$$3x = 3$$
$$x = 1.$$

$3(1) + 2y = 7$ *Let x = 1 in (4)*
$$2y = 4$$
$$y = 2$$

$1 + 2 + z = 2$ *Let x = 1, y = 2 in (1)*

$$z = -1$$

Check by substituting x = 1, y = 2, and z = -1 in the three original equations.

Solution set: $\{(1, 2, -1)\}$

3. $x + 3y + 4z = 14$ (1)
    $2x - 3y + 2z = 10$ (2)
    $3x - y + z = 9$ (3)

Eliminate y first. Add equations (1) and (2) to get

$$3x + 6z = 24. \quad (4)$$

Multiply equation (3) by 3 and add the result to equation (1).

$$x + 3y + 4z = 14$$
$$\underline{9x - 3y + 3z = 27}$$
$$10x + \qquad 7z = 41$$

This gives the new system

$$3x + 6z = 24 \quad (4)$$
$$10x + 7z = 41. \quad (5)$$

Multiply equation (4) by 10 and equation (5) by -3 and add.

$$30x + 60z = 240$$
$$\underline{-30x - 21z = -123}$$
$$39z = 117$$
$$z = 3$$

$3x + 6(3) = 24$ *Let z = 3 in (4)*
$$3x = 6$$
$$x = 2$$

$2 + 3y + 4(3) = 14$ *Let x = 2, z = 3 in (1)*
$$3y = 0$$
$$y = 0$$

Solution set: $\{(2, 0, 3)\}$

5. $x + 2y + 3z = 8$ (1)
    $3x - y + 2z = 5$ (2)
    $-2x - 4y - 6z = 5$ (3)

Eliminate x first. Multiply equation (1) by 2 and add the result to equation (2).

$$2x + 4y + 6z = 16$$
$$\underline{-2x - 4y - 6z = 5}$$
$$0 = 21$$

All three variables have been eliminated, leaving the false statement $0 = 21$, so there is no solution.

Solution set: $\emptyset$

7. $x + 4y - z = 6$ (1)
   $2x - y + z = 3$ (2)
   $3x + 2y + 3z = 16$ (3)

Eliminate z first. Add equations (1) and (2) to get

$$3x + 3y = 9 \text{ or}$$
$$x + y = 3. \quad (4)$$

Multiply equation (1) by 3 and add the result to equation (3).

$$3x + 12y - 3z = 18$$
$$\underline{3x + 2y + 3z = 16}$$
$$6x + 14y = 34 \text{ or}$$
$$3x + 7y = 17 \quad (5)$$

The new system is

$$x + y = 3 \quad (4)$$
$$3x + 7y = 17. \quad (5)$$

Multiply equation (3) by -3 and add the result to equation (7).

$$-3x - 3y = -9$$
$$\underline{3x + 7y = 17}$$
$$4y = 8$$
$$y = 2$$

$x + 2 = 3$ *Let y = 2 in (4)*
  $x = 1$

$1 + 4(2) - z = 6$ *Let x = 1,*
                    y = 2 in (1)
  $9 - z = 6$
      $z = 3$

Solution set: $\{(1, 2, 3)\}$

9. $5x + y - 3z = -6$ (1)
   $2x + 3y + z = 5$ (2)
   $-3x - 2y + 4z = 3$ (3)

Eliminate y first.
Multiply equation (1) by 2 and add to equation (3).

$$10x + 2y - 6z = -12$$
$$\underline{-3x - 2y + 4z = 3}$$
$$7x - 2z = -9 \quad (4)$$

Multiply equation (1) by -3 and add to equation (2).

$$-15x - 3y + 9z = 18$$
$$\underline{2x + 3y + z = 5}$$
$$-13x + 10z = 23 \quad (5)$$

This gives the new system

$$7x - 2z = -9 \quad (4)$$
$$-13x + 10z = 23. \quad (5)$$

Multiply equation (4) by 5 and add to equation (5).

$$35x - 10z = -45$$
$$\underline{-13x + 10z = 23}$$
$$22x = -22$$
$$x = -1$$

$7(-1) - 2z = -9$ *Let x = -1 in (4)*
     $-2z = -2$
       $z = 1$

$5(-1) + y - 3(1) = -6$ *Let* $x = -1$,
$\qquad\qquad\qquad\qquad$ *z = 1 in* (1)

$\qquad\qquad y - 8 = -6$

$\qquad\qquad\qquad y = 2$

Solution set: $\{(-1,\ 2,\ 1)\}$

11. $\quad x - 3y - 2z = -3 \qquad (1)$
$\quad 3x + 2y - z = 12 \qquad (2)$
$\quad -x - y + 4z = 3 \qquad (3)$

Eliminate x first. Add equations (1) and (3).

$\qquad\quad -4 + 2z = 0 \qquad\quad (4)$

Multiply equation (3) by 3 and add to equation (2).

$\qquad -3x - 3y + 12z = 9$
$\qquad \underline{\ 3x + 2y - \ \ z = 12}$
$\qquad\qquad -2y + 11z = 21 \qquad (5)$

This gives the new system

$\qquad\quad -4y + 2z = 0 \qquad (4)$
$\qquad\quad -y + 11z = 21. \qquad (5)$

Multiply equation (5) by -4 and add to equation (4).

$\qquad\quad -4y + 2z = \quad 0$
$\qquad\quad \underline{\ 4y - 44z = -84}$
$\qquad\qquad -42z = -84$
$\qquad\qquad\quad z = 2$

$\quad -4y + 2(2) = 0$ *Let* $z = 2$ *in* (4)
$\qquad\qquad -4y = -4$
$\qquad\qquad\quad y = 1$

$x - 3(1) - 2(2) = -3$ *Let* $y = 1$,
$\qquad\qquad\qquad\qquad$ *z = 2 in* (1)

$\qquad\quad x - 7 = -3$

$\qquad\qquad x = 4$

Solution set: $\{(4,\ 1,\ 2)\}$

13. $\quad 2x + 6y - z = 6 \qquad (1)$
$\quad 4x - 3y + 5z = -5 \qquad (2)$
$\quad 6x + 9y - 2z = 11 \qquad (3)$

Eliminate y first. Add equation (1) to 2 times equation (2).

$\qquad 2x + 6y - \ \ z = \quad 6$
$\qquad \underline{8x - 6y + 10z = -10}$
$\qquad 10x \qquad + 9z = -4 \qquad (4)$

Add equation (3) to 3 times equation (2).

$\qquad 6x + 9y - \ \ 2z = \quad 11$
$\qquad \underline{12x - 9y + 15z = -15}$
$\qquad 18x \qquad + 13z = -4 \qquad (5)$

This gives the new system

$\qquad 10x + 9z = -4 \qquad (4)$
$\qquad 18x + 13z = -4. \qquad (5)$

Multiply equation (4) by 9 and equation (5) by -5 and add the results.

$\qquad\quad 90x + 81z = -36$
$\qquad\quad \underline{-90x - 65z = \ \ 20}$
$\qquad\qquad\quad 16z = -16$
$\qquad\qquad\qquad z = -1$

$10x + 9(-1) = -4$ *Let* $z = -1$ *in* (4)
$\qquad\quad 10x = 5$
$\qquad\qquad x = \dfrac{1}{2}$

$2\left(\dfrac{1}{2}\right) + 6y - (-1) = 6$ *Let* $x = 1/2$,
$\qquad\qquad\qquad\qquad\qquad$ *z = -1 in* (1)

$\qquad\quad 6y + 2 = 6$

$\qquad\qquad 6y = 4$

$\qquad\qquad y = \dfrac{2}{3}$

Solution set: $\left\{\left(\dfrac{1}{2},\ \dfrac{2}{3},\ -1\right)\right\}$

15.

$$\frac{1}{x} + \frac{1}{y} - \frac{1}{z} = \frac{1}{4} \qquad (1)$$

$$\frac{2}{x} - \frac{1}{y} + \frac{3}{z} = \frac{9}{4} \qquad (2)$$

$$-\frac{1}{x} - \frac{2}{y} + \frac{4}{z} = 1 \qquad (3)$$

Let $\frac{1}{x} = t$, $\frac{1}{y} = u$, $\frac{1}{z} = v$.

Making these substitutions, we get the following system.

$$t + u - v = \frac{1}{4} \qquad (4)$$

$$2t - u + 3v = \frac{9}{4} \qquad (5)$$

$$-t - 2u + 4v = 1 \qquad (6)$$

To eliminate fractions, multiply equations (1) and (2) by 4, and then work with the resulting system.

$$4t + 4u - 4v = 1 \qquad (7)$$
$$8t - 4u + 12v = 9 \qquad (8)$$
$$-t - 2u + 4v = 1 \qquad (6)$$

Eliminate u first. Add equations (7) and (8).

$$12t \qquad + 8v = 10$$

Multiply equation (6) by −2 and add the result to equation (8).

$$8t - 4u + 12v = 9$$
$$\underline{2t + 4u - 8v = -2}$$
$$10t \qquad + 4v = 7$$

This gives the new system

$$12t + 8v = 10 \qquad (9)$$
$$10t + 4v = 7. \qquad (10)$$

Multiply equation (10) by −2 and add the result to equation (9).

$$12t + 8v = 10$$
$$\underline{-20t - 8v = -14}$$
$$-8t \qquad = -4$$

$$t = \frac{1}{2}$$

To find v, substitute t = 1/2 in equation (9).

$$12\left(\frac{1}{2}\right) + 8v = 10$$
$$6 + 8v = 10$$
$$8v = 4$$
$$v = \frac{1}{2}$$

To find u, substitute t = 1/2 and v = 1/2 in equation (7).

$$4\left(\frac{1}{2}\right) + 4u - 4\left(\frac{1}{2}\right) = 1$$
$$4u = 1$$
$$u = \frac{1}{4}$$

Finally, find x, y, and z.

$$\frac{1}{x} = t \qquad \frac{1}{y} = u \qquad \frac{1}{z} = v$$

$$\frac{1}{x} = \frac{1}{2} \qquad \frac{1}{y} = \frac{1}{4} \qquad \frac{1}{z} = \frac{1}{2}$$

$$x = 2 \qquad y = 4 \qquad z = 2$$

Solution set: $\{(2, 4, 2)\}$

17.

$$\frac{2}{x} - \frac{2}{y} + \frac{1}{z} = -1 \qquad (1)$$

$$\frac{4}{x} + \frac{1}{y} - \frac{2}{z} = -9 \qquad (2)$$

$$\frac{1}{x} + \frac{1}{y} - \frac{3}{z} = -9 \qquad (3)$$

Let $\dfrac{1}{x} = t$, $\dfrac{1}{y} = u$, $\dfrac{1}{z} = v$.

Making these substitutions, we get the following system.

$$2t - 2u + v = -1 \quad (4)$$
$$4t + u - 2v = -9 \quad (5)$$
$$t + u - 3v = -9 \quad (6)$$

Eliminate u first. Multiply equation (5) by 2 and add the result to equation (4).

$$
\begin{array}{r}
2t - 2u + v = -1 \\
8t + 2u - 4v = -18 \\
\hline
10t \quad\;\; - 3v = -19
\end{array}
$$

Multiply equation (6) by -1 and add the result to equation (5).

$$
\begin{array}{r}
4t + u - 2v = -9 \\
-t - u + 3v = 9 \\
\hline
3t \quad\;\; + v = 0
\end{array}
$$

This gives the new system

$$10t - 3v = -19 \quad (7)$$
$$3t + v = 0. \quad (8)$$

Multiply equation (8) by 3 and add the result to equation (7).

$$
\begin{array}{r}
10t - 3v = -19 \\
9t + 3v = 0 \\
\hline
19t \quad\;\; = -19 \\
t = -1
\end{array}
$$

To find v, substitute $t = -1$ in equation (8).

$$3(-1) + v = 0$$
$$v = 3$$

To find u, substitute $t = -1$ and $v = 3$ in equation (6).

$$-1 + u - 3(3) = -9$$
$$-1 + u - 9 = -9$$
$$-10 + u = -9$$
$$u = 1$$

Finally, find x, y, and z.

$$\frac{1}{x} = t \qquad \frac{1}{y} = u \qquad \frac{1}{z} = v$$

$$\frac{1}{x} = -1 \qquad \frac{1}{y} = 1 \qquad \frac{1}{z} = -3$$

$$x = -1 \qquad y = 1 \qquad z = \frac{1}{3}$$

Solution set: $\left\{-1,\ 1,\ \dfrac{1}{3}\right\}$

19.

$$x + y + z = 4 \quad (1)$$

(a)
$$x + 2y + z = 5 \quad (2)$$
$$2x - y + 3z = 4 \quad (3)$$

Equations (1), (2), and (3) form a system having exactly one solution, namely $(4, 1, -1)$.
(There are other equations that would do the same.)

(b)
$$x + y + z = 5 \quad (4)$$
$$2x - y + 3z = 4 \quad (5)$$

Equations (1), (4), and (5) form a system having no solution, since no ordered triple can satisfy equations (1) and (4) simultaneously.
(There are other equations that would do the same.)

(c) $2x + 2y + 2z = 8$ (6)

   $2x - y + 3z = 4$ (7)

Equations (1), (6), and (7) form a system having infinitely many solutions, since all the ordered triples that satisfy equation (1) will also satisfy equation (6).
(There are other equations that would do the same.)

21. For example, the ceiling and two perpendicular walls of a standard room meet at a single point.

23. $x - 2y + 3z = 6$ (1)

   $2x - y + 2z = 5$ (2)

First eliminate y. Multiply equation (2) by -2 and add the result to equation (1).

$x - 2y + 3z = 6$

$\underline{-4x + 2y - 4z = -10}$

$-3x - z = -4$ (3)

Now solve equation (3) for z in terms of x.

$$-z = 3x - 4$$

$$z = -3x + 4 \quad (4)$$

Now, express y also in terms of x by solving equation (2) for y and substituting $-3x + 4$ for z in the result.

$2x - y + 2z = -5$

$-y = -2x - 2z + 5$

$y = 2x + 2z - 5$

$y = 2x + 2(-3x + 4) - 5$

$y = 2x - 6x + 8 - 5$

$y = -4x + 3$ (5)

Equations (4) and (5) are used to write the solution with x arbitrary.

Solution set:

$$\{(x, -4x + 3, -3x + 4)\}$$

25. $5x - 4y + z = 9$ (1)

   $x + y = 15$ (2)

Add equation (1) to 4 times equation (2).

$5x - 4y + z = 9$

$\underline{4x + 4y = 60}$

$9x + z = 69$

$z = -9x + 69$

Now express y in terms of x by solving equation (2) for y.

$$y = -x + 15$$

Solution set:

$$\{(x, -x + 15, -9x + 69)\}$$

27. $3x - 5y - 4z = -7$ (1)

   $y - z = -13$ (2)

Add 5 times equation (2) to equation (1).

$3x - 5y - 4z = -7$

$\underline{5y - 5z = -65}$

$3x - 9z = -72$

$9z = 3x + 72$

$z = \dfrac{24 + x}{3}$

Now express y in terms of x by solving equation (2) for y and substituting $\frac{24 + x}{3}$ for z.

$$y = z - 13$$

$$y = \frac{24 + x}{3} - 13$$

$$y = \frac{-15 + x}{3}$$

Solution set:

$$\left\{ \left(x, \frac{-15 + x}{3}, \frac{24 + x}{3} \right) \right\}$$

29. The graph $y = ax^2 + bx + c$ passes through $(2, 3)$, $(-1, 0)$, and $(-2, 2)$.

If $x = 2$, then $y = 3$, so

$$3 = a(2)^2 + b(2) + c$$
$$3 = 4a + 2b + c$$

If $x = -1$, then $y = 0$, or

$$0 = a(-1)^2 + b(-1) + c$$
$$0 = a - b + c.$$

If $x = -2$, then $y = 2$, so

$$2 = a(-2)^2 + b(-2) + c$$
$$2 = 4a - 2b + c.$$

This gives the system

$$4a + 2b + c = 3 \quad (1)$$
$$a - b + c = 0 \quad (2)$$
$$4a - 2b + c = 2. \quad (3)$$

Eliminate b first. Add equations (1) and (3) to get

$$8a + 2c = 5.$$

Multiply equation (2) by 2 and add to equation (1).

$$4a + 2b + c = 3$$
$$\underline{2a - 2b + 2c = 0}$$
$$6a \quad\;\; + 3c = 3$$

or $2a + c = 1$

This gives the new system

$$8a + 2c = 5 \quad (4)$$
$$2a + c = 1. \quad (5)$$

Multiply equation (5) by -2 and add to equation (4).

$$8a + 2c = 5$$
$$\underline{-4a - 2c = -2}$$
$$4a \qquad = 3$$

$$a = \frac{3}{4}$$

$2\left(\frac{3}{4}\right) + c = 1$ *Substitute in (5)*

$$\frac{3}{2} + c = 1$$

$$c = -\frac{1}{2}$$

$\frac{3}{4} - b - \frac{1}{2} = 0$ *Substitute in (2)*

$$\frac{1}{4} - b = 0$$

$$b = \frac{1}{4}$$

Thus, the graph of $y = ax^2 + bx + c$ passes through $(2, 3)$, $(-1, 0)$, and $(-2, 2)$ if $a = \frac{3}{4}$, $b = \frac{1}{4}$, and $c = -\frac{1}{2}$.

31. The graph of

$$x^2 + y^2 + ax + by + c = 0$$

passes through $(2, 1)$, $(-1, 0)$, and $(3, 3)$.

Substitute these values for x and y to get three equations in a, b, and c.

$2^2 + 1^2 + 2a + b + c = 0$
     Let x = 2, y = 1

  $2a + b + c = -5$

$(-1)^2 + 0^2 - a + 0b + c = 0$
     Let x = -1, y = 0

   $-a + c = -1$

$3^2 + 3^2 + 3a + 3b + c = 0$
     Let x = 3, y = 3

  $3a + 3b + c = -18$

We have the system

  $2a + b + c = -5$ *(1)*

   $-a + c = -1$ *(2)*

 $3a + 3b + c = -18.$ *(3)*

Add -3 times equation (1) to equation (3) to eliminate b.

  $-6a - 3b - 3c = 15$

  $\underline{3a + 3b + c = -18}$

  $-3a - 2c = -3$ *(4)*

Then we have the system

   $-a + c = -1$ *(2)*

  $-3a - 2c = -3.$ *(4)*

Multiply equation (2) by -3 and add to equation (4).

   $3a - 3c = 3$

  $\underline{-3a - 2c = -3}$

    $-5c = 0$

     $c = 0$

  $-a + 0 = -1$ *Let c = 0 in (2)*

   $a = 1$

$2(1) + b + 0 = -5$ *Let a = 1,*
       c = 0 in (1)

   $b = -7$

Thus, a = 1, b = -7, and c = 0, so the equation of the circle is

$$x^2 + y^2 + x - 7y = 0.$$

33. Let x = the number of cents;
  y = the number of nickels;
  z = the number of quarters.

z = x - 8 may be rewritten as x - z = 8.

    $x - z = 8$ *(1)*

   $x + y + z = 29$ *(2)*

 $.01x + .05y + .25z = 1.77$ *(3)*

First eliminate z.

  $x - z = 8$ *(1)*

  $\underline{x + y + z = 29}$ *(2)*

  $2x + y = 37$ *(4)*

 $.25x - .25z = 2$ *Multiply*
         (1) by .25

 $\underline{.01x + .05y + .25z = 1.77}$ *(3)*

 $.26x + .05y = 3.77$ *(5)*

Use equations (4) and (5) to solve for x.

 $-.1x - .05y = -1.85$ *Multiply (4)*
         by -.05

 $\underline{.26x + .05y = 3.77}$ *(5)*

 $.16x = 1.92$

    $x = 12$

Now solve for y and z.

  $12 - z = 8$ *Let x = 12 in (1)*

   $-z = -4$

    $z = 4$

$12 + y + 4 = 29$ *Let $x = 12$ and*
$z = 4$ *in (2)*

$y = 13$

There are 12 cents, 13 nickels, and
4 quarters.

35. Let x = the number of barrels of
$150 glue;

y = the number of barrels of
$190 glue;

z = the number of barrels of
$120 glue.

Organize the given information in a
chart.

Price of glue	Number of barrels	Value
$100	150	15,000
$150	x	150x
$190	y	190y
$120	z	120z

$x = y$ *(1) Same number of barrels*

$z = 150 + x + y$ *(2)*
*$120 glue is mixture
of other three*

$15,000 + 150x + 190y = 120z$ *(3)*
Total value from chart

Substitute x for y in equation (2).

$z = 150 + x + x$

$z = 150 + 2x$

Substitute x for y and $150 + 2x$ for
z in equation (3).

$15,000 + 150x + 190x = 120(150 + 2x)$

$15,000 + 340x = 18,000 + 240x$

$100x = 30,000$

$x = 30$

$y = 30$

$z = 150 + 2(30)$

$= 210$

30 barrels each of $150 and $190
glue are needed.   210 barrels of
$120 glue will be produced.

37. Let x = the length of the shortest
side;

y = the length of the medium
side;

z = the length of the longest
side.

$z = y + 11$

$y = x + 3$

$x + y + z = 59$

Rewrite these equations.

$-y + z = 11$ *(1)*

$-x + y = 3$ *(2)*

$x + y + z = 59$ *(3)*

First eliminate x.

$-x + y \qquad = 3$ *(2)*

$\underline{x + y + z = 59}$ *(3)*

$2y + z = 62$ *(4)*

Use equations (1) and (4) to solve
for z.

$-2y + 2z = 22$ *Multiply (1) by 2*

$\underline{2y + z = 62}$ *(4)*

$3z = 84$

$z = 28$

Now solve for x and y.

$-y + 28 = 11$ *Let z = 28 in (1)*

$-y = -17$

$y = 17$

$-x + 17 = 3$ *Let y = 17 in (2)*

$-x = -14$

$x = 14$

The lengths of the sides of the triangle are 14 in, 17 in, and 28 in.

39. Let x = the amount invested at 5%;

y = the amount invested at 4.5%;

z = the amount invested at 3.75%.

$z = x + y - 20,000$ may be rewritten as $x + y - z = 20,000$.

$$\begin{array}{llll} x + & y - & z = 20,000 & (1) \\ x + & y + & z = 100,000 & (2) \\ .05x + & .045y + & .0375z = 4450 & (3) \end{array}$$

First eliminate z. Add equations (1) and (2).

$$\begin{array}{lr} x + y - z = 20,000 & (1) \\ \underline{x + y + z = 100,000} & (2) \\ 2x + 2y = 120,000 & \\ \text{or} \quad x + y = 60,000 & (4) \end{array}$$

Multiply equation (1) by .0375 and add the result to equation (3).

$$\begin{array}{l} .0375x + .0375y - .0375z = 750 \\ \underline{.05x + .045y + .0375z = 4450} \\ .0875x + .0825y = 5200 \quad (5) \end{array}$$

Use equations (4) and (5) to solve for x.

Multiply equation (4) by −.0825 and add the result to equation (5).

$$\begin{array}{l} -.0825x - .0825y = -4950 \\ \underline{.0875x + .0825y = 5200} \\ .005x = 250 \\ x = 50,000 \end{array}$$

Now solve for y and z.

$50,000 + y = 60,000$ *Let x = 50,000 in (4)*

$y = 10,000$

$50,000 + 10,000 + z = 100,000$

Let x = 50,000 and y = 10,000 in (2)

$z = 40,000$

The amounts invested were $50,000 at 5%, $10,000 at 4.5%, and $40,000 at 3.75%.

Section 7.3

1. $\begin{bmatrix} 2 & 4 \\ 4 & 7 \end{bmatrix}$; −2 times row 1 added to row 2

Using the third row transformation, the matrix is changed to

$$\begin{bmatrix} 2 & 4 \\ 4 + (-2)(2) & 7 + (-2)(4) \end{bmatrix}$$

$$= \begin{bmatrix} 2 & 4 \\ 0 & -1 \end{bmatrix}.$$

3. $\begin{bmatrix} 1 & 5 & 6 \\ -2 & 3 & -1 \\ 4 & 7 & 0 \end{bmatrix}$; 2 times row 1 added to row 2

Using the third row transformation, the matrix is changed to

$$\begin{bmatrix} 1 & 5 & 6 \\ -2+2(1) & 3+2(5) & -1+2(6) \\ 4 & 7 & 0 \end{bmatrix}$$

$$= \begin{bmatrix} 1 & 5 & 6 \\ 0 & 13 & 11 \\ 4 & 7 & 0 \end{bmatrix}.$$

5. $\begin{bmatrix} -3 & 1 & -4 \\ 2 & 1 & 3 \\ -7 & 5 & 2 \end{bmatrix}$; $\begin{array}{l} -5 \text{ times row 2} \\ \text{added to row 3} \end{array}$

Using the third row transformation, the matrix is changed to

$$\begin{bmatrix} -3 & 1 & -4 \\ 2 & 1 & 3 \\ -7+(-5)(2) & 5+(-5)(1) & 2+(-5)(3) \end{bmatrix}$$

$$= \begin{bmatrix} -3 & 1 & -4 \\ 2 & 1 & 3 \\ -17 & 0 & -13 \end{bmatrix}.$$

7. $2x + 3y = 11$
    $x + 2y = 8$

The augmented matrix is

$$\begin{bmatrix} 2 & 3 & | & 11 \\ 1 & 2 & | & 8 \end{bmatrix}.$$

9. $x + 5y = 6$
    $x + 2y = 8$

The augmented matrix is

$$\begin{bmatrix} 1 & 5 & | & 6 \\ 1 & 2 & | & 8 \end{bmatrix}.$$

11. $2x + y + z = 3$
    $3x - 4y + 2z = -7$
    $x - y + z = 2$

has the augmented matrix

$$\begin{bmatrix} 2 & 1 & 1 & | & 3 \\ 3 & -4 & 2 & | & -7 \\ 1 & 1 & 1 & | & 2 \end{bmatrix}.$$

13. $x + y = 2$
    $2y + z = -4$
    $z = 2$

has the augmented matrix

$$\begin{bmatrix} 1 & 1 & 0 & | & 2 \\ 0 & 2 & 1 & | & -4 \\ 0 & 0 & 1 & | & 2 \end{bmatrix}.$$

15. $\begin{bmatrix} 2 & 1 & | & 1 \\ 3 & -2 & | & -9 \end{bmatrix}$

is associated with the system

$$2x + y = 1$$
$$3x - 2y = -9.$$

17. $\begin{bmatrix} 1 & 0 & 0 & | & 2 \\ 0 & 1 & 0 & | & 3 \\ 0 & 0 & 1 & | & -2 \end{bmatrix}$

is associated with the system

$$x = 2$$
$$y = 3$$
$$z = -2.$$

19. $\begin{bmatrix} 3 & 2 & 1 & | & 1 \\ 0 & 2 & 4 & | & 22 \\ -1 & -2 & 3 & | & 15 \end{bmatrix}$

is associated with the system

$$3x + 2y + z = 1$$
$$2y + 4z = 22$$
$$-x - 2y + 3x = 15.$$

21. $x + y = 5$
    $x - y = -1$

has the augmented matrix

$$\begin{bmatrix} 1 & 1 & | & 5 \\ 1 & -1 & | & -1 \end{bmatrix}.$$

$$\begin{bmatrix} 1 & 1 & | & 5 \\ 0 & -2 & | & -6 \end{bmatrix} \quad -1R1 + R2$$

$$\begin{bmatrix} 1 & 1 & | & 5 \\ 0 & 1 & | & 3 \end{bmatrix} \quad -\tfrac{1}{2}R2$$

$$\begin{bmatrix} 1 & 0 & | & 2 \\ 0 & 1 & | & 3 \end{bmatrix} \quad -1R2 + R1$$

Solution set: $\{(2, 3)\}$

23. $x + y = -3$
$2x - 5y = -6$

$$\begin{bmatrix} 1 & 1 & | & -3 \\ 2 & -5 & | & -6 \end{bmatrix}$$

$$\begin{bmatrix} 1 & 1 & | & -3 \\ 0 & -7 & | & 0 \end{bmatrix} \quad -2R1 + R2$$

$$\begin{bmatrix} 1 & 1 & | & -3 \\ 0 & 1 & | & 0 \end{bmatrix} \quad -\tfrac{1}{7}R2$$

$$\begin{bmatrix} 1 & 0 & | & -3 \\ 0 & 1 & | & 0 \end{bmatrix} \quad -1R2 + R1$$

Solution set: $\{(-3, 0)\}$

25. $2x - 3y = 10$
$2x + 2y = 5$

$$\begin{bmatrix} 2 & -3 & | & 10 \\ 2 & 2 & | & 5 \end{bmatrix}$$

$$\begin{bmatrix} 2 & -3 & | & 10 \\ 0 & 5 & | & -5 \end{bmatrix} \quad -1R + R2$$

$$\begin{bmatrix} 1 & -\tfrac{3}{2} & | & 5 \\ 0 & 1 & | & -1 \end{bmatrix} \quad \begin{matrix} \tfrac{1}{2}R1 \\ \tfrac{1}{5}R2 \end{matrix}$$

$$\begin{bmatrix} 1 & 0 & | & \tfrac{7}{2} \\ 0 & 1 & | & -1 \end{bmatrix} \quad \tfrac{3}{2}R2 + R1$$

Solution set: $\{(\tfrac{7}{2}, -1)\}$

27. $2x - 3y = 2$
$4x - 6y = 1$

$$\begin{bmatrix} 2 & -3 & | & 2 \\ 4 & -6 & | & 1 \end{bmatrix}$$

$$\begin{bmatrix} 2 & -3 & | & 2 \\ 0 & 0 & | & -3 \end{bmatrix} \quad -2R1 + R2$$

It is impossible to go further.
Since the second row corresponds
to the equation

$$0x + 0y = -3,$$

which has no solution, the system is
inconsistent.

Solution set: Ø

29. $6x - 3y = 1$ (1)
$-12x + 6y = -2$ (2)

$$\begin{bmatrix} 6 & -3 & | & 1 \\ -12 & 6 & | & -2 \end{bmatrix}$$

$$\begin{bmatrix} 6 & -3 & | & 1 \\ 0 & 0 & | & 0 \end{bmatrix} \quad 2R1 + R2$$

Since every entry in one of the rows
is zero, the equations of the system
are dependent, and the system has
infinitely many solutions. Solve
equation (1) for x in terms of y.

$$6x - 3y = 1$$
$$6x = 3y + 1$$
$$x = \frac{3y + 1}{6}$$

Solution set: $\left\{ \left(\frac{3y + 1}{6}, y \right) \right\}$

31. $x + y = -1$

$y + z = 4$

$x + z = 1$

$$\begin{bmatrix} 1 & 1 & 0 & | & -1 \\ 0 & 1 & 1 & | & 4 \\ 1 & 0 & 1 & | & 1 \end{bmatrix}$$

$$\begin{bmatrix} 1 & 1 & 0 & | & -1 \\ 0 & 1 & 1 & | & 4 \\ 0 & -1 & 1 & | & 2 \end{bmatrix} \quad -1R1 + R3$$

$$\begin{bmatrix} 1 & 1 & 0 & | & -1 \\ 0 & 1 & 1 & | & 4 \\ 0 & 0 & 2 & | & 6 \end{bmatrix} \quad R2 + R3$$

$$\begin{bmatrix} 1 & 1 & 0 & | & -1 \\ 0 & 1 & 1 & | & 4 \\ 0 & 0 & 1 & | & 3 \end{bmatrix} \quad \tfrac{1}{2}R3$$

$$\begin{bmatrix} 1 & 1 & 0 & | & -1 \\ 0 & 1 & 0 & | & 1 \\ 0 & 0 & 1 & | & 3 \end{bmatrix} \quad -1R3 + R2$$

$$\begin{bmatrix} 1 & 0 & 0 & | & -2 \\ 0 & 1 & 0 & | & 1 \\ 0 & 0 & 1 & | & 3 \end{bmatrix} \quad -1R2 + R1$$

Solution set: $\{(-2, 1, 3)\}$

33. $x + y - z = 6$

$2x - y + z = -9$

$x - 2y + 3z = 1$

$$\begin{bmatrix} 1 & 1 & -1 & | & 6 \\ 2 & -1 & 1 & | & -9 \\ 1 & -2 & 3 & | & 1 \end{bmatrix}$$

$$\begin{bmatrix} 1 & 1 & -1 & | & 6 \\ 0 & -3 & 3 & | & -21 \\ 0 & -3 & 4 & | & -5 \end{bmatrix} \quad \begin{array}{l} -2R1 + R2 \\ -1R1 + R3 \end{array}$$

$$\begin{bmatrix} 1 & 1 & -1 & | & 6 \\ 0 & -3 & 3 & | & -21 \\ 0 & 0 & 1 & | & 16 \end{bmatrix} \quad -1R2 + R3$$

$$\begin{bmatrix} 1 & 1 & -1 & | & 6 \\ 0 & 1 & -1 & | & 7 \\ 0 & 0 & 1 & | & 16 \end{bmatrix} \quad -\tfrac{1}{3}R2$$

$$\begin{bmatrix} 1 & 1 & 0 & | & 22 \\ 0 & 1 & 0 & | & 23 \\ 0 & 0 & 1 & | & 16 \end{bmatrix} \quad \begin{array}{l} R3 + R1 \\ R3 + R2 \end{array}$$

$$\begin{bmatrix} 1 & 0 & 0 & | & -1 \\ 0 & 1 & 0 & | & 23 \\ 0 & 0 & 1 & | & 16 \end{bmatrix} \quad -1R2 + R1$$

Solution set: $\{(-1, 23, 16)\}$

35. $-x + y = -1$

$y - z = 6$

$x + z = -1$

$$\begin{bmatrix} -1 & 1 & 0 & | & -1 \\ 0 & 1 & -1 & | & 6 \\ 1 & 0 & 1 & | & -1 \end{bmatrix}$$

$$\begin{bmatrix} -1 & 1 & 0 & | & -1 \\ 0 & 1 & -1 & | & 6 \\ 0 & 1 & 1 & | & -2 \end{bmatrix} \quad R1 + R3$$

$$\begin{bmatrix} 1 & -1 & 0 & | & 1 \\ 0 & 1 & -1 & | & 6 \\ 0 & 1 & 1 & | & -2 \end{bmatrix} \quad -1R1$$

$$\begin{bmatrix} 1 & 0 & -1 & | & 7 \\ 0 & 1 & -1 & | & 6 \\ 0 & 1 & 1 & | & -2 \end{bmatrix} \quad R2 + R1$$

$$\begin{bmatrix} 1 & 0 & -1 & | & 7 \\ 0 & 1 & -1 & | & 6 \\ 0 & 0 & 2 & | & -8 \end{bmatrix} \quad -1R2 + R3$$

$$\begin{bmatrix} 1 & 0 & -1 & | & 7 \\ 0 & 1 & -1 & | & 6 \\ 0 & 0 & 1 & | & -4 \end{bmatrix} \quad \tfrac{1}{2}R3$$

$$\begin{bmatrix} 1 & 0 & 0 & | & 3 \\ 0 & 1 & 0 & | & 2 \\ 0 & 0 & 1 & | & -4 \end{bmatrix} \quad \begin{array}{l} R3 + R1 \\ R3 + R2 \end{array}$$

Solution set: $\{(3, 2, -4)\}$

37. $2x - y + 3z = 0$

$x + 2y - z = 5$

$2y + z = 1$

$$\begin{bmatrix} 2 & -1 & 3 & | & 0 \\ 1 & 2 & -1 & | & 5 \\ 0 & 2 & 1 & | & 1 \end{bmatrix}$$

$$\begin{bmatrix} 1 & -\frac{1}{2} & \frac{3}{2} & | & 0 \\ 1 & 2 & -1 & | & 5 \\ 0 & 2 & 1 & | & 1 \end{bmatrix} \quad \frac{1}{2}R1$$

$$\begin{bmatrix} 1 & -\frac{1}{2} & \frac{3}{2} & | & 0 \\ 0 & \frac{5}{2} & -\frac{5}{2} & | & 5 \\ 0 & 2 & 1 & | & 1 \end{bmatrix} \quad -1R1 + R2$$

$$\begin{bmatrix} 1 & -\frac{1}{2} & \frac{3}{2} & | & 0 \\ 1 & 1 & -1 & | & 2 \\ 0 & 2 & 1 & | & 1 \end{bmatrix} \quad \frac{2}{5}R2$$

$$\begin{bmatrix} 1 & -\frac{1}{2} & \frac{3}{2} & | & 0 \\ 0 & 1 & -1 & | & 2 \\ 0 & 0 & 3 & | & -3 \end{bmatrix} \quad -2R2 + R3$$

$$\begin{bmatrix} 1 & -\frac{1}{2} & \frac{3}{2} & | & 0 \\ 0 & 1 & -1 & | & 2 \\ 0 & 0 & 1 & | & -1 \end{bmatrix} \quad \frac{1}{3}R3$$

$$\begin{bmatrix} 1 & 0 & 1 & | & 1 \\ 0 & 1 & -1 & | & 2 \\ 0 & 0 & 1 & | & -1 \end{bmatrix} \quad \frac{1}{2}R2 + R1$$

$$\begin{bmatrix} 1 & 0 & 0 & | & 2 \\ 0 & 1 & 0 & | & 1 \\ 0 & 0 & 1 & | & -1 \end{bmatrix} \quad \begin{matrix} -1R3 + R1 \\ R3 + R2 \end{matrix}$$

Solution set: $\{(2, 1, -1)\}$

41. $3x + 2y - w = 0$

$2x + z + 2w = 5$

$x + 2y - z = -2$

$2x - y + z + w = 2$

$$\begin{bmatrix} 3 & 2 & 0 & -1 & | & 0 \\ 2 & 0 & 1 & 2 & | & 5 \\ 1 & 2 & -1 & 0 & | & -2 \\ 2 & -1 & 1 & 1 & | & 2 \end{bmatrix}$$

$$\begin{bmatrix} 3 & 2 & 0 & -1 & | & 0 \\ 0 & -\frac{4}{3} & 1 & \frac{8}{3} & | & 5 \\ 0 & \frac{4}{3} & -1 & \frac{1}{3} & | & -2 \\ 0 & -\frac{7}{3} & 1 & \frac{5}{3} & | & 2 \end{bmatrix} \quad \begin{matrix} -\frac{2}{3}R1 + R2 \\ -\frac{1}{3}R1 + R3 \\ -\frac{2}{3}R1 + R4 \end{matrix}$$

$$\begin{bmatrix} 1 & \frac{2}{3} & 0 & -\frac{1}{3} & | & 0 \\ 0 & 1 & -\frac{3}{4} & -2 & | & -\frac{15}{4} \\ 0 & 1 & -\frac{3}{4} & \frac{1}{4} & | & -\frac{3}{2} \\ 0 & 1 & -\frac{3}{7} & -\frac{5}{7} & | & -\frac{6}{7} \end{bmatrix} \quad \begin{matrix} \frac{1}{3}R1 \\ -\frac{3}{4}R2 \\ \frac{3}{4}R3 \\ -\frac{3}{7}R4 \end{matrix}$$

$$\begin{bmatrix} 1 & \frac{2}{3} & 0 & -\frac{1}{3} & | & 0 \\ 0 & 1 & -\frac{3}{4} & -2 & | & -\frac{15}{4} \\ 0 & 0 & 0 & \frac{9}{4} & | & \frac{9}{4} \\ 0 & 0 & \frac{9}{28} & \frac{9}{7} & | & \frac{81}{28} \end{bmatrix} \quad \begin{matrix} -1R2 + R3 \\ -1R2 + R4 \end{matrix}$$

$$\begin{bmatrix} 1 & \frac{2}{3} & 0 & -\frac{1}{3} & | & 0 \\ 0 & 1 & -\frac{3}{4} & -2 & | & -\frac{15}{4} \\ 0 & 0 & \frac{9}{28} & \frac{9}{7} & | & \frac{81}{28} \\ 0 & 0 & 0 & \frac{9}{4} & | & \frac{9}{4} \end{bmatrix} \quad R3 \leftrightarrow R4$$

Note: R3 $\leftrightarrow$ R4 means "interchange row 3 and row 4."

$$\begin{bmatrix} 1 & \frac{2}{3} & 0 & -\frac{1}{3} & \Big| & 0 \\ 0 & 1 & -\frac{3}{4} & -2 & \Big| & -\frac{15}{4} \\ 0 & 0 & 1 & 4 & \Big| & 9 \\ 0 & 0 & 0 & 1 & \Big| & 1 \end{bmatrix} \begin{matrix} \\ \\ \frac{28}{9}R3 \\ \frac{4}{9}R4 \end{matrix}$$

$$\begin{bmatrix} 1 & \frac{2}{3} & 0 & 0 & \Big| & \frac{1}{3} \\ 0 & 1 & -\frac{3}{4} & 0 & \Big| & -\frac{7}{4} \\ 0 & 0 & 1 & 0 & \Big| & 5 \\ 0 & 0 & 0 & 1 & \Big| & 1 \end{bmatrix} \begin{matrix} \frac{1}{3}R4 + R1 \\ 2R4 + R2 \\ -4R4 + R3 \\ \\ \end{matrix}$$

$$\begin{bmatrix} 1 & \frac{2}{3} & 0 & 0 & \Big| & \frac{1}{3} \\ 0 & 1 & 0 & 0 & \Big| & 2 \\ 0 & 0 & 1 & 0 & \Big| & 5 \\ 0 & 0 & 0 & 1 & \Big| & 1 \end{bmatrix} \begin{matrix} \\ \frac{3}{4}R3 + R2 \\ \\ \\ \end{matrix}$$

$$\begin{bmatrix} 1 & 0 & 0 & 0 & \Big| & -1 \\ 0 & 1 & 0 & 0 & \Big| & 2 \\ 0 & 0 & 1 & 0 & \Big| & 5 \\ 0 & 0 & 0 & 1 & \Big| & 1 \end{bmatrix} \begin{matrix} -\frac{2}{3}R2 + R1 \\ \\ \\ \\ \end{matrix}$$

Solution set: $\{(-1, 2, 5, 1)\}$

43. Let x = the amount invested at 3%;

    y = the amount invested at 4%;

    z = the amount invested at 4.5%.

One equation is $y = \frac{1}{3}x$, or

$x - 3y = 0$.

$$\begin{array}{ll} x - 3y = 0 & (1) \\ x + y + z = 10{,}000 & (2) \\ .03x + .04y + .045z = 400 & (3) \end{array}$$

Write the augmented matrix and solve the system using the Gauss–Jordan method.

$$\begin{bmatrix} 1 & -3 & 0 & \Big| & 0 \\ 1 & 1 & 1 & \Big| & 10{,}000 \\ .03 & .04 & .045 & \Big| & 400 \end{bmatrix}$$

$$\begin{bmatrix} 1 & -3 & 0 & \Big| & 0 \\ 0 & 4 & 1 & \Big| & 10{,}000 \\ 0 & .13 & .045 & \Big| & 400 \end{bmatrix} \begin{matrix} \\ -1R1 + R2 \\ -.03R1 + R3 \end{matrix}$$

$$\begin{bmatrix} 1 & -3 & 0 & \Big| & 0 \\ 0 & 1 & .25 & \Big| & 2500 \\ 0 & .13 & .045 & \Big| & 400 \end{bmatrix} \begin{matrix} \\ .25R2 \\ \\ \end{matrix}$$

$$\begin{bmatrix} 1 & 0 & .75 & \Big| & 7500 \\ 0 & 1 & .25 & \Big| & 2500 \\ 0 & 0 & .0125 & \Big| & 75 \end{bmatrix} \begin{matrix} 3R2 + R1 \\ \\ -.13R2 + R3 \end{matrix}$$

$$\begin{bmatrix} 1 & 0 & .75 & \Big| & 7500 \\ 0 & 1 & .25 & \Big| & 2500 \\ 0 & 0 & 1 & \Big| & 6000 \end{bmatrix} \begin{matrix} \\ \\ \frac{1}{.0125}R3 \end{matrix}$$

$$\begin{bmatrix} 1 & 0 & 0 & \Big| & 3000 \\ 0 & 1 & 0 & \Big| & 1000 \\ 0 & 0 & 1 & \Big| & 6000 \end{bmatrix} \begin{matrix} -.75R3 + R1 \\ \\ -.25R3 + R2 \end{matrix}$$

Thus, x = 3000, y = 1000, and z = 6000.

The amount invested at 3% is $3000, at 4% is $1000, and at 4.5% is $6000.

45. Let x, y, and z be the prices per gallon respectively of premium gasoline, regular gasoline, and super gasoline.

From the information given in the exercise, we have the system

$$\begin{array}{ll} 400x + 150y + 130z = 909 & (1) \\ 380x + 170y + 150z = 931 & (2) \\ z - y = \frac{1}{2}(x - y). & (3) \end{array}$$

Equation (3) may be written without fractions as

$$\begin{array}{ll} 2(z - y) = x - y & \textit{Multiply by 2} \\ 2z - 2y = x - y & \\ x + y - 2z = 0. & \end{array}$$

Write the augmented matrix and solve
by the Gauss–Jordan method.

$$\begin{bmatrix} 1 & 1 & -2 & 0 \\ 400 & 150 & 130 & 909 \\ 380 & 170 & 150 & 931 \end{bmatrix}$$

$$\begin{bmatrix} 1 & 1 & -2 & 0 \\ 0 & -250 & 930 & 909 \\ 0 & -210 & 910 & 931 \end{bmatrix} \begin{matrix} \\ -400R1 + R2 \\ -380R1 + R3 \end{matrix}$$

$$\begin{bmatrix} 1 & 1 & -2 & 0 \\ 0 & 1 & -3.72 & -3.636 \\ 0 & -210 & 910 & 931 \end{bmatrix} \begin{matrix} \\ -\frac{1}{250}R2 \\ \\ \end{matrix}$$

$$\begin{bmatrix} 1 & 0 & 1.72 & 3.636 \\ 0 & 1 & -3.72 & -3.636 \\ 0 & 0 & 128.8 & 167.44 \end{bmatrix} \begin{matrix} -1R2 + R1 \\ \\ 210R2 + R3 \end{matrix}$$

$$\begin{bmatrix} 1 & 0 & 1.72 & 3.636 \\ 0 & 1 & -3.72 & -3.636 \\ 0 & 0 & 1 & 1.30 \end{bmatrix} \begin{matrix} \\ \\ \frac{1}{128.8}R3 \end{matrix}$$

$$\begin{bmatrix} 1 & 0 & 0 & 1.40 \\ 0 & 1 & 0 & 1.20 \\ 0 & 0 & 1 & 1.30 \end{bmatrix} \begin{matrix} -1.72R3 + R1 \\ 3.72R3 + R2 \\ \end{matrix}$$

$$x = 1.40$$
$$y = 1.20$$
$$z = 1.30$$

Premium costs $1.40 per gallon,
regular costs $1.20 per gallon,
and super costs $1.30 per gallon.

47. $x - 3y + 2z = 10$
$2x - y - z = 8$

$$\begin{bmatrix} 1 & -3 & 2 & 10 \\ 2 & -1 & -1 & 8 \end{bmatrix}$$

$$\begin{bmatrix} 1 & -3 & 2 & 10 \\ 0 & 5 & -5 & -12 \end{bmatrix} \begin{matrix} \\ -2R1 + R2 \end{matrix}$$

$$\begin{bmatrix} 1 & -3 & 2 & 10 \\ 0 & 1 & -1 & -\frac{12}{5} \end{bmatrix} \begin{matrix} \\ \frac{1}{5}R2 \end{matrix}$$

$$\begin{bmatrix} 1 & 0 & -1 & \frac{14}{5} \\ 0 & 1 & -1 & -\frac{12}{5} \end{bmatrix} \begin{matrix} 3R2 + R1 \\ \\ \end{matrix}$$

It is not possible to go further
with the Gauss–Jordan method. The
equations that correspond to the
final matrix are

$$x - z = \frac{14}{5} \quad \text{and} \quad y - z = -\frac{12}{5}.$$

Solve these equations for x and y,
respectively.

$$x - z = \frac{14}{5} \qquad\qquad y - z = -\frac{12}{5}$$
$$x = z + \frac{14}{5} \qquad\qquad y = z - \frac{12}{5}$$
$$x = \frac{5z + 14}{5} \qquad\qquad y = \frac{5z - 12}{5}$$

Solution set (with z arbitrary):

$$\left\{ \left(\frac{5z + 14}{5}, \frac{5z - 12}{5}, z \right) \right\}$$

49. $x + 2y - z = 0$ (1)
$3x - y + z = 6$ (2)
$-2x - 4y + 2z = 0$ (3)

$$\begin{bmatrix} 1 & 2 & -1 & 0 \\ 3 & -1 & 1 & 6 \\ -2 & -4 & 2 & 0 \end{bmatrix}$$

$$\begin{bmatrix} 1 & 2 & -1 & 0 \\ 0 & -7 & 4 & 6 \\ 0 & 0 & 0 & 0 \end{bmatrix} \begin{matrix} \\ -3R1 + R2 \\ 2R1 + R3 \end{matrix}$$

The third row gives no information,
so we drop it. This occurs because
equations (1) and (3) are dependent.

$$\begin{bmatrix} 1 & 0 & \frac{1}{7} & \frac{12}{7} \\ 0 & -7 & 4 & 6 \end{bmatrix} \begin{matrix} \frac{2}{7}R2 + R1 \\ \\ \end{matrix}$$

$$\begin{bmatrix} 1 & 0 & \frac{1}{7} & \frac{12}{7} \\ 0 & 1 & -\frac{4}{7} & -\frac{6}{7} \end{bmatrix} \begin{matrix} \\ -\frac{1}{7}R2 \end{matrix}$$

Then,

$$x + \frac{1}{7}z = \frac{12}{7}$$

$$x = \frac{12 - z}{7}$$

and

$$y - \frac{4}{7}z = -\frac{6}{7}$$

$$y = \frac{4z - 6}{7}.$$

Solution set (with z arbitrary):

$$\left\{ \left(\frac{12 - z}{7}, \frac{4z - 6}{7}, z \right) \right\}$$

51. $x - 2y + z = 5$
$-2x + 4y - 2z = 2$
$2x + y - z = 2$

$$\begin{bmatrix} 1 & -2 & 1 & | & 5 \\ -2 & 4 & -2 & | & 2 \\ 2 & 1 & -1 & | & 2 \end{bmatrix}$$

$$\begin{bmatrix} 1 & -2 & 1 & | & 5 \\ 0 & 0 & 0 & | & 12 \\ 2 & 1 & -1 & | & 2 \end{bmatrix} \quad 2R1 + R2$$

The second row corresponds to the equation $0x + 0y + 0z = 12$, which is impossible, so there is no solution.

Solution set: Ø

53. (a) The number of cars coming into intersection D must equal the number of cars leaving intersection D.

$x_3 + x_4 = 200 + 400$
$x_3 + x_4 = 600$ (4)

(b) $$\begin{bmatrix} 1 & 0 & 0 & 1 & | & 1000 \\ 1 & 1 & 0 & 0 & | & 1100 \\ 0 & 1 & 1 & 0 & | & 700 \\ 0 & 0 & 1 & 1 & | & 600 \end{bmatrix}$$

$$\begin{bmatrix} 1 & 0 & 0 & 1 & | & 1000 \\ 0 & 1 & 0 & -1 & | & 100 \\ 0 & 1 & 1 & 0 & | & 700 \\ 0 & 0 & 1 & 1 & | & 600 \end{bmatrix} \quad -1R1 + R2$$

$$\begin{bmatrix} 1 & 0 & 0 & 1 & | & 1000 \\ 0 & 1 & 0 & -1 & | & 100 \\ 0 & 0 & 1 & 1 & | & 600 \\ 0 & 0 & 1 & 1 & | & 600 \end{bmatrix} \quad -1R2 + R3$$

Add -1 times row 3 to row 4 to get a row of zeros. Equations (3) and (4) are dependent and the system has no unique solution.

(c) $x_1 + x_4 = 1000$, or
$$x_1 = 1000 - x_4,$$
   and $x_1 + x_2 = 100,$
   so
$(1000 - x_4) + x_2 = 1100$
$$x_2 - x_4 = 100.$$

(d) $x_1 + x_4 = 1000$
$$x_4 = 1000 - x_1$$

x_1 can be at most 1000 for x_4 to be nonnegative.

(e) $x_2 - x_4 = 100$
$$x_2 - 100 = x_4$$

x_2 must be at least 100 for x_4 to be nonnegative.

(f) $x_3 + x_4 = 600$
$$x_4 = 600 - x_3$$

x_3 must be at most 600 for x_4 to be nonnegative. x_4 can be at most 600 for x_3 to be nonnegative.

(g) $x_1 + x_4 = 1000$

        $x_1 + x_2 = 1100$

        $x_2 + x_3 = 700$

        $x_3 + x_4 = 600$

From part (f), x_4 can be at most 600. If $x_4 = 600$ in the given equations above, the other values x_1, x_2, and x_3 will be nonnegative. From part (f), $x_3 \leq 600$. If $x_3 = 600$, then $x_4 = 0$. From part (e), $x_2 = 100$, and from part (d), $x_1 = 1000$. All are nonnegative, so 600 is the maximum possible value of x_3.

From part (e) $x_2 - x_4 = 100$. Since $x_4 \leq 600$ we have

$$x_2 = 100 + x_4$$
$$x_2 \leq 100 + 600 = 700,$$

so x_2 can be at most 700. If x_2 is 700, $x_1 = 400$, $x_4 = 600$, $x_3 = 0$, all nonnegative, so 700 is the maximum value of x_2. Finally, since $x_1 + x_4 = 1000$, $x_1 \leq 1000$. If $x_1 = 1000$, then $x_4 = 0$, $x_2 = 100$, $x_3 = 600$, all nonnegative, so 1000 is the maximum possible value of x_1.

Section 7.4

For Exercises 1–7, see the answer sketches in the textbook.

1. a line and an ellipse; two points of intersection

Draw an ellipse and then draw a line that intersects the ellipse but it is not tangent to it.

3. a circle and an ellipse; one point of intersection

Draw a circle and then draw an ellipse that is tangent to the circle.

5. a line and a hyperbola; two points of intersection

Draw a hyperbola and then draw a line that intersects both branches of the hyperbola.

7. a parabola and an ellipse; four points of intersection

Draw an ellipse and, using a point in the exterior of the ellipse as the vertex, draw a parabola that opens toward the ellipse and intersects it in four points.

11. $y = x^2$ (1)

    $x + y = 2$ (2)

Use the substitution method. Solve equation (2) for y.

$$y = 2 - x$$

Substitute this result into equation (1).

$$2 - x = x^2$$
$$x^2 + x - 2 = 0$$
$$(x + 2)(x - 1) = 0$$
$$x = -2 \quad \text{or} \quad x = 1$$

If $x = -2$, then $y = (-2)^2 = 4$.
If $x = 1$, then $y = 1^2 = 1$.

Solution set: $\{(-2, 4), (1, 1)\}$

13. $y = (x - 1)^2$ (1)
$x - 3y = -1$ (2)

Substitute $(x - 1)^2$ for y in equation (2).

$$x - 3(x - 1)^2 = -1$$
$$x - 3(x^2 - 2x + 1) = -1$$
$$x - 3x^2 + 6x - 3 = -1$$
$$-3x^2 + 7x - 2 = 0$$
$$3x^2 - 7x + 2 = 0$$
$$(3x - 1)(x - 2) = 0$$
$$x = \frac{1}{3} \quad \text{or} \quad x = 2$$

If $x = \frac{1}{3}$, then

$$y = \left(\frac{1}{3} - 1\right)^2$$
$$= \left(-\frac{2}{3}\right)^2 = \frac{4}{9}.$$

If $x = 2$, then

$$y = (2 - 1)^2$$
$$= 1^2 = 1.$$

Solution set: $\left\{\left(\frac{1}{3}, \frac{4}{9}\right), (2, 1)\right\}$

15. $y = x^2 + 4x$ (1)
$2x - y = -8$ (2)

Substitute $x^2 + 4x$ for y in equation (2).

$$2x - (x^2 + 4x) = -8$$
$$2x - x^2 - 4x = -8$$
$$x^2 + 2x - 8 = 0$$
$$(x - 2)(x + 4) = 0$$
$$x = 2 \quad \text{or} \quad x = -4$$

If $x = 2$, then

$$y = 2^2 + 4(2)$$
$$= 4 + 8 = 12.$$

If $x = -4$, then

$$y = (-4)^2 + 4(-4)$$
$$= 0.$$

Solution set: $\{(2, 12), (-4, 0)\}$

17. $3x^2 + 2y^2 = 5$ (1)
$x - y = -2$ (2)

Solve equation (2) for x.

$$x = y - 2$$

Substitute $y - 2$ for x in equation (1).

$$3(y - 2)^2 + 2y^2 = 5$$
$$3(y^2 - 4y + 4) + 2y^2 = 5$$
$$2y^2 - 12y + 12 + 2y^2 = 5$$
$$5y^2 - 12y + 7 = 0$$
$$(5y - 7)(y - 1) = 0$$
$$y = \frac{7}{5} \quad \text{or} \quad y = 1$$

If $y = \frac{7}{5}$, then,

$$x = y - 2$$
$$= \frac{7}{5} - 2$$
$$= -\frac{3}{5}.$$

If $y = 1$, then

$$x = 1 - 2 = -1.$$

Solution set: $\left\{ \left(-\frac{3}{5}, \frac{7}{5}\right), (-1, 1) \right\}$

19. $x^2 + y^2 = 8$ (1)
$x^2 - y^2 = 0$ (2)

Use the addition method.
Add equations (1) and (2).

$$\begin{array}{r} x^2 + y^2 = 8 \\ \underline{x^2 - y^2 = 0} \\ 2x^2 \quad\quad = 8 \\ x^2 = 4 \\ x = \pm 2 \end{array}$$

If $x = 2$, then

$$2^2 - y^2 = 0$$
$$4 - y^2 = 0$$
$$y^2 = 4$$
$$y = \pm 2.$$

If $x = -2$, then

$$(-2)^2 - y^2 = 0$$
$$4 - y^2 = 0$$
$$y^2 = 4$$
$$y = \pm 2.$$

Solution set:
$\{(2, 2), (2, -2), (-2, 2), (-2, -2)\}$

21. $5x^2 - y^2 = 0$ (1)
$3x^2 + 4y^2 = 0$ (2)

Multiply equation (1) by 4 and add to equation (2).

$$\begin{array}{r} 20x^2 - 4y^2 = 0 \\ \underline{3x^2 + 4y^2 = 0} \\ 23x^2 \quad\quad = 0 \\ x = 0 \end{array}$$

If $x = 0$,

$$5(0)^2 - y^2 = 0,$$
$$y = 0.$$

Solution set: $\{(0, 0)\}$

23. $3x^2 + y^2 = 3$ (1)
$4x^2 + 5y^2 = 26$ (2)

Multiply equation (1) by -5 and add to equation (2).

$$\begin{array}{r} -15x^2 - 5y^2 = -15 \\ \underline{4x^2 + 5y^2 = 26} \\ -11x^2 \quad\quad = 11 \\ x^2 = -1 \\ x = \pm i \end{array}$$

If $x = \pm i$,

$$3(\pm i)^2 + y^2 = 3$$
$$-3 + y^2 = 3$$
$$y^2 = 6$$
$$y = \pm\sqrt{6}.$$

Solution set:
$\{(i, \sqrt{6}), (-i, \sqrt{6}), (i, -\sqrt{6}), (-i, -\sqrt{6})\}$

25. $2x^2 + 3y^2 = 5$ (1)

$3x^2 - 4y^2 = -1$ (2)

Multiply equation (1) by 4 and
equation (2) by 3 and add.

$$8x^2 + 12y^2 = 20$$
$$\underline{9x^2 - 12y^2 = -3}$$
$$17x^2 \quad\quad = 17$$
$$x^2 = 1$$
$$x = \pm1$$

If $x = \pm1$,

$$2(\pm1)^2 + 3y^2 = 5$$
$$3y^2 = 3$$
$$y^2 = 1$$
$$y = \pm1.$$

Solution set:
$$\left\{(1,\ 1),\ (1,\ -1),\ (-1,\ 1),\ (-1,\ -1)\right\}$$

27. $2x^2 + 2y^2 = 20$ (1)

$3x^2 + 3y^2 = 30$ (2)

Multiply equation (1) by 3 and
equation (2) by -2 and add.

$$6x^2 + 6y^2 = 60$$
$$\underline{-6x^2 - 6y^2 = -60}$$
$$0 = 0$$

The equations are dependent. Both
equations are equivalent to

$$x^2 + y^2 = 10$$

and their graphs are the same
circle.

To write the solution set as an
ordered pair, solve for y,

$$y^2 = 10 - x^2$$
$$y = \pm\sqrt{10 - x^2}$$

Solution set: $\left\{(x,\ \pm\sqrt{10 - x^2})\right\}$

29. $9x^2 + 4y^2 = 1$ (1)

$x^2 + y^2 = 1$ (2)

$$9x^2 + 4y^2 = 1$$
$$\underline{-4x^2 - 4y^2 = -4}\quad \textit{Multiply (2) by -4}$$
$$5x^2 \quad\quad = -3$$
$$x^2 = -\frac{3}{5}$$
$$x = \pm\frac{i\sqrt{15}}{5}$$
$$y^2 = 1 - x^2$$
$$= 1 - \frac{3}{5} = \frac{2}{5}$$
$$y = \pm\frac{i\sqrt{10}}{5}$$

Solution set:
$$\left\{\left(\frac{i\sqrt{15}}{5},\ \frac{2\sqrt{10}}{5}\right),\ \left(\frac{i\sqrt{15}}{5},\ -\frac{2\sqrt{10}}{5}\right)\right.$$
$$\left.\left(\frac{i\sqrt{15}}{5},\ \frac{2\sqrt{10}}{5}\right),\ \left(-\frac{i\sqrt{15}}{5},\ -\frac{2\sqrt{10}}{5}\right)\right\}$$

31. $xy = -15$ (1)

$4x + 3y = 3$ (2)

Solve equation (1) for y.

$$y = -\frac{15}{x}\quad (3)$$

Substitute into equation (2) and
solve for x.

$$4x + 3\left(-\frac{15}{x}\right) = 3$$
$$4x^2 - 45 = 3x$$
$$4x^2 - 3x - 45 = 0$$
$$(4x - 15)(x + 3) = 0$$
$$x = \frac{15}{4}\quad \text{or}\quad x = -3$$

Substitute each value of x into equation (3) to find the corresponding value of y.

If $x = \dfrac{15}{4}$,

$$y = -\dfrac{15}{\dfrac{15}{4}} = -4.$$

If $x = -3$,

$$y = \dfrac{-15}{-3} = 5.$$

Solution set: $\left\{\left(\dfrac{15}{4},\ -4\right),\ (-3,\ 5)\right\}$

33. $2xy + 1 = 0$ (1)

$x + 16y = 2$ (2)

Solve equation (2) for x.

$$x = -16y + 2 \quad (3)$$

Substitute into equation (1).

$$2(-16y + 2)(y) + 1 = 0$$
$$-32y^2 + 4y + 1 = 0$$
$$32y^2 - 4y - 1 = 0$$
$$(4y - 1)(8y + 1) = 0$$
$$y = \dfrac{1}{4} \quad \text{or} \quad y = -\dfrac{1}{8}$$

Substitute each value of y into equation (3) to find the corresponding value of x.

If $y = \dfrac{1}{4}$, then

$$x = -16\left(\dfrac{1}{4}\right) + 2$$
$$= -2.$$

If $y = -\dfrac{1}{8}$, then

$$x = -16\left(-\dfrac{1}{8}\right) + 2$$
$$= 4.$$

Solution set: $\left\{\left(-2,\ \dfrac{1}{4}\right),\ \left(4,\ -\dfrac{1}{8}\right)\right\}$

35. $x^2 + 4y^2 = 25$ (1)

$xy = 6$ (2)

Solve equation (2) for x.

$$x = \dfrac{6}{y} \quad (3)$$

Substitute into equation (2).

$$\left(\dfrac{6}{y}\right)^2 + 4y^2 = 25$$
$$\dfrac{36}{y^2} + 4y^2 = 25$$
$$36 + 4y^4 = 25y^2$$
$$4y^4 - 25y^2 + 36 = 0$$
$$(y^2 - 4)(4y^2 - 9) = 0$$
$$y^2 = 4 \quad \text{or} \quad y^2 = \dfrac{9}{4}$$
$$y = \pm 2 \quad \text{or} \quad y = \pm\dfrac{3}{2}$$

Substitute each value of x into equation (3) to find the corresponding value of y.

If $y = 2$, $x = \dfrac{6}{2} = 3.$

If $y = -2$, $x = \dfrac{6}{-2} = -3.$

If $y = \dfrac{3}{2}$, $y = \dfrac{6}{\dfrac{3}{2}} = 6\left(\dfrac{2}{3}\right) = 4.$

If $y = -\dfrac{3}{2}$, $x = \dfrac{6}{-\dfrac{3}{2}} = 6\left(-\dfrac{2}{3}\right) = -4.$

Solution set:

$$\left\{(3,\ 2),\ (-3,\ -2),\ \left(4,\ \dfrac{3}{2}\right),\ \left(-4,\ -\dfrac{3}{2}\right)\right\}$$

37. $x^2 + 2xy - y^2 = 14$ *(1)*

   $x^2 - y^2 = -16$ *(2)*

This system can be solved using a combination of the addition and substitution methods.

Multiply equation (2) by -1 and add to equation (1).

$$x^2 + 2xy - y^2 = 14$$
$$\underline{-x^2 \qquad\quad + y^2 = 16}$$
$$2xy \qquad\quad = 30$$
$$xy = 15 \quad (3)$$

Solve equation (3) for y.

$$y = \frac{15}{x} \quad (4)$$

Substitute into equation (2).

$$x^2 - \left(\frac{15}{x}\right)^2 = -16$$

$$x^2 - \frac{225}{x^2} = -16$$

$$x^4 + 16x^2 - 225 = 0$$

Solve this equation by factoring.

$$(x^2 - 9)(x^2 + 25) = 0$$
$$x^2 - 9 = 0$$
$$x = \pm 3$$
$$x^2 + 25 = 0$$
$$x = \pm 5i$$

Substitute each value of x into equation (4) to find the corresponding value of y.

If $x = 3$, then

$$y = \frac{15}{x} = \frac{15}{3} = 5.$$

If $x = -3$, then

$$y = \frac{15}{x} = \frac{15}{-3} = -5.$$

If $x = 5i$, then

$$y = \frac{15}{5i} = \frac{3}{i} = -3i.$$

If $x = -5i$, then

$$y = \frac{15}{-5i} = -\frac{3}{i} = 3i.$$

Solution set: $\{(3, 5), (-3, -5),$
$(5i, -3i), (-5i, 3i)\}$

39. $x^2 - xy - y^2 = 5$ *(1)*

   $2x^2 + xy - y^2 = 10$ *(2)*

Add equation (1) and (2).

$$3x^2 = 15$$
$$x^2 = 5$$
$$x = \pm\sqrt{5}$$

If $x = \sqrt{5}$, then

$$x^2 - xy + y^2 = 5$$
$$(\sqrt{5})^2 - (\sqrt{5})y + y^2 = 5$$
$$5 - y\sqrt{5} + y^2 = 5$$
$$y^2 - y\sqrt{5} = 0$$
$$y(y - \sqrt{5}) = 0$$
$$y = 0 \quad\text{or}\quad y = \sqrt{5}.$$

If $x = -\sqrt{5}$, then

$$(-\sqrt{5})^2 - y(-\sqrt{5}) + y^2 = 5$$
$$5 + y\sqrt{5} + y^2 = 5$$
$$y^2 + y\sqrt{5} = 0$$
$$y(y + \sqrt{5}) = 0$$
$$y = 0 \quad\text{or}\quad y = -\sqrt{5}.$$

Solution set:
$$\{(\sqrt{5}, 0), (\sqrt{5}, \sqrt{5}), (-\sqrt{5}, 0), (-\sqrt{5}, -\sqrt{5})\}$$

41. $x = |y|$ (1)

$x^2 + y^2 = 18$ (2)

If $x = |y|$, then

$$x^2 = y^2$$

since

$$|y|^2 = y^2.$$

Substitute $x^2 = y^2$ into equation (2).

$$y^2 + y^2 = 18$$
$$2y^2 = 18$$
$$y^2 = 9$$
$$y = \pm 3$$

If $y = 3$, then

$$x = |y| = |3|$$
$$= 3.$$

If $y = -3$, then

$$x = |y| = |-3|$$
$$= 3.$$

Solution set: $\{(3, 3), (3, -3)\}$

43. $y = |x - 1|$ (1)

$y = x^2 - 4$ (2)

Since $y = |x - 1|$, by the definition of absolute value,

$$y = x - 1$$
or $y = -(x - 1).$

Suppose $y = x - 1$.

Substitute $x - 1$ for y in equation (2).

$$x - 1 = x^2 - 4$$
$$0 = x^2 - x - 3$$

Solve by the quadratic formula.

$$x = \frac{1 \pm \sqrt{1 - 4(1)(-3)}}{2}$$
$$= \frac{1 \pm \sqrt{13}}{2}$$

$$y = x - 1$$
$$= \frac{1 \pm \sqrt{13}}{2} - \frac{2}{2}$$
$$= \frac{-1 \pm \sqrt{13}}{2}$$

y must be nonnegative since it is the absolute value of $x - 1$, so reject the pair $(\frac{1 - \sqrt{13}}{2}, \frac{-1 - \sqrt{13}}{2})$.

Now suppose $y = -(x - 1)$. Then

$$-(x - 1) = x^2 - 4$$
$$-x + 1 = x^2 - 4$$
$$0 = x^2 + x - 5.$$

Solve by the quadratic formula.

$$x = \frac{-1 \pm \sqrt{1 - 4(1)(-5)}}{2}$$
$$= \frac{-1 \pm \sqrt{21}}{2}$$

$$y = -(x - 1) = -(\frac{-1 \pm \sqrt{21}}{2} - \frac{2}{2})$$
$$= -(\frac{-3 \pm \sqrt{21}}{2})$$
$$= \frac{3 \pm \sqrt{21}}{2}$$

Since $\frac{3 - \sqrt{21}}{2}$ is negative, we reject the pair $(\frac{(-1 + \sqrt{21})}{2}, \frac{(3 - \sqrt{21})}{2})$.

Solution set:

$$\left\{ (\frac{1 + \sqrt{13}}{2}, \frac{-1 + \sqrt{13}}{2}), \right.$$
$$\left. (\frac{-1 - \sqrt{21}}{2}, \frac{3 + \sqrt{21}}{2}) \right\}$$

45. Let x and y represent the numbers.

$$\frac{x}{y} = \frac{9}{2} \quad (1)$$

$$xy = 162 \quad (2)$$

Rewrite (1) as $x = \frac{9}{2}y$, and substitute $\frac{9}{2}y$ for x in (2).

$$\left(\frac{9}{2}y\right)y = 162$$

$$\frac{9}{2}y^2 = 162$$

$$y^2 = 36$$

$$y = \pm 6$$

If $y = 6$, $x = \frac{9}{2}(6) = 27$.

If $y = -6$, $x = \frac{9}{2}(-6) = -27$.

The two numbers are either 6 and 27, or −6 and −27.

47. If the system

$$3x - 2y = 9 \quad (1)$$
$$x^2 + y^2 = 25 \quad (2)$$

has a solution, the line and the circle intersect.
Solve equation (1) for x.

$$3x = 9 + 2y$$

$$x = 3 + \frac{2}{3}y \quad (3)$$

Substitute into equation (2).

$$\left(3 + \frac{2}{3}y\right)^2 + y^2 = 25$$

$$9 + 4y + \frac{4}{9}y^2 + y^2 = 25$$

$$\frac{13}{9}y^2 + 4y - 16 = 0$$

$$13y^2 + 36y - 144 = 0$$

Use the quadratic formula.

$$y = \frac{-36 \pm \sqrt{(36)^2 - 4(13)(144)}}{2(13)}$$

$$= \frac{-36 \pm \sqrt{8784}}{26}$$

$$y \approx 2.2 \quad \text{or} \quad y \approx -4.989$$

Substitute into equation (3) to find x.

If $y = 2.22$,

$$x = 3 + \frac{2}{3}(2.22)$$

$$= 4.48.$$

If $y = -4.989$,

$$x = 3 + \frac{2}{3}(-4.989)$$

$$= -.326.$$

Thus, the circle and the line do intersect, in fact twice, at (4.48, 2.22) and at (−3.26, −4.99).

49. Write the equation of the line as $y = mx + b$.

$$y = mx + b \quad (1)$$
$$y = x^2 \quad (2)$$

Substitute x^2 for y in equation (1).

$$x^2 = mx + b$$
$$x^2 - mx - b = 0$$

Find the discriminant of this equation and set it equal to 0.

$$(-m)^2 - 4(1)(-b) = 0$$
$$m^2 + 4b = 0 \quad (3)$$

Since (2, 4) is a point on the line, substitute x = 2 and y = 4 in equation (1).

$$4 = m \cdot 2 + b \text{ or}$$

$$4 - 2m = b \qquad (4)$$

Substitute 4 - 2m for b in equation (3).

$$m^2 + 4(4 - 2m) = 0$$

$$m^2 + 16 - 8m = 0$$

$$m^2 - 8m + 16 = 0$$

$$(m - 4)(m - 4) = 0$$

$$m = 4$$

Substitute m = 4 in equation (4).

$$b = 4 - 2(4)$$

$$= -4$$

Thus m = 4 and b = -4, so the equation of the desired line is y = 4x - 4.

51.
$$x^2 + y^2 = 25 \qquad (1)$$

$$\frac{x^2}{a^2} + \frac{y^2}{25} = 1 \qquad (2)$$

The graph of equation (1) is a circle which passes through (0, 5) and (5, 0). The graph of equation (2) is an ellipse or circle which also passes through (0, 5) and (5, 0). Thus, for every nonzero value of a, the graphs have at least those two points in common.

However, if a = ±5, then equation (2) would be equivalent to equation (1) and the graphs would have infinitely many points in common. To have exactly two points in common, therefore, a can be any nonzero real number except -5 and 5.

Section 7.5

For Exercises 1-27, see the answer graphs in the textbook.

1. x + y ≤ 4

x - 2y ≥ 6

Graph x + y = 4 as a solid line with y-intercept 4 and x-intercept 4. Shade the region below this line. Graph x - 2y = 6 as a solid line with y-intercept -3 and x-intercept 6. Shade the region below this line. The solution set is the common region, which is shaded in the final graph.

3. 4x + 3y < 12

y + 4x > -4

Graph 4x + 3y = 12 as a dashed line with y-intercept 4 and x-intercept 3. Shade the region below the line. Graph y + 4x = -4 as a dashed line with y-intercept -4 and x-intercept -1. Shade the region above the line. The solution set is the common region, which is shaded in the final graph.

5. x + y ≤ 6

2x + 2y ≥ 12

Graph x + y = 6 as a solid line with y-intercept 6 and x-intercept 6. Shade the region below the line, including the boundary. Since 2x + 2y = 12 is equivalent to

$$x + y = 6,$$

the boundary line is the same as
before. Shade the region above
the line, including the boundary.
The solution set is the common part,
which is the line $x + y = 6$.

7. $x + 2y \leq 4$
    $y \geq x^2 - 1$

    Graph $x + 2y = 4$ as a solid line
    with y-intercept 2 and x-intercept
    4, as a solid line. Shade the
    region below the line.
    Graph the solid parabola $y = x^2 - 1$.
    Shade the region inside of the
    parabola. The solution set is the
    common region, which is shaded in
    the final graph.

9. $y \leq -x^2$
    $y \geq x^2 - 6$

    Graph $y = -x^2$ as a solid parabola.
    Shade the region below the curve.
    Graph $y = x^2 - 6$ using a solid para-
    bola. Shade the region above the
    curve. The solution set is the
    region between the two curves, which
    is shaded in the final graph.

11. $x^2 - y^2 < 1$
    $-1 < y < 1$

    Use a dashed curve to graph the
    hyperbola $x^2 - y^2 = 1$, which has
    x-intercepts 1 and −1, no y-inter-
    cepts, asymptotes $y = \pm x$, and
    branches opening to the left and
    right. Shade the region between

the branches. Graph the horizontal lines $y = -1$ and $y = 1$ as
dashed lines. Shade the region
between these lines. The solution
set is the intersection of these two
regions, which is shaded in the
final graph.

13. $2x^2 - y^2 > 4$
    $2y^2 - x^2 > 4$

    The inequalities may be rewritten in
    standard form by dividing both sides
    of each inequality by 4.

    $$\frac{x^2}{2} - \frac{y^2}{4} > 1$$

    $$\frac{y^2}{2} - \frac{x^2}{4} > 1$$

    Use a dashed curve to graph the
    hyperbola $\frac{x^2}{2} - \frac{y^2}{4} = 1$, which has
    x-intercepts $\sqrt{2} \approx 1.4$ and $-\sqrt{2} \approx$
    −1.4, no y-intercepts, asymptotes
    $y = \pm\frac{2}{\sqrt{2}}x$ or $y = \pm\sqrt{2}x$, and branches
    opening to the left and right.
    Shade the regions that are inside
    each of the branches of the hyper-
    bola. Use a dashed curve to graph
    the hyperbola $\frac{y^2}{2} - \frac{x^2}{4} = 1$, which
    has y-intercepts $\sqrt{2} \approx 1.4$ and
    $-\sqrt{2} \approx -1.4$, no x-intercepts, asymp-
    totes $y = \pm\frac{\sqrt{2}}{2}x$, and branches opening
    upward and downward. Shade the
    regions that are inside of each the
    branches of this hyperbola. The
    solution set is the common region,
    which is shaded in the final graph.

15. $\dfrac{x^2}{16} + \dfrac{y^2}{9} \le 1$

$\dfrac{x^2}{4} - \dfrac{y^2}{16} \ge 1$

Graph the solid ellipse $\dfrac{x^2}{16} + \dfrac{y^2}{9} = 1$, which has x-intercepts -4 and 4 and y-intercepts -3 and 3. Shade the region inside the ellipse.
Use a solid curve to graph the hyperbola $\dfrac{x^2}{4} - \dfrac{y^2}{16} = 1$, which has x-intercepts -2 and 2, no y-intercepts, asymptotes $y = \pm 2x$, and branches opening to the left and right. Shade the regions that are inside of each of the branches of the hyperbola.
The solution set is the common region, which is shaded.

17. $x + y \le 4$

$x - y \le 5$

$4x + y \le -4$

Graph $x + y = 4$ as a solid line and shade the region below it.
Graph $x - y = 5$ as solid line and shade the region above it. Graph $4x + y = -4$ as a solid line and shade the region below it. The solution set is the common region, which is shaded in the final graph.

19. $2y + x \le -5$

$y \ge 3 + x$

$x \ge 0$

$y \ge 0$

Graph $2y + x = -5$ as a solid line and shade the region below it.
Graph $y = 3 + x$ as a solid line and shade the region above it. $x = 0$ is the y-axis; shade the region to the right of it. $y = 0$ is the x-axis; shade the region above it. There are no points in the inter- section. No points satisfy all the conditions, so the system has no solution.

21. $\dfrac{x^2}{4} + \dfrac{y^2}{9} > 1$

$x^2 - y^2 \ge 1$

$-4 \le x \le 4$

Graph the solid vertical lines $x = -4$ and $x = 4$. Shade the region between them. Graph the ellipse $\dfrac{x^2}{4} + \dfrac{y^2}{9} = 1$ as a dashed curve with x-intercepts -2 and 2 and y-intercepts -3 and 3; shade the region outside the ellipse. Graph the hyperbola $x^2 - y^2 = 1$ as a solid curve with x-intercepts -1 and 1, no y-intercepts, asymptotes $y = \pm x$, and branches opening to the left and right. Shade the regions that are inside of each of the branches of the hyperbola. The solution set is the common region, which is shaded.

The four open circles in the final graph indicate that those four points are not included in the solution (due to the fact that the ellipse on which they lie,

$\frac{x^2}{4} + \frac{y^2}{9} > 1$, is not included in the system's solution).

23. $y \geq 3^x$

$y \geq 2$

Graph $y = 3^x$ using a solid curve passing through the points $\left(-2, \frac{1}{9}\right)$, $\left(-1, \frac{1}{3}\right)$, $(0, 1)$, $(1, 3)$, and $(2, 9)$; shade the region above it. Graph the solid horizontal line $y = 2$ and shade the region above it. The solution set is the common region, which is shaded.

25. $|x| \geq 2$

$|y| \geq 4$

$y < x^2$

$|x| \geq 2$ means $x \geq 2$ or $x \leq -2$. Graph the solid vertical line $x = 2$ and shade the region to the right, and graph the solid vertical line $x = -2$ and shade the region to the left.

$|y| \geq 4$ means $y \geq 4$ or $y \leq -4$. Graph the solid horizontal line $y = 4$ and shade the region above it, and graph the solid horizontal line $y = -4$ and shade the region below it. Graph the parabola $y = x^2$

using a dashed curve and shade the region below it. The solution set is the common region, which is shaded.

The two open circles are drawn because no points on the parabola are in the solution set.

27. $y \leq |x + 2|$

$\frac{x^2}{16} - \frac{y^2}{9} \leq 1$

Graph $y = |x + 2|$ using a solid curve. This is the same as the graph of $y = |x|$, but translated to the left 2 units. Shade the region below it. Graph the hyperbola $\frac{x^2}{16} - \frac{y^2}{9} = 1$, using a solid curve. This hyperbola has branches opening to the left and right. The vertices are $(-4, 0)$ and $(4, 0)$, and the asymptotes are $y = \pm\frac{3}{4}x$. Shade the region between the two branches. The solution set is the common region, which is shaded in the final graph.

29. (d) $x^2 + 4y^2 < 36$

$y < x$

The solution set of this system consists of all points inside the ellipse $x^2 + 4y^2 = 36$ and below the line $y = x$.

31. Find $x \geq 0$ and $y \geq 0$ such that

$$2x + 3y \leq 6$$
$$4x + y \leq 6$$

and $5x + 2y$ is maximized.

First, graph the solid lines $x = 0$, $y = 0$, $2x + 3y = 6$ and $4x + y = 6$.

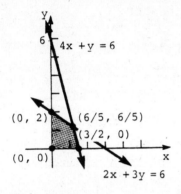

Notice that the vertices are

at $(0, 0)$, $(0, 2)$, $\left(\frac{3}{2}, 0\right)$, and

$\left(\frac{6}{5}, \frac{6}{5}\right)$, which is the intersection

of $2x + 3y = 6$ and $4x + y = 6$. This

intersection point result may be

obtained by solving the system

$$2x + 3y = 6$$
$$4x + y = 6$$

by either addition or substitution. Substitute the coordinates of each point into the expression $5x + 2y$ to see which gives the maximum value.

Point	Value = 5x + 2y	
$(0, 0)$	0	
$(0, 2)$	4	
$\left(\frac{3}{2}, 0\right)$	$\frac{15}{2}$	
$\left(\frac{6}{5}, \frac{6}{5}\right)$	$\frac{42}{5}$	← Maximum

The maximum value is $\frac{42}{5}$. It occurs

at $\left(\frac{6}{5}, \frac{6}{5}\right)$.

33. Find $x \geq 2$ and $y \geq 5$ such that

$$3x - y \geq 12$$
$$x + y \geq 15$$

and $2x + y$ is minimized.

The intersection of the given equation is the triangular region bounded by the lines $3x - y = 12$, $y = 5$, and $x + y = 15$. The vertices of the triangle can be obtained as follows. Substitute $y = 5$ into $3x - y = 12$.

$$3x - 5 = 12 \quad \text{or} \quad x = \frac{17}{3}$$

$\left(\frac{17}{3}, 5\right)$ is one vertex.

Substitute $y = 5$ into $x + y = 15$.

$$x + 5 = 15 \quad \text{or} \quad x = 10$$

$(10, 5)$ is another vertex.
To find the point of intersection of the lines $3x - y = 12$ and $x + y = 15$, solve this system by addition.

$$3x - y = 12 \quad (1)$$
$$\underline{x + y = 15} \quad (2)$$
$$4x \qquad = 27 \quad Add$$
$$x = \frac{27}{4}$$

$\frac{27}{4} + y = 15 \quad$ *Let* $x = 27/4$ *in* (2)

$$y = 15 - \frac{27}{4}$$
$$= \frac{33}{4}$$

$\left(\frac{27}{4}, \frac{33}{4}\right)$ is the third vertex.

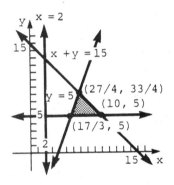

Substitute the coordinates of each vertex into the expression $2x + y$ to see where this value is minimized.

Point	Value = $2x + y$
$\left(\frac{17}{3}, 5\right)$	$\frac{49}{3}$ ← Minimum
$(10, 5)$	25
$\left(\frac{27}{4}, \frac{33}{4}\right)$	$\frac{87}{4}$

The minimum value is $\frac{49}{3}$ at $\left(\frac{17}{3}, 5\right)$.

35. Let P = number of pigs;
    G = number of geese.

$P + G \le 16$ *Total number*
        of animals

$G \le 12$ *No more than*
      12 geese

$50P + 20G \le 500$ *$500 available*
          to spend

Maximize the profit $80G + 40P$. Find the feasible region by graphing on a horizontal P-axis and a vertical G-axis.

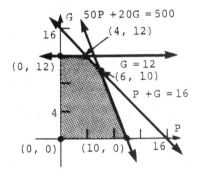

The boundaries of the graph are

$$P + G = 16$$
$$G = 12$$
$$50P + 20G = 500.$$

Also, $P \ge 0$ and $G \ge 0$, so the P- and G-axes are boundaries. The vertices are $(0, 0)$, $(0, 12)$, $(4, 12)$, $(10, 0)$, and $(6, 10)$, the intersection of $P + G = 16$ and $50P + 20G = 500$. Check the value of $80G + 40P$ at each vertex to find the maximum profit.

Point (P, G)	Profit = 80G + 40P
(0, 0)	0
(0, 12)	960
(4, 12)	1120 ← Maximum
(10, 0)	400
(6, 10)	1040

The maximum profit is $1120 with 4 pigs and 12 geese.

37. Let x = number of cabinet #1;

   y = number of cabinet #2.

The cost constraint is

$$10x + 20y \leq 140.$$

The space constraint is

$$6x + 8y \leq 72.$$

Since the numbers of cabinets cannot be negative, we also have

$$x \geq 0$$
$$y \geq 0.$$

We want to maximize the volume of files, given by $8x + 12y$.
Find the region of feasible solutions by graphing.

20x + 40y = 280 (shade below)
 6x + 8y = 72 (shade below)
    x = 0   (shade right)
    y = 0   (shade above)

The vertices are at (0, 7), (0, 0), (12, 0), and the intersection of 20x + 40y = 280 and 6x + 8y = 72, which is the point (8, 3).

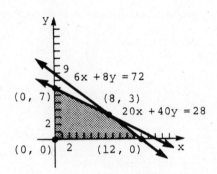

Find the value of $8x + 12y$ at each vertex.

Point	Value = 8x + 12y
(0, 7)	84
(0, 0)	0
(12, 0)	96
(8, 3)	100 ← Maximum

$8x + 12y$ is maximized at (8, 3). She should get 8 #1 cabinets and 3 #2 cabinets. This will correspond to maximum storage capacity of 100 cu ft.

39. Let x = number of barrels of gasoline;

   y = number of barrels of fuel oil.

The constraints are

$$x \geq 0,\ y \geq 0$$
$$x \geq 2y$$
$$y \geq 3{,}000{,}000$$
$$x \leq 6{,}400{,}000.$$

Maximize revenue, given by 1.9x + 1.5y.

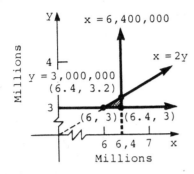

The boundaries are

$$x = 0$$
$$y = 0$$
$$x = 2y$$
$$y = 3,000,000$$
$$x = 6,400,000.$$

The vertices of the feasible region are

(6.4, 3), (6, 3) and (6.4, 3.2).

Testing these points in the expression to be maximized will show that (6.4, 3.2) will maximize revenue.

6.4 million barrels of gasoline and 3.2 million barrels of fuel oil should be produced for maximum revenue of

$$1.9(6.4) + 1.5(3.2) = 16.96$$

or $16,960,000.

Chapter 7   Review Exercises

1. $4x - 3y = -1$ (1)
    $3x + 5y = 50$ (2)

Solve equation (1) for x.

$$4x = -1 + 3y$$
$$x = \frac{-1 + 3y}{4} \quad (3)$$

Substitute this result into equation (2).

$$3\left(\frac{-1 + 3y}{4}\right) + 5y = 50$$
$$4 \cdot 3\left(\frac{-1 + 3y}{4}\right) + 4(5y) = 4 \cdot 50$$
$$3(-1 + 3y) + 20y = 200$$
$$-3 + 9y + 20y = 203$$
$$29y = 203$$
$$y = 7$$

Substitute y = 7 back into equation (3) to find x.

$$x = \frac{-1 + 3(7)}{4}$$
$$= \frac{20}{4} = 5$$

Solution set: $\{(5, 7)\}$

3. $7x - 10y = 11$ (1)
    $\frac{3x}{2} - 5y = 8$ (2)

Multiply both sides of equation (2) by 2 to eliminate the fraction.

$$3x - 10y = 16 \quad (3)$$

Solve equation (3) for x.

$$3x = 16 + 10y$$
$$x = \frac{16 + 10y}{3}$$

Substitute this result into equation (1).

$$7\left(\frac{16 + 10y}{3}\right) - 10y = 11$$

$$3 \cdot 7\left(\frac{16 + 10y}{3}\right) - 3(10y) = 3(11)$$

$$7(16 + 10y) - 30y = 33$$

$$112 + 70y - 30y = 33$$

$$112 + 40y = 33$$

$$40y = -79$$

$$y = -\frac{79}{40}$$

To find x, substitute $y = -\frac{79}{40}$ in equation (3).

$$3x - 10\left(-\frac{79}{40}\right) = 16$$

$$3x + \frac{79}{4} = 16$$

$$3x = \frac{64}{4} - \frac{79}{4}$$

$$= -\frac{15}{4}$$

$$x = \frac{1}{3}\left(-\frac{15}{4}\right) = -\frac{5}{4}$$

Solution set: $\left\{\left(-\frac{5}{4}, -\frac{79}{40}\right)\right\}$

5. $\dfrac{x}{2} - \dfrac{y}{5} = \dfrac{11}{10}$ (1)

$2x - \dfrac{4y}{5} = \dfrac{22}{5}$ (2)

To eliminate fractions, multiply equation (1) by 10 and equation (2) by 5.

$$5x - 2y = 11 \quad (3)$$

$$10x - 4y = 22 \quad (4)$$

Solve equation (3) for x.

$$5x - 2y = 11$$

$$5x = 2y + 11$$

$$x = \frac{2y + 11}{5}$$

Substitute into equation (4).

$$10\left(\frac{2y + 11}{5}\right) - 4y = 22$$

$$2(2y + 11) - 4y = 22$$

$$22 + 4y - 4y = 22$$

$$22 = 22$$

Thus, the equations are dependent.

Solution set: $\left\{\left(\dfrac{2y + 11}{5}, y\right)\right\}$

7. $3x - 5y = -18$ (1)

$2x + 7y = 19$ (2)

Multiply equation (1) by 2 and equation (2) by -3 and add.

$$6x - 10y = -36$$

$$\underline{-6x - 21y = -57}$$

$$-31y = -93$$

$$y = 3$$

To find x, substitute $y = 3$ in equation (1).

$$3x - 5(3) = -18$$

$$3x = -3$$

$$x = -1$$

Solution set: $\{(-1, 3)\}$

9. $\dfrac{2}{3}x - \dfrac{3}{4}y = 13$ (1)

$\dfrac{1}{2}x + \dfrac{2}{3}y = -5$ (2)

To eliminate fractions, multiply equation (1) by 12 and equation (2) by 6.

$$8x - 9y = 156 \quad (3)$$
$$3x + 4y = -30 \quad (4)$$

Multiply equation (3) by 4 and equation (4) by 9 and add.

$$
\begin{array}{r}
32x - 36y = 624 \\
27x + 36y = -270 \\
\hline
59x \qquad\;\; = 354 \\
x = 6
\end{array}
$$

Substitute 6 for x in equation (4).

$$3(6) + 4y = -30$$
$$18 + 4y = -30$$
$$4y = -48$$
$$y = -12$$

Solution set: $\{(6, -12)\}$

11. $\dfrac{1}{x} + \dfrac{1}{y} = \dfrac{7}{10}$ $\qquad (1)$

$\dfrac{3}{x} - \dfrac{5}{y} = \dfrac{1}{2}$ $\qquad (2)$

Let $\dfrac{1}{x} = t$, $\dfrac{1}{y} = u$.

Making these substitutions gives the following system:

$$t + u = \dfrac{7}{10} \quad (3)$$
$$3t - 5u = \dfrac{1}{2}. \quad (4)$$

To eliminate fractions, multiply equation (3) by 10 and equation (4) by 2.

$$10t + 10u = 7 \quad (5)$$
$$6t - 10u = 1 \quad (6)$$

Add equations (5) and (6).

$$
\begin{array}{r}
10t + 10u = 7 \\
6t - 10u = 1 \\
\hline
16t \qquad\;\; = 8 \\
t = \dfrac{1}{2}
\end{array}
$$

To find u, substitute $t = 1/2$ in equation (6).

$$6\left(\dfrac{1}{2}\right) - 10u = 1$$
$$3 - 10u = 1$$
$$-10u = -2$$
$$u = \dfrac{1}{5}$$

Now find x and y.

$$\dfrac{1}{x} = t \qquad\qquad \dfrac{1}{y} = u$$
$$\dfrac{1}{x} = \dfrac{1}{2} \qquad\qquad \dfrac{1}{y} = \dfrac{1}{5}$$
$$x = 2 \qquad\qquad y = 5$$

Solution set: $\{(2, 5)\}$

13. Let x = the number of 25¢ candy bars;

y = the number of 50¢ candy bars.

$$x + y = 22 \quad (1) \text{ Total number of bars purchased}$$

$$.25x + .50y = 8.50 \quad (2) \text{ Total money spent}$$

Multiply equation (2) by -4 and add to equation (1).

$$
\begin{array}{r}
x + y = 22 \\
-x - 2y = -34 \\
\hline
-y = -12 \\
y = 12
\end{array}
$$

To find x, substitute $y = 12$ into equation (1).

$$x + 12 = 22$$
$$x = 10$$

He bought 10 of the 25¢ bars and 12 of the 50¢ bars.

15. Let x = the amount invested at 3%;

y = the amount invested at 3.5%;

z = the amount invested at 4.5%.

$$y = 2x \qquad\qquad (1)$$
$$z = x + 10,000 \qquad (2)$$
$$x + y + z = 50,000 \qquad (3)$$
$$.03x + .035y + .045z = 1900 \quad (4)$$

Substitute 2x for y and x + 10,000 for z in equation (3) and solve for x.

$$x + 2x + (x + 10,000) = 50,000$$
$$4x = 40,000$$
$$x = 10,000$$

Substitute $x = 10,000$ in equation (1).

$$y = 2(10,000)$$
$$= 20,000$$

Substitute $x = 10,000$ in equation (2).

$$z = 10,000 + 10,000$$
$$= 20,000$$

$10,000 is invested at 3%, $20,000 at 3.5%, and $20,000 at 4.5%.

17. Let x = the number of grams of 12-carat gold;

y = the number of grams of 22-carat gold;

z = the number of grams of pure gold.

$$x + y = 25 \qquad (1) \; \textit{Total amount of mixture}$$
$$\frac{12}{24}x + \frac{22}{24}y = z \qquad (2) \; \textit{Total amount of pure gold}$$
$$z = \frac{15}{24} \cdot 25$$
$$z = \frac{375}{24} \qquad (3) \; \textit{Total amount of pure gold}$$
$$\frac{12}{24}x + \frac{22}{24}y = \frac{375}{24} \qquad (4) \; \textit{Let } z = 375/24 \textit{ in (2)}$$

$$\begin{array}{ll} 12x + 22y = 375 & \textit{Multiply (4) by 24} \\ \underline{-12x - 12y = -300} & \textit{Multiply (1)} \\ 10y = 75 & \textit{by } -12 \end{array}$$

$$y = \frac{75}{10} = 7.5$$

$$x + 7.5 = 25 \qquad \textit{Let } y = 7.5 \textit{ in (1)}$$
$$x = 17.5$$

17.5 g of 12-carat gold should be mixed with 7.5 g of 22-carat gold.

19.
$$2x - 3y + z = -5 \quad (1)$$
$$x + 4y + 2z = 13 \quad (2)$$
$$5x + 5y + 3z = 14 \quad (3)$$

Multiply equation (2) by -2 and add to equation (1).

$$\begin{array}{r} 2x - 3y + z = -5 \\ \underline{-2x - 8y - 4z = -26} \\ -11y - 3z = -31 \end{array}$$

or $\qquad 11y + 3z = 31$

Multiply equation (1) by 5 and equation (3) by -2, and add the results.

$$\begin{array}{r} 10x - 15y + 5z = -25 \\ \underline{-10x - 10y - 6z = -28} \\ -25y - z = -53 \end{array}$$

or $\qquad 25y + z = 53$

We now have the system

$$11y + 3z = 31 \quad (4)$$
$$25y + z = 53. \quad (5)$$

Multiply equation (4) by 25 and equation (5) by -11, and add the results.

$$275y + 75z = 775$$
$$\underline{-275y - 11z = -583}$$
$$64z = 192$$
$$z = 3$$

To find y, substitute $z = 3$ into equation (5).

$$25y + 3 = 53$$
$$25y = 50$$
$$y = 2$$

To find x, substitute $z = 3$ and $y = 2$ into equation (2).

$$x + 4y + 2z = 13$$
$$x + 4(2) + 2(3) = 13$$
$$x + 14 = 13$$
$$x = -1$$

Solution set: $\{(-1, 2, 3)\}$

21.
$$x + y - z = 5 \quad (1)$$
$$2x + y + 3z = 2 \quad (2)$$
$$4x - y + 2z = -1 \quad (3)$$

Add equation (1) to equation (3). Add equation (2) to equation (3). We get the system

$$5x + z = 4 \quad (4)$$
$$6x + 5z = 1. \quad (5)$$

Multiply equation (4) by -5 and add to equation (5).

$$-25x - 5z = -20$$
$$\underline{6x + 5z = 1}$$
$$-19x = -19$$
$$x = 1$$

To find z, substitute $x = 1$ into equation (4).

$$5(1) + z = 4$$
$$z = -1$$

To find y, substitute $x = 1$ and $z = -1$ into equation (5).

$$1 + y - (-1) = 5$$
$$y + 2 = 5$$
$$y = 3$$

Solution set: $\{(1, 3, -1)\}$

23.
$$3x - 4y + z = 2 \quad (1)$$
$$2x + y - 4z = 1 \quad (2)$$

Add 4 times equation (2) to equation (1).

$$3x - 4y + z = 2$$
$$\underline{8x + 4y - 16z = 4}$$
$$11x - 15z = 6$$

Solve for x in terms of z.

$$11x = 15z + 6$$
$$x = \frac{15z + 6}{11}$$

Substitute this result into equation (1).

$$3\left(\frac{15z + 6}{11}\right) - 4y + z = 2$$

Multiply both sides by 11, and solve for y in terms of z.

$$45z + 18 - 44y + 11z = 22$$
$$56z - 44y = 4$$
$$-44y = -56z + 4$$
$$y = \frac{56z - 4}{44}$$
$$= \frac{14z - 1}{11}$$

The solution, with z arbitrary, is

$$\left\{ \left(\frac{15z + 6}{11}, \ \frac{14z - 1}{11}, \ z \right) \right\}.$$

25.   Let x = the number of pounds of
$4.60 tea;

   y = the number of pounds of
$5.75 tea;

   z = the number of pounds of
$6.50 tea.

$$x + y + z = 20$$
   (1) *Total pounds*

$$4.6x + 5.75y + 6.5z = 20(5.25)$$
   (2) *Total value*

$$x = y + z$$
   (3) *Amount of $4.60*
   tea equals sum of
   other two

Substitute x for y + z in equation (1).

$$x + x = 20$$
$$2x = 20$$
$$x = 10$$

$$4.6(10) + 5.75y + 6.5z = 105$$
   Let x = 10 in (2)
$$5.75y + 6.5z = 105 - 46$$
$$5.75y + 6.5z = 59 \quad (4)$$
$$y + z = 10 \quad (5)$$
   Let x = 10 in (3)

$$5.75y + 6.5z = \ \ 59 \quad (4)$$
$$\underline{-6.5 y - 6.5z = -65} \quad \textit{Multiply (5)}$$
$$-.75y \qquad\qquad = \ -6 \quad \textit{by -6.5}$$
$$y = 8$$

$$8 + z = 10 \quad \textit{Let y = 8 in (5)}$$
$$z = 2$$

Use 10 lb of $4.60 tea, 8 lb of
$5.75 tea, and 2 lb of $6.50 tea.

27.   Let x = the number of fives;

   y = the number of tens;

   z = the number of twenties.

$$5x + 10y + 20z = 2480 \qquad (1)$$
$$x + y + z = 290 \qquad (2)$$
$$10y = 20z + 60 \qquad (3)$$

Divide both sides of equation (3) by 10 to obtain

$$y = 2z + 6. \qquad (4)$$

Substitute this result into equations (1) and (2).

$$5x + 10(2z + 6) + 20z = 2480$$
$$x + (2z + 6) + z = \ 290$$

or

$$5x + 40z = 2420 \quad (5)$$
$$x + \ 3z = \ 284 \quad (6)$$

Multiply equation (6) by −5 and add to equation (5).

$$5x + 40z = \ \ 2420$$
$$\underline{-5x - 15z = -1420}$$
$$25z = \ 1000$$
$$z = 40$$

To find y, substitute z = 40 in equation (4).

$$y = 2(40) + 6$$
$$= 86$$

To find x, substitute y = 86 and
z = 40 in equation (2).

$$x + 86 + 40 = 290$$
$$x = 164$$

There are 164 fives, 86 tens, and 40
twenties.

29. $2x + 3y = 10$
$$-3x + y = 18$$

$$\begin{bmatrix} 2 & 3 & | & 10 \\ -3 & 1 & | & 18 \end{bmatrix}$$

$$\begin{bmatrix} 1 & \frac{3}{2} & | & 5 \\ -3 & 1 & | & 18 \end{bmatrix} \frac{1}{2}R1$$

$$\begin{bmatrix} 1 & \frac{3}{2} & | & 5 \\ 0 & \frac{11}{2} & | & 33 \end{bmatrix} 3R1 + R2$$

$$\begin{bmatrix} 1 & \frac{3}{2} & | & 5 \\ 0 & 1 & | & 6 \end{bmatrix} \frac{2}{11}R2$$

$$\begin{bmatrix} 1 & 0 & | & -4 \\ 0 & 1 & | & 6 \end{bmatrix} -\frac{3}{2}R2 + R1$$

Solution set: $\{(-4, 6)\}$

31. $3x + y = -7$
$$x - y = -5$$

$$\begin{bmatrix} 3 & 1 & | & -7 \\ 1 & -1 & | & -5 \end{bmatrix}$$

$$\begin{bmatrix} 4 & 0 & | & -12 \\ 1 & -1 & | & -5 \end{bmatrix} R2 + R1$$

$$\begin{bmatrix} 1 & 0 & | & -3 \\ -1 & 1 & | & 5 \end{bmatrix} \begin{matrix} \frac{1}{4}R2 \\ -1R2 \end{matrix}$$

$$\begin{bmatrix} 1 & 0 & | & -3 \\ 0 & 1 & | & 2 \end{bmatrix} R1 + R2$$

Solution set: $\{(-3, 2)\}$

33. $2x - y + 4z = -1$
$$-3x + 5y - z = 5$$
$$2x + 3y + 2z = 3$$

$$\begin{bmatrix} 2 & -1 & 4 & | & -1 \\ -3 & 5 & -1 & | & 5 \\ 2 & 3 & 2 & | & 3 \end{bmatrix}$$

$$\begin{bmatrix} 2 & -1 & 4 & | & -1 \\ 0 & \frac{7}{2} & 5 & | & \frac{7}{2} \\ 0 & 4 & -2 & | & 4 \end{bmatrix} \begin{matrix} \frac{3}{2}R1 + R2 \\ -1R1 + R3 \end{matrix}$$

$$\begin{bmatrix} 1 & -\frac{1}{2} & 2 & | & -\frac{1}{2} \\ 0 & 1 & \frac{10}{7} & | & 1 \\ 0 & 4 & -2 & | & 4 \end{bmatrix} \begin{matrix} \frac{1}{2}R1 \\ \frac{2}{7}R2 \end{matrix}$$

$$\begin{bmatrix} 1 & -\frac{1}{2} & 2 & | & -\frac{1}{2} \\ 0 & 1 & \frac{10}{7} & | & 1 \\ 0 & 0 & -\frac{54}{7} & | & 0 \end{bmatrix} -4R2 + R3$$

$$\begin{bmatrix} 1 & -\frac{1}{2} & 0 & | & -\frac{1}{2} \\ 0 & 1 & \frac{10}{7} & | & 1 \\ 0 & 0 & 1 & | & 0 \end{bmatrix} -\frac{7}{54}R3$$

$$\begin{bmatrix} 1 & -\frac{1}{2} & 0 & | & -\frac{1}{2} \\ 0 & 1 & 0 & | & 1 \\ 0 & 0 & 1 & | & 0 \end{bmatrix} \begin{matrix} -2R3 + R1 \\ -\frac{10}{7}R3 + R2 \end{matrix}$$

$$\begin{bmatrix} 1 & 0 & 0 & | & 0 \\ 0 & 1 & 0 & | & 1 \\ 0 & 0 & 1 & | & 0 \end{bmatrix} \frac{1}{2}R2 + R1$$

Solution set: $\{(0, 1, 0)\}$

35. $y = x^2 - 1$ (1)

$x + y = 1$ (2)

Substitute $x^2 - 1$ for y in equation (2).

$$x + x^2 - 1 = 1$$
$$x^2 + x - 2 = 0$$
$$(x - 1)(x + 2) = 0$$
$$x = 1 \text{ or } x = -2$$

If $x = 1$, then

$$y = x^2 - 1$$
$$= 1^2 - 1 = 0.$$

If $x = -2$, then

$$y = (-2)^2 - 1$$
$$= 4 - 1 = 3.$$

Solution set: $\{(1, 0), (-2, 3)\}$

37. $x^2 + 2y^2 = 22$ (1)

$2x^2 - y^2 = -1$ (2)

Multiply equation (1) by -2 and add.

$$-2x^2 - 4y^2 = -44$$
$$\underline{2x^2 - y^2 = -1}$$
$$-5y^2 = -45$$
$$y^2 = 9$$
$$y = \pm 3$$

To find x, substitute into equation (1).

$$x^2 + 2y^2 = 22$$
$$x^2 + 2(\pm 3)^2 = 22$$
$$x^2 + 18 = 22$$
$$x^2 = 4$$
$$x = \pm 2$$

Solution set:

$\{(2, 3), (2, -3), (-2, 3), (-2, -3)\}$

39. $xy = 4$ (1)

$x - 6y = 2$ (2)

Solve equation (2) for x.

$$x = 6y + 2 \quad (3)$$

Substitute $6y + 2$ for x in equation (1) and solve for y.

$$(6y + 2)y = 4$$
$$6y^2 + 2y = 4$$
$$6y^2 + 2y - 4 = 0$$
$$3y^2 + y - 2 = 0$$
$$(3y - 2)(y + 1) = 0$$
$$y = \frac{2}{3} \text{ or } y = -1$$

Substitute each value of y into equation (3) to find the corresponding value of x.

If $y = \frac{2}{3}$, then

$$x = 6\left(\frac{2}{3}\right) + 2 = 6.$$

If $y = -1$, then

$$x = 6(-1) + 2 = -4.$$

Solution set: $\left\{\left(6, \frac{2}{3}\right), (-4, -1)\right\}$

41. $x^2 + y^2 = 144$ (1)

$x + 2y = 8$ (2)

Solve equation (2) for x.

$$x = -2y + 8$$

Substitute $-2y + 8$ for x in equation (1).

$$(-2y + 8)^2 + y^2 = 144$$
$$4y^2 - 32y + 64 + y^2 = 144$$
$$5y^2 - 32y - 80 = 0$$

Use the quadratic formula.

$$y = \frac{-(-32) \pm \sqrt{(-32)^2 - 4(5)(-80)}}{2(5)}$$

$$= \frac{32 \pm \sqrt{2624}}{10}$$

$$= \frac{32 \pm 2\sqrt{656}}{10}$$

$$= \frac{16 \pm \sqrt{656}}{5}$$

$$= \frac{16 \pm 4\sqrt{41}}{5}$$

Substitute each value of y into equation (3) to find the corresponding value of x.

If

$$y = \frac{16 + 4\sqrt{41}}{5}, \text{ then}$$

$$x = -2y + 8$$

$$= -2\left(\frac{16 + 4\sqrt{41}}{5}\right) + 8$$

$$= \frac{-32 - 8\sqrt{41}}{5} + \frac{40}{5}$$

$$= \frac{8 - 8\sqrt{41}}{5}.$$

If

$$y = \frac{16 - 4\sqrt{41}}{5}, \text{ then}$$

$$x = -2y + 8$$

$$= -2\left(\frac{16 - 4\sqrt{41}}{5}\right) + 8$$

$$= \frac{8 + 8\sqrt{41}}{5}.$$

The circle and line intersect in two points,

$$\left(\frac{8 - 8\sqrt{41}}{5}, \frac{16 + 4\sqrt{41}}{5}\right) \text{ and}$$

$$\left(\frac{8 + 8\sqrt{41}}{5}, \frac{16 - 4\sqrt{41}}{5}\right).$$

43. See the sketches in the answer section of the textbook.

45. $x - 3y \geq 6$

$y^2 \leq 16 - x^2$

Graph the line $x - 3y = 6$ as a solid line. Shade the region below the line. Graph the circle $x^2 + y^2 = 16$ as a solid curve.

Shade the interior of the circle. The solution set is the intersection of the two regions which is shaded in the final graph.

See the answer graph in the textbook.

47. Find $x \geq 0$ and $y \geq 0$ such that

$$3x + 2y \leq 12$$
$$5x + y \geq 5$$

and $2x + 4y$ is maximized.

Graph the solid lines $x = 0$, $y = 0$, $3x + 2y = 12$, and $5x + y = 5$.

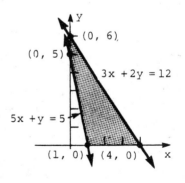

The vertices are $(1, 0)$, $(4, 0)$, $(0, 5)$, and $(0, 6)$.

Point	Value = 2x + 4y
(1, 0)	2(1) + 4(0) = 2
(4, 0)	2(4) + 4(0) = 8
(0, 5)	2(0) + 4(5) = 20
(0, 6)	2(0) + 4(6) = 24 ← Maximum

The maximum value is 24, which occurs at (0, 6).

49. Let x = the number of batches of cakes;

y = the number of batches of cookies.

Find $x \geq 0$ and $y \geq 0$ such that

$$2x + \frac{3}{2}y \leq 16$$

$$3x + \frac{2}{3}y \leq 12$$

and 30x + 20y is maximized.

Graph the lines x = 0, y = 0,

$2x + \frac{3}{2}y = 16$, and $3x + \frac{2}{3}y = 12$.

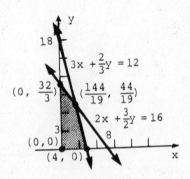

The vertices are (0, 0), (4, 0),

$\left(\frac{144}{19}, \frac{44}{19}\right)$, and $\left(0, \frac{32}{3}\right)$.

Point	Value = 30x + 20y
(0, 0)	30(0) + 20(0) = 0
(4, 0)	30(4) + 20(0) = 120
$\left(\frac{44}{19}, \frac{144}{19}\right)$	$30\left(\frac{44}{19}\right) + 20\left(\frac{144}{19}\right) \approx 221.05$
$\left(9, \frac{32}{3}\right)$	$30(0) + 20\left(\frac{32}{3}\right) \approx 213.33$

For a maximum profit of approximately $220, bake $\frac{44}{19} \approx 2$ batches of cakes and $\frac{144}{19} \approx 8$ batches of cookies.

Chapter 7   Test

1.   x − 3y = −5   (1)
      2x + y = 4   (2)

Solve equation (1) for x.

x = 3y − 5   (3)

Substitute this result into equation (2) and solve for y.

2(3y − 5) + y = 4
6y − 10 + y = 4
7y = 14
y = 2

Substitute y = 2 back into equation (3) to find x.

x = 3(2) − 5
x = 1

Solution set: {(1, 2)}

2. $3m - n = 9$ *(1)*

    $m + 2n = 10$ *(2)*

Solve equation (2) for m.

$$m = 10 - 2n \quad (3)$$

Substitute this result into equation (1) and solve for n.

$$3(10 - 2n) - n = 9$$
$$30 - 6n - n = 9$$
$$-7n = -21$$
$$n = 3$$

Substitute n = 3 back into equation (3) to find m.

$$m = 10 - 2(3)$$
$$m = 4$$

Solution set: $\{(4, 3)\}$

3. $6x + 9y = -21$ *(1)*

   $4x + 6y = -14$ *(2)*

Solve equation (1) for x.

$$6x = -9y - 21$$
$$x = \frac{-9y - 21}{6}$$
$$x = \frac{-3y - 7}{2} \quad (3)$$

Substitute this result into equation (2).

$$4\left(\frac{-3y - 7}{2}\right) + 6y = -14$$
$$2(-3y - 7) + 6y = -14$$
$$-6y - 14 + 6y = -14$$
$$-14 = -14$$

The equations are dependent.

The solution set is an infinite set of ordered pairs, and the graphs of the equations are the same line.

Solution set: $\left\{\left(\frac{-3y - 7}{2}, y\right)\right\}$

4. $4a + 6b = 31$ *(1)*

   $3a + 4b = 22$ *(2)*

Multiply equation (1) by 3 and equation (2) by −4 and add the resulting equations.

$$\begin{aligned} 12a + 18b &= 93 \\ -12a - 16b &= -88 \\ \hline 2b &= 5 \\ b &= \frac{5}{2} \end{aligned}$$

Substitute $b = \frac{5}{2}$ in equation (1) and solve for a.

$$4a + 6\left(\frac{5}{2}\right) = 31$$
$$4a + 15 = 31$$
$$4a = 16$$
$$a = 4$$

Solution set: $\left\{\left(4, \frac{5}{2}\right)\right\}$

5. $\frac{1}{4}x - \frac{1}{3}y = -\frac{5}{12}$ *(1)*

   $\frac{1}{10}x + \frac{1}{5}y = \frac{1}{2}$ *(2)*

To eliminate fractions, multiply equation (1) by 12 and equation (2) by 10.

$$3x + 4y = -5 \quad (3)$$
$$x + 2y = 5 \quad (4)$$

Multiply equation (4) by 2 and add the result to equation (3).

$$3x - 4y = -5$$
$$\underline{2x + 4y = 10}$$
$$5x \quad\quad = 5$$
$$x = 1$$

Substitute x = 1 in equation (4) to find y.

$$1 + 2y = 5$$
$$2y = 4$$
$$y = 2$$

Solution set: $\{(1, 2)\}$

6. $x - 2y = 4$ (1)
    $-2x + 4y = 6$ (2)

Multiply equation (1) by 2 and add the result to equation (2).

$$2x - 4y = 8$$
$$\underline{-2x + 4y = 6}$$
$$0 = 14$$

The system is inconsistent.
Solution set: Ø

8. Let x = the amount invested at 5%;
     y = the amount invested at 4%.

There is $1000 more invested at 5% than at 4%, so

$$x = 1000 + y. \quad\quad (1)$$

The total annual income is $698, so

$$.05x + .04y = 698. \quad (2)$$

Equations (1) and (2) make up a system, which can be solved by the substitution method. Substitute 1000 + y for x in equation (2).

$$.05(1000 + y) + .04y = 698$$
$$50 + .05y + .04y = 698$$
$$.09y = 648$$
$$y = 7200$$
$$x = 1000 + 7200$$
$$= 8200$$

Christopher has $7200 invested at 4% and $8200 invested at 5%.

9. $3a - 4b + 2c = 15$ (1)
    $2a - b + c = 13$ (2)
    $a + 2b - c = 5$ (3)

Eliminate c first. Add equations (2) and (3).

$$3a + b = 18 \quad\quad (4)$$

Multiply equation (3) by 2 and add the result to equation (1).

$$3a - 4b + 2c = 15$$
$$\underline{2a + 4b - 2c = 10}$$
$$5a \quad\quad\quad = 25$$
$$a = 5$$

To find b, substitute a = 5 in equation (4).

$$3(5) + b = 18$$
$$b = 3$$

To find c, substitute a = 5 and b = 3 in equation (3).

$$5 + 2(3) - c = 5$$
$$c = 6$$

Solution set: $\{(5, 3, 6)\}$

10.
$$2x + y + z = 3 \quad (1)$$
$$x + 2y - z = 3 \quad (2)$$
$$3x - y + z = 5 \quad (3)$$

Eliminate z first. Add equations (1) and (2).

$$3x + 3y = 6 \quad (4)$$

Add equations (2) and (3).

$$4x + y = 8 \quad (5)$$

Multiply equation (5) by −3 and add the result to equation (4).

$$3x + 3y = 6$$
$$\underline{-12x - 3y = -24}$$
$$-9x \qquad = -18$$
$$x = 2$$

Substitute x = 2 and in equation (5) to find y.

$$4(2) + y = 8$$
$$y = 0$$

Substitute x = 2 and y = 0 in equation (1) to find z.

$$2(2) + 0 + z = 3$$
$$z = -1$$

Solution set: $\{(2, 0, -1)\}$

11.
$$x - 2y + 3z = 2 \quad (1)$$
$$4x + y - z = 1 \quad (2)$$

To eliminate y, multiply equation (2) by 2 and add to equation (1).

$$x - 2y + 3z = 2$$
$$\underline{8x + 2y - 2z = 2}$$
$$9x \qquad + z = 4 \quad (3)$$

Solve equation (3) for x in terms of z.

$$9x = 4 - z$$
$$x = \frac{4 - z}{9}$$

To eliminate x, multiply equation (1) by −4 and add to equation (2).

$$-4x + 8y - 12z = -8$$
$$\underline{4x + y - z = 1}$$
$$9y - 13z = -7 \quad (4)$$

Solve equation (4) for y in terms of z.

$$9y = -7 + 13z$$
$$y = \frac{-7 + 13z}{9}$$

The solution set, with z arbitrary, is $\left\{ \left(\frac{4 - z}{9}, \frac{-7 + 13z}{9}, z \right) \right\}$.

12. Let x, y, and z be the three numbers.
The sum of the three numbers is 2, so

$$x + y + z = 2. \quad (1)$$

The first number is equal to the sum of the other two, so

$$x = y + z. \quad (2)$$

The third number is the result of subtracting the first from the second, so

$$z = y - x. \quad (3)$$

These three equations make up the following system:

$$x + y + z = 2 \qquad (1)$$
$$x - y - z = 0 \qquad (4)$$
$$x - y + z = 0. \qquad (5)$$

Add equations (1) and (4).

$$2x = 2$$
$$x = 1$$

Add equations (4) and (5).

$$2x - 2y = 0 \qquad (6)$$

Substitute $x = 1$ in equation (6) to find y.

$$2(1) - 2y = 0$$
$$y = 1$$

Substitute $x = 1$ and $y = 1$ in equation (1) to find z.

$$1 + 1 + z = 2$$
$$z = 0$$

The first number is 1, the second number is 1, the third number is 0.

13. $3a - 2b = 13$
    $4a - b = 19$

Write the augmented matrix.

$$\begin{bmatrix} 3 & -2 & \bigm| & 13 \\ 4 & -1 & \bigm| & 19 \end{bmatrix}$$

$$\begin{bmatrix} 1 & -\dfrac{2}{3} & \bigm| & \dfrac{13}{3} \\ 4 & -1 & \bigm| & 19 \end{bmatrix} \quad \dfrac{1}{3}R1$$

$$\begin{bmatrix} 1 & -\dfrac{2}{3} & \bigm| & \dfrac{13}{3} \\ 0 & \dfrac{5}{3} & \bigm| & \dfrac{5}{3} \end{bmatrix} \quad -4R1 + R2$$

$$\begin{bmatrix} 1 & -\dfrac{2}{3} & \bigm| & \dfrac{13}{3} \\ 0 & 1 & \bigm| & 1 \end{bmatrix} \quad \dfrac{3}{5}R2$$

$$\begin{bmatrix} 1 & 0 & \bigm| & 5 \\ 0 & 1 & \bigm| & 1 \end{bmatrix} \quad \dfrac{2}{3}R2 + R1$$

Solution set: $\{(5, 1)\}$

14. $2x + 3y - 6z = 1 \quad (1)$
     $x - y + 2z = 3 \quad (2)$
     $4x + y - 2z = 7 \quad (3)$

$$\begin{bmatrix} 2 & 3 & -6 & \bigm| & 1 \\ 1 & -1 & 2 & \bigm| & 3 \\ 4 & 1 & -2 & \bigm| & 7 \end{bmatrix}$$

$$\begin{bmatrix} 1 & \dfrac{3}{2} & -3 & \bigm| & \dfrac{1}{2} \\ 1 & -1 & 2 & \bigm| & 3 \\ 4 & 1 & -2 & \bigm| & 7 \end{bmatrix} \quad \dfrac{1}{2}R1$$

$$\begin{bmatrix} 1 & \dfrac{3}{2} & -3 & \bigm| & \dfrac{1}{2} \\ 0 & -\dfrac{5}{2} & 5 & \bigm| & \dfrac{5}{2} \\ 0 & -5 & 10 & \bigm| & 5 \end{bmatrix} \quad \begin{matrix} \\ -1R1 + R2 \\ -4R1 + R3 \end{matrix}$$

$$\begin{bmatrix} 1 & \dfrac{3}{2} & -3 & \bigm| & \dfrac{1}{2} \\ 0 & 1 & -2 & \bigm| & -1 \\ 0 & -5 & 10 & \bigm| & 5 \end{bmatrix} \quad -\dfrac{2}{5}R2$$

$$\begin{bmatrix} 1 & \dfrac{3}{2} & -3 & \bigm| & 1 \\ 0 & 1 & -2 & \bigm| & -1 \\ 0 & 0 & 0 & \bigm| & 0 \end{bmatrix} \quad 5R2 + R3$$

Since the third row contains only zeros, which indicates a system with dependent equations, the method cannot be coninued. We will write the solution with z arbitrary.

The second line of the last matrix corresponds to the equation.

$$y - 2z = -1.$$

Solve for y.

$$y = 2z - 1$$

Now substitute this expression in equation (1) and solve for x.

$$2x + 3(2z - 1) - 6z = 1$$
$$2x + 6z - 3 - 6z = 1$$
$$2x = 4$$
$$x = 2$$

The solution set, with z arbitrary, is $\{(2, 2z - 1, z)\}$.

15. $2x^2 + y^2 = 6$ (1)
    $x^2 - 4y^2 = -15$ (2)

    $8x^2 + 4y^2 = 24$ *Multiply (1) by 4*
    $\underline{x^2 - 4y^2 = -15}$ (2)
    $9x^2 \qquad = 9$
    $\qquad x^2 = 1$
    $\qquad x = \pm 1$

Substitute these values into equation (1) and solve for y.

If x = 1, $\quad 2(1)^2 + y^2 = 6$
$$2 + y^2 = 6$$
$$y^2 = 4$$
$$y = \pm 2.$$

Thus, (1, 2) and (1, -2) are solutions.

If x = -1, $\quad 2(-1)^2 + y^2 = 6$
$$2 + y^2 = 6$$
$$y^2 = 4$$
$$y = \pm 2.$$

Thus, (-1, 2) and (-1, -2) are solutions.

Solution set:

$$\{(1, 2), (-1, 2), (1, -2), (-1, -2)\}$$

16. $x^2 + y^2 = 25$ (1)
    $x + y = 7$ (2)

Solve equation (2) for x.

$$x = 7 - y$$

Substitute this result into equation (1).

$$(7 - y)^2 + y^2 = 25$$
$$49 - 14y + y^2 + y^2 = 25$$
$$2y^2 - 14y + 24 = 0$$
$$y^2 - 7y + 12 = 0$$
$$(y - 3)(y - 4) = 0$$
$$y = 3 \quad \text{or} \quad y = 4$$

If y = 3, x = 7 - 3 = 4.
If y = 4, x = 7 - 4 = 3.

Solution set: $\{(3, 4), (4, 3)\}$

17. The system will have exactly one solution if the line is tangent to the circle. See the answer sketch in the textbook.

18. Let x and y represent the numbers.

    $x + y = -1$ (1)
    $x^2 + y^2 = 61$ (2)

Rewrite equation (1) as y = -x - 1 and substitute -x - 1 for y in (2).

$$x^2 + (-x - 1)^2 = 61$$
$$x^2 + x^2 + 2x + 1 = 61$$
$$2x^2 + 2x - 60 = 0$$
$$x^2 + x - 30 = 0$$
$$(x + 6)(x - 5) = 0$$
$$x = -6 \quad \text{or} \quad x = 5$$

Substitute these values in (1) to find the corresponding values of y.

If x = -6, -6 + y = -1 or y = 5.

If x = 5, 5 + y = -1 or y = -6.

The same pair of numbers results from both cases.

The numbers are 5 and -6.

19. $9x^2 + 4y^2 \geq 36$

$y < x^2$

The boundary for the first curve has equation

$$9x^2 + 4y^2 = 36$$

or $$\frac{x^2}{4} + \frac{y^2}{9} = 1.$$

The graph of this equation is an ellipse centered at the origin with x-intercepts -2 and 2 and y-intercepts -3 and 3. Draw this ellipse as a solid curve and shade the region outside the ellipse.

The boundary for the second curve has equation

$$y = x^2.$$

Graph this parabola as a dotted curve and shade the region below the parabola.

The solution set for the system is the region common to the two shaded regions. This region is shaded in the final graph.

See the answer graph in the textbook.

20. Let x = the number of servings of A;

y = the number of servings of B.

Find $x \geq 0$ and $y \geq 0$ such that

$$3x + 2y \geq 15$$
$$2x + 4y \geq 15$$

and $.25x + .40y$ is minimized.

Graph the lines $x = 0$, $y = 0$,

$3x + 2y = 15$, and $2x + 4y = 15$.

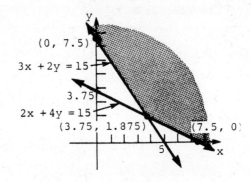

The vertices are $\left(0, 7\frac{1}{2}\right)$, $\left(3\frac{3}{4}, 1\frac{7}{8}\right)$,

and $\left(7\frac{1}{2}, 0\right)$.

Point	Value = .25x + .40y
$\left(0, 7\frac{1}{2}\right)$	$.25(0) + .40\left(7\frac{1}{2}\right) = 3.00$
$\left(3\frac{3}{4}, 1\frac{7}{8}\right)$	$.25\left(3\frac{3}{4}\right) + .40\left(1\frac{7}{8}\right) \approx 1.69$
$\left(7\frac{1}{2}, 0\right)$	$.25\left(7\frac{1}{2}\right) + .40(0) \approx 1.88$

Gwen can satisfy her daily requirements for a minimum cost of $1.69 by using $3\frac{3}{4}$ servings of Product A and $1\frac{7}{8}$ servings of Product B.

CHAPTER 8   MATRICES

Section 8.1

1. A 3×8 matrix has eight columns and three rows.

3. $\begin{bmatrix} -9 & 6 & 2 \\ 4 & 1 & 8 \end{bmatrix}$

   This matrix has 2 rows and 3 columns, so it is a 2×3 matrix.

5. $[8 \quad -2 \quad 4 \quad 6 \quad 3]$

   This matrix has 1 row and 5 columns, so it is a 1×5 row matrix.

7. $[-9]$

   This matrix has 1 row and 1 column, so it is a 1×1 square matrix. It is also a row matrix since it has only one row, and a column matrix because it has only one column.

9. $\begin{bmatrix} -4 & 2 \\ 3 & 5 \end{bmatrix}$

   This matrix has 2 rows and 2 columns, so it is a 2×2 square matrix.

11. $\begin{bmatrix} -5 \\ y \end{bmatrix} = \begin{bmatrix} -5 \\ 8 \end{bmatrix}$

    Two matrices can be equal only if they are the same size and corresponding elements are equal. These matrices will be equal if $y = 8$.

13. $\begin{bmatrix} 9 & 7 \\ r & 0 \end{bmatrix} = \begin{bmatrix} m - 3 & n + 5 \\ 8 & 0 \end{bmatrix}$

    The matrices are the same size, so they will be equal if corresponding elements are equal.

    $\begin{array}{ll} 9 = m - 3 & 7 = n + 5 \\ 12 = m & 2 = n \end{array}$

    $r = 8$

    Thus, $m = 12$, $n = 2$, and $r = 8$.

15. $[8 \quad p + 9 \quad q + 5] + [9 \quad -3 \quad 12]$
    $\quad = [k - 2 \quad 12 \quad 2q]$

    $[17 \quad p + 6 \quad q + 17] = [k - 2 \quad 12 \quad 2q]$

    $\begin{array}{ll} 17 = k - 2 & p + 6 = 12 \\ 19 = k & p = 6 \end{array}$

    $q + 17 = 2q$
    $\quad 17 = q$

    Thus, $k = 19$, $p = 6$, and $q = 17$.

17. $\begin{bmatrix} a + 2 & 3z + 1 & 5m \\ 4k & 0 & 3 \end{bmatrix} + \begin{bmatrix} 3a & 2z & 5m \\ 2k & 5 & 6 \end{bmatrix}$

    $= \begin{bmatrix} 10 & -14 & 80 \\ 10 & 5 & 9 \end{bmatrix}$

    $\begin{bmatrix} 4a + 2 & 5z + 1 & 10m \\ 6k & 5 & 9 \end{bmatrix}$

    $= \begin{bmatrix} 10 & -14 & 80 \\ 10 & 5 & 9 \end{bmatrix}$

    $\begin{array}{ll} 4a + 2 = 10 & 5z + 1 = -14 \\ a = 2 & z = -3 \end{array}$

    $\begin{array}{ll} 10m = 80 & 6k = 10 \\ m = 8 & k = \dfrac{5}{3} \end{array}$

    Thus, $a = 2$, $z = -3$, $m = 8$, and $k = \dfrac{5}{3}$.

21. $3\begin{bmatrix} 6 & -1 & 4 \\ 2 & 8 & -3 \\ -4 & 5 & 6 \end{bmatrix} + 5\begin{bmatrix} -2 & -8 & -6 \\ 4 & 1 & 3 \\ 2 & -1 & 5 \end{bmatrix}$

$\quad = \begin{bmatrix} 18 & -3 & 12 \\ 6 & 24 & -9 \\ -12 & 15 & 18 \end{bmatrix} + \begin{bmatrix} -10 & -40 & -30 \\ 20 & 5 & 15 \\ 10 & -5 & 25 \end{bmatrix}$

$\quad = \begin{bmatrix} 8 & -43 & -18 \\ 26 & 29 & 6 \\ -2 & 10 & 43 \end{bmatrix}$

23. $\begin{bmatrix} -8 & 4 & 0 \\ 2 & 5 & 0 \end{bmatrix} + \begin{bmatrix} 6 & 3 \\ 8 & 9 \end{bmatrix}$

This operation is not possible because the matrices are different sizes. Only matrices of the same size can be added.

25. $\begin{bmatrix} 9 & 4 & 1 & -2 \\ 5 & -6 & 3 & 4 \\ 2 & -5 & 1 & 2 \end{bmatrix} - \begin{bmatrix} -2 & 5 & 1 & 3 \\ 0 & 1 & 0 & 2 \\ -8 & 3 & 2 & 1 \end{bmatrix}$

$\quad + \begin{bmatrix} 2 & 4 & 0 & 3 \\ 4 & -5 & 1 & 6 \\ 2 & -3 & 0 & 8 \end{bmatrix}$

$\quad = \begin{bmatrix} 9 & 4 & 1 & -2 \\ 5 & -6 & 3 & 4 \\ 2 & -5 & 1 & 2 \end{bmatrix} + \begin{bmatrix} 2 & -5 & -1 & -3 \\ 0 & -1 & 0 & -2 \\ 8 & -3 & -2 & -1 \end{bmatrix}$

$\quad + \begin{bmatrix} 2 & 4 & 0 & 3 \\ 4 & -5 & 1 & 6 \\ 2 & -3 & 0 & 8 \end{bmatrix}$

$\quad = \begin{bmatrix} 13 & 3 & 0 & -2 \\ 9 & -12 & 4 & 8 \\ 12 & -11 & -1 & 9 \end{bmatrix}$

27. $\begin{bmatrix} -4x + 2y & -3x + y \\ 6x - 3y & 2x - 5y \end{bmatrix}$

$\quad + \begin{bmatrix} -8x + 6y & 2x \\ 3y - 5x & 6x + 4y \end{bmatrix}$

$\quad = \begin{bmatrix} -12x + 8y & -x + y \\ x & 8x - y \end{bmatrix}$

In Exercises 29–35,

$\qquad A = \begin{bmatrix} -2 & 4 \\ 0 & 3 \end{bmatrix}$ and $B = \begin{bmatrix} -6 & 2 \\ 4 & 0 \end{bmatrix}$.

29. $2A = 2\begin{bmatrix} -2 & 4 \\ 0 & 3 \end{bmatrix} = \begin{bmatrix} -4 & 8 \\ 0 & 6 \end{bmatrix}$

31. $-4A = -4\begin{bmatrix} -2 & 4 \\ 0 & 3 \end{bmatrix} = \begin{bmatrix} 8 & -16 \\ 0 & -12 \end{bmatrix}$

33. $2A - B = 2\begin{bmatrix} -2 & 4 \\ 0 & 3 \end{bmatrix} - \begin{bmatrix} -6 & 2 \\ 4 & 0 \end{bmatrix}$

$\quad = \begin{bmatrix} -4 & 8 \\ 0 & 6 \end{bmatrix} + \begin{bmatrix} 6 & -2 \\ -4 & 0 \end{bmatrix}$

$\quad = \begin{bmatrix} 2 & 6 \\ -4 & 6 \end{bmatrix}$

35. $3A - 11B = 3\begin{bmatrix} -2 & 4 \\ 0 & 3 \end{bmatrix} - 11\begin{bmatrix} -6 & 2 \\ 4 & 0 \end{bmatrix}$

$\quad = \begin{bmatrix} -6 & 12 \\ 0 & 9 \end{bmatrix} + \begin{bmatrix} 66 & -22 \\ -44 & 0 \end{bmatrix}$

$\quad = \begin{bmatrix} 60 & -10 \\ -44 & 9 \end{bmatrix}$

37. The given information may be written as the 3 × 2 matrix

$\qquad \begin{bmatrix} 7 & 2 \\ 9 & 0 \\ 8 & 6 \end{bmatrix},$

or as the 2 × 3 matrix.

$\qquad \begin{bmatrix} 7 & 9 & 8 \\ 2 & 0 & 6 \end{bmatrix}.$

39. The given information may be written as the 2 × 2 matrix

$$\begin{bmatrix} 5411 & 11,352 \\ 9371 & 15,956 \end{bmatrix},$$

or as the 2 × 2 matrix

$$\begin{bmatrix} 5411 & 9371 \\ 11,352 & 15,956 \end{bmatrix}.$$

41. $A + B = \begin{bmatrix} a + e & b + f \\ c + g & d + h \end{bmatrix}$

$B + A = \begin{bmatrix} e + a & f + b \\ g + c & h + d \end{bmatrix}$

$\quad = \begin{bmatrix} a + e & b + f \\ c + g & d + h \end{bmatrix}$

Commutative property of addition of real numbers

Therefore, A + B = B + A (commutative property) is true for the addition of 2 × 2 matrices.

It is also true for the addition of any two matrices of the same size.

43. Is there a matrix *0* such that
A + *0* = A and *0* + A = A for a
matrix $A = \begin{bmatrix} a & b \\ c & d \end{bmatrix}$?

Yes, there is such a matrix;
consider $\begin{bmatrix} 0 & 0 \\ 0 & 0 \end{bmatrix}$.

$\begin{bmatrix} a & b \\ c & d \end{bmatrix} + \begin{bmatrix} 0 & 0 \\ 0 & 0 \end{bmatrix} = \begin{bmatrix} a & b \\ c & d \end{bmatrix}$ and

$\begin{bmatrix} 0 & 0 \\ 0 & 0 \end{bmatrix} + \begin{bmatrix} a & b \\ c & d \end{bmatrix} = \begin{bmatrix} a & b \\ c & d \end{bmatrix}$

Moreover, there is a similar matrix *0* of any size, composed of zeros in every position.

45. Yes, all of these properties are valid for matrices that are not square.

Section 8.2

1. A is 2 × 2, B is 2 × 2

The product AB exists; its size is 2 × 2.
The product BA exists; its size is is also 2 × 2.

3. A is 4 × 2, B is 2 × 4

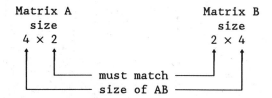

The product AB exists; its size is 4 × 4.

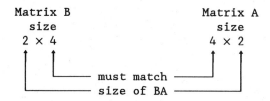

The product BA also exists; its size is 2 × 2.

5. A is 3 × 5, B is 5 × 2

The product AB exists; its size is 3 × 2.
The product BA cannot be found, since the number of columns of B (two), is not the same as the number of rows of A (three).

7. A is 4×2, B is 3×4

    The product AB cannot be found, since the number of columns of A is not the same as the number of rows of B.
    The product BA exists; its size is 3×2.

9. A is 4×3, B is 2×5

    AB cannot be found because the number of columns of A does not equal the number of rows of B.
    BA cannot be found because the number of columns of B does not equal the number of rows of A.

11. The product MN of two matrices can be found only if the number of columns of M equals the number of rows of N.

13. To find the product AB of matrices A and B, the first row, second column entry is found by multiplying the first row elements in A and the second column elements in B and then adding these products.

15. $\begin{bmatrix} -1 & 5 \\ 7 & 0 \end{bmatrix} \begin{bmatrix} 6 \\ 2 \end{bmatrix} = \begin{bmatrix} -1(6) + 5(2) \\ 7(6) + 0(2) \end{bmatrix} = \begin{bmatrix} 4 \\ 42 \end{bmatrix}$

17. $\begin{bmatrix} -6 & 3 & 5 \\ 2 & 9 & 1 \end{bmatrix} \begin{bmatrix} -2 \\ 0 \\ 3 \end{bmatrix} = \begin{bmatrix} -6(-2) + 3(0) + 5(3) \\ 2(-2) + 9(0) + 1(3) \end{bmatrix} = \begin{bmatrix} 27 \\ -1 \end{bmatrix}$

19. $\begin{bmatrix} -4 & 0 \\ 1 & 3 \end{bmatrix} \begin{bmatrix} -2 & 4 \\ 0 & 1 \end{bmatrix} = \begin{bmatrix} -4(-2) + 0(0) & -4(4) + 0(1) \\ 1(-2) + 3(0) & 1(4) + 3(1) \end{bmatrix} = \begin{bmatrix} 8 & -16 \\ -2 & 7 \end{bmatrix}$

21. $\begin{bmatrix} -9 & 2 & 1 \\ 3 & 0 & 0 \end{bmatrix} \begin{bmatrix} 2 \\ -1 \\ 4 \end{bmatrix} = \begin{bmatrix} -9(2) + 2(-1) + 1(4) \\ 3(2) + 0(-1) + 0(4) \end{bmatrix} = \begin{bmatrix} -16 \\ 6 \end{bmatrix}$

23. $\begin{bmatrix} -2 & -3 & -4 \\ 2 & -1 & 0 \\ 4 & -2 & 3 \end{bmatrix} \begin{bmatrix} 0 & 1 & 4 \\ 1 & 2 & -1 \\ 3 & 2 & -2 \end{bmatrix}$

$\begin{bmatrix} -2(0)+(-3)(1)+(-4)(-3) & -2(1)+(-3)(2)+(-4)(2) & -2(4)+(-3)(-1)+(-4)(-2) \\ 2(0)+(-1)(1)+0(3) & 2(1)+(-1)(2)+0(2) & 2(4)+(-1)(-1)+0(-2) \\ 4(0)+(-2)(1)+3(3) & 4(1)+(-2)(2)+3(2) & 4(4)+(-2)(-1)+3(-2) \end{bmatrix}$

$\begin{bmatrix} -15 & -16 & 3 \\ -1 & 0 & 9 \\ 7 & 6 & 12 \end{bmatrix}$

25. $\begin{bmatrix} 0 & 3 & -4 \end{bmatrix} \begin{bmatrix} -2 & 6 & 3 \\ 0 & 4 & 2 \\ -1 & 1 & 4 \end{bmatrix}$

$= \begin{bmatrix} 0(-2)+3(0)+(-4)(-1) & 0(6)+3(4)+(-4)(1) & 0(3)+3(2)+(-4)(4) \end{bmatrix}$

$= \begin{bmatrix} 4 & 8 & -10 \end{bmatrix}$

27. $\begin{bmatrix} -1 & 0 & 0 \\ 2 & 1 & 4 \end{bmatrix} \begin{bmatrix} 4 & -2 & 5 \\ 0 & 1 & 4 \\ 2 & -9 & 0 \end{bmatrix}$

$= \begin{bmatrix} -1(4)+0(0)+0(2) & -1(-2)+0(1)+0(-9) & -1(5)+0(4)+0(0) \\ 2(4)+1(0)+4(2) & 2(-2)+1(1)+4(-9) & 2(5)+1(4)+4(0) \end{bmatrix}$

$= \begin{bmatrix} -4 & 2 & -5 \\ 16 & -39 & 14 \end{bmatrix}$

29. $\begin{bmatrix} -1 & 2 & 4 & 1 \\ 0 & 2 & -3 & 5 \end{bmatrix} \begin{bmatrix} 1 & 2 & 4 \\ -2 & 5 & 1 \end{bmatrix}$

It is not possible to find this product because the number of columns of the first matrix (four) is not equal to the number of rows of the second matrix (two).

31. $[4 \quad 0 \quad 2]\begin{bmatrix} -5 \\ 1 \\ 6 \end{bmatrix}$

$= [4(-5) + 0(-1) + 2(6)]$

$= [-8]$

33. $\begin{bmatrix} -5 \\ 1 \\ 6 \end{bmatrix}[4 \quad 0 \quad 2]$

$= \begin{bmatrix} -5(4) & -5(0) & -5(2) \\ 1(4) & 1(0) & 1(2) \\ 6(4) & 6(0) & 6(2) \end{bmatrix}$

$= \begin{bmatrix} -20 & 0 & -10 \\ 4 & 0 & 2 \\ 24 & 0 & 12 \end{bmatrix}$

35. $BA = \begin{bmatrix} -2 & 1 \\ 3 & 6 \end{bmatrix}\begin{bmatrix} -2 & 4 \\ 1 & 3 \end{bmatrix}$

$= \begin{bmatrix} -2(-2) + 1(1) & -2(4) + 1(3) \\ 3(-2) + 6(1) & 3(4) + 6(3) \end{bmatrix}$

$= \begin{bmatrix} 5 & -5 \\ 0 & 30 \end{bmatrix}$

37. $CA = \begin{bmatrix} 5 & -2 & 1 \\ 0 & 3 & 7 \end{bmatrix}\begin{bmatrix} -2 & 4 \\ 1 & 3 \end{bmatrix}$

It is not possible to find this product because the number of columns of C (three) is not equal to the number of rows of A (two).

39. If P has size m × n and Q has size n × m, then PQ and QP both exist.

For example, if the size of P is 2 × 4 and the size of Q is 4 × 2, both products would exist: PQ would be a 2 × 2 matrix and QP would be 4 × 4 a matrix. Multiplying two

n × n matrices (square matrices of the same size) is a special case of the condition.

41.

A(B + C)

$= \begin{bmatrix} a & b \\ c & d \end{bmatrix}\left(\begin{bmatrix} e & f \\ g & h \end{bmatrix} + \begin{bmatrix} j & m \\ k & n \end{bmatrix}\right)$

$= \begin{bmatrix} a & b \\ c & d \end{bmatrix}\begin{bmatrix} e + j & f + m \\ g + k & h + n \end{bmatrix}$

$= \begin{bmatrix} a(e+j) + b(g+k) & a(f+m) + b(h+n) \\ c(e+j) + d(g+k) & c(f+m) + d(h+n) \end{bmatrix}$

$= \begin{bmatrix} ae + aj + bg + bk & af + am + bh + bn \\ ce + cj + dg + dk & cf + cm + dh + dn \end{bmatrix}$

AB + AC

$= \begin{bmatrix} a & b \\ c & d \end{bmatrix}\begin{bmatrix} e & f \\ g & h \end{bmatrix} + \begin{bmatrix} a & b \\ c & d \end{bmatrix}\begin{bmatrix} j & m \\ k & n \end{bmatrix}$

$= \begin{bmatrix} ae + bg & af + bh \\ ce + dg & cf + dh \end{bmatrix} + \begin{bmatrix} aj + bk & am + bn \\ cj + dk & cm + dn \end{bmatrix}$

$= \begin{bmatrix} ae + aj + bg + bk & af + am + bh + bn \\ ce + cj + dg + dk & cf + cm + dh + dn \end{bmatrix}$

Observe that A(B + C) = AB + AC for these three matrices A, B, and C. A similar property holds for any square matrices of the same size.

43. (k + p)A = kA + pA for any real numbers k and p

$(k + p)A = (k + p)\begin{bmatrix} a & b \\ c & d \end{bmatrix}$

$= \begin{bmatrix} (k+p)a & (k+p)b \\ (k+p)c & (k+p)d \end{bmatrix}$

$= \begin{bmatrix} ka + pa & kb + pb \\ kc + pc & kd + pd \end{bmatrix}$

Distributive property for real numbers

$$kA + pA = k\begin{bmatrix} a & b \\ c & d \end{bmatrix} + p\begin{bmatrix} a & b \\ c & d \end{bmatrix}$$

$$= \begin{bmatrix} ka & kb \\ kc & kd \end{bmatrix} + \begin{bmatrix} pa & pb \\ pc & pd \end{bmatrix}$$

$$= \begin{bmatrix} ka + pa & kb + pb \\ kc + pc & kd + pd \end{bmatrix}$$

Thus, $(k + p)A$, where k and p are any real numbers, is true when A is a 2 × 2 matrix. This property will also be true if A is any square matrix.

45. (a) The sales figure information may be written as the 3 × 3 matrix

$$\begin{bmatrix} 50 & 100 & 30 \\ 10 & 90 & 50 \\ 60 & 120 & 40 \end{bmatrix}.$$

(b) The income per gallon information may be written as the 3 × 1

matrix $\begin{bmatrix} 12 \\ 10 \\ 15 \end{bmatrix}.$

(If the matrix in part (a) had been written with its rows and columns interchanged, then this income per gallon information would be written instead as a 1 × 3 matrix.)

(c) $\begin{bmatrix} 50 & 100 & 30 \\ 10 & 90 & 50 \\ 60 & 120 & 40 \end{bmatrix}\begin{bmatrix} 12 \\ 10 \\ 15 \end{bmatrix} = \begin{bmatrix} 2050 \\ 1770 \\ 2520 \end{bmatrix}$

(This result may be written as a 1 × 3 matrix instead.)

(d) 2050 + 1770 + 2520 = 6340

The total daily income from the three locations is $6340.

Section 8.3

1. When an n × n matrix A is multiplied by the n × n identity matrix, the product is the matrix A.

3. $\begin{bmatrix} 2 & 3 \\ 1 & 1 \end{bmatrix}\begin{bmatrix} -1 & 3 \\ 1 & -2 \end{bmatrix}$

$$= \begin{bmatrix} 2(-1) + 3(1) & 2(3) + 3(-2) \\ 1(-1) + 1(1) & 1(3) + 1(-2) \end{bmatrix}$$

$$= \begin{bmatrix} 1 & 0 \\ 0 & 1 \end{bmatrix}$$

$$\begin{bmatrix} -1 & 3 \\ 1 & -2 \end{bmatrix}\begin{bmatrix} 2 & 3 \\ 1 & 1 \end{bmatrix}$$

$$= \begin{bmatrix} -1(2) + 3(1) & -1(3) + 3(1) \\ 1(2) + (-2)(1) & 1(3) + (-2)(1) \end{bmatrix}$$

$$= \begin{bmatrix} 1 & 0 \\ 0 & 1 \end{bmatrix}$$

Since the products obtained by multiplying the matrices in either order are both the 2 × 2 identity matrix, the given matrices are inverses of each other.

5. $\begin{bmatrix} 2 & 1 \\ 3 & 2 \end{bmatrix}\begin{bmatrix} 2 & 1 \\ -3 & 2 \end{bmatrix} = \begin{bmatrix} 1 & 4 \\ 0 & 7 \end{bmatrix}$

Since this product is not the 2 × 2 identity matrix, the given matrices are not inverses of each other.

7. $\begin{bmatrix} 1 & -2 & -3 \\ 2 & -2 & -5 \\ -1 & 1 & 4 \end{bmatrix}\begin{bmatrix} -1 & \frac{5}{3} & \frac{4}{3} \\ -1 & \frac{1}{3} & -\frac{1}{3} \\ 0 & \frac{1}{3} & \frac{2}{3} \end{bmatrix}$

$$= \begin{bmatrix} 1 & 0 & 0 \\ 0 & 1 & 0 \\ 0 & 0 & 1 \end{bmatrix}$$

Multiplication of the two given matrices in the opposite order will also result in the 3×3 identity matrix.

Therefore, the given matrices are inverses of each other.

9.
$$\begin{bmatrix} 1 & 2 & -1 \\ 0 & 1 & 3 \\ 2 & 1 & -2 \end{bmatrix} \begin{bmatrix} 1 & 1 & 2 \\ 1 & 1 & 1 \\ 2 & 3 & 4 \end{bmatrix}$$

$$= \begin{bmatrix} 1 & 0 & 0 \\ 7 & 10 & 13 \\ -1 & -3 & -3 \end{bmatrix}$$

This product is not the 3×3 identity matrix, so the given matrices are not inverses of each other.

11.   Find the inverse of $A = \begin{bmatrix} 1 & -1 \\ 2 & 0 \end{bmatrix}$, if it exists.

Form the augmented matrix $[A|I]$.

$$[A|I] = \begin{bmatrix} 1 & -2 & | & 1 & 0 \\ 2 & 0 & | & 0 & 1 \end{bmatrix}$$

Perform row transformations on $[A|I]$ to get a matrix of the form $[I|B]$.

$$\begin{bmatrix} 1 & -1 & | & 1 & 0 \\ 0 & 2 & | & -2 & 1 \end{bmatrix} \quad -2R1 + R2$$

$$\begin{bmatrix} 1 & -1 & | & 1 & 0 \\ 0 & 1 & | & -1 & \frac{1}{2} \end{bmatrix} \quad \frac{1}{2}R2$$

$$\begin{bmatrix} 1 & 0 & | & 0 & \frac{1}{2} \\ 0 & 1 & | & -1 & \frac{1}{2} \end{bmatrix} \quad R2 + R1$$

$$A^{-1} = B = \begin{bmatrix} 0 & \frac{1}{2} \\ -1 & \frac{1}{2} \end{bmatrix}$$

13.   Find the inverse of $A = \begin{bmatrix} -6 & 4 \\ -3 & 2 \end{bmatrix}$, if it exists.

$$[A|I] = \begin{bmatrix} -6 & 4 & | & 1 & 0 \\ -3 & 2 & | & 0 & 1 \end{bmatrix}$$

$$\begin{bmatrix} 1 & -\frac{2}{3} & | & -\frac{1}{6} & 0 \\ -3 & 2 & | & 0 & 1 \end{bmatrix} \quad -\frac{1}{6}R1$$

$$\begin{bmatrix} 1 & -\frac{2}{3} & | & -\frac{1}{6} & 0 \\ 0 & 0 & | & -\frac{1}{2} & 1 \end{bmatrix} \quad 3R1 + R2$$

At this point, the matrix should be changed so that the second-row, second-column element will be 1. Since that element is now 0, the desired transformation cannot be completed.

Therefore, the inverse of the given matrix does not exist.

15.   Find the inverse of $A = \begin{bmatrix} -1 & -2 \\ 3 & 4 \end{bmatrix}$, if it exists.

$$[A|I] = \begin{bmatrix} -1 & -2 & | & 1 & 0 \\ 3 & 4 & | & 0 & 1 \end{bmatrix}$$

$$\begin{bmatrix} -1 & -2 & | & 1 & 0 \\ 0 & -2 & | & 3 & 1 \end{bmatrix} \quad 3R1 + R2$$

$$\begin{bmatrix} 1 & 2 & | & -1 & 0 \\ 0 & -2 & | & 3 & 1 \end{bmatrix} \quad -1R1$$

$$\begin{bmatrix} 1 & 0 & | & 2 & 1 \\ 0 & -2 & | & 3 & 1 \end{bmatrix} \quad R2 + R1$$

$$\begin{bmatrix} 1 & 0 & | & 2 & 1 \\ 0 & 1 & | & -\frac{3}{2} & -\frac{1}{2} \end{bmatrix} \quad -\frac{1}{2}R2$$

$$A^{-1} = \begin{bmatrix} 2 & 1 \\ -\frac{3}{2} & -\frac{1}{2} \end{bmatrix}$$

17. Find the inverse of $A = \begin{bmatrix} .6 & .2 \\ .5 & .1 \end{bmatrix}$, if it exists.

$[A|I] = \begin{bmatrix} .6 & .2 & | & 1 & 0 \\ .5 & .1 & | & 0 & 1 \end{bmatrix}$

First, multiply each row by 10 to eliminate decimals.

$\begin{bmatrix} 6 & 2 & | & 10 & 0 \\ 5 & 1 & | & 0 & 10 \end{bmatrix}$ 10R1
 10R2

$\begin{bmatrix} 1 & 1 & | & 10 & -10 \\ 5 & 1 & | & 0 & 10 \end{bmatrix}$ -1R2 + R1

$\begin{bmatrix} 1 & 1 & | & 10 & -10 \\ 0 & -4 & | & -50 & 60 \end{bmatrix}$ -5R1 + R2

$\begin{bmatrix} 1 & 1 & | & 10 & -10 \\ 0 & 1 & | & 12.5 & -15 \end{bmatrix}$ $-.25$R2

$\begin{bmatrix} 1 & 0 & | & -2.5 & 5 \\ 0 & 1 & | & 12.5 & -15 \end{bmatrix}$ -1R2 + R1

$A^{-1} = \begin{bmatrix} -2.5 & 5 \\ 12.5 & -15 \end{bmatrix}$

19. Find the inverse of $A = \begin{bmatrix} 1 & 0 & 0 \\ 0 & -1 & 0 \\ 1 & 0 & 1 \end{bmatrix}$, if it exists.

$[A|I] = \begin{bmatrix} 1 & 0 & 0 & | & 1 & 0 & 0 \\ 0 & -1 & 0 & | & 0 & 1 & 0 \\ 1 & 0 & 1 & | & 0 & 0 & 1 \end{bmatrix}$

$\begin{bmatrix} 1 & 0 & 0 & | & 1 & 0 & 0 \\ 0 & -1 & 0 & | & 0 & 1 & 0 \\ 0 & 0 & 1 & | & -1 & 0 & 1 \end{bmatrix}$ -1R1 + R3

$\begin{bmatrix} 1 & 0 & 0 & | & 1 & 0 & 0 \\ 0 & 1 & 0 & | & 0 & -1 & 0 \\ 1 & 0 & 1 & | & -1 & 0 & 1 \end{bmatrix}$ -1R2

$A^{-1} = \begin{bmatrix} 1 & 0 & 0 \\ 0 & -1 & 0 \\ -1 & 0 & 1 \end{bmatrix}$

21. Find the inverse of $A = \begin{bmatrix} 3 & 6 & 3 \\ 6 & 4 & -2 \\ 0 & 1 & -1 \end{bmatrix}$, if it exists.

$[A|I] = \begin{bmatrix} 3 & 6 & 3 & | & 1 & 0 & 0 \\ 6 & 4 & -2 & | & 0 & 1 & 0 \\ 0 & 1 & -1 & | & 0 & 0 & 1 \end{bmatrix}$

$\begin{bmatrix} 1 & 2 & 1 & | & \frac{1}{3} & 0 & 0 \\ 6 & 4 & -2 & | & 0 & 1 & 0 \\ 0 & 1 & -1 & | & 0 & 0 & 1 \end{bmatrix}$ $\frac{1}{3}$R1

$\begin{bmatrix} 1 & 2 & 1 & | & \frac{1}{3} & 0 & 0 \\ 0 & -8 & -8 & | & -2 & 1 & 0 \\ 0 & 1 & -1 & | & 0 & 0 & 1 \end{bmatrix}$ -6R1 + R2

$\begin{bmatrix} 1 & 0 & 3 & | & \frac{1}{3} & 0 & -2 \\ 0 & -8 & -8 & | & -2 & 1 & 0 \\ 0 & 1 & -1 & | & 0 & 0 & 1 \end{bmatrix}$ -2R3 + R1

$\begin{bmatrix} 1 & 0 & 3 & | & \frac{1}{3} & 0 & -2 \\ 0 & 1 & 1 & | & \frac{1}{4} & -\frac{1}{8} & 0 \\ 0 & 1 & -1 & | & 0 & 0 & 1 \end{bmatrix}$ $-\frac{1}{8}$R2

$\begin{bmatrix} 1 & 0 & 3 & | & \frac{1}{3} & 0 & -2 \\ 0 & 1 & 1 & | & \frac{1}{4} & -\frac{1}{8} & 0 \\ 0 & 0 & -2 & | & -\frac{1}{4} & \frac{1}{8} & 1 \end{bmatrix}$ -1R2 + R3

$\begin{bmatrix} 1 & 0 & 3 & | & \frac{1}{3} & 0 & -2 \\ 0 & 1 & 1 & | & \frac{1}{4} & -\frac{1}{8} & 0 \\ 0 & 0 & 1 & | & \frac{1}{8} & -\frac{1}{16} & -\frac{1}{2} \end{bmatrix}$ $-\frac{1}{2}$R3

$\begin{bmatrix} 1 & 0 & 0 & | & -\frac{1}{24} & \frac{3}{16} & -\frac{1}{2} \\ 0 & 1 & 1 & | & \frac{1}{4} & -\frac{1}{8} & 0 \\ 0 & 0 & 1 & | & \frac{1}{8} & -\frac{1}{16} & -\frac{1}{2} \end{bmatrix}$ -3R3 + R1

$$\left[\begin{array}{ccc|ccc} 1 & 0 & 0 & -\dfrac{1}{24} & \dfrac{3}{16} & -\dfrac{1}{2} \\ 0 & 1 & 0 & \dfrac{1}{8} & -\dfrac{1}{16} & \dfrac{1}{2} \\ 0 & 0 & 1 & \dfrac{1}{8} & -\dfrac{1}{16} & -\dfrac{1}{2} \end{array}\right] \quad -1R3 + R2$$

$$A^{-1} = \left[\begin{array}{ccc} -\dfrac{1}{24} & \dfrac{3}{16} & -\dfrac{1}{2} \\ \dfrac{1}{8} & -\dfrac{1}{16} & \dfrac{1}{2} \\ \dfrac{1}{8} & -\dfrac{1}{16} & -\dfrac{1}{2} \end{array}\right]$$

23. Find the inverse of $A = \begin{bmatrix} -1 & -1 & -1 \\ 4 & 5 & 0 \\ 0 & 1 & -3 \end{bmatrix}$,

if it exists.

$$[A|I] = \left[\begin{array}{ccc|ccc} -1 & -1 & -1 & 1 & 0 & 0 \\ 4 & 5 & 0 & 0 & 1 & 0 \\ 0 & 1 & -3 & 0 & 0 & 1 \end{array}\right]$$

$$\left[\begin{array}{ccc|ccc} -1 & -1 & -1 & 1 & 0 & 0 \\ 0 & 1 & -4 & 4 & 1 & 0 \\ 0 & 1 & -3 & 0 & 0 & 1 \end{array}\right] \quad 4R1 + R2$$

$$\left[\begin{array}{ccc|ccc} 1 & 1 & 1 & -1 & 0 & 0 \\ 0 & 1 & -4 & 4 & 1 & 0 \\ 0 & 0 & 1 & -4 & -1 & 1 \end{array}\right] \quad \begin{array}{l}-1R1 \\ \\ -1R2 + R3\end{array}$$

$$\left[\begin{array}{ccc|ccc} 1 & 1 & 1 & -1 & 0 & 0 \\ 0 & 1 & 0 & -12 & -3 & 4 \\ 0 & 0 & 1 & -4 & -1 & 1 \end{array}\right] \quad 4R3 + R2$$

$$\left[\begin{array}{ccc|ccc} 1 & 0 & 1 & 11 & 3 & -4 \\ 0 & 1 & 0 & -12 & -3 & 4 \\ 0 & 0 & 1 & -4 & -1 & 1 \end{array}\right] \quad -1R2 + R1$$

$$\left[\begin{array}{ccc|ccc} 1 & 0 & 0 & 15 & 4 & -5 \\ 0 & 1 & 0 & -12 & -3 & 4 \\ 0 & 0 & 1 & -4 & -1 & 1 \end{array}\right] \quad -1R3 + R1$$

$$A^{-1} = \begin{bmatrix} 15 & 4 & -5 \\ -12 & -3 & 4 \\ -4 & -1 & 1 \end{bmatrix}$$

25. Find the inverse of $A = \begin{bmatrix} -.4 & .1 & .2 \\ 0 & .6 & .8 \\ .3 & 0 & -.2 \end{bmatrix}$,

if it exists.

$$[A|I] = \left[\begin{array}{ccc|ccc} -.4 & .1 & .2 & 1 & 0 & 0 \\ 0 & .6 & .8 & 0 & 1 & 0 \\ .3 & 0 & -.2 & 0 & 0 & 1 \end{array}\right]$$

$$\left[\begin{array}{ccc|ccc} -4 & 1 & 2 & 10 & 0 & 0 \\ 0 & 6 & 8 & 0 & 10 & 0 \\ 3 & 0 & -2 & 0 & 0 & 10 \end{array}\right] \quad \begin{array}{l}10R1 \\ 10R2 \\ 10R3\end{array}$$

$$\left[\begin{array}{ccc|ccc} -1 & 1 & 0 & 10 & 0 & 10 \\ 0 & 6 & 8 & 0 & 10 & 0 \\ 3 & 0 & -2 & 0 & 0 & 10 \end{array}\right] \quad R3 + R1$$

$$\left[\begin{array}{ccc|ccc} -1 & 1 & 0 & 10 & 0 & 0 \\ 0 & 6 & 8 & 0 & 10 & 0 \\ 0 & 3 & -2 & 30 & 0 & 40 \end{array}\right] \quad 3R1 + R3$$

$$\left[\begin{array}{ccc|ccc} 1 & -1 & 0 & -10 & 0 & -10 \\ 0 & 6 & 8 & 0 & 10 & 0 \\ 0 & 3 & -2 & 30 & 0 & 40 \end{array}\right] \quad -1R1$$

$$\left[\begin{array}{ccc|ccc} 1 & -1 & 0 & -10 & 0 & -10 \\ 0 & 0 & -12 & -60 & 10 & -80 \\ 0 & 3 & -2 & 30 & 0 & 40 \end{array}\right] \quad -2R3 + R2$$

$$\left[\begin{array}{ccc|ccc} 1 & -1 & 0 & -10 & 0 & -10 \\ 0 & 3 & -2 & 30 & 0 & 40 \\ 0 & 0 & 12 & -60 & 10 & -80 \end{array}\right] \quad R2 \leftrightarrow R3$$

(Interchange rows 2 and 3.)

$$\left[\begin{array}{ccc|ccc} 1 & -1 & 0 & -10 & 0 & -10 \\ 0 & 1 & -\dfrac{2}{3} & 10 & 0 & \dfrac{40}{3} \\ 0 & 0 & 12 & -60 & 10 & -80 \end{array}\right] \quad \dfrac{1}{3}R2$$

$$\left[\begin{array}{ccc|ccc} 1 & 0 & -\dfrac{2}{3} & 0 & 0 & \dfrac{10}{3} \\ 0 & 1 & -\dfrac{2}{3} & 10 & 0 & \dfrac{40}{3} \\ 0 & 0 & 12 & -60 & 10 & -80 \end{array}\right] \quad R2 + R1$$

$$\left[\begin{array}{ccc|ccc} 1 & 0 & -\dfrac{2}{3} & 0 & 0 & \dfrac{10}{3} \\ 0 & 1 & -\dfrac{2}{3} & 10 & 0 & \dfrac{40}{3} \\ 0 & 0 & 1 & -5 & \dfrac{5}{6} & -\dfrac{20}{3} \end{array}\right] \quad \dfrac{1}{12}R3$$

$$\left[\begin{array}{ccc|ccc} 1 & 0 & 0 & -\dfrac{10}{3} & \dfrac{5}{9} & -\dfrac{10}{9} \\ 0 & 1 & 0 & \dfrac{20}{3} & \dfrac{5}{9} & \dfrac{80}{8} \\ 0 & 0 & 1 & -5 & \dfrac{5}{6} & -\dfrac{20}{3} \end{array}\right] \quad \begin{array}{l}\dfrac{2}{3}R3 + R1 \\ \dfrac{2}{3}R3 + R2\end{array}$$

$$A^{-1} = \begin{bmatrix} -\dfrac{10}{3} & \dfrac{5}{9} & -\dfrac{10}{9} \\ \dfrac{20}{3} & \dfrac{5}{9} & \dfrac{80}{9} \\ -5 & \dfrac{5}{6} & -\dfrac{20}{3} \end{bmatrix}$$

27. $x + y = 8$

$2x - y = 4$

Matrix of coefficients: $A = \begin{bmatrix} 1 & 1 \\ 2 & -1 \end{bmatrix}$

Matrix of variables: $X = \begin{bmatrix} x \\ y \end{bmatrix}$

Matrix of constants: $B = \begin{bmatrix} 8 \\ 4 \end{bmatrix}$

29. $4x + 5y = 7$

$2x + 3y = 5$

$A = \begin{bmatrix} 4 & 5 \\ 2 & 3 \end{bmatrix}$

$X = \begin{bmatrix} x \\ y \end{bmatrix}$

$B = \begin{bmatrix} 7 \\ 5 \end{bmatrix}$

31. $x + y + z = 9$

$2x + y + 3z = 17$

$5x - 2y + 2z = 16$

$A = \begin{bmatrix} 1 & 1 & 1 \\ 2 & 1 & 3 \\ 5 & -2 & 2 \end{bmatrix}$

$X = \begin{bmatrix} x \\ y \\ z \end{bmatrix}$

$B = \begin{bmatrix} 9 \\ 17 \\ 16 \end{bmatrix}$

33. X is a column matrix of the variables of the system; A^{-1} is the inverse of matrix A, which is a matrix of the coefficients of the variables of the system; B is a column matrix of the constants of the system.

35. $x + y = 5$

$x - y = -1$

$A = \begin{bmatrix} 1 & 1 \\ 1 & -1 \end{bmatrix}$, $X = \begin{bmatrix} x \\ y \end{bmatrix}$, $B = \begin{bmatrix} 5 \\ -1 \end{bmatrix}$

Find A^{-1}.

$[A|I] = \begin{bmatrix} 1 & 1 & | & 1 & 0 \\ 1 & -1 & | & 0 & 1 \end{bmatrix}$

$\begin{bmatrix} 1 & 1 & | & 1 & 0 \\ 0 & -2 & | & -1 & 1 \end{bmatrix}$ $-1R1 + R2$

$\begin{bmatrix} 1 & 1 & | & 1 & 0 \\ 0 & 1 & | & \dfrac{1}{2} & -\dfrac{1}{2} \end{bmatrix}$ $-\dfrac{1}{2}R2$

$\begin{bmatrix} 1 & 0 & | & \dfrac{1}{2} & \dfrac{1}{2} \\ 0 & 1 & | & \dfrac{1}{2} & -\dfrac{1}{2} \end{bmatrix}$ $-1R2 + R1$

$A^{-1} = \begin{bmatrix} \dfrac{1}{2} & \dfrac{1}{2} \\ \dfrac{1}{2} & -\dfrac{1}{2} \end{bmatrix}$

$X = A^{-1}B$

$= \begin{bmatrix} \dfrac{1}{2} & \dfrac{1}{2} \\ \dfrac{1}{2} & -\dfrac{1}{2} \end{bmatrix} \begin{bmatrix} 5 \\ -1 \end{bmatrix} = \begin{bmatrix} 2 \\ 3 \end{bmatrix}$

Solution set: $\{(2, 3)\}$

37. $x + 3y = -12$

$2x - y = 11$

$A = \begin{bmatrix} 1 & 3 \\ 2 & -1 \end{bmatrix}$, $X = \begin{bmatrix} x \\ y \end{bmatrix}$, $B = \begin{bmatrix} -12 \\ 11 \end{bmatrix}$

Find A^{-1}.

$[A|I] = \begin{bmatrix} 1 & 3 & | & 1 & 0 \\ 2 & -1 & | & 0 & 1 \end{bmatrix}$

$\begin{bmatrix} 1 & 3 & | & 1 & 0 \\ 0 & -7 & | & -2 & 1 \end{bmatrix}$ $-2R1 + R2$

$\begin{bmatrix} 1 & 3 & | & 1 & 0 \\ 0 & 1 & | & \frac{2}{7} & -\frac{2}{7} \end{bmatrix}$ $-\frac{1}{7}R2$

$\begin{bmatrix} 1 & 0 & | & \frac{1}{7} & \frac{3}{7} \\ 0 & 1 & | & \frac{2}{7} & -\frac{1}{7} \end{bmatrix}$ $-3R2 + R1$

$A^{-1} = \begin{bmatrix} \frac{1}{7} & \frac{3}{7} \\ \frac{2}{7} & -\frac{1}{7} \end{bmatrix}$

$X = A^{-1}B$

$= \begin{bmatrix} \frac{1}{7} & \frac{3}{7} \\ \frac{1}{7} & -\frac{1}{7} \end{bmatrix} \begin{bmatrix} -12 \\ 11 \end{bmatrix} = \begin{bmatrix} 3 \\ -5 \end{bmatrix}$

Solution set: $\{(3, -5)\}$

39. $2x - 3y = 10$
$2x + 2y = 5$

$A = \begin{bmatrix} 2 & -3 \\ 2 & 2 \end{bmatrix}$, $X = \begin{bmatrix} x \\ y \end{bmatrix}$, $B = \begin{bmatrix} 10 \\ 5 \end{bmatrix}$

Find A^{-1}.

$[A|I] = \begin{bmatrix} 2 & -3 & | & 1 & 0 \\ 2 & 2 & | & 0 & 1 \end{bmatrix}$

$\begin{bmatrix} 1 & -\frac{3}{2} & | & \frac{1}{2} & 0 \\ 2 & 2 & | & 0 & 1 \end{bmatrix}$ $\frac{1}{2}R1$

$\begin{bmatrix} 1 & -\frac{3}{2} & | & \frac{1}{2} & 0 \\ 0 & 5 & | & -1 & 1 \end{bmatrix}$ $-2R1 + R2$

$\begin{bmatrix} 1 & -\frac{3}{2} & | & \frac{1}{2} & 0 \\ 0 & 1 & | & -\frac{1}{5} & \frac{1}{5} \end{bmatrix}$ $\frac{1}{5}R2$

$\begin{bmatrix} 1 & 0 & | & \frac{1}{5} & \frac{3}{10} \\ 0 & 1 & | & -\frac{1}{5} & \frac{1}{5} \end{bmatrix}$ $\frac{3}{2}R2 + R1$

$A^{-1} = \begin{bmatrix} \frac{1}{5} & \frac{3}{10} \\ -\frac{1}{5} & \frac{1}{5} \end{bmatrix}$

$X = A^{-1}B$

$= \begin{bmatrix} \frac{1}{5} & \frac{3}{10} \\ -\frac{1}{5} & \frac{1}{5} \end{bmatrix} \begin{bmatrix} 10 \\ 5 \end{bmatrix} = \begin{bmatrix} \frac{7}{2} \\ -1 \end{bmatrix}$

Solution set: $\{(\frac{7}{2}, -1)\}$

41. $2x - 3y = 2$ (1)
$4x - 6y = 1$ (2)

$A = \begin{bmatrix} 2 & -3 \\ 4 & -6 \end{bmatrix}$, $X = \begin{bmatrix} x \\ y \end{bmatrix}$, $B = \begin{bmatrix} 2 \\ 1 \end{bmatrix}$

$[A|I] = \begin{bmatrix} 2 & -3 & | & 1 & 0 \\ 4 & -6 & | & 0 & 1 \end{bmatrix}$

$\begin{bmatrix} 1 & -\frac{3}{2} & | & \frac{1}{2} & 0 \\ 4 & -6 & | & 0 & 1 \end{bmatrix}$ $\frac{1}{2}R1$

$\begin{bmatrix} 1 & -\frac{3}{2} & | & \frac{1}{2} & 0 \\ 0 & 0 & | & -2 & 1 \end{bmatrix}$ $-4R1 + R2$

The inverse of A does not exist.
Since equation (2) added to -2 times
equation (1) yields the equation
$0 = -3$, this system has no solution.

Solution set: Ø

43. $x + 3y + 3z = 11$

$$-x = -2$$

$$-4x - 4y - 3z = -10$$

$$A = \begin{bmatrix} 1 & 3 & 3 \\ -1 & 0 & 0 \\ -4 & -4 & -3 \end{bmatrix}, \quad X = \begin{bmatrix} x \\ y \\ z \end{bmatrix},$$

$$B = \begin{bmatrix} 11 \\ -2 \\ -10 \end{bmatrix}$$

From Exercise 20,

$$A^{-1} = \begin{bmatrix} 0 & -1 & 0 \\ -1 & 3 & -1 \\ \frac{4}{3} & -\frac{8}{3} & 1 \end{bmatrix}.$$

$X = A^{-1}B$

$$= \begin{bmatrix} 0 & -1 & 0 \\ -1 & 3 & -1 \\ \frac{4}{3} & -\frac{8}{3} & 1 \end{bmatrix} \begin{bmatrix} 11 \\ -2 \\ -10 \end{bmatrix}$$

$$= \begin{bmatrix} 2 \\ -7 \\ 10 \end{bmatrix}$$

Solution set: $\{(2, -7, 10)\}$

45. $2x + 4z = 14$

$$3x + y + 5z = 19$$

$$-x + y - 2z = -7$$

$$A = \begin{bmatrix} 2 & 0 & 4 \\ 3 & 1 & 5 \\ -1 & 1 & -2 \end{bmatrix}, \quad X = \begin{bmatrix} x \\ y \\ z \end{bmatrix},$$

$$B = \begin{bmatrix} 14 \\ 19 \\ -7 \end{bmatrix}.$$

From Exercise 24,

$$A^{-1} = \begin{bmatrix} -\frac{7}{2} & 2 & -2 \\ \frac{1}{2} & 0 & 1 \\ 2 & -1 & 1 \end{bmatrix}.$$

$X = A^{-1}B$

$$= \begin{bmatrix} -\frac{7}{2} & 2 & -2 \\ \frac{1}{2} & 0 & 1 \\ 2 & -1 & 1 \end{bmatrix} \begin{bmatrix} 14 \\ 19 \\ -7 \end{bmatrix} = \begin{bmatrix} 3 \\ 0 \\ 2 \end{bmatrix}$$

Solution set: $\{(3, 0, 2)\}$

47. $x + y - z = 6$

$$2x - y + z = -9$$

$$x - 2y + 3z = 1$$

$$A = \begin{bmatrix} 1 & 1 & -1 \\ 2 & -1 & 1 \\ 1 & -2 & 3 \end{bmatrix}, \quad X = \begin{bmatrix} x \\ y \\ z \end{bmatrix},$$

$$B = \begin{bmatrix} 6 \\ -9 \\ 1 \end{bmatrix}$$

Find A^{-1}.

$$[A|I] = \begin{bmatrix} 1 & 1 & -1 & | & 1 & 0 & 0 \\ 2 & -1 & 1 & | & 0 & 1 & 0 \\ 1 & -2 & 3 & | & 0 & 0 & 1 \end{bmatrix}$$

$$\begin{bmatrix} 1 & 1 & -1 & | & 1 & 0 & 0 \\ 0 & -3 & 3 & | & -2 & 1 & 0 \\ 0 & -3 & 4 & | & -1 & 0 & 1 \end{bmatrix} \begin{matrix} \\ -2R1 + R2 \\ -1R1 + R3 \end{matrix}$$

$$\begin{bmatrix} 1 & 1 & -1 & | & 1 & 0 & 0 \\ 0 & 1 & -1 & | & \frac{2}{3} & -\frac{1}{3} & 0 \\ 0 & -3 & 4 & | & -1 & 0 & 1 \end{bmatrix} \begin{matrix} \\ -\frac{1}{3}R2 \\ \\ \end{matrix}$$

$$\begin{bmatrix} 1 & 0 & 0 & | & \frac{1}{3} & \frac{1}{3} & 0 \\ 0 & 1 & -1 & | & \frac{2}{3} & -\frac{1}{3} & 0 \\ 0 & 0 & 1 & | & 1 & -1 & 1 \end{bmatrix} \begin{matrix} -1R2 + R1 \\ \\ 3R2 + R3 \end{matrix}$$

$$\begin{bmatrix} 1 & 0 & 0 & | & \frac{1}{3} & \frac{1}{3} & 0 \\ 0 & 1 & 0 & | & \frac{5}{3} & -\frac{4}{3} & 1 \\ 0 & 0 & 1 & | & 1 & -1 & 1 \end{bmatrix} \begin{matrix} \\ R3 + R2 \\ \\ \end{matrix}$$

$$A^{-1} = \begin{bmatrix} \frac{1}{3} & \frac{1}{3} & 0 \\ \frac{5}{3} & -\frac{4}{3} & 1 \\ 1 & -1 & 1 \end{bmatrix}$$

$$X = A^{-1}B$$

$$= \begin{bmatrix} \frac{1}{3} & \frac{1}{3} & 0 \\ \frac{5}{3} & -\frac{4}{3} & 1 \\ 1 & -1 & 1 \end{bmatrix}\begin{bmatrix} 6 \\ -9 \\ 1 \end{bmatrix} = \begin{bmatrix} -1 \\ 23 \\ 16 \end{bmatrix}$$

Solution set: $\{(-1, 23, 16)\}$

49. Let x = the number of model 201 bicycles;

   y = the number of model 301 bicycles.

The system of equations is

$$2x + 3y = 34$$
$$25x + 30y = 365.$$

$$A = \begin{bmatrix} 2 & 3 \\ 25 & 30 \end{bmatrix}, \quad X = \begin{bmatrix} x \\ y \end{bmatrix}, \quad B = \begin{bmatrix} 34 \\ 365 \end{bmatrix}$$

Find A^{-1}.

$$[A|I] = \begin{bmatrix} 2 & 3 & | & 1 & 0 \\ 25 & 30 & | & 0 & 1 \end{bmatrix}$$

$$\begin{bmatrix} 1 & \frac{3}{2} & | & \frac{1}{2} & 0 \\ 25 & 30 & | & 0 & 1 \end{bmatrix} \quad \frac{1}{2}R1$$

$$\begin{bmatrix} 1 & \frac{3}{2} & | & \frac{1}{2} & 0 \\ 0 & -\frac{15}{2} & | & -\frac{25}{2} & 1 \end{bmatrix} \quad -25R1 + R2$$

$$\begin{bmatrix} 1 & \frac{3}{2} & | & \frac{1}{2} & 0 \\ 0 & 1 & | & \frac{5}{3} & -\frac{2}{15} \end{bmatrix} \quad -\frac{2}{15}R2$$

$$\begin{bmatrix} 1 & 0 & | & -2 & \frac{1}{5} \\ 0 & 1 & | & \frac{5}{3} & -\frac{2}{15} \end{bmatrix} \quad -\frac{3}{2}R2 + R1$$

$$A^{-1} = \begin{bmatrix} -2 & \frac{1}{5} \\ \frac{5}{3} & -\frac{2}{15} \end{bmatrix}$$

$$X = A^{-1}B$$

$$= \begin{bmatrix} -2 & \frac{1}{5} \\ \frac{5}{3} & -\frac{2}{15} \end{bmatrix}\begin{bmatrix} 34 \\ 365 \end{bmatrix} = \begin{bmatrix} 5 \\ 8 \end{bmatrix}$$

5 model 201 bicycles and 8 model 301 bicycles can be made in a day.

51. For $A = \begin{bmatrix} a & b \\ c & d \end{bmatrix}$, show that $AA^{-1} = I$ and $A^{-1}A = I$.

Note that the 2 × 2 identity matrix is $\begin{bmatrix} 1 & 0 \\ 0 & 1 \end{bmatrix}$.

Begin by finding A^{-1}.

$$[A|I] = \begin{bmatrix} a & b & | & 1 & 0 \\ c & d & | & 0 & 1 \end{bmatrix}$$

$$\begin{bmatrix} 1 & \frac{b}{a} & | & \frac{1}{a} & 0 \\ c & d & | & 0 & 1 \end{bmatrix} \quad \frac{1}{a}R1$$

$$\begin{bmatrix} 1 & \frac{b}{a} & | & \frac{1}{a} & 0 \\ 0 & \frac{ad-bc}{a} & | & -\frac{c}{a} & 1 \end{bmatrix} \quad -cR1 + R2$$

$$\begin{bmatrix} 1 & \frac{b}{a} & | & \frac{1}{a} & 0 \\ 0 & 1 & | & \frac{-c}{ad-bc} & \frac{a}{ad-bc} \end{bmatrix} \quad \frac{a}{ad-bc}R2$$

$$\begin{bmatrix} 1 & 0 & | & \frac{d}{ad-bc} & \frac{-b}{ad-bc} \\ 0 & 1 & | & \frac{-c}{ad-bc} & \frac{a}{ad-bc} \end{bmatrix} \quad -\frac{b}{a}R2 + R1$$

$$A^{-1} = \begin{bmatrix} \dfrac{d}{ad-bc} & \dfrac{-b}{ad-bc} \\ \dfrac{-c}{ad-bc} & \dfrac{a}{ad-bc} \end{bmatrix}$$

$$AA^{-1} = \begin{bmatrix} a & b \\ c & d \end{bmatrix} \begin{bmatrix} \dfrac{d}{ad-bc} & \dfrac{-b}{ad-bc} \\ \dfrac{-c}{ad-bc} & \dfrac{a}{ad-bc} \end{bmatrix}$$

$$= \begin{bmatrix} 1 & 0 \\ 0 & 1 \end{bmatrix} = I$$

and

$$A^{-1}A = \begin{bmatrix} \dfrac{d}{ad-bc} & \dfrac{-b}{ad-bc} \\ \dfrac{-c}{ad-bc} & \dfrac{a}{ad-bc} \end{bmatrix} \begin{bmatrix} a & b \\ c & d \end{bmatrix}$$

$$= \begin{bmatrix} 1 & 0 \\ 0 & 1 \end{bmatrix} = I$$

53. For $A = \begin{bmatrix} a & b \\ c & d \end{bmatrix}$, show that $AI = A$.

Note that the 2×2 identity matrix is $\begin{bmatrix} 1 & 0 \\ 0 & 1 \end{bmatrix}$.

$$AI = \begin{bmatrix} a & b \\ c & d \end{bmatrix} \begin{bmatrix} 1 & 0 \\ 0 & 1 \end{bmatrix} = \begin{bmatrix} a & b \\ c & d \end{bmatrix} = A$$

Section 8.4

1. $\begin{vmatrix} 5 & 8 \\ 2 & -4 \end{vmatrix} = 5(-4) - 2(8)$
$$= -20 - 16$$
$$= -36$$

3. $\begin{vmatrix} -1 & -2 \\ 5 & 3 \end{vmatrix} = -1(3) - 5(-2)$
$$= -3 + 10 = 7$$

5. $\begin{vmatrix} 9 & 3 \\ -3 & -1 \end{vmatrix} = 9(-1) - (-3)(3)$
$$= -9 + 9 = 0$$

7. $\begin{vmatrix} 3 & 4 \\ 5 & -2 \end{vmatrix} = 3(-2) - 5(4)$
$$= -6 - 20 = -26$$

9. $\begin{vmatrix} 0 & 4 \\ 4 & 0 \end{vmatrix} = 0(0) - 4(4)$
$$= -16$$

11. $\begin{vmatrix} 8 & 3 \\ 8 & 3 \end{vmatrix} = 8(3) - 8(3) = 0$

13. $\begin{vmatrix} x & 4 \\ 8 & 2 \end{vmatrix} = x(2) - 8(4)$
$$= 2x - 32$$

15. $\begin{vmatrix} y & 2 \\ 8 & y \end{vmatrix} = y(y) - 8(2)$
$$= y^2 - 16$$

17. $\begin{vmatrix} x & y \\ y & x \end{vmatrix} = x(x) - y(y)$
$$= x^2 - y^2$$

19. $\begin{vmatrix} 1.4 & 2.5 \\ 3.7 & 6.2 \end{vmatrix} = (1.4)(6.2) - (3.7)(2.5)$
$$= 8.68 - 9.25 = -.57$$

23. $\begin{vmatrix} -2 & 0 & 1 \\ 3 & 2 & -1 \\ 1 & 0 & 2 \end{vmatrix}$

Cofactor of 3: $(-1)\begin{vmatrix} 0 & 1 \\ 0 & 2 \end{vmatrix} = -1(0)$
$$= 0$$

Cofactor of 2: $1\begin{vmatrix} -2 & 1 \\ 1 & 2 \end{vmatrix} = 1(-5)$
$$= -5$$

Cofactor of -1: $(-1)\begin{vmatrix} -2 & 0 \\ 1 & 0 \end{vmatrix} = -1(0)$
$$= 0$$

25. $\begin{vmatrix} 1 & 2 & -1 \\ 2 & 3 & -2 \\ -1 & 4 & 1 \end{vmatrix}$

Cofactor of 2: $(-1)\begin{vmatrix} 2 & -1 \\ 4 & 1 \end{vmatrix} = -1(6)$

$\qquad\qquad\qquad = -6$

Cofactor of 3: $1\begin{vmatrix} 1 & -1 \\ -1 & 1 \end{vmatrix} = 1(0) = 0$

Cofactor of -2: $(-1)\begin{vmatrix} 1 & 2 \\ -1 & 4 \end{vmatrix} = (-1)(6)$

$\qquad\qquad\qquad\qquad = -6$

27. $\begin{vmatrix} 1 & 0 & 0 \\ 0 & -1 & 0 \\ 1 & 0 & 1 \end{vmatrix}$

Expand by minors about the first row.
(Expanding by minors about any other
row or column gives the same value.)

$= 1\begin{vmatrix} -1 & 0 \\ 0 & 1 \end{vmatrix} - 0\begin{vmatrix} 0 & 0 \\ 1 & 1 \end{vmatrix} + 0\begin{vmatrix} 0 & -1 \\ 1 & 0 \end{vmatrix}$

$= 1[-1(1) - 0(0)] - 0 + 0$

$= -1$

29. $\begin{vmatrix} -2 & 0 & 0 \\ 4 & 0 & 1 \\ 3 & 4 & 2 \end{vmatrix}$

Expand by minors about the second
column.

$= -0\begin{vmatrix} 4 & 1 \\ 3 & 2 \end{vmatrix} + 0\begin{vmatrix} -2 & 0 \\ 3 & 2 \end{vmatrix} - 4\begin{vmatrix} -2 & 0 \\ 4 & 1 \end{vmatrix}$

$= -0 + 0 - 4[-2(1) - 4(0)]$

$= -4(-2)$

$= 8$

31. $\begin{vmatrix} 1 & 2 & 0 \\ -1 & 2 & -1 \\ 0 & 1 & 4 \end{vmatrix}$

Expand by minors about the first
column.

$= 1\begin{vmatrix} 2 & -1 \\ 1 & 4 \end{vmatrix} - (-1)\begin{vmatrix} 2 & 0 \\ 1 & 4 \end{vmatrix} + 0\begin{vmatrix} 2 & 0 \\ 2 & -1 \end{vmatrix}$

$= 1[2(4) - 1(-1)] + 1[2(4) - 1(0)]$

$\quad + 0$

$= 9 + 8 + 0$

$= 17$

33. $\begin{vmatrix} 10 & 2 & 1 \\ -1 & 4 & 3 \\ -3 & 8 & 10 \end{vmatrix}$

Expand by minors about the first
row.

$= 10\begin{vmatrix} 4 & 3 \\ 8 & 10 \end{vmatrix} - 2\begin{vmatrix} -1 & 3 \\ -3 & 10 \end{vmatrix} + 1\begin{vmatrix} -1 & 4 \\ -3 & 8 \end{vmatrix}$

$= 10(40 - 24) - 2(-10 + 9)$

$\quad + 1(-8 + 12)$

$= 10(16) + 2 + 4$

$= 166$

35. $\begin{vmatrix} 1 & -2 & 3 \\ 0 & 0 & 0 \\ 1 & 10 & -12 \end{vmatrix}$

Expand by minors about the second
row.
(The arithmetic is easier if you
choose a row or column with the
most zeros and/or small numbers.
In this case, the second row is
the easiest.)

$= -0\begin{vmatrix} -2 & 3 \\ 10 & -12 \end{vmatrix} + 0\begin{vmatrix} 1 & 3 \\ 1 & -12 \end{vmatrix}$

$\quad - 0\begin{vmatrix} 1 & -2 \\ 1 & 10 \end{vmatrix}$

$= -0 + 0 - 0$

$= 0$

37.
$$\begin{vmatrix} 3 & 3 & -1 \\ 2 & 6 & 0 \\ -6 & -6 & 2 \end{vmatrix}$$

Expand by minors about the second row.

$$= -2\begin{vmatrix} 3 & -1 \\ -6 & 2 \end{vmatrix} + 6\begin{vmatrix} 3 & -1 \\ -6 & 2 \end{vmatrix}$$

$$- \; 0\begin{vmatrix} 3 & 3 \\ -6 & -6 \end{vmatrix}$$

$$= -2(6 - 6) + 6(6 - 6) - 0$$
$$= -2(0) + 6(0) - 0$$
$$= 0$$

39.
$$\begin{vmatrix} 3 & 2 & 0 \\ 0 & 1 & x \\ 2 & 0 & 0 \end{vmatrix}$$

Expand by minors about the third column. (The third row would be an equally good choice.)

$$= +0\begin{vmatrix} 0 & 1 \\ 2 & 0 \end{vmatrix} - x\begin{vmatrix} 3 & 2 \\ 2 & 0 \end{vmatrix} + 0\begin{vmatrix} 3 & 2 \\ 0 & 1 \end{vmatrix}$$

$$= 0 - x(0 - 4) + 0 = 4x$$

43. To solve the equation

$$\begin{vmatrix} -2 & 0 & 1 \\ -1 & 3 & x \\ 5 & -2 & 0 \end{vmatrix} = 3,$$

expand by minors about the first row.

$$-2\begin{vmatrix} 3 & x \\ -2 & 0 \end{vmatrix} - 0\begin{vmatrix} -1 & x \\ 5 & 0 \end{vmatrix} + 1\begin{vmatrix} -1 & 3 \\ 5 & -2 \end{vmatrix} = 3$$

$$-2(0 + 2x) - 0 + 1(2 - 15) = 3$$
$$-4x - 13 = 3$$
$$-4x = 16$$
$$x = -4$$

Solution set: $\{-4\}$

45.
$$\begin{vmatrix} 5 & 3x & -3 \\ 0 & 2 & -1 \\ 4 & -1 & x \end{vmatrix} = -7$$

Expand about the second row.

$$-0\begin{vmatrix} 3x & -3 \\ -1 & x \end{vmatrix} + 2\begin{vmatrix} 5 & -3 \\ 4 & x \end{vmatrix} - (-1)\begin{vmatrix} 5 & 3x \\ 4 & -1 \end{vmatrix} = -7$$

$$2(5x + 12) + (-5 - 12x) = -7$$
$$10x + 24 - 5 - 12x = -7$$
$$-2x + 19 = 2x$$
$$26 = 2x$$
$$13 = x$$

Solution set: $\{13\}$

47.
$$\begin{vmatrix} 4 & 0 & 0 & 2 \\ -1 & 0 & 3 & 0 \\ 2 & 4 & 0 & 1 \\ 0 & 0 & 1 & 2 \end{vmatrix}$$

Expand about the second column since it has the most zeros.

$$= -0\begin{vmatrix} -1 & 3 & 0 \\ 2 & 0 & 1 \\ 0 & 1 & 2 \end{vmatrix} + 0\begin{vmatrix} 4 & 0 & 2 \\ 2 & 0 & 1 \\ 0 & 1 & 2 \end{vmatrix}$$

$$-4\begin{vmatrix} 4 & 0 & 2 \\ -1 & 3 & 0 \\ 0 & 1 & 2 \end{vmatrix} + 0\begin{vmatrix} 4 & 0 & 2 \\ -1 & 3 & 0 \\ 2 & 0 & 1 \end{vmatrix}$$

$$= -4\begin{vmatrix} 4 & 0 & 2 \\ -1 & 3 & 0 \\ 0 & 1 & 2 \end{vmatrix}$$

Expand this determinant about the third row.

$$= -4\left[0\begin{vmatrix} 0 & 2 \\ 3 & 0 \end{vmatrix} - 1\begin{vmatrix} 4 & 2 \\ -1 & 0 \end{vmatrix} + 2\begin{vmatrix} 4 & 0 \\ -1 & 3 \end{vmatrix}\right]$$

$$= -4[0 - 1(2) + 2(12)]$$
$$= -4(-2 + 24)$$
$$= -4(22)$$
$$= -88$$

49. $\begin{vmatrix} 1 & 1 & 0 & 1 \\ 2 & 1 & 0 & 2 \\ 0 & 1 & -1 & 1 \\ 1 & -1 & 1 & 1 \end{vmatrix}$

Expand about the third column since there are more zeros there than in any other row or column.

$= 0\begin{vmatrix} 2 & 1 & 2 \\ 0 & 1 & 1 \\ 1 & -1 & 1 \end{vmatrix} - 0\begin{vmatrix} 1 & 1 & 1 \\ 0 & 1 & 1 \\ 1 & -1 & 1 \end{vmatrix}$

$\quad + (-1)\begin{vmatrix} 1 & 1 & 1 \\ 2 & 1 & 2 \\ 1 & -1 & 1 \end{vmatrix} - 1\begin{vmatrix} 1 & 1 & 1 \\ 2 & 1 & 2 \\ 0 & 1 & 1 \end{vmatrix}$

$= -\begin{vmatrix} 1 & 1 & 1 \\ 2 & 1 & 2 \\ 1 & -1 & 1 \end{vmatrix} - \begin{vmatrix} 1 & 1 & 1 \\ 2 & 1 & 2 \\ 0 & 1 & 1 \end{vmatrix}$

Expand the first determinant about the first row, and expand the second determinant about the third row.

$= -\left[1\begin{vmatrix} 1 & 2 \\ -1 & 1 \end{vmatrix} - 1\begin{vmatrix} 2 & 2 \\ 1 & 1 \end{vmatrix} + 1\begin{vmatrix} 2 & 1 \\ 1 & -1 \end{vmatrix}\right]$

$\quad -\left[0\begin{vmatrix} 1 & 1 \\ 1 & 2 \end{vmatrix} - 1\begin{vmatrix} 1 & 1 \\ 2 & 2 \end{vmatrix} + 1\begin{vmatrix} 1 & 1 \\ 2 & 1 \end{vmatrix}\right]$

$= -[1(3) - 1(0) + 1(-3)]$

$\quad - [0 - 1(0) + 1(-1)]$

$= -(3 - 0 - 3) - (0 - 0 - 1)$

$= 0 + 1 = 1$

51. For $A = \begin{bmatrix} a_{11} & a_{12} & a_{13} \\ a_{21} & a_{22} & a_{23} \\ a_{31} & a_{32} & a_{33} \end{bmatrix}$,

find $|A|$ by expansion about row 3. First get the minors of each element in the third row.

$M_{31} = \begin{vmatrix} a_{12} & a_{13} \\ a_{22} & a_{23} \end{vmatrix} = a_{12}a_{23} - a_{13}a_{22}$

$M_{32} = \begin{vmatrix} a_{11} & a_{13} \\ a_{21} & a_{23} \end{vmatrix} = a_{11}a_{23} - a_{13}a_{21}$

$M_{33} = \begin{vmatrix} a_{11} & a_{12} \\ a_{21} & a_{22} \end{vmatrix} = a_{11}a_{22} - a_{12}a_{21}$

Now find the cofactor of each of these minors.

$A_{31} = (-1)^{3+1}M_{31}$

$\quad = 1(a_{12}a_{23} - a_{13}a_{22})$

$\quad = a_{12}a_{23} - a_{13}a_{22}$

$A_{32} = (-1)^{3+2}M_{32}$

$\quad = (-1)(a_{11}a_{23} - a_{13}a_{21})$

$\quad = a_{13}a_{21} - a_{11}a_{23}$

$A_{33} = (-1)^{3+3}M_{33}$

$\quad = 1(a_{11}a_{22} - a_{12}a_{21})$

$\quad = a_{11}a_{22} - a_{12}a_{21}$

The determinant is found by multiplying each cofactor by its corresponding element in the matrix and finding the sum of these products.

$\begin{vmatrix} a_{11} & a_{12} & a_{13} \\ a_{21} & a_{22} & a_{23} \\ a_{31} & a_{32} & a_{33} \end{vmatrix}$

$= a_{31}A_{31} + a_{32}A_{32} + a_{33}A_{33}$

$= a_{31}(a_{12}a_{23} - a_{13}a_{22})$

$\quad + a_{32}(a_{13}a_{21} - a_{11}a_{23})$

$\quad + a_{33}(a_{11}a_{22} - a_{12}a_{21})$

$= a_{12}a_{23}a_{31} - a_{13}a_{22}a_{31}$

$\quad + a_{13}a_{21}a_{32} - a_{11}a_{23}a_{32}$

$\quad + a_{11}a_{22}a_{33} - a_{12}a_{21}a_{33}$

This result may be written in the form

$(a_{11}a_{22}a_{33} + a_{12}a_{23}a_{31} + a_{13}a_{21}a_{32})$

$- (a_{31}a_{22}a_{13} + a_{32}a_{23}a_{11} + a_{33}a_{21}a_{12})$, which is what was given in the definition of the determinant of a 3 × 3 matrix.

53. $\begin{vmatrix} x & y & 1 \\ 2 & 3 & 1 \\ -1 & 4 & 1 \end{vmatrix} = 0$

Expand about the third column.

$$1\begin{vmatrix} 2 & 3 \\ -1 & 4 \end{vmatrix} - 1\begin{vmatrix} x & y \\ -1 & 4 \end{vmatrix} + 1\begin{vmatrix} x & y \\ 2 & 3 \end{vmatrix} = 0$$

$$1(11) - 1(4x + y) + 1(3x - 2y) = 0$$

$$11 - 4x - y + 3x - 2y = 0$$

$$-x - 3y + 11 = 0$$

$$x - 3y = -11$$

$$x + 3y = 11$$

$$\text{or } x + 3y - 11 = 0$$

The slope of the line through (2, 3) and (−1, 4) is

$$\frac{y_2 - y_1}{x_2 - x_1} = \frac{4 - 3}{-1 - 2} = -\frac{1}{3}.$$

Using the point-slope form, we can find the equation of the line.

$$y - 3 = -\frac{1}{3}(x - 2)$$

$$3(y - 3) = -1(x - 2)$$

$$3y - 9 = -x + 2$$

$$x + 3y = 11$$

$$\text{or } x + 3y - 11 = 0$$

Thus, the given determinant is the equation of the given line.

55. The inverse of matrix A = $\begin{bmatrix} a & b \\ c & d \end{bmatrix}$

is

$$A^{-1} = \begin{bmatrix} \dfrac{d}{ad - bc} & \dfrac{-b}{ad - bc} \\ \dfrac{-c}{ad - bc} & \dfrac{a}{ad - bc} \end{bmatrix}.$$

By the definition of a determinant of a 2 × 2 matrix,

$$|A| = \begin{vmatrix} a & b \\ c & d \end{vmatrix} = ad - bc.$$

If $|A| = 0$ then $ad - bc = 0$, so each of the denominators in the elements of A^{-1} would be 0. Therefore, each element of A^{-1} would not exist. Thus, if $|A| = 0$, A^{-1} does not exist.

If A is the coefficient matrix of a linear system and $|A| = 0$, then the system has either no solution, or an infinite number of solutions, since A^{-1} does not exist.

Section 8.5

1. $\begin{vmatrix} 2 & 3 \\ 2 & 3 \end{vmatrix} = 0$

This statement is true by Property 5, since the two rows of the matrix

$\begin{bmatrix} 2 & 3 \\ 2 & 3 \end{bmatrix}$ are identical.

3. $\begin{vmatrix} 2 & 0 \\ 3 & 0 \end{vmatrix} = 0$

This statement is true by Property 1, since every element of column 2

of the matrix $\begin{bmatrix} 2 & 0 \\ 3 & 0 \end{bmatrix}$ is 0.

5. $\begin{vmatrix} -1 & 2 & 4 \\ 4 & -8 & -16 \\ 3 & 0 & 5 \end{vmatrix} = 0$

Row 1 of the matrix for this determinant equals row 2 multiplied by $-1/4$, so, by Property 4, we have

$$\begin{vmatrix} -1 & 2 & 4 \\ 4 & -8 & -16 \\ 3 & 0 & 5 \end{vmatrix} = -\frac{1}{4}\begin{vmatrix} 4 & -8 & -16 \\ 4 & -8 & -16 \\ 3 & 0 & 5 \end{vmatrix}.$$

By Property 5, the determinant of a matrix with two identical rows equals 0, so the value of the original determinant is

$$-\frac{1}{4}(0) = 0.$$

7. $\begin{vmatrix} 3 & 6 & 6 \\ 2 & 0 & 4 \\ 1 & 4 & 2 \end{vmatrix} = 0$

Column 3 of the matrix for this determinant equals column 1 multiplied by 2, so, by Property 4, we have

$$\begin{vmatrix} 3 & 6 & 6 \\ 2 & 0 & 4 \\ 1 & 4 & 2 \end{vmatrix} = 2\begin{vmatrix} 3 & 6 & 3 \\ 2 & 0 & 2 \\ 1 & 4 & 1 \end{vmatrix}.$$

By Property 5, the determinant of a matrix with two identical rows equals 0, so the value of the original determinant is

$$2(0) = 0.$$

9. $\begin{vmatrix} m & 2 & 2m \\ 3n & 1 & 6n \\ 5p & 6 & 10p \end{vmatrix} = 0$

Column 3 equals column 1 multiplied by 2, so by Property 4,

$$\begin{vmatrix} m & 2 & 2m \\ 3n & 1 & 6n \\ 5p & 6 & 10p \end{vmatrix} = 2\begin{vmatrix} m & 2 & m \\ 3n & 1 & 3n \\ 5p & 6 & 5p \end{vmatrix}.$$

By Property 5, the determinant of a matrix with two identical columns equals 0, so the value of the original determinant is

$$2(0) = 0.$$

13. $\begin{vmatrix} 4 & -2 \\ 3 & 8 \end{vmatrix} = \begin{vmatrix} 4 & 3 \\ -2 & 8 \end{vmatrix}$

Since corresponding rows and columns of the matrix are interchanged, Property 2 says that the determinant is not changed.

15. $\begin{vmatrix} -1 & 8 & 9 \\ 0 & 2 & 1 \\ 3 & 2 & 0 \end{vmatrix} = -\begin{vmatrix} 8 & -1 & 9 \\ 2 & 0 & 1 \\ 2 & 3 & 0 \end{vmatrix}$

By Property 3, interchanging two columns of a matrix reverses the sign of the determinant. Note that columns 1 and 2 have been interchanged here.

17. $-\frac{1}{2}\begin{vmatrix} 5 & -8 & 2 \\ 3 & -6 & 9 \\ 2 & 4 & 4 \end{vmatrix} = \begin{vmatrix} 5 & 4 & 2 \\ 3 & 3 & 9 \\ 2 & -2 & 4 \end{vmatrix}$

Property 4 says that if every element of column 2 of the matrix

$\begin{bmatrix} 5 & -8 & 2 \\ 3 & -6 & 9 \\ 2 & 4 & 4 \end{bmatrix}$ is multiplied by $-1/2$,

then the determinant of the new

matrix, $\begin{bmatrix} 5 & 4 & 2 \\ 3 & 3 & 9 \\ 2 & -2 & 4 \end{bmatrix}$, is $-\frac{1}{2}\begin{vmatrix} 5 & -8 & 2 \\ 3 & -6 & 9 \\ 2 & 4 & 4 \end{vmatrix}$.

19. $\begin{vmatrix} -1 & 6 \\ 3 & -5 \end{vmatrix} = \begin{vmatrix} -1 & 6 \\ 2 & 1 \end{vmatrix}$

By Property 6, if a multiple of a row of a matrix is added to the corresponding elements of another row, the value of the determinant is unchanged.

In this case, row 1 of the matrix $\begin{bmatrix} -1 & 6 \\ 3 & -5 \end{bmatrix}$ was added to row 2 to obtain the matrix $\begin{bmatrix} -1 & 6 \\ 2 & 1 \end{bmatrix}$.

21. $\begin{vmatrix} 13 & 5 \\ 6 & 1 \end{vmatrix} = \begin{vmatrix} -2 & 5 \\ 3 & 1 \end{vmatrix}$

Multiply column 2 of $\begin{bmatrix} 13 & 5 \\ 6 & 1 \end{bmatrix}$ by -3 and add the result to column 1 to obtain $\begin{bmatrix} -2 & 5 \\ 3 & 1 \end{bmatrix}$. By Property 6, the determinants are equal.

23. $2\begin{bmatrix} 4 & 2 & -1 \\ m & 2n & 3p \\ 5 & 2 & 0 \end{bmatrix} = \begin{bmatrix} 4 & 2 & -1 \\ 2m & 4n & 6p \\ 5 & 1 & 0 \end{bmatrix}$

Multiply row 2 of $\begin{vmatrix} 4 & 2 & -1 \\ m & 2n & 3p \\ 5 & 1 & 0 \end{vmatrix}$ by

2 to obtain $\begin{bmatrix} 4 & 2 & -1 \\ 2m & 4n & 6p \\ 5 & 1 & 0 \end{bmatrix}$.

By Property 4, the determinants are equal.

25. $\begin{bmatrix} -4 & 2 & 1 \\ 3 & 0 & 5 \\ -1 & 4 & -2 \end{bmatrix} = \begin{bmatrix} -4 & 2 & 1 + (-4)k \\ 3 & 0 & 5 + 3k \\ -1 & 4 & -2 + (-1)k \end{bmatrix}$

Multiply column 1 of $\begin{bmatrix} -4 & 2 & 1 \\ 3 & 0 & 5 \\ -1 & 4 & -2 \end{bmatrix}$ by k and add it to column 3. By Property 6, these determinants are equal.

27. $\begin{vmatrix} -5 & 10 \\ 6 & -12 \end{vmatrix}$

Multiply column 1 of $\begin{bmatrix} -5 & 10 \\ 6 & -12 \end{bmatrix}$ by 2, and add to column 2 to get $\begin{bmatrix} -5 & 0 \\ 6 & 0 \end{bmatrix}$.

By Property 1, the value of $\begin{bmatrix} -5 & 0 \\ 6 & 0 \end{bmatrix}$ is 0, so $\begin{vmatrix} -5 & 10 \\ 6 & -12 \end{vmatrix} = 0$.

29. $\begin{vmatrix} 6 & 8 & -12 \\ -1 & 16 & 2 \\ 4 & 0 & -8 \end{vmatrix}$

Multiply column 1 of the matrix for this determinant by 2 and add the result to column 3.

Thus,

$\begin{vmatrix} 6 & 8 & -12 \\ -1 & 16 & 2 \\ 4 & 0 & -8 \end{vmatrix} = \begin{vmatrix} 6 & 8 & 0 \\ -1 & 16 & 0 \\ 4 & 0 & 0 \end{vmatrix}$

Property 6

$= 0$ Property 1

31. $\begin{vmatrix} -2 & 2 & 3 \\ 0 & 2 & 1 \\ -1 & 4 & 0 \end{vmatrix} = \begin{vmatrix} -2 & -6 & 3 \\ 0 & 2 & 1 \\ -1 & 0 & 0 \end{vmatrix}$

Multiply column 1 of the corresponding matrix by 4 and add to column 2.

Expand the determinant about row 3.

$$= -1 \begin{vmatrix} -6 & 3 \\ 2 & 1 \end{vmatrix} - 0 \begin{vmatrix} -2 & 3 \\ 0 & 1 \end{vmatrix} + 0 \begin{vmatrix} -2 & -6 \\ 0 & 2 \end{vmatrix}$$

$$= -1(-6 - 6) = 12$$

33. $\begin{vmatrix} 6 & 3 & 2 \\ 1 & 0 & 2 \\ -1 & 4 & 1 \end{vmatrix} = \begin{vmatrix} 6 & 3 & 2 \\ 1 & 0 & 2 \\ 0 & 4 & 3 \end{vmatrix}$

Add row 2 of the corresponding matrix to row 3.

Expand about column 1.

$$= 6 \begin{vmatrix} 0 & 2 \\ 4 & 3 \end{vmatrix} - 1 \begin{vmatrix} 3 & 2 \\ 4 & 3 \end{vmatrix} + 0 \begin{vmatrix} 3 & 2 \\ 0 & 2 \end{vmatrix}$$

$$= 6(-8) - 1(1) = -49$$

35. $\begin{vmatrix} 2 & -1 & 1 & 0 \\ 1 & 1 & 0 & 1 \\ 0 & -1 & 1 & 1 \\ 1 & 2 & 1 & 2 \end{vmatrix} = \begin{vmatrix} 2 & 0 & 0 & -1 \\ 1 & 1 & 0 & 1 \\ 0 & -1 & 1 & 1 \\ 1 & 2 & 1 & 2 \end{vmatrix}$

Multiply row 3 of the corresponding matrix by -1 and add to row 1.

$$= \begin{vmatrix} 2 & 0 & 0 & -1 \\ 1 & 1 & 0 & 1 \\ 0 & -1 & 1 & 1 \\ 1 & 3 & 0 & 1 \end{vmatrix}$$

Multiply row 3 of the corresponding matrix by -1 and add to row 4.

Expand the determinant about column 3.

$$= 0 \begin{vmatrix} 1 & 1 & 1 \\ 0 & -1 & 1 \\ 1 & 3 & 1 \end{vmatrix} - 0 \begin{vmatrix} 2 & 0 & -1 \\ 0 & -1 & 1 \\ 1 & 3 & 1 \end{vmatrix}$$

$$+ 1 \begin{vmatrix} 2 & 0 & -1 \\ 1 & 1 & 1 \\ 1 & 3 & 1 \end{vmatrix} - 0 \begin{vmatrix} 2 & 0 & -1 \\ 1 & 1 & 1 \\ 0 & -1 & 1 \end{vmatrix}$$

$$= \begin{vmatrix} 2 & 0 & -1 \\ 1 & 1 & 1 \\ 1 & 3 & 1 \end{vmatrix}$$

Multiply row 2 by -2 and add to row 1.

$$= \begin{vmatrix} 0 & -2 & -3 \\ 1 & 1 & 1 \\ 1 & 3 & 1 \end{vmatrix}$$

Multiply row 2 by -1 and add to row 3.

$$= \begin{vmatrix} 0 & -2 & -3 \\ 1 & 1 & 1 \\ 0 & 2 & 0 \end{vmatrix}$$

Expand about column 1.

$$= 0 \begin{vmatrix} 1 & 1 \\ 2 & 0 \end{vmatrix} - 1 \begin{vmatrix} -2 & -3 \\ 2 & 0 \end{vmatrix} + 0 \begin{vmatrix} -2 & -3 \\ 1 & 1 \end{vmatrix}$$

$$= -1(6) = -6$$

37. $\begin{vmatrix} a & b & c \\ d & e & f \\ g & h & j \end{vmatrix}$

Expand about the first row.

$$= a \begin{vmatrix} e & f \\ h & j \end{vmatrix} - b \begin{vmatrix} d & f \\ g & j \end{vmatrix} + c \begin{vmatrix} d & e \\ g & h \end{vmatrix}$$

$$= a(ej - fh) - b(dj - fg)$$
$$+ c(dh - eg)$$

$$= aej - afh - bdj + bfg$$
$$+ cdh - ceg$$

Find the value of the determinant formed by interchanging rows and columns of the matrix $\begin{bmatrix} a & b & c \\ d & e & f \\ g & h & j \end{bmatrix}$ to prove Property 2.

Thus, $\begin{vmatrix} a & d & g \\ b & e & h \\ c & f & j \end{vmatrix}$ is to be evaluated.

Expand about the first column.

$$= a \begin{vmatrix} e & h \\ f & j \end{vmatrix} - b \begin{vmatrix} d & g \\ f & j \end{vmatrix} + c \begin{vmatrix} d & g \\ e & h \end{vmatrix}$$

= a(ej − fh) + b(dj − gf)

+ c(dh − eg)

= aej − afg − bdj − bfg

+ cdh − ceg

Since the values of the two determinants are the same, Property 2 is true for 3 × 3 determinants.

39. To prove Property 4, we must show that multiplying a row or column of a matrix by k multiplies the determinant by k.

If the first row of

$$\begin{bmatrix} a & b & c \\ d & e & f \\ g & h & j \end{bmatrix}$$

is multiplied by k, the result is

$$\begin{bmatrix} ka & kb & kc \\ d & e & f \\ g & h & j \end{bmatrix}.$$

Find its determinant by expanding by minors about the first row.

$$\begin{vmatrix} ka & kb & kc \\ d & e & f \\ g & h & j \end{vmatrix}$$

$$= ka\begin{vmatrix} e & f \\ h & j \end{vmatrix} - kb\begin{vmatrix} d & f \\ g & j \end{vmatrix} + kc\begin{vmatrix} d & e \\ g & h \end{vmatrix}$$

= ka(ej − fh) − kb(dj − gf)

+ kc(dh − eg)

= kaej − kafh − kbdj − kbgf

+ kcdh − kceg

= k(aej − afh − bdj + bfg

+ cdh − ceg)   (1)

By the definition of the value of a 3 × 3 determinant,

$$\begin{vmatrix} a & b & c \\ d & e & f \\ g & h & j \end{vmatrix}$$

= aej − afh − bdj + bfg + cdh

− ceg.   (2)

Notice that expression (1) is k times expression (2); thus Property (4) is true.

If a different row or column were multiplied by k and the determinant calculated, the result would be the same.

Section 8.6

1.   x + y = 4

2x − y = 2

$$D = \begin{vmatrix} 1 & 1 \\ 2 & -1 \end{vmatrix} = -3$$

To find D_x, replace the first column of $\begin{vmatrix} 1 & 1 \\ 2 & -1 \end{vmatrix}$ with $\begin{matrix} 4 \\ 2 \end{matrix}$.

$$D_x = \begin{vmatrix} 4 & 1 \\ 2 & -1 \end{vmatrix} = -6$$

To form D_y, replace the second column of D with $\begin{matrix} 4 \\ 2 \end{matrix}$.

$$D_y = \begin{vmatrix} 1 & 4 \\ 2 & 2 \end{vmatrix} = -6$$

$$x = \frac{D_x}{D} = \frac{-6}{-3} = 2$$

$$y = \frac{D_y}{D} = \frac{-6}{-3} = 2$$

Solution set: {(2, 2)}

3. $4x + 3y = -7$
$2x + 3y = -11$

$$D = \begin{vmatrix} 4 & 3 \\ 2 & 3 \end{vmatrix} = 6$$

To form D_x, replace the first column
of $\begin{vmatrix} 4 & 3 \\ 2 & 3 \end{vmatrix}$ with $\begin{matrix} -7 \\ -11 \end{matrix}$.

$$D_x = \begin{vmatrix} -7 & 3 \\ -11 & 3 \end{vmatrix} = 12$$

To form D_y, replace the second column
of D with $\begin{matrix} -7 \\ -11 \end{matrix}$.

$$D_y = \begin{vmatrix} 4 & -7 \\ 2 & -11 \end{vmatrix} = -30$$

$$x = \frac{D_x}{D} = \frac{12}{6} = 2$$

$$y = \frac{D_y}{D} = \frac{-30}{6} = -5$$

Solution set: $\{(2, -5)\}$

5. $5x + 4y = 10$
$3x - 7y = 6$

$$D = \begin{vmatrix} 5 & 4 \\ 3 & -7 \end{vmatrix} = -35 - 12 = -47$$

$$D_x = \begin{vmatrix} 10 & 4 \\ 6 & -7 \end{vmatrix} = -70 - 24 = -94$$

$$D_y = \begin{vmatrix} 5 & 10 \\ 3 & 6 \end{vmatrix} = 30 - 30 = 0$$

$$x = \frac{D_x}{D} = \frac{-94}{-47} = 2$$

$$y = \frac{D_y}{D} = \frac{0}{-47} = 0$$

Solution set: $\{(2, 0)\}$

7. $2x - 3y = -5$
$x + 5y = 17$

$$D = \begin{vmatrix} 2 & -3 \\ 1 & 5 \end{vmatrix} = 13$$

$$D_x = \begin{vmatrix} -5 & -3 \\ 17 & 5 \end{vmatrix} = 26$$

$$D_y = \begin{vmatrix} 2 & -5 \\ 1 & 17 \end{vmatrix} = 39$$

$$x = \frac{D_x}{D} = \frac{26}{13} = 2$$

$$y = \frac{D_y}{D} = \frac{39}{13} = 3$$

Solution set: $\{(2, 3)\}$

9. $3x + 2y = 4$ (1)
$6x + 4y = 8$ (2)

$$D = \begin{vmatrix} 3 & 2 \\ 6 & 4 \end{vmatrix} = 0$$

Since $D = 0$, Cramer's rule does not
apply. To determine whether the
system is inconsistent or contains
dependent equations, use the addition method. Multiply equation (1)
by -2 and add the result to equation
(2).

$$\begin{array}{r} -6x - 4y = -8 \\ \underline{6x + 4y = 8} \\ 0 = 0 \quad True \end{array}$$

This shows that equations (1) and
(2) are dependent.
To write the solution as an ordered
pair, solve equation (1) for x in
terms of y.

$$3x + 2y = 4$$
$$3x = 4 - 2y$$
$$x = \frac{4 - 2y}{3}$$

Solution set: $\left\{ \left(\frac{4 - 2y}{3}, y \right) \right\}$

11. $12x + 8y = 3$ (1)
     $15x + 10y = 9$ (2)

$$D = \begin{vmatrix} 12 & 8 \\ 15 & 10 \end{vmatrix} = 0$$

Since D = 0, Cramer's rule does not apply. Use the addition method. Multiply equation (1) by 5 and equation (2) by −4, and then add the resulting equations.

$$\begin{array}{r} 60x + 40y = 15 \\ -60x - 40y = -36 \\ \hline 0 = -21 \end{array}$$

This shows that the system is inconsistent.

Solution set: ∅

13. $4x - y + 3z = -3$
     $3x + y + z = 0$
     $2x - y + 4z = 0$

$$D = \begin{vmatrix} 4 & -1 & 3 \\ 3 & 1 & 1 \\ 2 & -1 & 4 \end{vmatrix}$$

Add row 1 to row 2; then add −1 times row 1 to row 3.

$$D = \begin{vmatrix} 4 & -1 & 3 \\ 7 & 0 & 4 \\ -2 & 0 & 1 \end{vmatrix}$$

Expand about column 2 to get

$$D = -(-1) \begin{vmatrix} 7 & 4 \\ -2 & 1 \end{vmatrix}$$
$$= 1(15)$$
$$= 15.$$

Replace the first column of D with $\begin{smallmatrix} -3 \\ 0 \\ 0 \end{smallmatrix}$ to find D_x.

$$D_x = \begin{vmatrix} -3 & -1 & 3 \\ 0 & 1 & 1 \\ 0 & -1 & 4 \end{vmatrix}$$

Expand about column 1 to get

$$D_x = -3 \begin{vmatrix} 1 & 1 \\ -1 & 4 \end{vmatrix} = -3(5)$$
$$= -15.$$

Replace the second column of D with $\begin{smallmatrix} -3 \\ 0 \\ 0 \end{smallmatrix}$ to get

$$D_y = \begin{vmatrix} 4 & -3 & 3 \\ 3 & 0 & 1 \\ 2 & 0 & 4 \end{vmatrix} = -(-3) \begin{vmatrix} 3 & 1 \\ 2 & 4 \end{vmatrix}$$
$$= 3(10)$$
$$= 30.$$

To find D_z, replace the third column of D with $\begin{smallmatrix} -3 \\ 0 \\ 0 \end{smallmatrix}$.

$$D_z = \begin{vmatrix} 4 & -1 & -3 \\ 3 & 1 & 0 \\ 2 & -1 & 0 \end{vmatrix}$$

Expand about column 3 to get

$$D_z = -3 \begin{vmatrix} 3 & 1 \\ 2 & -1 \end{vmatrix}$$
$$= -3(-5) = 15.$$

$$x = \frac{D_x}{D} = \frac{-15}{15} = -1$$

$$y = \frac{D_y}{D} = \frac{30}{15} = 2$$

$$z = \frac{D_z}{D} = \frac{15}{15} = 1$$

Solution set: $\{(-1,\ 2,\ 1)\}$

15. $2x - y + 4z = -2$
     $3x + 2y - z = -3$
     $x + 4y + 2z = 17$

$$D = \begin{vmatrix} 2 & -1 & 4 \\ 3 & 2 & -1 \\ 1 & 4 & 2 \end{vmatrix}$$

Multiply row 1 by 2 and add to row 2; then multiply row 1 by 4 and add to row 3.

$$D = \begin{vmatrix} 2 & -1 & 4 \\ 7 & 0 & 7 \\ 9 & 0 & 18 \end{vmatrix}$$

Expand about column 2.

$$D = -(-1)\begin{vmatrix} 7 & 7 \\ 9 & 18 \end{vmatrix} = 63$$

$$D_x = \begin{vmatrix} -2 & -1 & 4 \\ -3 & 2 & -1 \\ 17 & 4 & 2 \end{vmatrix}$$

Multiply row 1 by 2 and add to row 2; then multiply row 1 by 4 and add to row 3.

$$D_x = \begin{vmatrix} -2 & -1 & 4 \\ -7 & 0 & 7 \\ 9 & 0 & 18 \end{vmatrix}$$

Expand about column 2.

$$D_x = -(-1)\begin{vmatrix} -7 & 7 \\ 9 & 18 \end{vmatrix} = -189$$

$$D_y = \begin{vmatrix} 2 & -2 & 4 \\ 3 & -3 & -1 \\ 1 & 17 & 2 \end{vmatrix}$$

Add column 2 to column 1.

$$D_y = \begin{vmatrix} 0 & -2 & 4 \\ 0 & -3 & -1 \\ 18 & 17 & 2 \end{vmatrix}$$

Expand about column 1.

$$D_y = 18\begin{vmatrix} -2 & 4 \\ -3 & -1 \end{vmatrix} = 18(14)$$
$$= 252$$

$$D_z = \begin{vmatrix} 2 & -1 & -2 \\ 3 & 2 & -3 \\ 1 & 4 & 17 \end{vmatrix}$$

Add column 3 to column 1.

$$D_z = \begin{vmatrix} 0 & -1 & -2 \\ 0 & 2 & -3 \\ 18 & 4 & 17 \end{vmatrix}$$

Expand about column 1.

$$D_z = 18\begin{vmatrix} -1 & -2 \\ 2 & -3 \end{vmatrix}$$
$$= 18(7) = 126$$

$$x = \frac{D_x}{D} = \frac{-189}{63} = -3$$

$$y = \frac{D_y}{D} = \frac{252}{63} = 4$$

$$z = \frac{D_z}{D} = \frac{126}{63} = 2$$

Solution set: $\{(-3,\ 4,\ 2)\}$

17. $4x - 3y + z = -1$
     $5x - 7y + 2z = -2$
     $3x - 5y + z = 1$

$$D = \begin{vmatrix} 4 & -3 & 1 \\ 5 & 7 & 2 \\ 3 & -5 & -1 \end{vmatrix}$$

Multiply row 3 by 2 and add to row 2. Add row 1 to row 3.

$$D = \begin{vmatrix} 4 & -3 & 1 \\ 11 & -3 & 0 \\ 7 & -8 & 0 \end{vmatrix}$$

Expand about column 3 to get

$$D = 1\begin{vmatrix} 11 & -3 \\ 7 & -8 \end{vmatrix} = -67.$$

$$D_x = \begin{vmatrix} -1 & -3 & 1 \\ -2 & 7 & 2 \\ 1 & -5 & -1 \end{vmatrix}$$

Add column 1 to get column 3 to get

$$D_x = \begin{vmatrix} -1 & 3 & 0 \\ -2 & 7 & 0 \\ 1 & -5 & 0 \end{vmatrix} = 0,$$

since it has a column of zeros.

$$D_y = \begin{vmatrix} 4 & -1 & 1 \\ 5 & -2 & 2 \\ 3 & 1 & -1 \end{vmatrix}$$

Add column 2 to column 3 to get

$$D_y = \begin{vmatrix} 4 & -1 & 0 \\ 5 & -2 & 0 \\ 3 & 1 & 0 \end{vmatrix} = 0,$$

since it has a column of zeros.

$$D_z = \begin{vmatrix} 4 & -3 & -1 \\ 5 & 7 & -2 \\ 3 & -5 & 1 \end{vmatrix}$$

Add row 3 to row 1. Add twice row 3 to row 2.

$$D_z = \begin{vmatrix} 7 & -8 & 0 \\ 11 & -3 & 0 \\ 3 & -5 & 1 \end{vmatrix}$$

Expand about column 3 to get

$$D_z = 1\begin{vmatrix} 7 & -8 \\ 11 & -3 \end{vmatrix} = 67.$$

$$x = \frac{D_x}{D} = \frac{0}{-67} = 0$$

$$y = \frac{D_y}{D} = \frac{0}{-67} = 0$$

$$z = \frac{D_z}{D} = \frac{67}{-67} = -1$$

Solution set: $\{(0, 0, -1)\}$

19.
$$\begin{aligned} x + 2y + 3z &= 4 \quad (1) \\ 4x + 3y + 2z &= 1 \quad (2) \\ -x - 2y - 3z &= 0 \quad (3) \end{aligned}$$

$$D = \begin{vmatrix} 1 & 2 & 3 \\ 4 & 3 & 2 \\ -1 & -2 & -3 \end{vmatrix}$$

Add row 1 to row 3.

$$D = \begin{vmatrix} 1 & 2 & 3 \\ 4 & 3 & 2 \\ 0 & 0 & 0 \end{vmatrix} = 0$$

Since D = 0, Cramer's rule does not apply. Use the addition method.
Add equations (1) and (3).

$$\begin{aligned} x + 2y + 3z &= 4 \\ -x - 2y - 3z &= 0 \\ \hline 0 &= 4 \quad \textit{False} \end{aligned}$$

The system is inconsistent.
Solution set: Ø

21.
$$\begin{aligned} -2x - 2y + 3z &= 4 \quad (1) \\ 5x + 7y - z &= 2 \quad (2) \\ 2x + 2y - 3z &= -4 \quad (3) \end{aligned}$$

$$D = \begin{vmatrix} -2 & -2 & 3 \\ 5 & 7 & -1 \\ 2 & 2 & -3 \end{vmatrix}$$

Add row 1 to row 3.

$$D = \begin{vmatrix} -2 & -2 & 3 \\ 5 & 7 & -1 \\ 0 & 0 & 0 \end{vmatrix} = 0,$$

since there is a row of zeros.
Since D = 0, Cramer's rule does not apply. Use the addition method.

Add equations (1) and (3).

$$-2x - 2y + 3z = 4$$
$$\underline{2x + 2y - 3z = -4}$$
$$0 = 0$$

Equations (1) and (3) are dependent. Solve the system made up of equations (2) and (3) in terms of the arbitrary variable z.

To eliminate x, multiply equation (2) by -2 and equation (3) by 5 and add the results.

$$-10x - 14y + 2z = -4$$
$$\underline{10x + 10y - 15z = -20}$$
$$-4y - 13z = -24$$

Solve for y in terms of z.

$$-4y = -24 + 13z$$
$$y = \frac{24 - 13z}{4}$$

Now, express x also in terms of z by solving equation (3) for x and substituting $\frac{24 - 13z}{4}$ for y in the result.

$$2x + 2y - 3z = -4$$
$$2x = -2y + 3z - 4$$
$$x = \frac{-2y + 13z - 4}{2}$$
$$= \frac{-2\left(\frac{24 - 13z}{2}\right) + 3z - 4}{2}$$
$$= \frac{\frac{-24 + 13z}{2} + 3z - 4}{2}$$
$$= \frac{-24 + 13z + 6z - 8}{4}$$
$$x = \frac{-32 + 19z}{4}$$

Solution set (with z arbitrary):

$$\left\{\left(\frac{-32 + 19z}{4}, \frac{24 - 13z}{4}, z\right)\right\}$$

23. $2x + 3y = 13$
$2y - z = 5$
$x + 2z = 4$

$$D = \begin{vmatrix} 2 & 3 & 0 \\ 0 & 2 & -1 \\ 1 & 0 & 2 \end{vmatrix}$$

Add twice row 2 to row 3.

$$D = \begin{vmatrix} 2 & 3 & 0 \\ 0 & 2 & -1 \\ 1 & 4 & 0 \end{vmatrix}$$

Expand about column 3.

$$D = -(-1)\begin{vmatrix} 2 & 3 \\ 1 & 4 \end{vmatrix} = 5$$

$$D_x = \begin{vmatrix} 13 & 3 & 0 \\ 5 & 2 & -1 \\ 4 & 0 & 2 \end{vmatrix}$$

Add 1/2 row 3 to row 2.

$$D_x = \begin{vmatrix} 13 & 3 & 0 \\ 7 & 2 & 0 \\ 4 & 0 & 2 \end{vmatrix}$$

Expand about column 3.

$$D_x = 2\begin{vmatrix} 13 & 3 \\ 7 & 2 \end{vmatrix} = 10$$

$$D_y = \begin{vmatrix} 2 & 13 & 0 \\ 0 & 5 & -1 \\ 1 & 4 & 2 \end{vmatrix}$$

Add 5 times column 3 to column 2.

$$D_y = \begin{vmatrix} 2 & 13 & 0 \\ 0 & 0 & -1 \\ 1 & 14 & 2 \end{vmatrix}$$

Expand about row 2.

$$D_y = -(-1)\begin{vmatrix} 2 & 13 \\ 1 & 14 \end{vmatrix} = 15$$

$$D_z = \begin{vmatrix} 2 & 3 & 13 \\ 0 & 2 & 5 \\ 1 & 0 & 4 \end{vmatrix}$$

Multiply row 3 by -2 and add to row 1.

$$D_z = \begin{vmatrix} 0 & 3 & 5 \\ 0 & 2 & 5 \\ 1 & 0 & 4 \end{vmatrix}$$

Expand about column 1.

$$D_z = 1 \begin{vmatrix} 3 & 5 \\ 2 & 5 \end{vmatrix} = 5$$

$$x = \frac{D_x}{D} = \frac{10}{5} = 2$$

$$y = \frac{D_y}{D} = \frac{15}{5} = 3$$

$$z = \frac{D_z}{D} = \frac{5}{5} = 1$$

Solution set: $\{(2,\ 3,\ 1)\}$

25. $5x - y = -4$

$3x + 2z = 4$

$4y + 3z = 22$

$$D = \begin{vmatrix} 5 & -1 & 0 \\ 3 & 0 & 2 \\ 0 & 4 & 3 \end{vmatrix}$$

Add 4 times row 1 to row 3.

$$D = \begin{vmatrix} 5 & -1 & 0 \\ 3 & 0 & 2 \\ 20 & 0 & 3 \end{vmatrix}$$

Expand about column 2.

$$D = -(-1) \begin{vmatrix} 3 & 2 \\ 20 & 3 \end{vmatrix} = -31$$

$$D_x = \begin{vmatrix} -4 & -1 & 0 \\ 4 & 0 & 2 \\ 22 & 4 & 3 \end{vmatrix}$$

Add 4 times row 1 to row 3.

$$D_x = \begin{vmatrix} -4 & -1 & 0 \\ 4 & 0 & 2 \\ 6 & 0 & 3 \end{vmatrix}$$

Expand about column 2.

$$D_x = -(-1) \begin{vmatrix} 4 & 2 \\ 6 & 3 \end{vmatrix} = 0$$

$$D_y = \begin{vmatrix} 5 & -4 & 0 \\ 3 & 4 & 2 \\ 0 & 22 & 3 \end{vmatrix}$$

Add column 2 to column 1.

$$D_y = \begin{vmatrix} 1 & -4 & 0 \\ 7 & 4 & 2 \\ 22 & 22 & 3 \end{vmatrix}$$

Add 4 times column 1 to column 2.

$$D_y = \begin{vmatrix} 1 & 0 & 0 \\ 7 & 32 & 2 \\ 22 & 110 & 3 \end{vmatrix}$$

Expand about row 1.

$$D_y = 1 \begin{vmatrix} 32 & 2 \\ 110 & 3 \end{vmatrix}$$

$$= 96 - 220 = -124$$

$$D_z = \begin{vmatrix} 5 & -1 & -4 \\ 3 & 0 & 4 \\ 0 & 4 & 22 \end{vmatrix}$$

Add 4 times row 1 to row 3.

$$D_z = \begin{vmatrix} 5 & -1 & -4 \\ 3 & 0 & 4 \\ 20 & 0 & 6 \end{vmatrix}$$

Expand about column 2.

$$D_z = -(-1) \begin{vmatrix} 3 & 4 \\ 20 & 6 \end{vmatrix}$$

$$= 1(18 - 80) = -62$$

$$x = \frac{D_x}{D} = \frac{0}{-31} = 0$$

$$y = \frac{D_y}{D} = \frac{-124}{-31} = 4$$

$$z = \frac{D_z}{D} = \frac{-62}{-31} = 2$$

Solution set: $\{(0,\ 4,\ 2)\}$

27.
$$x + 2y = 10$$
$$3x + 4z = 7$$
$$-y - z = 1$$

$$D = \begin{vmatrix} 1 & 2 & 0 \\ 3 & 0 & 4 \\ 0 & -1 & -1 \end{vmatrix}$$

Multiply row 1 by −3 and add to row 2.

$$D = \begin{vmatrix} 1 & 2 & 0 \\ 0 & -6 & 4 \\ 0 & -1 & -1 \end{vmatrix}$$

Expand about column 1.

$$D = 1 \begin{vmatrix} -6 & 4 \\ -1 & -1 \end{vmatrix} = 10$$

$$D_x = \begin{vmatrix} 10 & 2 & 0 \\ 7 & 0 & 4 \\ 1 & -1 & -1 \end{vmatrix}$$

Add column 1 to column 2 and to column 3.

$$D_x = \begin{vmatrix} 10 & 12 & 10 \\ 7 & 7 & 11 \\ 1 & 0 & 0 \end{vmatrix}$$

Expand about row 3.

$$D_x = 1 \begin{vmatrix} 12 & 10 \\ 7 & 11 \end{vmatrix} = 62$$

$$D_y = \begin{vmatrix} 1 & 10 & 0 \\ 3 & 7 & 4 \\ 0 & 1 & -1 \end{vmatrix}$$

Add column 2 to column 3.

$$D_y = \begin{vmatrix} 1 & 10 & 10 \\ 3 & 7 & 11 \\ 0 & 1 & 0 \end{vmatrix}$$

Expand about row 3.

$$D_y = -1 \begin{vmatrix} 1 & 10 \\ 3 & 11 \end{vmatrix} = 19$$

$$D_z = \begin{vmatrix} 1 & 2 & 10 \\ 3 & 0 & 7 \\ 0 & -1 & 1 \end{vmatrix}$$

Add column 3 to column 2.

$$D_z = \begin{vmatrix} 1 & 12 & 10 \\ 3 & 7 & 7 \\ 0 & 0 & 1 \end{vmatrix}$$

Expand about row 3.

$$D_z = 1 \begin{vmatrix} 1 & 12 \\ 3 & 7 \end{vmatrix} = -29$$

$$x = \frac{D_x}{D} = \frac{62}{10} = \frac{31}{5}$$

$$y = \frac{D_y}{D} = \frac{19}{10}$$

$$z = \frac{D_z}{D} = \frac{-29}{10} = -\frac{29}{10}$$

Solution set: $\left\{ \left(\frac{31}{5}, \frac{19}{10}, -\frac{29}{10} \right) \right\}$

31.
$$x + 3y - 2z - w = 9 \quad (1)$$
$$4x + y + z + 2w = 2 \quad (2)$$
$$-3x - y + z - w = -5 \quad (3)$$
$$x - y - 3z - 2w = 2 \quad (4)$$

$$D = \begin{vmatrix} 1 & 3 & -2 & -1 \\ 4 & 1 & 1 & 2 \\ -3 & -1 & 1 & -1 \\ 1 & -1 & -3 & -2 \end{vmatrix}$$

Expand about column 1.

$$= 1 \begin{vmatrix} -1 & 1 & 2 \\ -1 & 1 & -1 \\ -1 & -3 & -2 \end{vmatrix} - 4 \begin{vmatrix} 3 & -2 & -1 \\ -1 & 1 & -1 \\ -1 & -3 & -2 \end{vmatrix}$$

$$- 3 \begin{vmatrix} 3 & -2 & -1 \\ 1 & 1 & 2 \\ -1 & -3 & -2 \end{vmatrix} - 1 \begin{vmatrix} 3 & -2 & -1 \\ 1 & 1 & 2 \\ -1 & 1 & 1 \end{vmatrix}$$

$= 1[-1(-2-3)+(-2+6)-1(-1-2)]$

$\quad -4[3(-2-3)+1(4-3)-1(2+1)]$

$\quad -3[3(-2+6)-1(4-3)-1(-4+1)]$

$\quad -1[3(1-2)-1(-2+1)-1(-4+1)]$

$= 12 + 68 - 42 - 1$

$= 37$

$$D_x = \begin{vmatrix} 9 & 3 & -2 & -1 \\ 2 & 1 & 1 & 2 \\ -5 & -1 & 1 & -1 \\ 2 & -1 & -3 & -2 \end{vmatrix}$$

$$= -3\begin{vmatrix} 2 & 1 & 2 \\ -5 & 1 & -1 \\ 2 & -3 & -2 \end{vmatrix} + 1\begin{vmatrix} 9 & -2 & -1 \\ -5 & 1 & -1 \\ 2 & -3 & -2 \end{vmatrix}$$

$$+ 1\begin{vmatrix} 9 & -2 & -1 \\ 2 & 1 & 2 \\ 2 & -3 & -2 \end{vmatrix} - 1\begin{vmatrix} 9 & -2 & -1 \\ -2 & 1 & 2 \\ -5 & 1 & -1 \end{vmatrix}$$

$= -3[2(-2-3)-1(10+2)+2(15-2)]$

$\quad +1[-1(15-2)+1(-27+4)-2(9-10)]$

$\quad +1[-1(6-2)-2(-27+4)-2(9+4)]$

$\quad -1[-1(2+5)-2(9-10)-1(9+4)]$

$\quad = -12 - 34 + 28 + 18 = 0$

$$x = \frac{D_x}{D} = \frac{0}{37} = 0$$

Substitute 0 for x in equations (1), (2), and (3).

The resulting system of 3 equations in 3 variables may be solved by Cramer's rule.

$3y - 2z - w = 9$ (4)

$y + z + 2w = 2$ (5)

$-y + z - w = -5$ (6)

Find D, D_x, D_y, and D_z for this system.

$$D = \begin{vmatrix} 3 & -2 & -1 \\ 1 & 1 & 2 \\ -1 & -1 & -1 \end{vmatrix} = -9$$

$$D_y = \begin{vmatrix} 9 & -2 & -1 \\ 2 & 1 & 2 \\ -5 & 1 & -1 \end{vmatrix} = -18$$

$$D_z = \begin{vmatrix} 3 & 9 & -1 \\ 1 & 2 & 2 \\ -1 & -5 & -1 \end{vmatrix} = 18$$

$$D_w = \begin{vmatrix} 3 & -2 & 9 \\ 1 & 1 & 2 \\ -1 & 1 & -5 \end{vmatrix} = -9$$

$$y = \frac{D_y}{D} = \frac{-18}{-9} = 2$$

$$z = \frac{D_z}{D} = \frac{18}{-9} = -2$$

$$w = \frac{D_w}{D} = \frac{-9}{-9} = 1$$

Solution set: $\{(0, 2, -2, 1)\}$

33. $x + y - z + w = 2$

$\quad x - y + z + w = 4$

$-2x + y + 2z - w = -5$

$\quad x + 3z + 2w = 5$

$$D = \begin{vmatrix} 1 & 1 & -1 & 1 \\ 1 & -1 & 1 & 1 \\ -2 & 1 & 2 & -1 \\ 1 & 0 & 3 & 2 \end{vmatrix}$$

Expand about row 1.

$$= 1\begin{vmatrix} -1 & 1 & 1 \\ 1 & 2 & -1 \\ 0 & 3 & 2 \end{vmatrix} - 1\begin{vmatrix} 1 & 1 & 1 \\ -2 & 2 & -1 \\ 1 & 3 & 2 \end{vmatrix}$$

$$- 1\begin{vmatrix} 1 & -1 & 1 \\ -2 & 1 & -1 \\ 1 & 0 & 2 \end{vmatrix} - 1\begin{vmatrix} 1 & -1 & 1 \\ -2 & 1 & 2 \\ 1 & 0 & 3 \end{vmatrix}$$

$$= 1\left[-1\begin{vmatrix} 2 & -1 \\ 3 & 2 \end{vmatrix} - 1\begin{vmatrix} 1 & -1 \\ 0 & 2 \end{vmatrix} + 1\begin{vmatrix} 1 & 2 \\ 0 & 3 \end{vmatrix}\right]$$

$$- 1\left[1\begin{vmatrix} 2 & -1 \\ 3 & 2 \end{vmatrix} - 1\begin{vmatrix} -2 & -1 \\ 1 & 2 \end{vmatrix} + 1\begin{vmatrix} -2 & 2 \\ 1 & 3 \end{vmatrix}\right]$$

$$- 1\left[1\begin{vmatrix} 1 & -1 \\ 0 & 2 \end{vmatrix} + 1\begin{vmatrix} -2 & -1 \\ 1 & 2 \end{vmatrix} + 1\begin{vmatrix} -2 & 1 \\ 1 & 0 \end{vmatrix}\right]$$

$$- 1\left[1\begin{vmatrix} 1 & 2 \\ 0 & 3 \end{vmatrix} + 1\begin{vmatrix} -2 & 2 \\ 1 & 3 \end{vmatrix} + 1\begin{vmatrix} -2 & 1 \\ 1 & 0 \end{vmatrix}\right]$$

$$= 1[-1(4 + 3) - 1(2) + 1(3)]$$
$$\quad - 1[1(4 + 3) - 1(-4 + 1) + 1(-6 - 2)]$$
$$\quad - 1[1(2) + 1(-4 + 1) + 1(-1)]$$
$$\quad - 1[1(3) + 1(-6 - 2) + 1(-1)]$$
$$= 1(-7 - 2 + 3) - 1(7 + 3 - 8)$$
$$\quad - 1(2 - 3 - 1) - 1(3 - 8 - 1)$$
$$= -6 - 2 + 2 + 6$$
$$= 0$$

Cramer's rule does not apply since $D = 0$.

35. $bx + y = a^2$

$ax + y = b^2$

$$D = \begin{vmatrix} b & 1 \\ a & 1 \end{vmatrix} = b - a$$

$$D_x = \begin{vmatrix} a^2 & 1 \\ b^2 & 1 \end{vmatrix} = a^2 - b^2$$

$$D_y = \begin{vmatrix} b & a^2 \\ a & b^2 \end{vmatrix} = b^3 - a^3$$

$$x = \frac{D_x}{D} = \frac{a^2 - b^2}{b - a}$$

$$= \frac{(a + b)(a - b)}{(b - a)}$$
$$\quad \textit{Factor numerator as differ-}$$
$$\quad \textit{ence of two squares}$$

$$= -(a + b) = -a - b$$

$$y = \frac{D_y}{D} = \frac{a^3 - a^3}{b - a}$$

$$= \frac{(b - a)(b^2 + ab + a^2)}{b - a}$$
$$\quad \textit{Factor numerator as differ-}$$
$$\quad \textit{ence of two cubes}$$

$$= b^2 + ab + a^2$$

Solution set:
$$\{(-a - b, \ a^2 + ab + b^2)\}$$

37. $b^2x + a^2y = b^2$

$ax + \ by = a$

$$D = \begin{vmatrix} b^2 & a^2 \\ a & b \end{vmatrix} = b^3 - a^3$$

Note that for Cramer's rule to apply, $b^3 \neq a^3$, which is equivalent to $b \neq a$ or $b - a \neq 0$.

$$D_x = \begin{vmatrix} b^2 & a^2 \\ a & b \end{vmatrix} = b^3 - a^3$$

$$D_y = \begin{vmatrix} b^2 & b^2 \\ a & a \end{vmatrix} = ab^2 - ab^2 = 0$$

$$x = \frac{D_x}{D} = \frac{b^3 - a^3}{b^3 - a^3} = 1$$

$$y = \frac{D_y}{D} = \frac{0}{b^3 - a^3} = 0$$

Solution set: $\{(1, 0)\}$

39. $a_1x + b_1y = c_1 \quad (1)$

$a_2x + b_2y = c_2 \quad (2)$

If $D_y = 0$, then

$$\begin{vmatrix} a_1 & c_1 \\ a_2 & c_2 \end{vmatrix} = 0,$$

or

$a_1c_2 - a_2c_1 = 0.$

If $D_x = 0$, then

$$\begin{vmatrix} c_1 & b_1 \\ c_2 & b_2 \end{vmatrix} = 0,$$

or

$c_1b_2 - c_2b_1 = 0.$

$$D = \begin{vmatrix} a_1 & b_1 \\ a_2 & b_2 \end{vmatrix}$$

$$= a_1b_2 - b_1a_2$$

Show that

$a_1b_2 - b_1a_2 = 0$ if $c_1c_2 \neq 0$.

From

$$a_1c_2 - a_2c_1 = 0$$

we get

$$a_1c_2 = a_2c_1. \qquad (3)$$

From

$$c_1b_2 - c_1b_1 = 0$$

we get

$$b_2c_1 = b_1c_2. \qquad (4)$$

Multiplying the left sides and the right sides of equations (3) and (4), we get

$$(a_1c_2)(b_2c_1) = (a_2c_1)(b_1c_2)$$
$$a_1b_2c_1c_2 = a_2b_1c_2. \qquad (5)$$

Since $c_1c_2 \neq 0$, we can divide both sides of equation (5) by c_1c_2 to get

$$a_1b_2 = a_2b_1 \qquad (6)$$

or

$$a_1b_2 - a_2b_1 = 0$$

or $\qquad D = 0$.

To see that the equations are dependent, multiply equation (1) by a_2 and equation (2) by $-a_1$ and add.

$$\begin{array}{rcl} a_2a_1x + b_1a_2y &=& a_2c_1 \\ -a_1a_2x - a_1b_2y &=& -a_1c_2 \\ \hline (b_1a_2 - a_1b_2)y &=& a_2c_1 - a_1c_2 \end{array}$$

But $b_1a_2 - a_1b_2 = 0$ by equation (6) and $a_2c_1 - a_1c_2 = 0$ by equation (3).

This gives us

$$0y = 0$$
$$0 = 0.$$

Therefore, the equations are dependent.

Chapter 8 Review Exercises

1. $\begin{bmatrix} 2 & z & 1 \\ m & 9 & -7 \end{bmatrix} = \begin{bmatrix} x & 5 & 1 \\ -8 & y & p \end{bmatrix}$

All elements in corresponding positions must be equal, so $m = -8$, $p = -7$, $x = 2$, $y = 9$, and $z = 5$.

3. $\begin{bmatrix} 5 & x + 2 \\ -6y & z \end{bmatrix} = \begin{bmatrix} a & 3x - 1 \\ 5y & 9 \end{bmatrix}$

$a = 5 \qquad x + 2 = 3x - 1$
$\qquad\qquad\quad 3 = 2x$
$\qquad\qquad\quad \dfrac{3}{2} = x$

$-6y = 5y \qquad z = 9$
$\quad 0 = 11y$
$\quad 0 = y$

Thus, $a = 5$, $x = \dfrac{3}{2}$, $y = 0$, and $z = 9$.

5. $\begin{bmatrix} 3 & -4 & 2 \\ 5 & -1 & 6 \end{bmatrix} + \begin{bmatrix} -3 & 2 & 5 \\ 1 & 0 & 4 \end{bmatrix}$

$= \begin{bmatrix} 0 & -2 & 7 \\ 6 & -1 & 10 \end{bmatrix}$

7. $\begin{bmatrix} 2 & 5 & 8 \\ 1 & 9 & 2 \end{bmatrix} - \begin{bmatrix} 3 & 4 \\ 7 & 1 \end{bmatrix}$

This operation is not possible because one matrix is 2 × 3 and the other 2 × 2. Matrices of different sizes cannot be added or subtracted.

9. $-1\begin{bmatrix} 3 & -5 & 2 \\ 1 & 7 & -4 \end{bmatrix} + 5\begin{bmatrix} 0 & 2 \\ -1 & 3 \end{bmatrix} = \begin{bmatrix} -3 & 5 & -2 \\ -1 & -7 & 4 \end{bmatrix} + \begin{bmatrix} 0 & 10 \\ -5 & 15 \end{bmatrix}$

This operation is not possible because matrices of different sizes cannot be adde

11. The sum of two m × n matrices A and B is found by adding corresponding elements.

13. The given information may be written as the 3 × 4 matrix.

$$\begin{bmatrix} 606 & 354 & 4434 & 28 \\ 397 & 238 & 2981 & 18 \\ 479 & 263 & 3359 & 21 \end{bmatrix},$$

or as the 4 × 3 matrix

$$\begin{bmatrix} 606 & 397 & 479 \\ 354 & 238 & 263 \\ 4434 & 2981 & 3359 \\ 28 & 18 & 21 \end{bmatrix}.$$

15. $\begin{bmatrix} 3 & 2 & -1 \\ 4 & 0 & 6 \end{bmatrix}\begin{bmatrix} -2 & 0 \\ 0 & 2 \\ 3 & 1 \end{bmatrix}$

$= \begin{bmatrix} 3(-2) + 2(0) + (-1)(3) & 3(0) + 2(2) + (-1)(1) \\ 4(-2) + 0(0) + 6(3) & 4(0) + 0(2) + 6(1) \end{bmatrix}$

$= \begin{bmatrix} -9 & 3 \\ 10 & 6 \end{bmatrix}$

17. $\begin{bmatrix} 1 & 2 & 5 \\ -3 & 4 & 7 \\ 0 & 2 & -1 \end{bmatrix}\begin{bmatrix} 4 & 2 & 3 \\ 10 & -5 & 6 \end{bmatrix}$

It is not possible to find this product because the number of columns of the firs matrix (three) is not equal to the number of rows of the second matrix (two).

19. $[3 \quad -1 \quad 0]\begin{bmatrix} 1 & 3 & 2 \\ 2 & -4 & 0 \\ 5 & 7 & 3 \end{bmatrix} = [3 - 2 + 0 \quad 9 + 4 + 0 \quad 6 + 0 + 0]$

$= [1 \quad 13 \quad 6]$

21. $\begin{bmatrix} 2 & -3 \\ 1 & -2 \end{bmatrix}\begin{bmatrix} 2 & -3 \\ 1 & -2 \end{bmatrix} = \begin{bmatrix} 1 & 0 \\ 0 & 1 \end{bmatrix}$

The product is the 2 × 2 identity matrix. It is not necessary to multiply in the opposite order since the two matrices are the same. The given matrices are inverses of each other, that is, $\begin{bmatrix} 2 & -3 \\ 1 & -2 \end{bmatrix}$ is its own inverse.

23. $\begin{bmatrix} 2 & 0 & 6 \\ 0 & 1 & 0 \\ 1 & 0 & 1 \end{bmatrix}\begin{bmatrix} -1 & 0 & \frac{3}{2} \\ 0 & 1 & 0 \\ \frac{1}{4} & 0 & -1 \end{bmatrix}$

$= \begin{bmatrix} -\frac{1}{2} & 0 & -3 \\ 0 & 1 & 0 \\ -\frac{3}{4} & 0 & \frac{1}{2} \end{bmatrix}$

Since this product is not the 3 × 3 identity matrix, the given matrices are not inverses of each other.

25. Find the inverse of A = $\begin{bmatrix} 2 & 1 \\ 5 & 3 \end{bmatrix}$, if it exists.

$[A|I] = \begin{bmatrix} 2 & 1 & | & 1 & 0 \\ 5 & 3 & | & 0 & 1 \end{bmatrix}$

$\begin{bmatrix} 1 & \frac{1}{2} & | & \frac{1}{2} & 0 \\ 5 & 3 & | & 0 & 1 \end{bmatrix} \frac{1}{2}R1$

$\begin{bmatrix} 1 & \frac{1}{2} & | & \frac{1}{2} & 0 \\ 0 & \frac{1}{2} & | & -\frac{5}{2} & 1 \end{bmatrix} -5R1 + R2$

$\begin{bmatrix} 1 & \frac{1}{2} & | & \frac{1}{2} & 0 \\ 0 & 1 & | & -5 & 2 \end{bmatrix} 2R2$

$\begin{bmatrix} 1 & 0 & | & 3 & -1 \\ 0 & 1 & | & -5 & 2 \end{bmatrix} -\frac{1}{2}R2 + R1$

$A^{-1} = \begin{bmatrix} 3 & -1 \\ -5 & 2 \end{bmatrix}$

27. Find the inverse of A = $\begin{bmatrix} 2 & 0 \\ -1 & 5 \end{bmatrix}$, if it exists.

$[A|I] = \begin{bmatrix} 2 & 0 & | & 1 & 0 \\ -1 & 5 & | & 0 & 1 \end{bmatrix}$

$\begin{bmatrix} 1 & 0 & | & \frac{1}{2} & 0 \\ -1 & 5 & | & 0 & 1 \end{bmatrix} \frac{1}{2}R1$

$\begin{bmatrix} 1 & 0 & | & \frac{1}{2} & 0 \\ 0 & 5 & | & \frac{1}{2} & 1 \end{bmatrix} R1 + R2$

$\begin{bmatrix} 1 & 0 & | & \frac{1}{2} & 0 \\ 0 & 1 & | & \frac{1}{10} & \frac{1}{5} \end{bmatrix} \frac{1}{5}R2$

$A^{-1} = \begin{bmatrix} \frac{1}{2} & 0 \\ \frac{1}{10} & \frac{1}{5} \end{bmatrix}$

29. Find the inverse of A = $\begin{bmatrix} 2 & -1 & 0 \\ 1 & 0 & 1 \\ 1 & -2 & 0 \end{bmatrix}$, if it exists.

$[A|I] = \begin{bmatrix} 2 & -1 & 0 & | & 1 & 0 & 0 \\ 1 & 0 & 1 & | & 0 & 1 & 0 \\ 1 & -2 & 0 & | & 0 & 0 & 1 \end{bmatrix}$

$\begin{bmatrix} 1 & -1 & -1 & | & 1 & -1 & 0 \\ 1 & 0 & 1 & | & 0 & 1 & 0 \\ 1 & -2 & 0 & | & 0 & 0 & 1 \end{bmatrix} -1R2 + R1$

$\begin{bmatrix} 1 & -1 & -1 & | & 1 & -1 & 0 \\ 0 & 1 & 2 & | & -1 & 2 & 0 \\ 1 & -2 & 0 & | & 0 & 0 & 1 \end{bmatrix} -1R1 + R2$

Chapter 8 Matrices

$$\begin{bmatrix} 1 & -1 & -1 & | & 1 & -1 & 0 \\ 0 & 1 & 2 & | & -1 & 2 & 0 \\ 0 & -1 & 1 & | & -1 & 1 & 1 \end{bmatrix} \begin{array}{l} \\ \\ -1R1 + R3 \end{array}$$

$$\begin{bmatrix} 1 & -1 & -1 & | & 1 & -1 & 0 \\ 0 & 1 & 2 & | & -1 & 2 & 0 \\ 0 & 0 & 3 & | & -2 & 3 & 1 \end{bmatrix} \begin{array}{l} \\ \\ R2 + R3 \end{array}$$

$$\begin{bmatrix} 1 & -1 & -1 & | & 1 & -1 & 0 \\ 0 & 1 & 2 & | & -1 & 2 & 0 \\ 0 & 0 & 1 & | & -\frac{2}{3} & 1 & \frac{1}{3} \end{bmatrix} \begin{array}{l} \\ \\ \frac{1}{3}R3 \end{array}$$

$$\begin{bmatrix} 1 & -1 & -1 & | & 1 & -1 & 0 \\ 0 & 1 & 0 & | & \frac{1}{3} & 0 & -\frac{2}{3} \\ 0 & 0 & 1 & | & -\frac{2}{3} & 1 & \frac{1}{3} \end{bmatrix} \begin{array}{l} \\ -2R3 + R2 \\ \end{array}$$

$$\begin{bmatrix} 1 & 0 & -1 & | & \frac{4}{3} & -1 & -\frac{2}{3} \\ 0 & 1 & 0 & | & \frac{1}{3} & 0 & -\frac{2}{3} \\ 0 & 0 & 1 & | & -\frac{2}{3} & 1 & \frac{1}{3} \end{bmatrix} \begin{array}{l} R2 + R1 \\ \\ \end{array}$$

$$\begin{bmatrix} 1 & 0 & 0 & | & \frac{2}{3} & 0 & -\frac{1}{3} \\ 0 & 1 & 0 & | & \frac{1}{3} & 0 & -\frac{2}{3} \\ 0 & 0 & 1 & | & -\frac{2}{3} & 1 & \frac{1}{3} \end{bmatrix} \begin{array}{l} R3 + R1 \\ \\ \end{array}$$

$$A^{-1} = \begin{bmatrix} \frac{2}{3} & 0 & -\frac{1}{3} \\ \frac{1}{3} & 0 & -\frac{2}{3} \\ -\frac{2}{3} & 1 & \frac{1}{3} \end{bmatrix}$$

31. $\quad x + y = 4$
$\quad 2x + 3y = 10$

$A = \begin{bmatrix} 1 & 1 \\ 2 & 3 \end{bmatrix}$, $X = \begin{bmatrix} x \\ y \end{bmatrix}$, $B = \begin{bmatrix} 4 \\ 10 \end{bmatrix}$

Find A^{-1}.

$[A|I] = \begin{bmatrix} 1 & 1 & | & 1 & 0 \\ 2 & 3 & | & 0 & 1 \end{bmatrix}$

$\begin{bmatrix} 1 & 1 & | & 1 & 0 \\ 0 & 1 & | & -2 & 1 \end{bmatrix} \begin{array}{l} \\ -2R1 + R2 \end{array}$

$\begin{bmatrix} 1 & 0 & | & 3 & -1 \\ 0 & 1 & | & -2 & 1 \end{bmatrix} \begin{array}{l} -1R2 + R1 \\ \end{array}$

$A^{-1} = \begin{bmatrix} 3 & -1 \\ -2 & 1 \end{bmatrix}$

$X = A^{-1} = B = \begin{bmatrix} 3 & -1 \\ -2 & 1 \end{bmatrix} \begin{bmatrix} 4 \\ 10 \end{bmatrix} = \begin{bmatrix} 2 \\ 2 \end{bmatrix}$

Solution set: $\{(2, 2)\}$

33. $\quad 2x + y = 5$
$\quad 3x - 2y = 4$

$A = \begin{bmatrix} 2 & 1 \\ 3 & -2 \end{bmatrix}$, $X = \begin{bmatrix} x \\ y \end{bmatrix}$, $B = \begin{bmatrix} 5 \\ 4 \end{bmatrix}$

Find A^{-1}.

$[A|I] = \begin{bmatrix} 2 & 1 & | & 1 & 0 \\ 3 & -2 & | & 0 & 1 \end{bmatrix}$

$\begin{bmatrix} -1 & 3 & | & 1 & -1 \\ 3 & -2 & | & 0 & 1 \end{bmatrix} \begin{array}{l} -1R2 + R1 \\ \end{array}$

$\begin{bmatrix} 1 & -3 & | & -1 & 1 \\ 3 & -2 & | & 0 & 1 \end{bmatrix} \begin{array}{l} -1R1 \\ \end{array}$

$\begin{bmatrix} 1 & -3 & | & -1 & 1 \\ 0 & 7 & | & 3 & -2 \end{bmatrix} \begin{array}{l} -3R1 + R2 \\ \end{array}$

$\begin{bmatrix} 1 & -3 & | & -1 & 1 \\ 0 & 1 & | & \frac{3}{7} & -\frac{2}{7} \end{bmatrix} \begin{array}{l} \\ \frac{1}{7}R2 \end{array}$

$\begin{bmatrix} 1 & 0 & | & \frac{2}{7} & \frac{1}{7} \\ 0 & 1 & | & \frac{3}{7} & -\frac{2}{7} \end{bmatrix} \begin{array}{l} 3R2 + R1 \\ \end{array}$

$A^{-1} = \begin{bmatrix} \frac{2}{7} & \frac{1}{7} \\ \frac{3}{7} & -\frac{2}{7} \end{bmatrix}$

$X = A^{-1}B = \begin{bmatrix} \frac{2}{7} & \frac{1}{7} \\ \frac{3}{7} & -\frac{2}{7} \end{bmatrix} \begin{bmatrix} 5 \\ 4 \end{bmatrix} = \begin{bmatrix} 2 \\ 1 \end{bmatrix}$

Solution set: $\{(2, 1)\}$

$$3y + z = 2$$

$$A = \begin{bmatrix} 1 & 1 & 1 \\ 2 & -1 & 0 \\ 0 & 3 & 1 \end{bmatrix}, \quad X = \begin{bmatrix} x \\ y \\ z \end{bmatrix}, \quad B = \begin{bmatrix} 1 \\ -2 \\ 2 \end{bmatrix}$$

Find A^{-1}.

$$[A|I] = \left[\begin{array}{ccc|ccc} 1 & 1 & 1 & 1 & 0 & 0 \\ 2 & -1 & 0 & 0 & 1 & 0 \\ 0 & 3 & 1 & 0 & 0 & 1 \end{array}\right]$$

$$\left[\begin{array}{ccc|ccc} 1 & 1 & 1 & 1 & 0 & 0 \\ 0 & -3 & -2 & -2 & 1 & 0 \\ 0 & 3 & 1 & 0 & 0 & 1 \end{array}\right] \quad -2R1 + R2$$

$$\left[\begin{array}{ccc|ccc} 1 & 1 & 1 & 1 & 0 & 0 \\ 0 & -3 & -2 & -2 & 1 & 0 \\ 0 & 0 & -1 & -2 & 1 & 1 \end{array}\right] \quad R2 + R3$$

$$\left[\begin{array}{ccc|ccc} 1 & 1 & 1 & 1 & 0 & 0 \\ 0 & 1 & \frac{2}{3} & \frac{2}{3} & -\frac{1}{3} & 0 \\ 0 & 0 & -1 & -2 & 1 & 1 \end{array}\right] \quad -\frac{1}{3}R2$$

$$\left[\begin{array}{ccc|ccc} 1 & 0 & \frac{1}{3} & \frac{1}{3} & \frac{1}{3} & 0 \\ 0 & 1 & \frac{2}{3} & \frac{2}{3} & -\frac{1}{3} & 0 \\ 0 & 0 & -1 & -2 & 1 & 1 \end{array}\right] \quad -1R2 + R1$$

$$\left[\begin{array}{ccc|ccc} 1 & 0 & \frac{1}{3} & \frac{1}{3} & \frac{1}{3} & 0 \\ 0 & 1 & \frac{2}{3} & \frac{2}{3} & -\frac{1}{3} & 0 \\ 0 & 0 & 1 & 2 & -1 & -1 \end{array}\right] \quad -1R3$$

$$\left[\begin{array}{ccc|ccc} 1 & 0 & 0 & -\frac{1}{3} & \frac{2}{3} & \frac{1}{3} \\ 0 & 1 & \frac{2}{3} & \frac{2}{3} & -\frac{1}{3} & 0 \\ 0 & 0 & 1 & 2 & -1 & -1 \end{array}\right] \quad -\frac{1}{3}R3 + R1$$

$$\left[\begin{array}{ccc|ccc} 1 & 0 & 0 & -\frac{1}{3} & \frac{2}{3} & \frac{1}{3} \\ 0 & 1 & 0 & -\frac{2}{3} & \frac{1}{3} & \frac{2}{3} \\ 0 & 0 & 1 & 2 & -1 & -1 \end{array}\right] \quad -\frac{2}{3}R3 + R2$$

$$A^{-1} = \begin{bmatrix} -\frac{1}{3} & \frac{2}{3} & \frac{1}{3} \\ -\frac{2}{3} & \frac{1}{3} & \frac{2}{3} \\ 2 & -1 & -1 \end{bmatrix}$$

$$X = A^{-1}B$$

$$= \begin{bmatrix} -\frac{1}{3} & \frac{2}{3} & \frac{1}{3} \\ -\frac{2}{3} & \frac{1}{3} & \frac{1}{3} \\ 2 & -1 & -1 \end{bmatrix} \begin{bmatrix} 1 \\ -2 \\ 2 \end{bmatrix} = \begin{bmatrix} -1 \\ 0 \\ 2 \end{bmatrix}$$

Solution set: $\{(-1,\ 0,\ 2)\}$

39. $\begin{vmatrix} -1 & 8 \\ 2 & 9 \end{vmatrix} = -1(9) - 2(8)$

$\qquad\qquad\quad = -9 - 16$

$\qquad\qquad\quad = -25$

41. $\begin{vmatrix} -2 & 4 & 1 \\ 3 & 0 & 2 \\ -1 & 0 & 3 \end{vmatrix}$

Expand by minors about the second column.

$= -4\begin{vmatrix} 3 & 2 \\ -1 & 3 \end{vmatrix} + 0\begin{vmatrix} -2 & 1 \\ -1 & 3 \end{vmatrix} - 0\begin{vmatrix} -2 & 1 \\ 3 & 2 \end{vmatrix}$

$= -4(9 + 2) + 0 - 0$

$= -4(11) = -44$

43. $\begin{vmatrix} -3 & 2 \\ 1 & x \end{vmatrix} = 5$

$\qquad -3x - 2 = 5$

$\qquad\qquad -3x = 7$

$\qquad\qquad\quad x = -\dfrac{7}{3}$

Solution set: $\left\{-\dfrac{7}{3}\right\}$

45. $\begin{vmatrix} 2 & 5 & 0 \\ 1 & 3x & -1 \\ 0 & 2 & 0 \end{vmatrix} = 4$

Expand about the third column.

$0\begin{vmatrix} 1 & 3x \\ 0 & 2 \end{vmatrix} - (-1)\begin{vmatrix} 2 & 5 \\ 0 & 2 \end{vmatrix} + 0\begin{vmatrix} 2 & 5 \\ 1 & 3x \end{vmatrix} = 4$

$\qquad\qquad\quad 0 + 1(4 - 0) = 4$

$\qquad\qquad\qquad\qquad\quad 4 = 4$

$\qquad\qquad\qquad\qquad\qquad$ *True*

The given determinant equation is true for any number x.

Solution set: {all real numbers}

49. $\begin{vmatrix} 4 & 6 \\ 3 & 5 \end{vmatrix} = \begin{vmatrix} 4 & 3 \\ 6 & 5 \end{vmatrix}$

By Property 2, if corresponding rows and columns of a matrix are interchanged, the value of the corresponding determinant is not changed.

51. $\begin{vmatrix} 4 & 6 & 2 \\ -3 & 8 & -5 \\ 4 & 6 & 2 \end{vmatrix} = 0$

By Property 5, a determinant with two identical rows equals 0.

53. $\begin{vmatrix} 8 & 2 & -5 \\ -3 & 1 & 4 \\ 2 & 0 & 5 \end{vmatrix} = -\begin{vmatrix} 8 & -5 & 2 \\ -3 & 4 & 1 \\ 2 & 5 & 0 \end{vmatrix}$

By Property 3, interchanging two rows (or columns) of a matrix reverses the sign of the determinant.

55. $\begin{vmatrix} 8 & 2 & -5 \\ -3 & 1 & 4 \\ 2 & 0 & 5 \end{vmatrix}$

Multiplying row 2 by −2 and adding to row 1 we have

$\begin{vmatrix} 8 & 2 & -5 \\ -3 & 1 & 4 \\ 2 & 0 & 5 \end{vmatrix}$

$= \begin{vmatrix} 14 & 0 & -13 \\ -3 & 1 & 4 \\ 2 & 0 & 5 \end{vmatrix}$ −2R2 + R1

Expand about column 2.

$\begin{vmatrix} 14 & 0 & -13 \\ -3 & 1 & 4 \\ 2 & 0 & 5 \end{vmatrix}$

$= -0\begin{vmatrix} -3 & 4 \\ 2 & 5 \end{vmatrix} + 1\begin{vmatrix} 14 & -13 \\ 2 & 5 \end{vmatrix} - 0\begin{vmatrix} 14 & -13 \\ -3 & 4 \end{vmatrix}$

$= 70 + 26$

$= 96$

57. $3x + y = -1$

$5x + 4y = 10$

$$D = \begin{vmatrix} 3 & 1 \\ 5 & 4 \end{vmatrix} = 7$$

$$D_x = \begin{vmatrix} -1 & 1 \\ 10 & 4 \end{vmatrix} = -14$$

$$D_y = \begin{vmatrix} 3 & -1 \\ 5 & 10 \end{vmatrix} = 35$$

$$x = \frac{D_x}{D} = \frac{-14}{7} = -2$$

$$y = \frac{D_y}{D} = \frac{35}{7} = 5$$

Solution set: $\{(-2, 5)\}$

59. $2x - 5y = 8$

$3x + 4y = 10$

$$D = \begin{vmatrix} 2 & -5 \\ 3 & 4 \end{vmatrix} = (2)(4) - (3)(-5)$$
$$= 23$$

$$D_x = \begin{vmatrix} 8 & -5 \\ 10 & 4 \end{vmatrix} = (8)(4) - (10)(-5)$$
$$= 82$$

$$D_y = \begin{vmatrix} 2 & 8 \\ 3 & 10 \end{vmatrix} = (2)(10) - (3)(8)$$
$$= -4$$

$$x = \frac{D_x}{D} = \frac{82}{23}$$

$$y = \frac{D_y}{D} = -\frac{4}{23}$$

Solution set: $\left\{ \left(\frac{82}{23}, -\frac{4}{23} \right) \right\}$

61. $5x - 2y - z = 8$ (1)

$-5x + 2y + z = -8$ (2)

$x - 4y - 2z = 0$ (3)

$$D = \begin{vmatrix} 5 & -2 & -1 \\ -5 & 2 & 1 \\ 1 & -4 & -2 \end{vmatrix} = 0$$

Cramer's rule cannot be used since $D = 0$.

Adding the first and second equations results in the equality $0 = 0$. The system has dependent equations.

Chapter 8   Test

1. $\begin{bmatrix} 5 & x + 6 \\ 0 & 4 \end{bmatrix} = \begin{bmatrix} y - 2 & 4 - x \\ 0 & w + 7 \end{bmatrix}$

All corresponding elements, position by position, of the two matrices must be equal.

$5 = y - 2$ $x + 6 = 4 - x$ $4 = w + 7$

$7 = y$ $2x = -2$ $-3 = w$

             $x = -1$

Thus, $x = -1$, $y = 7$, and $w = -3$.

2. $3\begin{bmatrix} 2 & 3 \\ 1 & -4 \\ 5 & 9 \end{bmatrix} - \begin{bmatrix} -2 & 6 \\ 3 & -1 \\ 0 & 8 \end{bmatrix}$

$$= \begin{bmatrix} 6 & 9 \\ 3 & -12 \\ 15 & 27 \end{bmatrix} + \begin{bmatrix} 2 & -6 \\ -3 & 1 \\ 0 & -8 \end{bmatrix}$$

$$\begin{bmatrix} 8 & 3 \\ 0 & -11 \\ 5 & 19 \end{bmatrix}$$

3. $\begin{bmatrix} 1 \\ 2 \end{bmatrix} + \begin{bmatrix} 4 \\ -6 \end{bmatrix} + \begin{bmatrix} 2 & 8 \\ -7 & 5 \end{bmatrix}$

The first two matrices are 2×1 and the third is 2×2. Only matrices of the same size can be added, so it is not possible to find this sum.

4. Use the first row to represent the number of women in each class and the second row to represent the number of men in each class.

Then the first column will represent the number of freshman women and men, the second the sophomore women and men, and so on.

$$\begin{bmatrix} 108 & 112 & 123 & 110 \\ 142 & 98 & 117 & 130 \end{bmatrix}$$

5. $\begin{bmatrix} 2 & 1 & -3 \\ 4 & 0 & 5 \end{bmatrix} \begin{bmatrix} 1 & 3 \\ 2 & 4 \\ 3 & -2 \end{bmatrix}$

$= \begin{bmatrix} 2(1)+1(2)+(-3)(3) & 2(3)+1(4)+(-3)(-2) \\ 4(1)+0(2)+5(3) & 4(3)+0(4)+5(-2) \end{bmatrix}$

$= \begin{bmatrix} -5 & 16 \\ 19 & 2 \end{bmatrix}$

6. $\begin{bmatrix} 2 & -4 \\ 3 & 5 \end{bmatrix} \begin{bmatrix} 4 \\ 2 \\ 7 \end{bmatrix}$

The first matrix is 2×2 and the second is 1×3. The product of two matrices can be found only if the number of columns of the first matrix is the same as the number of rows of the second matrix. The first matrix has two columns and the second has one row, so it is not possible to find this product.

7. (a) There are associative, distributive, and identity properties that apply to multiplication of matrices, but matrix multiplication is not commutative.

8. $\begin{bmatrix} 1 & 0 & 1 \\ -1 & 0 & 1 \\ 0 & 1 & 0 \end{bmatrix} \begin{bmatrix} \frac{1}{2} & -\frac{1}{2} & 0 \\ 0 & 0 & 1 \\ \frac{1}{2} & \frac{1}{2} & 0 \end{bmatrix}$

$= \begin{bmatrix} \frac{1}{2}+0+\frac{1}{2} & -\frac{1}{2}+0+\frac{1}{2} & 0+0+0 \\ -\frac{1}{2}+0+\frac{1}{2} & \frac{1}{2}+0+\frac{1}{2} & 0+0+0 \\ 0+0+0 & 0+0+0 & 0+1+0 \end{bmatrix}$

$= \begin{bmatrix} 1 & 0 & 0 \\ 0 & 1 & 0 \\ 0 & 0 & 1 \end{bmatrix}$

The product of the two given 3×3 matrices is the 3×3 identity matrix.

It can be shown that the product obtained by multiplying the two given matrices in the opposite order is also the 3×3 identity matrix. Therefore, the given matrices are inverses.

9. Find the inverse of $A = \begin{bmatrix} -8 & 5 \\ 3 & -2 \end{bmatrix}$, if it exists.

Form the augmented matrix $[A|I]$.

$[A|I] = \begin{bmatrix} -8 & 5 & 1 & 0 \\ 3 & -2 & 0 & 1 \end{bmatrix}$

Perform row transformations on $[A|I]$ until a matrix of the form $[I|B]$ is obtained.

$[A|I] = \begin{bmatrix} 1 & -\frac{5}{8} & -\frac{1}{8} & 0 \\ 3 & -2 & 0 & 1 \end{bmatrix}$ $\quad -\frac{1}{8}R1$

$\begin{bmatrix} 1 & -\frac{5}{8} & -\frac{1}{8} & 0 \\ 0 & -\frac{1}{8} & \frac{3}{8} & 1 \end{bmatrix}$ $\quad -3R1 + R2$

$\begin{bmatrix} 1 & 0 & -2 & -5 \\ 0 & -\frac{1}{8} & \frac{3}{8} & 1 \end{bmatrix}$ $\quad -5R2 + R1$

$$[I|B] = \begin{bmatrix} 1 & 0 & | & -2 & -5 \\ 0 & 1 & | & -3 & -8 \end{bmatrix} \quad -8R2$$

$$A^{-1} = B = \begin{bmatrix} -2 & -5 \\ -3 & -8 \end{bmatrix}$$

$$\begin{bmatrix} 1 & 0 & 1 & | & -5 & 0 & -3 \\ 0 & 1 & 0 & | & -2 & 1 & 0 \\ 0 & 0 & 1 & | & -4 & -1 & 1 \end{bmatrix} \quad -1R2 + R3$$

$$\begin{bmatrix} 1 & 0 & 0 & | & -9 & 1 & -4 \\ 0 & 1 & 0 & | & -2 & 1 & 0 \\ 0 & 0 & 1 & | & 4 & -1 & 1 \end{bmatrix} \quad -1R3 + R1$$

10. Find the inverse of $A = \begin{bmatrix} 4 & 12 \\ 2 & 6 \end{bmatrix}$, if it exists.

$$[A|I] = \begin{bmatrix} 4 & 12 & | & 1 & 0 \\ 2 & 6 & | & 0 & 1 \end{bmatrix}$$

$$\begin{bmatrix} 1 & 3 & | & \frac{1}{4} & 0 \\ 2 & 6 & | & 0 & 1 \end{bmatrix} \quad \frac{1}{4}R1$$

$$\begin{bmatrix} 1 & 3 & | & \frac{1}{4} & 0 \\ 0 & 0 & | & -\frac{1}{2} & 1 \end{bmatrix} \quad -2R1 + R2$$

The second row, second column element is now 0, so the desired transformation cannot be completed.
Therefore, the inverse of the given matrix does not exist.

11. Find the inverse of

$$A = \begin{bmatrix} 1 & 3 & 4 \\ 2 & 7 & 8 \\ -2 & -5 & -7 \end{bmatrix}, \text{ if it exists.}$$

$$[A|I] = \begin{bmatrix} 1 & 3 & 4 & | & 1 & 0 & 0 \\ 2 & 7 & 8 & | & 0 & 1 & 0 \\ -2 & -5 & -7 & | & 0 & 0 & 1 \end{bmatrix}$$

$$\begin{bmatrix} 1 & 3 & 4 & | & 1 & 0 & 0 \\ 0 & 1 & 0 & | & -2 & 1 & 0 \\ -2 & -5 & -7 & | & 0 & 0 & 1 \end{bmatrix} \quad -2R1 + R2$$

$$\begin{bmatrix} 1 & 3 & 4 & | & 1 & 0 & 0 \\ 0 & 1 & 0 & | & -2 & 1 & 0 \\ 0 & 1 & 1 & | & 2 & 0 & 1 \end{bmatrix} \quad 2R1 + R3$$

$$\begin{bmatrix} 1 & 0 & 1 & | & -5 & 0 & -3 \\ 0 & 1 & 0 & | & -2 & 1 & 0 \\ 2 & 1 & 1 & | & 2 & 0 & 1 \end{bmatrix} \quad -3R3 + R1$$

$$A^{-1} = \begin{bmatrix} -9 & 1 & -4 \\ -2 & 1 & 0 \\ 4 & -1 & 1 \end{bmatrix}$$

12. $2x + y = -6$
$3x - y = -29$

Represent the system as a matrix equation as follows.

Let $A = \begin{bmatrix} 2 & 1 \\ 3 & -1 \end{bmatrix}$, $X = \begin{bmatrix} x \\ y \end{bmatrix}$,

$B = \begin{bmatrix} -6 \\ -29 \end{bmatrix}$.

Then $AX = B$.
Find A^{-1}.

$$\begin{bmatrix} 2 & 1 & | & 1 & 0 \\ 3 & -1 & | & 0 & 1 \end{bmatrix}$$

$$\begin{bmatrix} 1 & \frac{1}{2} & | & \frac{1}{2} & 0 \\ 3 & -1 & | & 0 & 1 \end{bmatrix} \quad \frac{1}{2}R1$$

$$\begin{bmatrix} 1 & \frac{1}{2} & | & \frac{1}{2} & 0 \\ 0 & -\frac{5}{2} & | & -\frac{3}{2} & 1 \end{bmatrix} \quad -3R1 + R2$$

$$\begin{bmatrix} 1 & 0 & | & \frac{1}{5} & \frac{1}{5} \\ 0 & -\frac{5}{2} & | & -\frac{3}{2} & 1 \end{bmatrix} \quad \frac{1}{5}R2 + R1$$

$$\begin{bmatrix} 1 & 0 & | & \frac{1}{5} & \frac{1}{5} \\ 0 & 1 & | & \frac{3}{5} & -\frac{2}{5} \end{bmatrix} \quad -\frac{2}{5}R2$$

Thus,

$$A^{-1} = \begin{bmatrix} \dfrac{1}{5} & \dfrac{1}{5} \\ \dfrac{3}{5} & -\dfrac{2}{5} \end{bmatrix}.$$

$$A^{-1}B = \begin{bmatrix} \dfrac{1}{5} & \dfrac{1}{5} \\ \dfrac{3}{5} & -\dfrac{2}{5} \end{bmatrix} \begin{bmatrix} -6 \\ -29 \end{bmatrix}$$

$$= \begin{bmatrix} -\dfrac{6}{5} + \left(-\dfrac{29}{5}\right) \\ -\dfrac{18}{5} + \dfrac{58}{5} \end{bmatrix} = \begin{bmatrix} -\dfrac{35}{5} \\ \dfrac{40}{5} \end{bmatrix}$$

$$= \begin{bmatrix} -7 \\ 8 \end{bmatrix}$$

Since $X = A^{-1}B$,

$$X = \begin{bmatrix} -7 \\ 8 \end{bmatrix}.$$

Solution set: $\{(-7, 8)\}$

13. $x + y = 5$
      $y - 2z = 23$
      $x + 3z = -27$

Let $A = \begin{bmatrix} 1 & 1 & 0 \\ 0 & 1 & -2 \\ 1 & 0 & 3 \end{bmatrix},$

$X = \begin{bmatrix} x \\ y \\ z \end{bmatrix},$

$B = \begin{bmatrix} 5 \\ 23 \\ -27 \end{bmatrix}.$

Then $AX = B$.
Find A^{-1}.

$$\left[\begin{array}{ccc|ccc} 1 & 1 & 0 & 1 & 0 & 0 \\ 0 & 1 & -2 & 0 & 1 & 0 \\ 1 & 0 & 3 & 0 & 0 & 1 \end{array}\right]$$

$$\left[\begin{array}{ccc|ccc} 1 & 1 & 0 & 1 & 0 & 0 \\ 0 & 1 & -2 & 0 & 1 & 0 \\ 0 & -1 & 3 & -1 & 0 & 1 \end{array}\right] \quad -R1 + R3$$

$$\left[\begin{array}{ccc|ccc} 1 & 0 & 3 & 0 & 0 & 1 \\ 0 & 1 & -2 & 0 & 1 & 0 \\ 0 & -1 & 3 & -1 & 0 & 1 \end{array}\right] \quad R3 + R1$$

$$\left[\begin{array}{ccc|ccc} 1 & 0 & 3 & 0 & 0 & 1 \\ 0 & 1 & -2 & 0 & 1 & 0 \\ 0 & 0 & 1 & -1 & 1 & 1 \end{array}\right] \quad R2 + R3$$

$$\left[\begin{array}{ccc|ccc} 1 & 0 & 0 & 3 & -3 & -2 \\ 0 & 1 & -2 & 0 & 1 & 0 \\ 0 & 0 & 1 & -1 & 1 & 1 \end{array}\right] \quad -3R1 + R1$$

$$\left[\begin{array}{ccc|ccc} 1 & 0 & 0 & 3 & -3 & -2 \\ 0 & 1 & 0 & -2 & 3 & 2 \\ 0 & 0 & 1 & -1 & 1 & 1 \end{array}\right] \quad 2R3 + R2$$

Thus,

$$A^{-1} = \begin{bmatrix} 3 & -3 & -2 \\ -2 & 3 & 2 \\ -1 & 1 & 1 \end{bmatrix}.$$

$$A^{-1}B = \begin{bmatrix} 3 & -3 & -2 \\ -2 & 3 & 2 \\ -1 & 1 & 1 \end{bmatrix} \begin{bmatrix} 5 \\ 23 \\ -27 \end{bmatrix}$$

$$= \begin{bmatrix} 0 \\ 5 \\ -9 \end{bmatrix} = X$$

Solution set: $\{(0, 5, -9)\}$

14. $\begin{vmatrix} 6 & 8 \\ 2 & -7 \end{vmatrix} = 6(-7) - 2(8)$

$= -58$

15. $\begin{vmatrix} 2 & 0 & 8 \\ -1 & 7 & 9 \\ 12 & 5 & -3 \end{vmatrix}$

This determinant may be evaluated by expanding about any row or any column. Choose the first row or second column because they contain a 0. We will expand by minors about the first row.

$$\begin{vmatrix} 2 & 0 & 8 \\ -1 & 7 & 9 \\ 12 & 5 & -3 \end{vmatrix}$$

$$= 2\begin{vmatrix} 7 & 9 \\ 5 & -3 \end{vmatrix} - 0\begin{vmatrix} -1 & 9 \\ 12 & -3 \end{vmatrix} + 8\begin{vmatrix} -1 & 7 \\ 12 & 5 \end{vmatrix}$$

$$= 2[7(-3) - 5(9) - 0$$
$$\quad + 8[(-1)(5) - 12(7)]$$

$$= 2(-21 - 45) + 8(-5 - 84)$$

$$= 2(-66) + 8(-89)$$

$$= -132 - 712$$

$$= -844$$

16. $\begin{vmatrix} 1 & 3x & 3 \\ -2 & 0 & 7 \\ 4 & x & 5 \end{vmatrix} = -202$

Evaluate the determinant by expanding by minors about the second column.

$$\begin{vmatrix} 1 & 3x & 3 \\ -2 & 0 & 7 \\ 4 & x & 5 \end{vmatrix}$$

$$= -(3x)\begin{vmatrix} -2 & 7 \\ 4 & 5 \end{vmatrix} + 0\begin{vmatrix} 1 & 3 \\ 4 & 5 \end{vmatrix}$$

$$\quad - 1(x)\begin{vmatrix} 1 & 3 \\ -2 & 7 \end{vmatrix}$$

$$= -3x(-38) - x(13)$$

$$= 114x - 13x$$

$$= 101x$$

To solve the determinant equation, solve

$$101x = -202$$
$$x = -2.$$

Solution set: $\{-2\}$

17. $\begin{vmatrix} 6 & 7 \\ -5 & 2 \end{vmatrix} = -\begin{vmatrix} -5 & 2 \\ 6 & 7 \end{vmatrix}$

Rows 1 and 2 have been interchanged. Interchanging two rows of a matrix reverses the sign of the determinant (Property 3).

18. $\begin{vmatrix} 7 & 2 & -1 \\ 5 & -4 & 3 \\ 6 & -2 & 1 \end{vmatrix} = \begin{vmatrix} 7 & 2 & -1 \\ -7 & 0 & 1 \\ 13 & 0 & 0 \end{vmatrix}$

-2 times row 1 has been added to row 2, and row 1 has been added to row 3. The determinant of a matrix is unchanged if a multiple of a row of the matrix is added to the corresponding elements of another row (Property 6).

19. $2x - 3y = -33$
$4x + 5y = 11$

$$D = \begin{vmatrix} 2 & -3 \\ 4 & 5 \end{vmatrix} = 2(5) - 4(-3) = 22$$

$$D_x = \begin{vmatrix} -33 & -3 \\ 11 & 5 \end{vmatrix} = -33(5) - 11(-3)$$
$$= -132$$

$$D_y = \begin{vmatrix} 2 & -33 \\ 4 & 11 \end{vmatrix} = 2(11) - 4(33) = 154$$

$$x = \frac{D_x}{D} = \frac{-132}{22} = -6$$

$$y = \frac{D_y}{D} = \frac{154}{22} = 7$$

Solution set: $\{(-6, 7)\}$

20. $x + y - z = -4$
$2x - 3y - z = 5$
$x + 2y + 2z = 3$

Expand about row 1.

$$D = \begin{vmatrix} 1 & 1 & -1 \\ 2 & -3 & -1 \\ 1 & 2 & 2 \end{vmatrix}$$

$$= 1\begin{vmatrix} -3 & -1 \\ 2 & 2 \end{vmatrix} - 1\begin{vmatrix} 2 & -1 \\ 1 & 2 \end{vmatrix}$$

$$+ (-1)\begin{vmatrix} 2 & -3 \\ 1 & 2 \end{vmatrix}$$

$$= -4 - 1(5) - 1(7)$$

$$= -16$$

Expand about row 1.

$$D_x = \begin{vmatrix} -4 & 1 & -1 \\ 5 & -3 & -1 \\ 3 & 2 & 2 \end{vmatrix}$$

$$= 4\begin{vmatrix} -3 & -1 \\ 2 & 2 \end{vmatrix} - 1\begin{vmatrix} 5 & -1 \\ 3 & 2 \end{vmatrix}$$

$$+ (-1)\begin{vmatrix} 5 & -3 \\ 3 & 2 \end{vmatrix}$$

$$= -4(-4) - 1(13) - 1(19)$$

$$= -16$$

Expand about row 1.

$$D_y = \begin{vmatrix} 1 & -4 & -1 \\ 2 & 5 & -1 \\ 1 & 3 & 2 \end{vmatrix}$$

$$= 1\begin{vmatrix} 5 & -1 \\ 3 & 2 \end{vmatrix} - (-4)\begin{vmatrix} 2 & -1 \\ 1 & 2 \end{vmatrix}$$

$$+ (-1)\begin{vmatrix} 2 & 5 \\ 1 & 3 \end{vmatrix}$$

$$= 13 + 4(5) - 1(1)$$

$$= 32$$

Expand about row 1.

$$D_z = \begin{vmatrix} 1 & 1 & -4 \\ 2 & -3 & 5 \\ 1 & 2 & 3 \end{vmatrix}$$

$$= 1\begin{vmatrix} -3 & 5 \\ 2 & 3 \end{vmatrix} - 1\begin{vmatrix} 2 & 5 \\ 1 & 3 \end{vmatrix}$$

$$+ (-4)\begin{vmatrix} 2 & -3 \\ 1 & 2 \end{vmatrix}$$

$$= 1(-19) - 1(1) - 4(7)$$

$$= -48$$

$$x = \frac{D_x}{D} = \frac{-16}{-16} = 1$$

$$x = \frac{D_y}{D} = \frac{32}{-16} = -2$$

$$z = \frac{D_z}{D} = \frac{-48}{-16} = 3$$

Solution set: $\{(1, -2, 3)\}$

CHAPTER 9   SEQUENCES AND SERIES;
PROBABILITY

Section 9.1

1. $a_n = 6n + 4$

   To find a_1, let $n = 1$ in
   $6n + 4$: $a_1 = 6(1) + 4 = 10$.
   To find a_2, let $n = 2$ in
   $6n + 4$: $a_2 = 6(2) + 4 = 16$.
   To find a_3, let $n = 3$ in
   $6n + 4$: $a_3 = 6(3) + 4 = 32$.
   To find a_4, let $n = 4$ in
   $6n + 4$: $a_4 = 6(4) + 4 = 28$.
   To find a_5, let $n = 5$ in
   $6n + 4$: $a_5 = 6(5) + 4 = 34$.

   The first five terms are

   $$10, \ 16, \ 22, \ 28, \ \text{and } 34.$$

3. $a_n = 2^n$

   $a_1 = 2^1 = 2$
   $a_2 = 2^2 = 4$
   $a_3 = 2^3 = 8$
   $a_4 = 2^4 = 16$
   $a_5 = 2^5 = 32$

   The first five terms are

   $$2, \ 4, \ 8, \ 16, \ \text{and } 32.$$

5. $a_n = (-1)^{n+1}$

   $a_1 = (-1)^{1+1}$
   $\quad = (-1)^2 = 1$
   $a_2 = (-1)^{2+1}$
   $\quad = (-1)^3 = -1$
   $a_3 = (-1)^{3+1}$
   $\quad = (-1)^4 = 1$

   $a_4 = (-1)^{4+1}$
   $\quad = (-1)^5 = -1$
   $a_5 = (-1)^{5+1}$
   $\quad = (-1)^6 = 1$

   The first five terms are

   $$1, \ -1, \ 1, \ -1, \ \text{and } 1.$$

7. $a_n = \dfrac{2n}{n + 3}$

   $a_1 = \dfrac{2(1)}{1 + 3} = \dfrac{2}{4} = \dfrac{1}{2}$

   $a_2 = \dfrac{2(2)}{2 + 3} = \dfrac{4}{5}$

   $a_3 = \dfrac{2(3)}{3 + 3} = \dfrac{6}{6} = 1$

   $a_4 = \dfrac{2(4)}{4 + 3} = \dfrac{8}{7}$

   $a_5 = \dfrac{2(5)}{5 + 3} = \dfrac{10}{8} = \dfrac{5}{4}$

   The first five terms are

   $$\frac{1}{2}, \ \frac{4}{5}, \ 1, \ \frac{8}{7}, \ \text{and } \frac{5}{4}.$$

9. $a_n = \dfrac{8n - 4}{2n + 1}$

   $a_1 = \dfrac{8(1) - 4}{2(1) + 1} = \dfrac{4}{3}$

   $a_2 = \dfrac{8(2) - 4}{2(2) + 1} = \dfrac{12}{5}$

   $a_3 = \dfrac{8(3) - 4}{2(3) + 1} = \dfrac{20}{7}$

   $a_4 = \dfrac{8(4) - 4}{2(4) + 1} = \dfrac{28}{9}$

   $a_5 = \dfrac{8(5) - 4}{2(5) + 1} = \dfrac{36}{11}$

   The first five terms are

   $$\frac{4}{3}, \ \frac{12}{5}, \ \frac{20}{7}, \ \frac{28}{9}, \ \text{and } \frac{36}{11}.$$

11. $a_n = (-2)^n(n)$

$a_1 = (-2)^1(1) = -2$

$a_2 = (-2)^2(2) = 4(2) = 8$

$a_3 = (-2)^3(3) = -8(3) = -24$

$a_4 = (-2)^4(4) = 16(4) = 64$

$a_5 = (-2)^5(5) = -32(5) = -160$

The first five terms are

   -2, 8, -24, 64, and -160.

13. $a_n = x^n$

$a_1 = x^1 = x$

$a_2 = x^2$

$a_3 = x^3$

$a_4 = x^4$

$a_5 = x^5$

The first five terms are

   x, x^2, x^3, x^4, and x^5.

15. $a_1 = 4$; for $n \geq 2$, $a_n = a_{n-1} + 5$

The equation $a_n = a_{n-1} + 5$ means that each term after the first is found by adding 5 to the previous term. We know that $a_1 = 4$. To find a_2, use $a_2 = a_{2-1} + 5$.

$a_2 = a_1 + 5 = 4 + 5 = 9$

$a_3 = a_{3-1} + 5 = a_2 + 5$

$\quad = 9 + 5 = 14$

$a_4 = a_3 + 5 = 14 + 5 = 19$

$a_5 = a_4 + 5 = 19 + 5 = 24$

$a_6 = a_5 + 5 = 24 + 5 = 29$

$a_7 = a_6 + 5 = 29 + 5 = 34$

$a_8 = a_7 + 5 = 34 + 5 = 39$

$a_9 = a_8 + 5 = 39 + 5 = 44$

$a_{10} = a_9 + 5 = 44 + 5 = 49$

The first ten terms are

4, 9, 14, 19, 24, 29, 34, 39, 44, and 49.

17. $a_1 = 2$; for $n \geq 2$, $a_n = 2 \cdot a_{n-1}$

The equation $a_n = 2 \cdot a_{n-1}$ means that each term after the first is found by multiplying the previous term by 2.

$a_2 = 2 \cdot a_1 = 2 \cdot 2 = 4$

$a_3 = 2 \cdot a_2 = 2 \cdot 4 = 8$

$a_4 = 2 \cdot a_3 = 2 \cdot 8 = 16$

$a_5 = 2 \cdot a_4 = 2 \cdot 16 = 32$

$a_6 = 2 \cdot a_5 = 2 \cdot 32 = 64$

$a_7 = 2 \cdot a_6 = 2 \cdot 64 = 128$

$a_8 = 2 \cdot a_7 = 2 \cdot 128 = 256$

$a_9 = 2 \cdot a_8 = 2 \cdot 256 = 512$

$a_{10} = 2 \cdot a_9 = 2 \cdot 512 = 1024$

The first ten terms are

2, 4, 8, 16, 32, 64, 128, 256, 512, and 1024.

19. $a_1 = 1$, $a_2 = 1$; for $n \geq 3$,

$a_n = a_{n-1} + a_{n-2}$

The equation $a_n = a_{n-1} + a_{n-2}$ means that each term after the second is found by adding the two previous terms.

$a_3 = a_{3-1} + a_{3-2} = a_2 + a_1$

$\quad = 1 + 1 = 2$

$a_4 = a_3 + a_2 = 2 + 1 = 3$

$a_5 = a_4 + a_3 = 3 + 2 = 5$

$a_6 = a_5 + a_4 = 5 + 3 = 8$

$a_7 = a_6 + a_5 = 8 + 5 = 13$

$a_8 = a_7 + a_6 = 13 + 8 = 21$

$a_9 = a_8 + a_7 = 21 + 13 = 34$

$a_{10} = a_9 + a_8 = 34 + 21 = 55$

The first ten terms are

1, 1, 2, 3, 5, 8, 13, 21, 34, and 55.

21. $a_1 = 4$, $d = 2$, $n = 5$

$a_2 = a_1 + d = 4 + 2 = 6$

$a_3 = a_2 + d = 6 + 2 = 8$

$a_4 = a_3 + d = 8 + 2 = 10$

$a_5 = a_4 + d = 10 + 2 = 12$

The first five terms are

4, 6, 8, 10, and 12.

23. $a_2 = 9$, $d = -2$, $n = 4$

First, find a_1.

$a_2 = a_1 - 2$

$9 = a_1 - 2$

$a_1 = 11$

$a_2 = 9$

$a_3 = a_2 + d = 9 - 2 = 7$

$a_4 = a_3 - 2 = 7 - 2 = 5$

The first four terms are

11, 9, 7, and 5.

25. $a_3 = -2$, $d = -4$, $n = 4$

Since $a_3 = a_2 + d$, substituting -2 for a_3 and -4 for d gives

$-2 = a_2 - 4$ or $a_2 = 2$.

Since $a_2 = a_1 + d$, we have

$2 = a_1 - 4$, or $a_1 = 6$.

$a_4 = a_3 + d = -2 - 4 = -6$.

The first four terms are

6, 2, -2, and -6.

27. $a_3 = 6$, $a_4 = 12$, $n = 6$

First, find d.

$a_4 = a_3 + d$

$d = a_4 - a_3 = 12 - 6 = 6$

Then

$a_2 = a_3 - d = 6 - 6 = 0$

$a_1 = a_2 - d = 0 - 6 = -6$

$a_5 = a_4 + d = 12 + 6 = 18$

$a_6 = a_5 + d = 18 + 6 = 24$.

The first six terms are

-6, 0, 6, 12, 18, and 24.

29. 12, 17, 22, 27, 32, 37, ... is arithmetic since there is a common difference.

$$d = 17 - 12 = 5$$

To find a_n, use the formula for the nth term of an arithmetic sequence with $a_1 = 12$ and $d = 5$.

$$a_n = a_1 + (n - 1)d$$
$$a_n = 12 + (n - 1)(5)$$
$$= 7 + 5n$$

31. -19, -12, -5, 2, 9, ... is arithmetic since there is a common difference.

$$d = -12 - (-19) = 7$$

Since $a_1 = -19$,

$$a_n = a_1 + (n - 1)d$$
$$a_n = -19 + (n - 1)(7)$$
$$= -19 + 7n - 7$$
$$a_n = -26 + 7n.$$

33. $x, x + m, x + 2m, x + 3m, x + 4m,$ ... is arithmetic since the difference between adjacent terms is m, or

$$d = m.$$

Since $a_1 = x$,

$$a_n = a_1 + (n - 1)d$$
$$a_n = x + (n - 1)m$$
$$= x + nm - m.$$

35. $2z + m, 2z, 2z - m, 2z - 2m,$ $2z - 3m,$ ... is arithmetic since the difference between adjacent terms is

$$d = 2z - (2z + m)$$
$$= -m.$$

Since $a_1 = 2z + m$,

$$a_n = a_1 + (n - 1)d$$
$$a_n = (2z + m) + (n - 1)(-m)$$
$$= 2z + m - mn + m$$
$$a_n = 2z + 2m - mn.$$

37. $a_1 = 5, d = 2$

$$a_8 = a_1 + 7d$$
$$= 5 + 7(2)$$
$$a_8 = 19$$

$$a_n = a_1 + (n - 1)d$$
$$= 5 + (n - 1)(2)$$
$$a_n = 3 + 2n$$

39. $a_3 = 2, d = 1$

First, find a_1.

$$a_3 = a_1 + 2d$$
$$2 = a_1 + 2(1)$$
$$a_1 = 0$$

$$a_8 = a_1 + 7d$$
$$a_8 = 0 + 7(1) = 7$$

$$a_n = a_1 + (n - 1)d$$
$$= 0 + (n - 1)(1)$$
$$a_n = n - 1$$

41. $a_1 = 8, a_2 = 6$

First, find d.

$$a_2 = a_1 + d$$
$$6 = 8 + d$$
$$d = -2$$

$$a_8 = a_1 + 7d$$
$$a_8 = 8 + 7(-2) = -6$$

$$a_n = a_1 + (n - 1)d$$
$$= 8 + (n - 1)(-2)$$
$$= 10 - 2n$$

43. $a_{10} = 6, a_{12} = 15$

First, find a_1 and d.

$$a_n = a_1 + (n - 1)d$$
$$a_{10} = a_1 + 9d$$
$$6 = a_1 + 9d \quad (1)$$
$$a_{12} = a_1 + 11d$$
$$15 = a_1 + 11d \quad (2)$$

Solve the system formed by equations (1) and (2) by the substitution method. Solve equation (1) for a_1.

$$a_1 = 6 - 9d \quad (3)$$

Substitute $6 - 9d$ for a_1 in equation (2).

$$15 = (6 - 9d) + 11d$$
$$15 = 6 + 2d$$
$$9 = 2d$$
$$\frac{9}{2} = d$$

Now substitute $d = \frac{9}{2}$ into equation (3) to find a_1.

$$a_1 = 6 - 9\left(\frac{9}{2}\right)$$
$$= \frac{12}{2} - \frac{81}{2}$$
$$a_1 = -\frac{69}{2}$$

Now find a_8 and a_n.

$$a_8 = a_1 + 7d$$
$$= -\frac{69}{2} + 7\left(\frac{9}{2}\right)$$
$$= -\frac{69}{2} + \frac{63}{2}$$
$$a_8 = -\frac{6}{2} = -3$$

$$a_n = a_1 + (n - 1)(d)$$
$$= -\frac{69}{2} + (n - 1)\left(\frac{9}{2}\right)$$
$$= -\frac{69}{2} + \frac{9}{2}n - \frac{9}{2}$$
$$= -\frac{78}{2} + \frac{9}{2}n$$
$$a_n = -39 + \frac{9}{2}n$$

45. $a_1 = x$, $a_2 = x + 3$

$$d = (x + 3) - x = 3$$
$$a_8 = a_1 + 7d$$
$$= x + 7(3)$$
$$a_8 = x + 21$$

$$a_n = a_1 + (n - 1)d$$
$$= x + (n - 1)(3)$$
$$a_n = x + 3n - 3$$

47. $a_6 = 2m$, $a_7 = 3m$

Find d and a_1.

$$d = 3m - 2m = m$$
$$a_6 = a_1 + 5d$$
$$2m = a_1 + 5m$$
$$a_1 = -3m$$

Now find a_8 and a_n.

$$a_8 = a_1 + 7d$$
$$= -3m + 7m$$
$$a_8 = 4m$$

$$a_n = a_1 + (n - 1)d$$
$$= -3m + (n - 1)m$$
$$= -3m + nm - m$$
$$= -4m + nm$$
$$a_n = mn - 4m$$

49. Consider the arithmetic sequence 6, 8, 10, 12, 14, 16, ... This sequence has $a_1 = 6$, $d = 2$, and, as specified, $a_4 = 12$. (There are many other arithmetic sequences that also have $a_4 = 12$.)

51. (a) 4, 6, 8, 10, ... is an arithmetic sequence with $d = 2$.

(b) -2, 6, 14, 22, ... is an arithmetic sequence with $d = 8$.

(c) $\frac{1}{2}$, 1, $\frac{3}{2}$, 2, ...

is an arithmetic sequence with

$d = \frac{1}{2}$.

(d) 5, 10, 20, 40, ...

is not an arithmetic sequence since there is not a common difference.

Thus, (d) is the only one of the given sequences that is not arithmetic.

53. $a_9 = 47$, $a_{15} = 77$

Use the formula

$a_n = a_1 + (n - 1)d$.

$a_9 = a_1 + 8d$

$47 = a_1 + 8d$ (1)

$a_{15} = a_1 + 14d$

$77 = a_1 + 14d$ (2)

The system formed by equations (1) and (2) may be solved by the substitution method. Solve equation (1) for a_1.

$a_1 = 47 - 8d$ (3)

Substitute $47 - 8d$ for a_1 in equation (2).

$77 = (47 - 8d) + 14d$

$77 = 47 + 6d$

$30 = 6d$

$5 = d$

Now substitute $d = 5$ into equation (3) to find a_1.

$a_1 = 47 - 8(5)$

$a_1 = 7$

55. $a_{15} = 168$, $a_{16} = 180$

$d = a_{16} - a_{15} = 180 - 168 = 12$

Use the formula $a_n = a_1 + (n - 1)d$ to find a_1.

$a_{16} = a_1 + 15d$

$180 = a_1 + 15(12)$

$180 = a_1 + 180$

$0 = a_1$

Section 9.2

1. $a_1 = 2$, $r = 3$, $n = 4$

$a_1 = 2$

$a_2 = a_1 r = 2(3) = 6$

$a_3 = a_2 r = 6(3) = 18$

$a_4 = a_3 r = 18(3) = 54$

The first four terms are

2, 6, 18, and 54.

3. $a_1 = \frac{1}{2}$, $r = 4$, $n = 4$

$a_1 = \frac{1}{2}$

$a_2 = a_1 r = \frac{1}{2}(4) = 2$

$a_3 = a_2 r = 2(4) = 8$

$a_4 = a_3 r = 8(4) = 32$

The first four terms are

$\frac{1}{2}$, 2, 8, and 32.

5. $a_1 = -2$, $r = -3$, $n = 4$

$\quad a_1 = -2$

$\quad a_2 = a_1 r = -2(-3) = 6$

$\quad a_3 = a_2 r = 6(-3) = -18$

$\quad a_4 = a_3 r = -18(-3) = 54$

The first four terms are

$\quad -2$, 6, -18, and 54.

7. $a_1 = 3125$, $r = \frac{1}{5}$, $n = 7$

$\quad a_1 = 3125$

$\quad a_2 = a_1 r = 3125\left(\frac{1}{5}\right) = 625$

$\quad a_3 = a_2 r = 625\left(\frac{1}{5}\right) = 125$

$\quad a_4 = a_3 r = 125\left(\frac{1}{5}\right) = 25$

$\quad a_5 = a_4 r = 25\left(\frac{1}{5}\right) = 5$

$\quad a_6 = a_5 r = 5\left(\frac{1}{5}\right) = 1$

$\quad a_7 = a_6 r = 1\left(\frac{1}{5}\right) = \frac{1}{5}$

The first seven terms are

3125, 625, 125, 25, 5, 1, and $\frac{1}{5}$.

9. $a_1 = -1$, $r = -1$, $n = 6$

$\quad a_1 = -1$

$\quad a_2 = a_1 r = -1(-1) = 1$

$\quad a_3 = a_2 r = 1(-1) = -1$

$\quad a_4 = a_3 r = -1(-1) = 1$

$\quad a_5 = a_4 r = 1(-1) = -1$

$\quad a_6 = a_5 r = -1(-1) = 1$

The first six terms are

$\quad -1$, 1, -1, 1, -1, and 1.

11. $a_3 = 6$, $a_4 = 12$, $n = 5$

Since $a_4 = a_3 \cdot r$,

$\quad 12 = 6r$

$\quad\quad 2 = r$.

Since $a_3 = a_1 r^2$,

$\quad\quad 6 = a_1(4)$

$\quad\quad a_1 = \frac{3}{2}$.

$\quad a_2 = a_1 r = \frac{3}{2}(2) = 3$

$\quad a_5 = a_4 r = 12(2) = 24$

The first five terms are

$\quad \frac{3}{2}$, 3, 6, 12, and 24.

13. $a_1 = 4$, $r = 3$

$\quad a_5 = a_1 r^{5-1}$

$\quad\quad = 4 \cdot 3^4 = 324$

$\quad a_n = a_1 r^{n-1}$

$\quad\quad = 4 \cdot 3^{n-1}$

15. $a_1 = -2$, $r = 3$

$\quad a_5 = a_1 r^4$

$\quad\quad = (-2)(3^4) = -162$

$\quad a_n = a_1 r^{n-1}$

$\quad\quad = -2 \cdot 3^{n-1}$

17. $a_1 = -3$, $r = -5$

$\quad a_5 = a_1 r^4$

$\quad\quad = -3(-5)^4$

$\quad\quad = -1875$

$\quad a_n = a_1 r^{n-1}$

$\quad\quad = -3(-5)^{n-1}$

19. $a_2 = 3$, $r = 2$

First find a_1.

$$a_2 = a_1 r$$
$$a_1 = \frac{a_2}{r} = \frac{3}{2}$$
$$a_5 = a_1 r^4$$
$$= \frac{3}{2}(2^4) = 24$$
$$a_n = a_1 r^{n-1}$$
$$= \left(\frac{3}{2}\right) 2^{n-1}$$

21. $a_4 = 64$, $r = -4$

First find a_1.

$$a_4 = a_1 r^3$$
$$64 = a_1 (-4)^3$$
$$64 = -64 a_1$$
$$a_1 = -1$$
$$a_5 = -1(-4)^4 = -256$$
$$a_n = a_1 r^{n-1}$$
$$= -1(-4)^{n-1}$$

23. 6, 12, 24, 48, ...

Each term (after the first) is found by multiplying the preceding term by 2, so this is a geometric sequence with $r = 2$.

$$a_n = a_1 r^{n-1}$$
$$= 6 \cdot 2^{n-1}$$

25. $\frac{3}{4}$, $\frac{3}{2}$, 3, 6, 12, ...

Each term (after the first) is found by multiplying the preceding term by 2, so this is a geometric sequence with $r = 2$.

$$a_n = a_1 r^{n-1}$$
$$= \left(\frac{3}{4}\right) 2^{n-1}$$

27. -4, 2, -1, $\frac{1}{2}$, ...

This is a geometric sequence.

$$r = \frac{2}{-4} = -\frac{1}{2}$$
$$a_n = a_1 r^{n-1}$$
$$= -4\left(-\frac{1}{2}\right)^{n-1}$$

29. 18, 20, 24, 32, 48, ...

$$\frac{20}{18} = \frac{10}{9}$$
$$\frac{24}{20} = \frac{6}{5}$$

This sequence is not geometric because it does not have a common ratio.

31. $a_3 = 9$, $r = 2$

This is a geometric sequence. First, find a_1.

$$a_n = a_1 r^{n-1}$$
$$a_3 = a_1 \cdot 2^2 = a_1 \cdot 4$$
$$9 = 4 a_1$$
$$a_1 = \frac{9}{4}$$
$$r = 2$$
$$a_n = \left(\frac{9}{4}\right) 2^{n-1}$$

33. $a_3 = -2$, $r = 3$

This is a geometric sequence.

$$a_3 = a_1 r^2$$
$$-2 = a_1 \cdot 9$$
$$a_1 = -\frac{2}{9}$$
$$r = 3$$
$$a_n = \left(-\frac{2}{9}\right) 3^{n-1}$$

35. Since John earns $.01 on the first day and his pay on each successive day is double that of the previous day, the sequence of his earnings on each day is geometric with $a_1 = .01$ and $r = 2$.

His earnings on the twentieth day would be represented by a_{20}.

$$a_n = a_1 r^{n-1}$$
$$a_n = .01(2)^{n-1}$$
$$a_{20} = .01(2)^{20-1}$$
$$= .01(2^{19})$$
$$= .01(524,288)$$
$$= 5242.88$$

He would earn $5242.88 on the twentieth day.

37. Consider the geometric sequence

$$\frac{3}{5}, \ 3, \ 15, \ 45, \ 135, \ \ldots \ .$$

This sequence has $a_1 = \frac{3}{5}$, $r = 5$, and, as specified, $a_3 = 15$. (There are many other geometric sequences that also have $a_3 = 15$.)

39. (a) $1, -2, 4, -8, \ldots$

is a geometric sequence with $r = -2$.

(b) $1, 10, 100, 1000, \ldots$

is a geometric sequence with $r = 10$.

(c) $3, \ \frac{3}{2}, \ \frac{3}{4}, \ \frac{3}{8}, \ \ldots$

is a geometric sequence with $r = \frac{1}{2}$.

(d) $1, 1, 2, 3, 5, 8, \ldots$

is not a geometric sequence since there is not a common ratio.

Thus, (d) is the only one of the given sequences that is not geometric.

41. $a_2 = 6$, $a_6 = 486$

Use the formula for the nth term of a geometric sequence,

$$a_n = a_1 r^{n-1}.$$

For $n = 2$,

$$a_2 = a_1 r = 6.$$

For $n = 6$,

$$a_6 = a_1 r^5 = 486.$$

These equations form a nonlinear system, which can be solved by the substitution method.

$$a_1 r = 6 \qquad (1)$$
$$a_1 r^5 = 486 \qquad (2)$$

Solve equation (2) for r.

$$a_1 = \frac{6}{r} \qquad (3)$$

Substitute $\frac{6}{r}$ for a_1 in equation (2).

$$\left(\frac{6}{r}\right)r^5 = 486$$

$$6r^4 = 486$$

$$r^4 = 81$$

$$r = \pm 3$$

Substitute these values for r in equation (3) to find the corresponding values of a_1.

If $r = 3$, $a_1 = \frac{6}{3} = 2$.

If $r = -3$, $a_1 = \frac{6}{-3} = -2$.

Therefore, there are two geometric sequences with $a_2 = 6$ and $a_6 = 486$. One sequence has $a_1 = 2$ and $r = 3$; the other has $a_1 = -2$ and $r = -3$.

43. $a_2 = 64$, $a_8 = 1$

For n = 2,

$$a_2 = a_1 r = 64.$$

For n = 8,

$$a_8 = a_1 r^7 = 1.$$

Use substitution to solve the nonlinear system

$$a_1 r = 64 \qquad (1)$$

$$a_1 r^7 = 1. \qquad (2)$$

Solve equation (1) for r.

$$a_1 = \frac{64}{r}$$

Substitute $\frac{64}{r}$ for a_1 in equation (2).

$$\left(\frac{64}{r}\right)r^7 = 1$$

$$64r^6 = 1$$

$$r^6 = \frac{1}{64}$$

$$r = \pm\frac{1}{2}$$

If $r = \frac{1}{2}$, $a_1 = \dfrac{64}{\frac{1}{2}} = 128$.

If $r = -\frac{1}{2}$, $a_1 = \dfrac{64}{-\frac{1}{2}} = -128$.

Therefore, there are two geometric sequences with $a_2 = 64$ and $a_8 = 1$. One has $a_1 = 128$ and $r = \frac{1}{2}$; the other has $a_1 = -128$ and $r = -\frac{1}{2}$.

Section 9.3

1. $\displaystyle\sum_{i=1}^{5} (2i + 1)$

Start with i = 1; end with i = 5.

$$\sum_{i=1}^{5} (2i + 1)$$

$$= (2 + 1) + (4 + 1) + (6 + 1)$$
$$+ (8 + 1) + (10 + 1)$$

$$= 3 + 5 + 7 + 9 + 11$$

$$= 35$$

3. $\displaystyle\sum_{i=1}^{4} \frac{1}{j}$

$= \frac{1}{1} + \frac{1}{2} + \frac{1}{3} + \frac{1}{4} = \frac{25}{12}$

5. $\displaystyle\sum_{i=1}^{4} i^i$

$= 1^1 + 2^2 + 3^3 + 4^4$

$= 1 + 4 + 27 + 256$

$= 288$

7. $\displaystyle\sum_{i=1}^{6} (-1)^k \cdot k$

$= (-1)^1 \cdot 1 + (-1)^2 \cdot 2 + (-1)^3 \cdot 3$

$\quad + (-1)^4 \cdot 4 + (-1)^5 \cdot 5 + (-1)^6 \cdot 6$

$= -1 + 2 - 3 + 4 - 5 + 6$

$= 3$

9. Find S_{10} if $a_1 = 8$ and $d = 3$.

S_{10} represents the sum of the first ten terms of a sequence.

To find S_{10}, substitute $n = 10$, $a_1 = 8$, and $d = 3$ into the formula

$S_n = \frac{n}{2}[2a_1 + (n-1)d].$

$S_{10} = \frac{10}{2}[2(8) + (9)(3)]$

$= 5(16 + 27) = 5(43)$

$= 215$

11. Find S_{10} if $a_3 = 5$ and $a_4 = 8$.

Find a_1 and d first.

$d = a_4 - a_3 = 8 - 5 = 3$

$a_3 = a_1 + 2d$

$5 = a_1 + 2 \cdot 3$

$-1 = a_1$

$S_n = \frac{n}{2}[2a_1 + (n-1)d$

$S_{10} = \frac{10}{2}[2(-1) + (9)(3)]$

$= 5(-2 + 27) = 5(25)$

$= 125$

13. To find S_{10} for the sequence 5, 9, 13, ..., substitute $n = 10$, $a_1 = 5$, and $d = 4$ into the formula

$S_n = \frac{n}{2}[2a_1 + (n-1)d].$

$S_{10} = \frac{10}{2}[2(5) + (9)(4)]$

$= 5(10 + 36) = 230$

15. $a_1 = 10$, $a_{10} = 5\frac{1}{2} = \frac{11}{2}$

$S_n = \frac{n}{2}(a_1 + a_n)$

$S_{10} = \frac{10}{2}\left(10 + \frac{11}{2}\right)$

$= 5\left(\frac{31}{2}\right) = \frac{155}{2}$

17. $S_{20} = 1090$, $a_{20} = 102$

The formula for the sum is

$S_n = \frac{n}{2}(a_1 + a_n)$, so

$S_{20} = \frac{20}{2}(a_1 + a_{20}).$

$1090 = 10(a_1 + 102)$

Solve for a_1.

$109 = a_1 + 102$

$7 = a_1$

Solve for d.

$$a_n = a_1 + (n - 1)d$$
$$a_{20} = a_1 + (20 - 1)d$$
$$102 = 7 + 19d$$
$$95 = 19d$$
$$5 = d$$

19. $S_{12} = -108$, $a_{12} = -19$

The formula for the sum is

$$S_n = \frac{n}{2}(a_1 + a_n), \text{ so}$$

$$S_{12} = \frac{12}{2}(a_1 + a_{12}).$$

$$-108 = 6(a_1 - 19)$$
$$-108 = 6a_1 - 114$$
$$6 = 6a_1$$
$$1 = a_1$$

Solve for d.

$$a_n = a_1 + (n - 1)d$$
$$a_{12} = a_1 + (12 - 1)d$$
$$-19 = 1 + 11d$$
$$-20 = 11d$$
$$-\frac{20}{11} = d$$

21. $\displaystyle\sum_{i=1}^{3} (i + 4)$

This is a sum of three terms having
a common difference of 1, so it is
the sum of the first three terms of
the arithmetic sequence having

$$a_1 = 1 + 4 = 5,$$
$$n = 3,$$

and

$$a_n = a_3 = 3 + 4 = 7.$$

Thus,

$$\sum_{i=1}^{3} (i + 4) = S_3 = \frac{3}{2}(a_1 + a_3)$$
$$= \frac{3}{2}(5 + 7)$$
$$= 18.$$

23. $\displaystyle\sum_{i=1}^{10} (2j + 3)$

There is a common difference of
$2 \cdot 1$ or 2, so this is the sum
of an arithmetic sequence with

$$a_1 = 2(1) + 3 = 5,$$
$$n = 10,$$

and

$$a_n = a_{10} = 2(10) + 3 = 23.$$

Thus,

$$\sum_{i=1}^{10} (2j + 3) = S_{10} = \frac{10}{2}(a_1 + a_{10})$$
$$= 5(5 + 23)$$
$$= 140.$$

25. $\displaystyle\sum_{i=1}^{12} (-5 - 8i)$

This is the sum of an arithmetic sum
sequence with

$$d = -8$$
$$a_1 = -5 - 8(1) = -13$$
$$n = 12$$
$$a_n = a_{12} = -5 - 8(12) = -101.$$

$$\sum_{i=1}^{12} (-5 - 8i)$$

$$= S_{12} = \frac{12}{2}[(-13) + (-101)]$$
$$= 6(-114)$$
$$= -684$$

27.
$$\sum_{i=1}^{1000} i$$

This is the sum of an arithmetic sequence with

$$d = 1$$
$$a_1 = 1$$
$$n = 1000$$
$$a_n = a_{1000} = 1000.$$

$$\sum_{i=1}^{1000} i = S_{1000} = \frac{1000}{2}(1 + 1000)$$
$$= 500(1001)$$
$$= 500,500$$

29. 3, 6, 12, 24, ...

This is a geometric sequence.

Use $S_n = \frac{a_1(1 - r^n)}{1 - r}$ with $a_1 = 3$,

$r = \frac{6}{3} = 2$, and $n = 5$.

$$S_5 = \frac{a_1(1 - r^5)}{1 - r}$$
$$= \frac{3(1 - 2^5)}{1 - 2}$$
$$= \frac{3(-31)}{-1}$$
$$= 93$$

31. 12, -6, 3, $-\frac{3}{2}$, ...

This is a geometric sequence.

Use $S_n = \frac{a_1(1 - r^n)}{1 - r}$ with $a_1 = 12$,

$r = \frac{-6}{12} = -\frac{1}{2}$, and $n = 5$.

$$S_5 = \frac{a_1(1 - r^5)}{1 - r}$$
$$= \frac{12\left(1 - \left(-\frac{1}{2}\right)^5\right)}{1 - \left(-\frac{1}{2}\right)}$$
$$= \frac{12\left(\frac{33}{32}\right)}{\frac{3}{2}}$$
$$= \frac{33}{4}$$

33. $a_1 = 4$, $r = 2$

Use $S_n = \frac{a_1(1 - r^n)}{1 - r}$ with $a_1 = 4$,

$r = 2$, and $n = 5$.

$$S_5 = \frac{4(1 - 2^5)}{1 - 2}$$
$$= \frac{4(-31)}{-1}$$
$$= 124$$

35. $a_2 = \frac{1}{3}$, $r = 3$

$$a_2 = a_1 \cdot r$$
$$\frac{1}{3} = a_1 \cdot 3$$
$$a_1 = \frac{1}{3}\left(\frac{1}{3}\right) = \frac{1}{9}$$

Use $S_n = \frac{a_1(1 - r^n)}{1 - r}$.

$$S_5 = \frac{\left(\frac{1}{9}\right)(1 - 3^5)}{1 - 3}$$
$$= \frac{1}{9}\left[\frac{1 - 243}{-2}\right]$$
$$= \frac{1}{9}(121)$$
$$= \frac{121}{9}$$

37. $\displaystyle\sum_{i=1}^{4} 2^i$

    Since this expression is the sum of a geometric sequence with $r = 2$, $a_1 = 2$, and $n = 4$, use

    $$S_n = \frac{a_1(1 - r^n)}{1 - r}.$$

    $$S_4 = \frac{2(1 - 2^4)}{1 - 2}$$

    $$= 2(15)$$

    $$= 30.$$

$$\sum_{i=1}^{6} 81\left(\frac{2}{3}\right)^i = S_6 = \frac{54\left[1 - \left(\frac{2}{3}\right)^6\right]}{1 - \frac{2}{3}}$$

$$= \frac{54\left(1 - \frac{64}{729}\right)}{1 - \frac{2}{3}}$$

$$= \frac{54\left(\frac{665}{729}\right)}{\frac{1}{3}}$$

$$= \frac{3 \cdot 54 \cdot 665}{729}$$

$$= \frac{1330}{9}$$

39. $\displaystyle\sum_{i=1}^{4} (-3)^i$

    Since this is the sum of a geometric sequence with $r = -3$, $a_1 = -3$, and $n = 4$, use

    $$S_n = \frac{a_1(1 - r^n)}{1 - r}.$$

    $$S_4 = \frac{(-3)[1 - (-3)^4]}{1 - (-3)}$$

    $$= \frac{(-3)(1 - 81)}{4}$$

    $$= 60$$

41. $\displaystyle\sum_{i=1}^{6} 81\left(\frac{2}{3}\right)^i$

    $a_1 = 81\left(\frac{2}{3}\right)^1 = 54$, $r = \frac{2}{3}$

    $$S_n = \frac{a_1(1 - r^n)}{1 - r}$$

43. $\displaystyle\sum_{i=3}^{6} 2^i$

    Since this is the sum of a geometric sequence with $r = 2$, $a_1 = 2^3$ or 8, and $n = 6$, use

    $$S_n = \frac{a_1(1 - r^n)}{1 - r}.$$

    (Note: This sum has four terms, not six.)

    $$S_4 = \frac{8(1 - 2^4)}{1 - 2}$$

    $$= \frac{8(-15)}{-1}$$

    $$= 120$$

45. $\displaystyle\sum_{i=1}^{4} 5\left(\frac{3}{5}\right)^{i-1}$

    Since this is the sum of a geometric sequence with $r = \frac{3}{5}$, $a_1 = 5$, and $n = 4$, use

$$S_n = \frac{a_1(1 - r^n)}{1 - r}.$$

$$S_4 = \frac{5[1 - \left(\frac{3}{5}\right)^4]}{1 - \frac{3}{5}}$$

$$= \frac{5\left(1 - \frac{81}{625}\right)}{\frac{2}{5}}$$

$$= \frac{272}{25}$$

47. Find the sum of all the integers from 51 to 71.

$$\sum_{i=51}^{71} i = \sum_{i=1}^{71} i - \sum_{i=1}^{50} i$$

$$= S_{71} - S_{50}$$

We know $a_1 = 1$, $d = 1$, $a_{50} = 50$, and $a_{71} = 71$. Thus,

$$S_{71} = \frac{71}{2}(1 + 71) = 71(36) = 2556,$$

$$S_{50} = \frac{50}{2}(1 + 50) = 25(51) = 1275.$$

Thus, the sum is

$$S_{71} - S_{50} = 2556 - 1275 = 1281.$$

49. In every 12 hour cycle, the clock will chime $1 + 2 + 3 + \ldots + 12$ times.
$a_1 = 1$, $n = 12$, $a_{12} = 12$

$$S_{12} = \frac{12}{2}(1 + 12) = 6(13) = 78$$

Since there are two 12 hour cycles in 1 day, every day the clock will chime $2(78) = 156$ times.

Since there are 30 days in this month, the clock will chime $156 \cdot 30 = 4680$ times.

51. The distances the sky diver falls during successive seconds is the arithmetic sequence 10, 20, 30, ... with $a_1 = 10$, $d = 10$. So during the tenth second he falls

$$a_{10} = a_1 + 9d$$
$$= 10 + 9(10)$$
$$= 100 \text{ m.}$$

The distance he falls in 10 seconds is

$$S_{10} = \frac{10}{2}(a_1 + a_{10})$$
$$= 5(10 + 100)$$
$$= 550 \text{ m.}$$

53. The distances the object falls during successive seconds form the arithmetic sequence 16, 48, 80, ... with $a_1 = 16$, $d = 32$. Thus, during the eighth second it falls

$$a_8 = a_1 + 7d$$
$$= 16 + 7(32)$$
$$= 240 \text{ ft.}$$

The total distance it falls in 8 sec is

$$S_8 = \frac{8}{2}(a_1 + a_8)$$
$$= 4(16 + 240)$$
$$= 4(256)$$
$$= 1024 \text{ ft.}$$

55. The longest support is 15 m long and the shortest support is 2 m long. Since the slide is of uniform slope, the sum of the lengths of these 20 supports can be thought of as the sum of an arithmetic sequence with $a_1 = 15$, $a_{20} = 2$, and $n = 20$.

$$S_{20} = \frac{20}{2}(a_1 + a_{20})$$
$$= 10(15 + 2)$$
$$= 10(17)$$
$$= 170 \text{ m}$$

57. The savings on the days in January form a geometric sequence with $a_1 = 1$, $r = 2$. On January 31, with $n = 31$, we would have

$$a_{31} = a_1 r^{30} = 1 \cdot 2^{30}$$
$$= \$2^{30}$$
$$\text{or} \quad \$1,073,741,824.$$

59. The total amount of income for the 6 yr forms a geometric sequence with $a_1 = 4,000,000$, $n = 6$, and $r = \frac{1}{2}$.

$$S_6 = \frac{a_1(1 - r^6)}{1 - r}$$

$$= \frac{4,000,000[1 - \left(\frac{1}{2}\right)^6]}{1 - \frac{1}{2}}$$

$$= \frac{4,000,000\left(1 - \frac{1}{64}\right)}{\frac{1}{2}}$$

$$= \$7,875,000$$

61. Since 100% of the original fixer is in the print at the start and 98% is removed with 15 minutes of washing, the amount left is 2%, or .02. Let $a_1 = .02$. Then the amounts left with each 15 minutes' washing form a geometric sequence with $r = .02$. After one hour, when $n = 4$, the amount left is

$$a_4 = a_1 r^3$$
$$= .02(.02)^3$$
$$\approx .0000002 = .00002\%.$$

Thus, essentially none of the original fixer is left after one hour.

63. Since the half life of this radioactive substance is 3 yr, the amount present forms a geometric sequence with $a_1 = 10^{15}$, $r = \frac{1}{2}$, and $n = 6$ (five 3-year cycles). Note that a_2 is the amount remaining after the first 3 yr, a_3 is the amount remaining after the second 3 yr, and so on.

$$a_6 = a_1 r^{n-1}$$
$$= 10^{15}\left(\frac{1}{2}\right)^{6-1}$$
$$= 10^{15}\left(\frac{1}{32}\right)$$
$$= \frac{1}{32} \times 10^{15} \text{ molecules}$$

65. The future value of an annuity uses the formula $S_n = \frac{a_1(1 - r^n)}{1 - r}$, where $r = 1 + \text{interest rate}$.

The payments are \$1000 for 9 yr at 8% compounded annually, so $a_1 = 1000$, $r = .08$, and $n = 9$.

$$S_9 = \frac{1000(1 - 1.08^9)}{1 - 1.08}$$

$$= \$12,487.56$$

67. The future value of an annuity uses the formula $S_n = \frac{a_1(1 - r^n)}{1 - r}$ where $r = 1 +$ interest rate.

The payments are \$2430 for 10 yr at 11% compounded annually, so $a_1 = 2430$, $r = .11$, and $n = 10$.

$$S_{10} = \frac{2430(1 - 1.11^{10})}{1 - 1.11}$$

$$= \$40,634.48$$

Section 9.4

1. 9, 18, 36, 72, 144, ...

$$r = \frac{18}{9} = 2$$

The sum would not exist because $|r| > 1$.

3. 10, 100, 1000, 10,000, ...

$$r = \frac{100}{10} = 10$$

The sum would not exist because $|r| > 1$.

5. 12, 6, 3, $\frac{3}{2}$, ...

$$r = \frac{6}{12} = \frac{1}{2}$$

The sum would exist because $-1 < r < 1$.

7. 1, 1.1, 1.2, 1.331, ...

$r = 1.1$

The sum would not exist because $|r| > 1$.

9. An infinite geometric sequence will have a sum so long as the common ratio satisfies $-1 < r < 1$.

11. 16 + 4 + 1 + ...

$a_1 = 16$ and $r = \frac{1}{4}$, so the sum exists and

$$S_\infty = \frac{a_1}{1 - r} = \frac{16}{1 - \frac{1}{4}} = \frac{16}{\frac{3}{4}} = \frac{64}{3}.$$

13. 100 + 10 + 1 + ...

$a_1 = 100$ and $r = \frac{1}{10}$, so the sum exists and

$$S_\infty = \frac{a_1}{1 - r} = \frac{100}{1 - \frac{1}{10}} = \frac{100}{\frac{9}{10}}$$

$$= \frac{1000}{9}.$$

15. 90 + 30 + 10 + ...

$a_1 = 90$ and $r = \frac{1}{3}$, so

$$S_\infty = \frac{a_1}{1 - r} = \frac{90}{1 - \frac{1}{3}} = \frac{90}{\frac{2}{3}} = 90 \cdot \frac{3}{2}$$

$$= 135.$$

17. $256 - 128 + 64 - 32 + 16 - \ldots$

$a_1 = 256$ and $r = -\frac{1}{2}$, so

$$S_\infty = \frac{256}{1 - \left(-\frac{1}{2}\right)} = \frac{256}{\frac{3}{2}} = 256 \cdot \frac{2}{3}$$

$$= \frac{512}{3}.$$

19. $108 - 36 + 12 - 4 + \ldots$

$a_1 = 108$, $r = -\frac{1}{3}$

$$S_\infty = \frac{108}{1 - \left(-\frac{1}{3}\right)} = \frac{108}{\frac{4}{3}} = 81$$

21. $\frac{3}{4} + \frac{3}{8} + \frac{3}{16} + \ldots$

$a_1 = \frac{3}{4}$, $r = \frac{1}{2}$

$$S_\infty = \frac{\frac{3}{4}}{1 - \frac{1}{2}} = \frac{\frac{3}{4}}{\frac{1}{2}} = \frac{3}{2}$$

23. $3 - \frac{3}{2} + \frac{3}{4} - \ldots$

$a_1 = 3$, $r = -\frac{1}{2}$

$$S_\infty = \frac{3}{1 - \left(-\frac{1}{2}\right)} = \frac{3}{\frac{3}{2}} = 2$$

25. $\frac{1}{3} - \frac{2}{9} + \frac{4}{27} - \frac{8}{21} + \ldots$

$a_1 = \frac{1}{3}$, $r = -\frac{2}{3}$

$$S_\infty = \frac{\frac{1}{3}}{1 - \left(-\frac{2}{3}\right)} = \frac{\frac{1}{3}}{\frac{5}{3}} = \frac{1}{5}$$

27. $\frac{1}{36} + \frac{1}{30} + \frac{1}{25} + \ldots$

$r = \frac{6}{5}$, $a_1 = \frac{1}{36}$

This sum does not exist since $|r| > 1$.

29. $\sum\limits_{i=1}^{\infty} \left(\frac{1}{4}\right)^i$ represents the sum of the infinite geometric sequence with

$$a_1 = \left(\frac{1}{4}\right)^1 = \frac{1}{4} \text{ and } r = \frac{1}{4}.$$

$$\sum\limits_{i=1}^{\infty} \left(\frac{1}{4}\right)^i = S_\infty = \frac{\frac{1}{4}}{1 - \frac{1}{4}} = \frac{\frac{1}{4}}{\frac{3}{4}} = \frac{1}{3}$$

31. $\sum\limits_{i=1}^{\infty} (1.2)^i$

$a_1 = 1.2$, $r = 1.2$

The sum does not exist since $|r| > 1$.

33. $\sum\limits_{i=1}^{\infty} \left(-\frac{1}{4}\right)^i$

$a_1 = -\frac{1}{4}$, $r = -\frac{1}{4}$

$$\sum\limits_{i=1}^{\infty} \left(-\frac{1}{4}\right)^i = S_\infty = \frac{-\frac{1}{4}}{1 - \left(-\frac{1}{4}\right)}$$

$$= \frac{-\frac{1}{4}}{\frac{5}{4}} = -\frac{1}{5}$$

35. $\displaystyle\sum_{i=1}^{\infty} \frac{1}{5i}$

$a_1 = \dfrac{1}{5}, \; r = \dfrac{1}{5}$

$\displaystyle\sum_{i=1}^{\infty} \frac{1}{5i} = S_\infty = \frac{\dfrac{1}{5}}{1 - \dfrac{1}{5}} = \frac{\dfrac{1}{5}}{\dfrac{4}{5}} = \frac{1}{4}$

37. $\displaystyle\sum_{i=1}^{\infty} 10^{-i}$

This represents the sum

$$\frac{1}{10} + \frac{1}{100} + \frac{1}{1000} + \cdots .$$

$a_1 = \dfrac{1}{10}, \; r = \dfrac{1}{10}$

$\displaystyle\sum_{i=1}^{\infty} 10^{-i} = S_\infty = \frac{\dfrac{1}{10}}{1 - \dfrac{1}{10}} = \frac{\dfrac{1}{10}}{\dfrac{9}{10}} = \frac{1}{9}$

39. $\displaystyle\sum_{i=1}^{\infty} \left(\frac{1}{2}\right)^{-i}$

This represents the sum

$$\left(\frac{1}{2}\right)^{-1} + \left(\frac{1}{2}\right)^{-2} + \left(\frac{1}{2}\right)^{-3} + \cdots$$

$$= 2 + 4 + 8 + \cdots .$$

$a_1 = 2, \; r = 2$

The sum does not exist since $|r| > 1$.

41. (b) $\displaystyle\sum_{i=1}^{\infty} \left(\frac{3}{2}\right)^{i}$

$r = \dfrac{3}{2}, \; a_1 = \dfrac{3}{2}$

This sum does not exist since $|r| > 1$.

43. .55555... can be written as
.5 + .05 + .005 + ..., which
is the sum of an infinite
geometric sequence having
$a_1 = .5$ and $r = .1$. The sum
of this sequence is given by

$$S_\infty = \frac{a_1}{1 - r}$$

$$= \frac{.5}{1 - .1} = \frac{.5}{.9} = \frac{5}{9}.$$

Thus, $.55555... = \dfrac{5}{9}$.

45. .121212... = .12 + .0012 + .000012
+ ..., which is the sum of an
infinite geometric sequence with

$$a_1 = .12 \quad \text{and} \quad r = .01.$$

$$S_\infty = \frac{.12}{1 - .01} = \frac{.12}{.99} = \frac{12}{99} = \frac{4}{33}$$

Thus, $.121212... = \dfrac{4}{33}$.

47. .313131... = .31 + .0031 + .000031
+ ...

$$a_1 = .31, \; r = .01$$

$$S_\infty = \frac{.31}{1 - .01} = \frac{.31}{.99} = \frac{31}{99}$$

Thus, $.313131... = \dfrac{31}{99}$.

49. 1.508508508... = 1 + .508 + .000508
+ .000000508 + ...

Beginning with the second term, this
is the sum of the terms of an infinite geometric sequence with

$$a_1 = .508 \text{ and } r = .001.$$

$$S_\infty = \frac{.508}{1 - .001} = \frac{.508}{.999} = \frac{508}{999}$$

Thus,

$$1.508508508\ldots = 1 + \frac{508}{999} = \frac{1507}{999}.$$

51. There are two sequences. A term of one is the distance the ball falls each time, and a term of the other is the distance the ball returns bouncing each time. We need the sum of both sequences.

Falling:

$$10 + 10\left(\frac{3}{4}\right) + 10\left(\frac{3}{4}\right)^2 + \ldots$$

$$= \frac{10}{1 - \frac{3}{4}}$$

$$= \frac{10}{\frac{1}{4}} = 40$$

Bouncing:

$$10\left(\frac{3}{4}\right) + 10\left(\frac{3}{4}\right)^2 + \ldots$$

$$= \frac{10\left(\frac{3}{4}\right)}{1 - \frac{3}{4}} = 30$$

Total distance = 40 + 30 = 70
The ball will travel 70 m before coming to rest.

53. Since the wheel initially rotates 400 times per minute, and then rotates only $\frac{3}{4}$ as many times as in the previous minute, $a_1 = 400$ and $r = \frac{3}{4}$.

$$\frac{a_1}{1 - r} = \frac{400}{1 - \frac{3}{4}} = \frac{400}{\frac{1}{4}} = 1600$$

55.

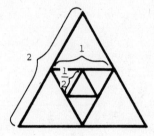

The first triangle has sides of length 2 m, so its perimeter is $3(2) = 6$ m. The second triangle has sides of length 1 m, so its perimeter is $3(1) = 3$ m.

$$3(2) + 3(1) + 3\left(\frac{1}{2}\right) + 3\left(\frac{1}{4}\right) + \ldots$$

$$= 6 + 3 + \frac{3}{2} + \frac{3}{4} + \ldots$$

This is the sum of an infinite geometric sequence with $a_1 = 6$ and $r = \frac{1}{2}$.

$$S_\infty = \frac{6}{1 - \frac{1}{2}} = \frac{6}{\frac{1}{2}} = 12$$

The total perimeter of all the triangles is 12 m.

Section 9.5

1. Prove that
$$2 + 4 + 6 + \ldots + 2n = n(n + 1)$$
is true for every positive integer n.

Step 1 Show that the statement is true for n = 1. S_1 is the statement

$$2 = 1(1 + 1),$$

which is true.

Step 2 Show that if S_k is true, then S_{k+1} is also true.

S_k is the statement

$$2 + 4 + 6 + \ldots + 2k = k(k + 1)$$

and S_{k+1} is the statement

$$2 + 4 + 6 + \ldots + 2k + 2(k + 1) = (k + 1)(k + 2).$$

Start with S_k:

$$2 + 4 + 6 + \ldots + 2k = k(k + 1).$$

Add the (k + 1)st term, 2(k + 1), to both sides:

$$2 + 4 + 6 + \ldots + 2k + 2(k + 1) = k(k + 1) + 2(k + 1).$$

Now factor out the common factor on the right to get

$$2 + 4 + 6 + \ldots + 2k + 2(k + 1) = (k + 1)(k + 2).$$

This result is the statement S_{k+1}. Thus, we have shown that if S_k is true, S_{k+1} is also true. The two steps required for a proof by mathematical induction have been completed, so the statement

$$2 + 4 + 6 + \ldots + 2n = n(n + 1)$$

is true for every positive integer n.

3. Let S_n be the statement

$$3 + 6 + 9 + \ldots + 3n = \frac{3n(n + 1)}{2}.$$

Prove that S_n is true for every positive integer n.

Step 1 S_1 is the statement

$$3 = \frac{3 \cdot 1(1 + 1)}{2}$$

$$3 = \frac{6}{2},$$

which is true.

Step 2 Show that if S_k is true, then S_{k+1} is also true.

S_k is the statement

$$3 + 6 + 9 + \ldots + 3k = \frac{3k(k + 1)}{2}.$$

Add the (k + 1)st term, 3(k + 1), to both sides.

$$3 + 6 + 9 + \ldots + 3k + 3(k + 1)$$
$$= \frac{3k(k + 1)}{2} + 3(k + 1)$$
$$= \frac{3k(k + 1)}{2} + \frac{6(k + 1)}{2}$$
$$= \frac{(k + 1)(3k + 6)}{2}$$

$$3 + 6 + 9 + \ldots + 3k + 3(k + 1)$$
$$= \frac{3(k + 1)(k + 2)}{2}$$

The final equation is the statement S_{k+1}. Thus, if S_k is true, S_{k+1} is also true.

Step 1 and 2 have been completed, so S_n is true for all positive integers n.

5. Let S_n be the statement

$$2 + 4 + 8 + \ldots + 2^n = 2^{n+1} - 2.$$

Prove that S_n is true for every positive integer n.

Step 1 S_1 is the statement

$$2 = 1^{1+1} - 2$$
$$2 = 4 - 2,$$

which is true.

Step 2   Show that if S_k is true, then S_{k+1} is also true.
S_k is the statement.

$$2 + 4 + 8 + \ldots + 2^k = 2^{k+1} - 2.$$

Add the (k + 1)st term, 2^{k+1}, to both sides.

$$2 + 4 + 8 + \ldots + 2^k + 2^{k+1}$$
$$= (2^{k+1} - 2) + 2^{k+1}$$
$$= 2 \cdot 2^{k+1} - 2$$
$$= 2^{k+2} - 2$$

$$2 + 4 + 8 + \ldots + 2^k + 2^{k+1}$$
$$= 2^{(k+1)+1} - 2$$

The final equation is the statement S_{k+1}. Thus, if S_k is true, S_{k+1} is also true.
Therefore, by mathematical induction, S_n is true for every positive integer n.

7. Let S_n be the statement

$$1^2 + 2^2 + 3^2 + \ldots + n^2$$
$$= \frac{n(n + 1)(2n + 1)}{6}.$$

Step 1 S_1 is the statement

$$1^2 = \frac{1(2)(3)}{6},$$

$$1 = \frac{6}{6},$$

which is true.

Step 2   Show that if S_k is true, S_{k+1} is also true. S_k is the statement

$$1^2 + 2^2 = 3^2 + \ldots + k^2$$
$$= \frac{k(k + 1)(2k + 1)}{6}.$$

Add the (k + 1)st term, $(k + 1)^2$, to both sides.

$$1^2 + 2^2 + 3^2 + \ldots + k^2 + (k + 1)^2$$
$$= \frac{k(k + 1)(2k + 1)}{6} + (k + 1)^2$$
$$= \frac{k(2k^2 + 3k + 1)}{6} + k^2 + 2k + 1$$
$$= \frac{k(2k^2 + 3k + 1)}{6} + \frac{6(k^2 + 2k + 1)}{6}$$
$$= \frac{2k^3 + 3k^2 + k + 6k^2 + 12k + 6}{6}$$
$$= \frac{2k^3 + 9k^2 + 13k + 6}{6}$$
$$= \frac{(k + 1)(2k^2 + 7k + 6)}{6}$$
$$= \frac{(k + 1)(k + 2)(2k + 3)}{6}$$
$$= \frac{(k + 1)(k + 2)[2(k + 1) + 1]}{6}$$

The final result is the statement S_{k+1}.

Thus, if S_k is true, S_{k+1} is also true. Therefore, by mathematical induction, S_n is true for every positive integer n.

9. Let S_n be the statement

$$5 \cdot 6 + 5 \cdot 6^2 + 5 \cdot 6^3 + \ldots + 5 \cdot 6^n$$
$$= 6(6^n - 1).$$

Step 1 S_1 is the statement

$$5 \cdot 6 = 6(6^1 - 1)$$
$$30 = 6 \cdot 5,$$

which is true.

Step 2 Show that if S_k is true, S_{k+1} is also true.

S_k is the statement

$$5 \cdot 6 + 5 \cdot 6^2 = 5 \cdot 6^3 + \ldots + 5 \cdot 6^k$$
$$= 6(6^k - 1).$$

Add the $(k + 1)$st term, $5 \cdot 6^{k+1}$, to both sides.

$$5 \cdot 6 + 5 \cdot 6^2 + \ldots + 5 \cdot 6^k + 5 \cdot 6^{k+1}$$
$$= 6(6^k - 1) + 5 \cdot 6^{k+1}$$
$$= 6 \cdot 6^k - 6 + 5 \cdot 6^{k+1}$$
$$= 6^{k+1} + 5 \cdot 6^{k+1} - 6$$
$$= 1 \cdot 6^{k+1} + 5 \cdot 6^{k+1} - 6$$
$$= 6 \cdot 6^{k+1} - 6$$
$$= 6(6^{k+1} - 1)$$

The final result is the statement S_{k+1}. Thus, if S_k is true, S_{k+1} is also true. Therefore, by mathematical induction, S_n is true for every positive integer n.

11. Let S_n be the statement

$$\frac{1}{1 \cdot 2} + \frac{1}{2 \cdot 3} + \frac{1}{3 \cdot 4} + \ldots + \frac{1}{n(n + 1)}$$
$$= \frac{n}{n + 1}.$$

Step 1 S_1 is the statement

$$\frac{1}{1 \cdot 2} = \frac{1}{1 + 1},$$

which is true.

Step 2 Show that if S_k is true, S_{k+1} is also true. S_k is the statement

$$\frac{1}{1 \cdot 2} + \frac{1}{2 \cdot 3} + \frac{1}{3 \cdot 4} + \ldots + \frac{1}{k(k + 1)}$$
$$= \frac{k}{k + 1}.$$

Add the $(k + 1)$st term,

$$\frac{1}{(k + 1)(k + 2)}, \text{ to both sides.}$$

$$\frac{1}{1 \cdot 2} + \frac{1}{2 \cdot 3} + \ldots$$
$$+ \frac{1}{k(k + 1)} + \frac{1}{(k + 1)(k + 2)}$$
$$= \frac{k}{k + 1} + \frac{1}{(k + 1)(k + 2)}$$
$$= \frac{k(k + 2) + 1}{(k + 1)(k + 2)}$$
$$= \frac{k^2 + 2k + 1}{(k + 1)(k + 2)}$$
$$= \frac{(k + 1)^2}{(k + 1)(k + 2)}$$
$$= \frac{k + 1}{k + 2}$$

The final result is the statement S_{k+1}. Thus, if S_k is true, S_{k+1} is also true. Therefore, by mathematical induction, S_n is true for every positive integer n.

13. Let S_n be the statement

$$\frac{1}{2} + \frac{1}{2^2} + \frac{1}{2^3} + \ldots + \frac{1}{2^n} = 1 - \frac{1}{2^n}.$$

Step 1 S_1 is the statement

$$\frac{1}{2} = 1 - \frac{1}{2},$$

which is true.

Step 2 Show that if S_k is true, S_{k+1} is also true.

S_k is the statement

$$\frac{1}{2} + \frac{1}{2^2} + \frac{1}{2^3} + \cdots + \frac{1}{2^k} = 1 - \frac{1}{2^k}.$$

Add the $(k + 1)$st term, $\frac{1}{2^{k+1}}$, to both sides.

$$\frac{1}{2} + \frac{1}{2^2} + \frac{1}{2^3} + \cdots + \frac{1}{2^k} + \frac{1}{2^{k+1}}$$

$$= 1 - \frac{1}{2^k} + \frac{1}{2^{k+1}}$$

$$= 1 - \frac{1}{2^k} + \frac{1}{2^1 2^k}$$

$$= 1 - \frac{1}{2^k} + \frac{1}{2}\left(\frac{1}{2^k}\right)$$

$$= 1 - \frac{1}{2}\left(\frac{1}{2^k}\right)$$

$$= 1 - \frac{1}{2^{k+1}}$$

The final result is the statement S_{k+1}. Thus, if S_k is true, S_{k+1} is also true. Therefore, by mathematical induction, S_n is true for every positive integer n.

15. Let S_n be the statement

$$x^{2n} + x^{2n-1}y + \cdots + xy^{2n-1} + y^{2n}$$
$$= \frac{x^{2n+1} - y^{2n+1}}{x - y}.$$

Step 1 S_1 is the statement

$$x^2 + xy + y^2 = \frac{x^3 - y^3}{x - y}.$$

To verify this statement, factor $x^3 - y^3$ as the difference of two cubes.

$$x^2 + xy + y^2 = \frac{(x - y)(x^2 + xy + y^2)}{x - y}$$

$$x^2 + xy + y^2 = x^2 + xy + y^2$$

Thus, S_1 is true.

Step 2 Show that if S_k is true, S_{k+1} is also true.

S_k is the statement

$$x^{2k} + x^{2k-1}y + \cdots + xy^{2k-1} + y^{2k}$$

$$= \frac{x^{2k+1} - y^{2k+1}}{x - y}.$$

S_{k+1} is the statement

$$x^{2(k+1)} + x^{2(k+1)-1}y + x^{2(k+1)-2}y^2$$
$$+ \cdots + x^2y^{2(k+1)-2} + xy^{2(k+1)-1}$$
$$+ y^{2(k+1)}$$

$$= \frac{x^{2(k+1)+1} - y^{2(k+1)+1}}{x - y}.$$

We will work with the left side of this equation and show that it is equal to the right side.

$$x^{2(k+1)} + x^{2(k+1)-1}y + x^{2(k+1)-2}y^2 + \cdots$$
$$+ x^2y^{2(k+1)-2} + xy^{2(k+1)-1} + y^{2(k+1)}$$
$$= x^{2(k+1)} + xy[x^{2k} + x^{2k-1} + \cdots$$
$$+ xy^{2k-1} + y^{2k}] + y^{2(k+1)}$$
$$= x^{2(k+1)} + xy\left(\frac{x^{2k+1} - y^{2k+1}}{x - y}\right)$$
$$+ y^{2(k+1)} \quad \text{By using the assumption that } S_k \text{ is true}$$

$$= \frac{x^{2k+2}(x-y)+x^{2k+2}y-xy^{2k+2}+y^{2k+2}(x-y)}{x - y}$$

$$= \frac{x^{2k+3}-yx^{2k+2}+yx^{2k+2}-xy^{2k+2}+xy^{2k+2}-y^{2k+3}}{x - y}$$

$$= \frac{x^{2k+3} - y^{2k+3}}{x - y}$$

$$= \frac{x^{2(k+1)+1} - y^{2(k+1)+1}}{x - y}$$

We have shown that S_1 is true, and that if S_k is true. S_{k+1} is also true. Therefore, by mathematical induction, S_n is true for every positive integer n.

17. Let S_n be the statement

$(a^m)^n = a^{mn}$. (Assume that a and m are constant.)

Step 1 S_1 is the statement

$(a^m)^1 = a^{m \cdot 1}$

$a^m = a^m$.

Thus, S_1 is true.

Step 2 Show that if S_k is true, then S_{k+1} is also true.
S_k is the statement

$(a^m)^k = a^{mk}$, and

S_{k+1} is the statement

$(a^m)^{k+1} = a^{m(k+1)}$.

$(a^m)^{k+1}$

$= (a^m)^k (a^m)^1$ *Using the property* $a^m \cdot a^n = a^{m+n}$

$= a^{mk} \cdot a^m$ *Using the assumption that S_k is true*

$= a^{mk+m}$ *Using the property* $a^m \cdot a^n = a^{m+n}$

$= a^{m(k+1)}$ *Factor mk + m*

Thus, we have shown that S_1 is true, and that if S_k is true, S_{k+1} is also true. Therefore, by mathematical induction, S_n is true for every positive integer n.

19. Let S_n be the statement:

If $a > 1$, then $a^n > 1$.

Step 1 S_1 is the statement:

If $a > 1$, then $a > 1$, which is obviously true.

Step 2 Show that if S_k is true, then S_{k+1} is also true.
S_k is the statement:

If $a > 1$, then $a^k > 1$.
Multiply both sides of the inequality

$a^k > 1$ by a.

Since $a > 1 > 0$, the direction of the inequality symbol will not be changed.

$a \cdot a^k > a \cdot 1$

$a^1 \cdot a^k > a$

$a^{k+1} > a$ *Using the property* $a^m \cdot a^n = a^{m+n}$

Since $a > 1$, we have

$a^{k+1} > a > 1$.

Therefore, if $a > 1$, $a^{k+1} > 1$, and S_{k+1} is true.

By Steps 1 and 2, S_n is true for every positive integer n.

21. Let S_n be the statement:
If $0 < a < 1$, then $a^n < a^{n-1}$.

Step 1 S_1 is the statement:

If $0 < a < 1$, then $a < a^0$.
Since $a^0 = 1$, this is equivalent to:

If $0 < a < 1$, then $a < 1$, which is obviously true.

Step 2   Show that S_k if true, then S_{k+1} is also true.

S_k is the statement:

If $0 < a < 1$, then $a^k < a^{k-1}$.

S_{k+1} is the statement:

If $0 < a < 1$, then

$$a^{k+1} < a^{(k+1)-1}$$

or $a^{k+1} < a^k$.

Multiply both sides of the inequality

$$a^k < a^{k-1} \text{ by } a.$$

Since $a > 0$, the direction of the inequality symbol will not change.

$$a \cdot a^k < a \cdot a^{k-1}$$
$$a^{k+1} < a^{1+(k-1)} \quad \textit{Using the property } a^m \cdot a^n = a^{m+n}$$
$$a^{k+1} < a^{(k+1)-1}$$
$$a^{k+1} < a^k$$

The final result is the statement S_{k+1}. Thus, if S_k is true, S_{k+1} is also true. By Steps 1 and 2, S_n is true for every positive integer n.

25.   Let S_n be the statement that the number of points of intersection of n lines is $\dfrac{n^2 - n}{2}$.

Since 2 is the smallest number of lines that can intersect, we need to prove this statement for every positive integer $n \geq 2$.

Step 1 S_2 is the statement that 2 non-parallel lines intersect in one point, since

$$\frac{2^2 - 2}{2} = \frac{2}{2} = 1.$$

It is obvious that S_2 is true.

Step 2   Show that if S_k is true, S_{k+1} is also true.

S_k is the statement:

k lines in a plane (no lines parallel, no three lines passing through the same point) intersect in $\dfrac{k^2 - k}{2}$ points.

If one more line is added to the set of k lines, this new $(k + 1)$st line will intersect each of the previous k lines in one point. Thus, there will be k additional intersection points. Thus, the number of intersection points for $k + 1$ lines is

$$\frac{k^2 - k}{2} + k$$
$$= \frac{k^2 - k}{2} + \frac{2k}{2}$$
$$= \frac{k^2 + k}{2}$$
$$= \frac{(k^2 + 2k + 1) - (k + 1)}{2}$$
$$= \frac{(k + 1)^2 - (k + 1)}{2}.$$

This result is the statement S_{k+1}. By Steps 1 and 2, S_n is true for every positive integer $n \geq 2$.

27. The number of sides of the nth figure is $3 \cdot 4^{n-1}$ (from Exercise 26). To see this, let a_n = the number of sides of the nth figure.

$a_1 = 3$

$a_2 = 3 \cdot 4$ since each side of the first figure will develop into 4 sides.

$a_3 = 3 \cdot 4^2$, and so on.

This gives a geometric sequence with $a_1 = 3$ and $r = 4$, so

$$a_n = 3 \cdot 4^{n-1}.$$

To find the perimeter of each figure, multiply the number of sides by the length of each side. In each figure, the lengths of the sides are 1/3 the lengths of the sides in the preceding figure.

Thus, if P_n = perimeter of nth figure,

$$P_1 = 3(1) = 3$$

$$P_2 = 3 \cdot 4\left(\frac{1}{3}\right) = 4$$

$$P_3 = 3 \cdot 4^2\left(\frac{1}{9}\right) = \frac{16}{3}, \text{ and so on.}$$

This gives a geometric sequence with

$$P_1 = 3 \text{ and } r = \frac{4}{3}.$$

Thus, $P_n = a_1 r^{n-1}$

$$P_n = 3\left(\frac{4}{3}\right)^{n-1}.$$

The result may also be written as

$$P_n = \frac{3^1 \cdot 4^{n-1}}{3^{n-1}}$$

$$= \frac{4^{n-1}}{3^{-1} \cdot 3^{n-1}}$$

$$P_n = \frac{4^{n-1}}{3^{n-2}}.$$

Section 9.6

1. $P(7, 7) = \dfrac{7!}{(7 - 7)!}$

          $= \dfrac{7!}{0!}$

          $= \dfrac{7!}{1}$

          $= 7 \cdot 6 \cdot 5 \cdot 4 \cdot 3 \cdot 2 \cdot 1$

          $= 5040$

3. $P(6, 5) = \dfrac{6!}{(6 - 5)!}$

          $= \dfrac{6!}{1!}$

          $= \dfrac{6!}{1}$

          $= 6!$

          $= 6 \cdot 5 \cdot 4 \cdot 3 \cdot 2 \cdot 1$

          $= 720$

5. $P(10, 2) = \dfrac{10!}{(10 - 2)!}$

          $= \dfrac{10!}{8!}$

          $= \dfrac{10 \cdot 9 \cdot 8!}{8!}$

          $= 10 \cdot 9$

          $= 90$

7. $P(8, 3) = \dfrac{8!}{(8 - 3)!}$

          $= \dfrac{8!}{5!}$

          $= \dfrac{8 \cdot 7 \cdot 6 \cdot 5!}{5!}$

          $= 8 \cdot 7 \cdot 6$

          $= 336$

9. $P(7, 1) = \dfrac{7!}{(7-1)!}$

$= \dfrac{7!}{6!}$

$= \dfrac{7 \cdot 6!}{6!}$

$= 7$

11. $P(9, 0) = \dfrac{9!}{(9-0)!}$

$= \dfrac{9!}{9!}$

$= 1$

15. $5 \cdot 3 \cdot 2 = 30$

There are 30 different homes available if a builder offers a choice of 5 basic plans, 3 roof styles, and 2 exterior finishes.

17. (a) The first letter can be one of 2.
The second letter can be one of 25.
The third letter can be one of 24.
The fourth letter can be one of 23.

Since $2 \cdot 25 \cdot 24 \cdot 23 = 27{,}600$, there are 27,600 different call letters without repeats.

(b) With repeats, the count is

$2 \cdot 26 \cdot 26 \cdot 26 = 35{,}152.$

(c) The first letter can be one of 2. The second letter can be one of 24, since it cannot repeat the first letter or be R.

The third letter can be one of 23, since it cannot repeat either of the first two letters or be R.
The fourth letter can be one of 1, since it must be R.
Since $2 \cdot 24 \cdot 23 \cdot 1 = 1104$, there are 1104 different such call letters.

19. Use the multiplication principle of counting.

$$3 \cdot 5 = 15$$

There are 15 different first and middle name arrangements.

21. (a) The first three positions could each be any one of 26 letters, and the second three positions could each be any one of 10 numbers.

$26 \cdot 26 \cdot 26 \cdot 10 \cdot 10 \cdot 10$
$= 17{,}576{,}000$

17,576,000 license plates were possible.

(b) $10 \cdot 10 \cdot 10 \cdot 26 \cdot 26 \cdot 26$
$= 17{,}576{,}000$

17,576,000 additional license plates were made possible by the reversal.

(c) $26 \cdot 10 \cdot 10 \cdot 10 \cdot 26 \cdot 26 \cdot 26$
$= 456{,}976{,}000$

456,976,000 plates were provided by prefixing the previous pattern with an additional letter.

23. The number of ways in which 6 people can be seated in 6 seats in a row is given by

$$P(6, 6) = \frac{6!}{(6 - 6)!}$$

$$= \frac{6!}{0!}$$

$$= 6!$$

$$= 720.$$

25. He has 6 choices for the first course, 5 choices for the second, and 4 choices for the third.

$$6 \cdot 5 \cdot 4 = 120$$

27. The number of ways in which the 3 officers can be chosen from the 15 members is given by

$$P(15, 3) = \frac{15!}{(15 - 3)!}$$

$$= \frac{15!}{12!}$$

$$= \frac{15 \cdot 14 \cdot 13 \cdot 12!}{12!}$$

$$= 15 \cdot 14 \cdot 13$$

$$= 2730.$$

29. 5 players can be assigned the 5 positions in

$$P(5, 5) = 5! = 120 \text{ ways.}$$

10 players can be assigned 5 positions in

$$P(10, 5) = \frac{10!}{(10 - 5)!}$$

$$= \frac{10!}{5!}$$

$$= 10 \cdot 9 \cdot 8 \cdot 7 \cdot 6$$

$$= 30,240 \text{ ways.}$$

Section 9.7

1. $\binom{6}{5} = \frac{6!}{(6 - 5)!5!}$

$$= \frac{6!}{1!5!}$$

$$= \frac{6 \cdot 5!}{5!}$$

$$= 6$$

3. $\binom{8}{5} = \frac{8!}{(8 - 5)!5!}$

$$= \frac{8!}{3!5!}$$

$$= \frac{8 \cdot 7 \cdot 6 \cdot 5!}{3 \cdot 2 \cdot 1 \cdot 5!}$$

$$= 56$$

5. $\binom{15}{4} = \frac{15!}{(15 - 4)!4!}$

$$= \frac{15!}{11!4!}$$

$$= \frac{15 \cdot 14 \cdot 13 \cdot 12 \cdot 11!}{11! \cdot 4!}$$

$$= \frac{15 \cdot 14 \cdot 13 \cdot 12}{4 \cdot 3 \cdot 2 \cdot 1}$$

$$= \frac{15 \cdot 14 \cdot 13 \cdot 12}{2 \cdot 12}$$

$$= \frac{15 \cdot 14 \cdot 13}{2}$$

$$= \frac{2730}{2}$$

$$= 1365$$

7. $\binom{10}{7} = \frac{10!}{(10 - 7)!7!}$

$$= \frac{10!}{3!7!}$$

$$= \frac{10 \cdot 9 \cdot 8 \cdot 7!}{3!7!}$$

$$= \frac{720}{6}$$

$$= 120$$

9. $\binom{14}{1} = \dfrac{14!}{(14 - 1)!1!}$

$= \dfrac{14!}{13!}$

$= \dfrac{14 \cdot 13!}{13!}$

$= 14$

11. $\binom{18}{0} = \dfrac{18!}{(18 - 0)!0!}$

$= \dfrac{18!}{18!}$

$= 1$

17. We want to choose 4 committee members out of 30 and the order is not important. The number of possible committees is

$\binom{30}{4} = \dfrac{30!}{(30 - 4)!4!}$

$= \dfrac{30!}{26!4!}$

$= \dfrac{30 \cdot 29 \cdot 28 \cdot 27 \cdot 26!}{4!26!}$

$= \dfrac{30 \cdot 29 \cdot 28 \cdot 27}{24}$

$= 27,405.$

19. $\binom{6}{3} = \dfrac{6!}{(6 - 3)!3!}$

$= \dfrac{6!}{3!3!}$

$= \dfrac{6 \cdot 5 \cdot 4 \cdot 3!}{3! \cdot 3 \cdot 2 \cdot 1}$

$= 20$

20 different kinds of hamburgers can be made.

21. This problem involves choosing 2 members from a set of 5 members. There are $\binom{5}{2}$ such subsets.

$\binom{5}{2} = \dfrac{5!}{3!2!}$

$= \dfrac{5 \cdot 4 \cdot 3!}{3! \cdot 2}$

$= \dfrac{5 \cdot 4}{2}$

$= 10$

There are 10 different 2-card combinations.

23. Since 2 blue marbles are to be chosen and there are 8 blue marbles, this problem involves choosing 2 members from a set of 8 members. There are $\binom{8}{2}$ ways of doing this.

$\binom{8}{2} = \dfrac{8!}{(8 - 2)!2!}$

$= \dfrac{8!}{6!2!}$

$= \dfrac{8 \cdot 7 \cdot 6!}{6!2!}$

$= \dfrac{8 \cdot 7}{2}$

$= 28$

28 samples of 2 marbles can be drawn in which both marbles are blue.

25. There are 5 liberals and 4 conservatives, giving a total of 9 members. Three members are chosen as delegates to a convention.

(a) There are $\binom{9}{3}$ ways of doing this.

$$\binom{9}{3} = \frac{9!}{(9 - 3)!3!}$$

$$= \frac{9!}{6!3!}$$

$$= \frac{9 \cdot 8 \cdot 7 \cdot 6!}{6!3!}$$

$$= \frac{9 \cdot 8 \cdot 7}{6}$$

$$= 84$$

84 delegations are possible.

(b) To get all liberals, we must choose 3 members from a set of 5, which can be done $\binom{5}{3}$ ways.

$$\binom{5}{3} = \frac{5!}{(5 - 3)!3!}$$

$$= \frac{5!}{2!3!}$$

$$= \frac{5 \cdot 4 \cdot 3!}{2!3!}$$

$$= \frac{5 \cdot 4}{2}$$

$$= 10$$

10 delegations could have all liberals.

(c) To get 2 liberals and 1 conservative involves two independent events. First select the liberals. The number of ways to do this is

$$\binom{5}{2} = \frac{5!}{3!2!}$$

$$= \frac{5 \cdot 4 \cdot 3!}{3! \cdot 2 \cdot 1}$$

$$= 10.$$

Now select the conservative. The number of ways to do this is

$$\binom{4}{1} = \frac{4!}{(4 - 1)!1!}$$

$$= \frac{4!}{3!}$$

$$= 4.$$

To find the number of delegations, use the fundamental principal of counting. The number of delegations with 2 liberals and 1 conservative is

$$10 \cdot 4 = 40.$$

(d) If one particular person must be on the delegation, then there are 2 people left to choose from a set consisting of 8 members.

$$\binom{8}{2} = \frac{8!}{(8 - 2)!2!}$$

$$= \frac{8!}{6!2!}$$

$$= \frac{8 \cdot 7 \cdot 6!}{6!2!}$$

$$= \frac{8 \cdot 7}{2}$$

$$= 28$$

28 delegations are possible which include the mayor.

27. Since order is important, this is a permutation problem.

$$P(5, 5) = \frac{5!}{(5 - 5)!}$$

$$= \frac{5!}{0!}$$

$$= 120 \text{ ways}$$

29. We are choosing a subset of 3 pineapples from a set containing 12 pineapples, so this is a combination problem.

$$\binom{12}{3} = \frac{12!}{(12-3)!3!}$$

$$= \frac{12!}{9!3!}$$

$$= \frac{12 \cdot 11 \cdot 10 \cdot 9!}{9!3!}$$

$$= \frac{12 \cdot 11 \cdot 10}{6}$$

$$= 220$$

220 samples can be drawn.

31. Order is important since the secretaries will be assigned to 3 different managers, so this is a permutation problem.

$$P(7, 3) = \frac{7!}{(7-3)!}$$

$$= \frac{7!}{4!}$$

$$= \frac{7 \cdot 6 \cdot 5 \cdot 4!}{4!}$$

$$= 7 \cdot 6 \cdot 5$$

$$= 210$$

The assignments can be made in 210 ways.

33. Order is unimportant, so these are combination problems. There are a total of 6 + 3 + 2 = 11 plants.

(a) At random, select 4 of the 11 plants.

$$\binom{11}{4} = \frac{11!}{(11-4)!4!}$$

$$= \frac{11!}{7!4!}$$

$$= \frac{11 \cdot 10 \cdot 9 \cdot 8 \cdot 7!}{7!4!}$$

$$= \frac{11 \cdot 10 \cdot 9 \cdot 8}{4 \cdot 3 \cdot 2 \cdot 1}$$

$$= 330$$

This can be done in 330 ways.

(b) If 2 of the 4 plants to be selected must be 2 of the 6 wheat plants, then the other 2 plants must be chosen from the remaining 11 - 6 = 5 plants.
Since selecting the wheat plants and selecting the other plants are independent events, the multiplication principle of counting can be used.

$$\binom{6}{2} \cdot \binom{5}{2} = \frac{6!}{(6-2)!2!} \cdot \frac{5!}{(5-2)!2!}$$

$$= \frac{6!}{4!2!} \cdot \frac{5!}{3!2!}$$

$$= \frac{6 \cdot 5 \cdot 4!}{4!2!} \cdot \frac{5 \cdot 4 \cdot 3!}{3!2!}$$

$$= \frac{6 \cdot 5}{2} \cdot \frac{5 \cdot 4}{2}$$

$$= 15 \cdot 10$$

$$= 150$$

This can be done in 150 ways.

35. Order is important since each person will have a different responsibility, so this is a permutation problem.

$$P(10, 4) = \frac{10!}{(10 - 4)!}$$

$$= \frac{10!}{6!}$$

$$= \frac{10 \cdot 9 \cdot 8 \cdot 7 \cdot 6!}{6!}$$

$$= 10 \cdot 9 \cdot 8 \cdot 7$$

$$= 5040$$

The committee can be formed in 5040 ways.

Section 9.8

1. Let h = heads, t = tails.

    The only possible outcome is a head. Hence, the sample space is

    $$S = \{h\}.$$

3. Since each coin can be a head or a tail and there are 3 coins, the sample space is

    $$S = \{hhh, hht, hth, thh, htt,$$
    $$tht, tth, ttt\}.$$

5. Let c = correct answer, w = wrong answer.
    There are 2 possible answers for each question and 3 questions, so the sample space is

    $$S = \{ccc, ccw, cwc, wcc, wwc,$$
    $$wcw, cww, www\}.$$

7. (a) "The result is heads" is the event $E_1 = \{h\}$. This event is certain to occur, so
    $P(E_1) = 1.$

    (b) "The result is tails" is the event $E_2 = \emptyset$. This event is an impossible event, so
    $P(E_2) = 0.$

9. The sample space is

    $$S = \{ccc, ccw, cwc, wcc, wwc, wcw,$$
    $$cww, www\}.$$
    $$n(S) = 8$$

    (a) The event, all three answers correct, is

    $$E = \{ccc\}.$$
    $$n(E) = 1$$
    $$P(e) = \frac{n(E)}{n(S)} = \frac{1}{8}$$

    (b) The event, all three answers wrong, is

    $$E = \{www\}.$$
    $$n(E) = 1$$
    $$P(E) = \frac{1}{8}$$

    (c) The event, exactly two answers correct, is

    $$E = \{ccw, cwc, wcc\}.$$
    $$n(E) = 3$$
    $$P(E) = \frac{3}{8}$$

(d) The event, at least one answer correct, is

$$E = \{ccc, ccw, cwc, wcc, wwc, wcw, cww\}.$$
$$n(E) = 7$$
$$P(E) = \frac{7}{8}$$

13. There are 15 marbles, so $n(S) = 15$.

(a) E_1: yellow marble is drawn
There are 3 yellow marbles, so $n(E_1) = 3$.

$$P(E_1) = \frac{3}{15} = \frac{1}{5}$$

(b) E_2: black marble is drawn
There are no black marbles, so $n(E_2) = 0$.

$$P(E_2) = \frac{0}{15} = 0$$

(c) E_3: yellow or white marble is drawn
There are 3 yellow and 4 white marbles, so $n(E_3) = 3 + 4 = 7$.

$$P(E_3) = \frac{7}{15}$$

(d) E_4: yellow marble is drawn
There are 3 yellow marbles, so $n(E_4) = 3$.

$$P(E_4) = \frac{3}{15} = \frac{1}{5}$$

E_4': yellow marble is not drawn
There are 12 non-yellow marbles, so $n(E_4') = 12$.

$$P(E_4') = \frac{12}{15} = \frac{4}{5}$$

The odds in favor of drawing a yellow marble are

$$= \frac{P(E_4)}{P(E_4')}$$
$$= \frac{\dfrac{1}{5}}{\dfrac{4}{5}}$$
$$= \frac{1}{4} \quad \text{or} \quad 1 \text{ to } 4.$$

(e) From part (b), the probability that a blue marble is drawn is

$$P(E_2) = \frac{8}{15}.$$

The probability that a blue marble is not drawn is

$$P(E_2') = 1 - P(E_2) = \frac{7}{15}.$$

The odds against drawing a blue marble are

$$= \frac{P(E_2')}{P(E_2)}$$
$$= \frac{\dfrac{7}{15}}{\dfrac{8}{15}}$$
$$= \frac{7}{8} \quad \text{or} \quad 7 \text{ to } 8.$$

15. E: sum of numbers is 5
$$= \{1 \text{ and } 4, \ 2 \text{ and } 3\}$$

From Exercise 4 or Exercise 9, $n(S) = 10$.
Therefore,

$$P(E) = \frac{2}{10} = \frac{1}{5}$$

and $P(E') = 1 - P(E) = \frac{4}{5}.$

Odds in favor of sum being 5

$$= \frac{P(E)}{P(E')} = \frac{\frac{1}{5}}{\frac{4}{5}} = \frac{1}{4} \quad \text{or} \quad 1 \text{ to } 4.$$

17. Let E be the event "candidate wins."
Then E′ is the event "candidate
loses."
If the odds in favor of the candidate winning are 3 to 2, then

$$P(E) = \frac{3}{3+2} = \frac{3}{5}.$$

Therefore,

$$P(E') = 1 - P(E) = 1 - \frac{3}{5} = \frac{2}{5}.$$

The probability that the candidate
will lose is $\frac{2}{5}$.

19. The sample space is S represented by
$\{M, U, U, B, B, B, C, C, C, C\}$, so
n(S) = 10.
The probability of alternate events
E and F is represented by

P(E or F) = P(E) + P(F) − P(E ∩ F).

(a) $P(U) = \frac{2}{10} = \frac{1}{5}$,

$$P(B) = \frac{3}{10},$$

P(U ∩ B) = 0, so

$$P(U \text{ or } B) = \frac{1}{5} + \frac{3}{10} - 0$$

$$= \frac{5}{10} = \frac{1}{2}.$$

(b) $P(B) = \frac{3}{10}$,

$$P(C) = \frac{4}{10}$$

$$= \frac{2}{5},$$

P(B ∩ C) = 0, so

$$P(B \text{ or } C) = \frac{3}{10} + \frac{4}{10} - 0$$

$$= \frac{7}{10}.$$

(c) $P(B) = \frac{3}{10}$,

$$P(M) = \frac{1}{10},$$

P(B ∩ M) = 0, so

$$P(B \text{ or } M) = \frac{3}{10} + \frac{1}{10} - 0$$

$$= \frac{4}{10} = \frac{2}{5}.$$

21. (a) P(less than $20)
= P($5–$19.99 or below $5)
= P($5–$19.99) + P(below $5)
= .37 + .25
= .62

(b) P($40 or more)
= P($40–$69.99 or $70–$99.99
or $100–$149.99 or $150 or
more)
= .09 + .07 + .08 + .03
= .27

(c) P(more than $99.99)
= P($100–$149.99 or $150 or
more)
= .08 + .03
= .11

(d) P(less than $100)

$\quad$ = P(below $5 or $5–$19.99 or
$\quad\quad$ $20–$39.99 or $40–$69.99
$\quad\quad$ or $70–$99.99)

$\quad$ = .25 + .37 + .11 + .09 + .07

$\quad$ = .89

23. There are four choices for the incorrect suit. The probability of
picking the incorrect card in a
given suit is 12/13. The probability of picking the correct card
in each of the other three suits is
1/13. Therefore, the probability of
getting three picks correct and
winning $200 is

$$4 \cdot \frac{12}{13} \cdot \frac{1}{13} \cdot \frac{1}{13} \cdot \frac{1}{13}$$

$$= \frac{48}{28,561} \approx .001681.$$

25. P(5 or more years) = .30
P(female) = .28
P(retirement plan contributor) = .65
P(retirement plan contributor and

female) = $\frac{1}{2}$(.28) = .14

(a) P(male) = 1 – P(female)

$\quad\quad$ = 1 – .28

$\quad\quad$ = .72

(b) P(less than 5 years)

$\quad$ = 1 – P(5 or more years)

$\quad$ = 1 – .30

$\quad$ = .70

(c) P(retirement plan contributor or
female)

$\quad$ = P(contributor) + P(female)

$\quad\quad$ – P(contributor and female)

$\quad$ = .65 + .28 – .14

$\quad$ = .79

Chapter 9 Review Exercises

1. $a_n = \dfrac{n}{n + 1}$

$a_1 = \dfrac{1}{1 + 1} = \dfrac{1}{2}$

$a_2 = \dfrac{2}{2 + 1} = \dfrac{2}{3}$

$a_3 = \dfrac{3}{3 + 1} = \dfrac{3}{4}$

$a_4 = \dfrac{4}{4 + 1} = \dfrac{4}{5}$

$a_5 = \dfrac{5}{5 + 1} = \dfrac{5}{6}$

The first five terms are

$$\frac{1}{2}, \frac{2}{3}, \frac{3}{4}, \frac{4}{5}, \text{ and } \frac{5}{6}.$$

3. $a_n = 2(n + 3)$

$a_1 = 2(1 + 3) = 8$

$a_2 = 2(2 + 3) = 10$

$a_3 = 2(3 + 3) = 12$

$a_4 = 2(4 + 3) = 14$

$a_5 = 2(5 + 3) = 16$

The first five terms are

$$8, 10, 12, 14, \text{ and } 16.$$

5. $a_1 = 5$; for $n \geq 2$, $a_n = a_{n-1} - 3$

$a_2 = a_{2-1} - 3 = a_1 - 3 = 5 - 3 = 2$

$a_3 = a_2 - 3 = 2 - 3 = -1$

$a_4 = a_3 - 3 = -1 - 3 = -4$

$a_5 = a_4 - 3 = -4 - 3 = -7$

The first five terms are

$5, 2, -1, -4$, and -7.

7. $a_1 = 5$, $a_2 = 3$; for $n \geq 3$,
$a_n = a_{n-1} - a_{n-2}$

$a_3 = a_2 - a_1 = 3 - 5 = -2$

$a_4 = a_3 - a_2 = -2 - 3 = -5$

$a_5 = a_4 - a_3 = -5 - (-2) = -3$

The first five terms are

$5, 3, -2, -5$, and -3.

9. Arithmetic, $a_1 = 6$, $d = -4$

$a_2 = a_1 + d = 6 + (-4) = 2$

$a_3 = a_1 + 2d = 6 + 2(-4) = -2$

$a_4 = a_1 + 3d = 6 + 3(-4) = -6$

$a_5 = a_1 + 4d = 6 + 4(-4) = -10$

The first five terms are

$6, 2, -2, -6$, and -10.

11. Arithmetic, $a_1 = 3 - \sqrt{5}$, $a_2 = 4$

$d = a_2 - a_1 = 4 - (3 - \sqrt{5})$
$\qquad = 1 + \sqrt{5}$

$a_3 = a_2 + d = 4 + (1 + \sqrt{5}) = 5 + \sqrt{5}$

$a_4 = a_3 + d$
$\quad = (5 + \sqrt{5}) + (1 + \sqrt{5})$
$\quad = 6 + 2\sqrt{5}$

$a_5 = a_4 + d$
$\quad = (6 + 2\sqrt{5}) + (1 + \sqrt{5})$
$\quad = 7 + 3\sqrt{5}$

The first five terms are

$3 - \sqrt{5}, \ 4, \ 5 + \sqrt{5}, \ 6 + 2\sqrt{5}$,
and $7 + 3\sqrt{5}$.

13. Find a_1 and a_{20} for the arithmetic sequence with $a_6 = -4$ and $a_{17} = 51$.

To find a_1 and d, use the formula

$a_n = a_1 + (n - 1)d$.

$a_6 = a_1 + 5d$

$-4 = a_1 + 5d$ $\quad (1)$

$a_{17} = a_1 + 16d$

$51 = a_1 + 16d$ $\quad (2)$

Solve the system formed by equations (1) and (2) by substitution.

Solve equation (1) for a_1.

$a_1 = -4 - 5d$ $\quad (3)$

Substitute $-4 - 5d$ for a_1 in equation (2).

$51 = (-4 - 5d) + 16d$

$51 = -4 + 11d$

$55 = 11d$

$5 = d$

Now substitute $d = 5$ into equation (3) to find a_1.

$a_1 = -4 - 5(5)$

$a_1 = -29$

Finally, find a_{20}.

$a_{20} = a_1 + 19d$

$\qquad = -29 + 19(5)$

$a_{20} = 66$

15. $a_1 = -4$, $d = 3$

$a_n = a_1 + (n - 1)d$

$a_8 = -4 + (8 - 1)3$

$\quad = -4 + 21$

$\quad = 17$

17. $\quad a_3 = 11m$, $a_5 = 7m - 4$

$2d = (7m - 4) - 11m$

$2d = -4m - 4$

$d = -2m - 2$

$a_n = a_1 + (n - 1)d$

$a_3 = a_1 + 2(-2m - 2)$

$11m = a_1 - 4m - 4$

$15m + 4 = a_1$

$a_n = a_1 + (n - 1)d$

$a_8 = (15m + 4) + 7(-2m - 2)$

$\quad = 15m + 4 - 14m - 14$

$\quad = m - 10$

19. Geometric, $a_4 = 8$, $r = \dfrac{1}{2}$

$a_n = a_1 r^{n-1}$

$a_4 = a_1 r^{4-1}$

$8 = a_1 \left(\dfrac{1}{2}\right)^3$

$8 = a_1 \left(\dfrac{1}{8}\right)$

$64 = a_1$

The first five terms are 64, 32, 16, 8, and 4.

21. Geometric, $a_3 = 8$, $a_5 = 72$

$r^2 = \dfrac{72}{8}$

$r^2 = 9$

$r = \pm 3$

$a_n = a_1 r^{n-1}$

$a_3 = a_1 r^{n-1}$

$8 = a_1 (\pm 3)^{3-1}$

$\quad = a_1 \cdot 9$

$\dfrac{8}{9} = a_1$

There are two geometric sequences that satisfy the given conditions.

If $r = 3$, the first five terms are

$\dfrac{8}{9}$, $\dfrac{8}{3}$, 8, 24, and 72.

If $r = -3$, the first five terms are

$\dfrac{8}{9}$, $-\dfrac{8}{3}$, 8, -24, and 72.

These two answers may be written together by listing the first five terms as

$\dfrac{8}{9}$, $\pm\dfrac{8}{3}$, 8, ± 24, and 72.

23. $a_1 = 3$, $r = 2$

$a_n = a_1 r^{n-1}$

$a_5 = 3 \cdot 2^4$

$\quad = 48$

25. $a_1 = 5x$, $a_2 = x^2$

$r = \dfrac{x^2}{5x}$

$r = \dfrac{x}{5}$

$a_n = a_1 r^{n-1}$

$a_5 = 5x\left(\dfrac{x}{5}\right)^4$

$\quad = 5x \cdot \dfrac{x^4}{25 \cdot 25}$

$\quad = \dfrac{x^5}{125}$

29. $a_2 = 6$, $d = 10$

$a_1 = 6 - 10 = -4$

$S_n = \dfrac{n}{2}[2a_1 + (n - 1)d]$

$S_{12} = \dfrac{12}{2}[2(-4) + (11)10]$

$\quad = 6(-8 + 110)$

$\quad = 6(102)$

$\quad = 612$

31. $a_1 = 1$, $r = 2$

$S_n = \dfrac{a_1(1 - r^n)}{1 - r}$

$S_4 = \dfrac{1(1 - 2^4)}{1 - 2}$

$\quad = \dfrac{-15}{-1}$

$\quad = 15$

33. $a_1 = 2k$, $a_2 = -4k$

$r = \dfrac{-4k}{2k}$

$r = -2$

$S_n = \dfrac{a_1(1 - r^n)}{1 - r}$

$S_4 = \dfrac{2k[1 - (-2)^4]}{1 - (-2)}$

$\quad = \dfrac{2k(-15)}{3}$

$\quad = \dfrac{-30k}{4}$

$\quad = -10k$

35. $\displaystyle\sum_{i=1}^{7} (-1)^{i+1} \cdot 6$

This is the sum of a geometric sequence with $a_1 = 6$ and $r = -1$.

$S_n = \dfrac{a_1(1 - r^n)}{1 - r}$

$\displaystyle\sum_{i=1}^{7} (-1)^{i+1} = S_7 = \dfrac{6[1 - (-1)^7]}{1 - (-1)}$

$\quad = \dfrac{6(2)}{2}$

$\quad = 6$

37. $\displaystyle\sum_{i=1}^{6} i(i + 2) = \sum_{i=1}^{6} i(i^2 + 2i)$

$\quad = \displaystyle\sum_{i=1}^{6} i^2 + \sum_{i=1}^{6} 2i$

On the right, sum of arithmetic sequence

$= 1 + 4 + 9 + 16 + 25$

$\quad + 36 + \dfrac{6}{2}(2 + 12)$

$= 91 + 42$

$= 133$

39. $\displaystyle\sum_{i=1}^{4} 8 \cdot 2^i$

This is the sum of a geometric sequence with $a_1 = 16$ and $r = 2$.

$\displaystyle\sum_{i=1}^{4} 8 \cdot 2^i = S_4 = \dfrac{16(1 - 2^4)}{1 - 2}$

$\quad = \dfrac{16(-15)}{-1}$

$\quad = 240$

41. Gale's payments form an arithmetic sequence with $n = 30$, $a_1 = 260$, and $d = -2$.

$S_n = \dfrac{n}{2}[2a_1 + (n - 1)d]$

$S_{30} = \dfrac{30}{2}[2(260) + 29(-2)]$

$S_{30} = 15(520 - 58)$

$S_{30} = 15(462)$

$S_{30} = 6930$

Gale will pay $6930 to pay off the loan plus the interest.

43. This situation may be represented by a geometric sequence where a_1 represents the number of grams present at the start, a_2 represents the number of grams present after 20 yr, etc., so a_6 will represent the amount present after $5 \cdot 20 = 100$ yr.

$$a_1 = 600, \ r = \frac{1}{2}$$

$$a_6 = a_1 r^5$$

$$= 600\left(\frac{1}{2}\right)^5$$

$$= 600 \cdot \frac{1}{32}$$

$$= 18.75$$

There will be 18.75 g left after 100 yr.

45. $20 + 15 + \dfrac{45}{4} + \dfrac{135}{16} + \cdots$

$$r = \frac{15}{20} = \frac{3}{4}$$

$$S_\infty = \frac{a_1}{1 - r}$$

$$= \frac{20}{1 - \dfrac{3}{4}}$$

$$= \frac{20}{\dfrac{1}{4}}$$

$$= 80$$

47. $.8 + .08 + .008 + .0008 + \cdots$

$$r = \frac{.08}{.8} = .1$$

$$S_\infty = \frac{a_1}{1 - r}$$

$$= \frac{.8}{1 - .1}$$

$$= \frac{.8}{.8}$$

$$= \frac{8}{9}$$

49. $\displaystyle\sum_{i=1}^{\infty} -10\left(\frac{5}{2}\right)^i$

This is the sum of a geometric sequence with $a_1 = -25$ and $a_2 = -\dfrac{125}{2}$. Hence, $r = \dfrac{5}{2}$. Note that $|r| > 1$, so the sum does not exist.

51. $.512512512\ldots$

$$= .5 + .0125 + .0000125 + \cdots$$

$$= .5 + \frac{.1025}{1 - .001}$$

$$= .50 + \frac{.0125}{.999}$$

$$= \frac{.512}{.999}$$

$$= \frac{512}{999}$$

53. Let S_n be the statement
$$2 + 6 + 10 + 14 + \ldots + (4n - 2) = 2n^2.$$

Step 1 S_1 is the statement

$$2 = 2(1)^2,$$

which is true.

Step 2 Show that if S_k is true, then S_{k+1} is also true, where S_k is the statement.

$2 + 6 + 10 + 14 + \ldots + (4k - 2) = 2k^2$,

and S_{k+1} is the statement

$2 + 6 + 10 + 14 + \ldots$
$+ (4k - 2) + [4(k + 1) - 2]$
$= 2(k + 1)^2$.

Start with S_k. Add $4(k + 1) - 2 = 4k + 2$ to both sides.

$2 + 6 + 10 + 14 + \ldots$
$+ (4k - 2) + [4(k + 1) - 2]$
$= 2k^2 + [4(k + 1) - 2]$
$= 2k^2 + 4k + 4 - 2$
$= 2k^2 + 4k + 2$
$= 2(k^2 + 2k + 1)$
$= 2(k + 1)^2$

Thus,

$2 + 6 + 10 + 14 + \ldots + (4n - 2) = 2n^2$ is true for every positive integer n.

55. Let S_n be the statement
$2 + 2^2 + 2^3 + \ldots + 2^n = 2(2^n - 1)$.
Step 1 S_1 is the statement

$$2 = 2(2^1 - 1)$$
$$2 = 2 \cdot 1,$$

which is true.

Step 2 Show that if S_k is true, then S_{k+1} is also true, where S_k is the statement

$2 + 2^2 + 2^3 + \ldots + 2^k = 2(2^k - 1)$,

and S_{k+1} is the statement

$2 + 2^2 + 2^3 + \ldots + 2^k + 2^{k+1}$
$= 2[2^{(k+1)} - 1]$.

Start with S_k. Add 2^{k+1} to both sides.

$2 + 2^2 + 2^3 + \ldots + 2^k + 2^{k+1}$
$= 2(2^k - 1) + 2^{k+1}$
$= 2^{k+1} - 2 + 2^{k+1}$
$= 2 \cdot 2^{k+1} - 2 \cdot 1$
$= 2(2^{k+1} - 1)$

Thus, $2 + 2^2 + 2^3 + \ldots + 2^n$ $= 2(2^n - 1)$ is true for every positive integer n.

57. $P(5, 5) = \dfrac{5!}{(5 - 5)!} = \dfrac{5!}{0!} = \dfrac{5!}{1} = 120$

59. $P(6, 0) = \dfrac{6!}{(6 - 0)!}$
$= \dfrac{6!}{6!} = 1$

61. $\dbinom{10}{5} = \dfrac{10!}{5!(10 - 5)!} = \dfrac{10!}{5!5!}$
$= \dfrac{10 \cdot 9 \cdot 8 \cdot 7 \cdot 6 \cdot 5!}{5 \cdot 4 \cdot 3 \cdot 2 \cdot 1 \cdot 5!}$
$= 252$

63. $2 \cdot 4 \cdot 3 \cdot 2 = 48$

48 different wedding arrangements are possible.

65. There are 4 choices for the first job, 3 choices for the second job, and so on.

$$4 \cdot 3 \cdot 2 \cdot 1 = 24$$

There are 24 ways in which the jobs can be assigned.

67. Order is important, so this is a permutation problem.

$$P(9, 3) = \frac{9!}{6!}$$
$$= 9 \cdot 8 \cdot 7$$
$$= 504$$

The winners can be determined 504 ways.

69. (a) Let E be the event "picking a green marble."

$n(E) = 4$

$n(S) = 4 + 5 + 6 = 15$
        (total number of marbles)

$$P(E) = \frac{P(E)}{n} = \frac{4}{15}$$

(b) Let E be the next event "picking a black marble."

$n(E) = 5$

$n(S) = 15$ (from part (a))

$$P(E) = \frac{5}{15} = \frac{1}{3}$$

The probability that the marble is not black is given by

$$P(E') = 1 - P(E)$$
$$= 1 - \frac{1}{3} = \frac{2}{3}.$$

(c) Let E be the event "picking a blue marble."

$n(E) = 0$ since there are no blue marbles.

$n(S) = 15$ (from part (a))

$$P(E) = \frac{n(E)}{n} = \frac{0}{15} = 0$$

71. $P(\text{black king}) = \dfrac{n \text{ (black kings)}}{n \text{ (cards in deck)}}$

$$= \frac{2}{52} = \frac{1}{26}$$

73. P(ace or diamond)

= P(ace) + P(diamond)

    − P(diamond and ace)

$$= \frac{4}{52} + \frac{13}{52} - \frac{1}{52}$$

$$= \frac{16}{52} = \frac{4}{13}$$

75. P(not a diamond)

= P(club or heart or spade)

$$= \frac{13}{52} + \frac{13}{52} + \frac{13}{52}$$

$$= \frac{1}{4} + \frac{1}{4} + \frac{1}{4} = \frac{3}{4}$$

P(not black) = P(red)

= P(diamond or heart)

$$= \frac{13}{52} + \frac{13}{52}$$

$$= \frac{1}{4} + \frac{1}{4} = \frac{1}{2}$$

P(not a diamond and not black)

= P(heart)

$$= \frac{13}{52}$$

$$= \frac{1}{4}$$

P(not a diamond or not black)

= P(not a diamond) + P(not black)

    − P(not a diamond and not black)

$$= \frac{3}{4} + \frac{1}{2} - \frac{1}{4}$$

$$= 1$$

77. P(at least 2)

= P(2 or 3 or 4 or 5)

= P(2) + P(3) + P(4) + P(5)
        Mutually exclusive events

= .18 + .12 + .08 + .06

= .44

Chapter 9 Test

1. $a_n = (-1)^n(n^2 + 1)$

   $a_1 = (-1)^1(1^2 + 1) = -2$

   $a_2 = (-1)^2(2^2 + 1) = 5$

   $a_3 = (-1)^3(3^2 + 1) = -10$

   $a_4 = (-1)^4(4^2 + 1) = 17$

   $a_5 = (-1)^5(5^2 + 1) = -26$

2. $a_1 = 5$; for $n \geq 2$, $a_n = n + a_{n-1}$

   $a_1 = 5$

   $a_2 = 2 + a_1 = 2 + 5 = 7$

   $a_3 = 3 + a_2 = 3 + 7 = 10$

   $a_4 = 4 + a_3 = 4 + 10 = 14$

   $a_5 = 5 + a_4 = 5 + 14 = 19$

3. $a_n = n + 1$ if n is odd and $a_n = a_{n-1} + 2$ if n is even.

   $a_1 = 1 + 1 = 2$

   $a_2 = a_1 + 2 = 2 + 2 = 4$

   $a_3 = 3 + 1 = 4$

   $a_4 = a_3 + 2 = 4 + 2 = 6$

   $a_5 = 5 + 1 = 6$

4. Arithmetic sequence with $a_2 = 1$, $a_4 = 25$

   $a_2 = a_1 + (2 - 1)d$

   $1 = a_1 + d$ *(1)*

   $a_4 = a_1 + (4 - 1)d$

   $25 = a_1 + 3d$ *(2)*

   Solve the system of equations (1) and (2) by the substitution method.

Solve equation (1) for a_1.

$1 - d = a_1$

Substitute $1 - d$ for a_1 in equation (2).

$25 = (1 - d) + 3d$

$25 = 1 + 2d$

$d = 12$

$a_1 = 1 - d$

$\quad = -11$

The first 5 terms of the sequence are:

$a_1 = -11$

$a_2 = -11 + 1(12) = 1$

$a_3 = -11 + 2(12) = 13$

$a_4 = -11 + 3(12) = 25$

$a_5 = -11 + 4(12) = 37.$

5. Geometric sequence with $a_1 = 81$, $r = \dfrac{2}{3}$

   $a_1 = 81$

   $a_2 = 81\left(\dfrac{2}{3}\right)^1 = 54$

   $a_3 = 81\left(\dfrac{2}{3}\right)^2 = 36$

   $a_4 = 81\left(\dfrac{2}{3}\right)^3 = 24$

   $a_5 = 81\left(\dfrac{2}{3}\right)^4 = 16$

6. An arithmetic sequence has $a_7 = -6$ and $a_{15} = -2$.

   Find a_{31}.

   Use the formula

   $a_n = a_1 + (n - 1)d.$

   $a_7 = a_1 + 6d$

   $-6 = a_1 + 6d$ *(1)*

$a_{15} = a_1 + 14d$

$-2 = a_1 + 14d$ (2)

Solve the system of equations (1) and (2) by the substitution method. Solve equation (1) for a_1.

$a_1 = -6 - 6d$ (3)

Substitute $-6 - 6d$ for a_1 in equation (2).

$-2 = (-6 - 6d) + 14d$

$-2 = -6 + 8d$

$4 = 8d$

$\frac{1}{2} = d$

To find a_1, substitute $d = \frac{1}{2}$ into equation (3).

$a_1 = -6 - 6\left(\frac{1}{2}\right)$

$= -6 - 3$

$a_1 = -9$

Now find a_{31}.

$a_{31} = a_1 + 30d$

$= -9 + 30\left(\frac{1}{2}\right)$

$= -9 + 15$

$a_{31} = 6$

7. Find S_{10} for the arithmetic sequence with $a_1 = 37$ and $d = 13$.

$S_n = \frac{n}{2}[2a_1 + (n - 1)d]$

$S_{10} = \frac{10}{2}[2(37) + 9(13)]$

$= 5(74 + 117)$

$= 5(191)$

$= 955$

8. A geometric sequence has $a_1 = 12$ and $a_6 = -\frac{3}{8}$. Find a_3.

First, find r.

$a_n = a_1 r^{n-1}$

$a_6 = a_1 r^5$

$-\frac{3}{8} = 12 r^5$

$\left(-\frac{3}{8}\right)\left(\frac{1}{12}\right) = r^5$

$-\frac{1}{32} = r^5$

$\sqrt[5]{-\frac{1}{32}} = r$

$r = -\frac{1}{2}$

$a_3 = a_1 r^2$

$= 12\left(-\frac{1}{2}\right)^2 = 12\left(\frac{1}{4}\right)$

$a_3 = 3$

9. Find a_7 for the geometric sequence with $a_1 = 2x^3$ and $a_3 = 18x^7$.

$a_n = a_1 r^{n-1}$

$a_3 = a_1 r^2$

$18x^7 = (2x^3)(r^2)$

$9x^4 = r^2$

$\pm 3x^2 = r$

$a_7 = a_1 r^6$

$= 2x^3(\pm 3x^2)^6$

$= 2x^3(729^{12})$

$a_7 = 1458x^{15}$

10. Find S_4 for the geometric sequence with $a_1 = 4$ and $r = \frac{1}{2}$.

$S_n = \frac{a_1(1 - r^n)}{1 - r}$

$$S_4 = \frac{4[1 - (\frac{1}{2})^4]}{1 - \frac{1}{2}}$$

$$= \frac{4(1 - \frac{1}{16})}{\frac{1}{2}}$$

$$= \frac{4(\frac{15}{16})}{\frac{1}{2}}$$

$$= 8 \cdot \frac{15}{16}$$

$$S_4 = \frac{15}{2}$$

11. $\displaystyle\sum_{i=1}^{30} (5i + 2)$

This sum represents the sum of the first 30 terms of the arithmetic sequence having

$a_1 = 5 \cdot 1 + 2 = 7$ and

$a_n = a_{30} = 5 \cdot 30 + 2 = 152.$

$$\sum_{i=1}^{30} (5i + 2) = S_{30}$$

$$= \frac{n}{2}(a_1 + a_n)$$

$$= \frac{30}{2}(a_1 + a_{30})$$

$$= 15(7 + 152)$$

$$= 2385$$

12. $\displaystyle\sum_{i=1}^{5} (-3 \cdot 2^i)$

This sum represents the sum of the first five terms of the geometric sequence having

$a_1 = -3 \cdot 2^1 = -6$ and $r = 2.$

$$\sum_{i=1}^{5} -3 \cdot 2^i = S_5$$

$$= \frac{a_1(1 - r^n)}{1 - r}$$

$$= \frac{-6(1 - 2^5)}{1 - 2}$$

$$= \frac{-6(1 - 32)}{-1}$$

$$= 6(-31)$$

$$= -186$$

13. $75 + 30 + 12 + \dfrac{24}{5} + \ldots$

$$r = \frac{a_2}{a_1} = \frac{30}{75} = \frac{2}{5}$$

Check that the ratios $\frac{a_3}{a_2}$ and $\frac{a_4}{a_3}$ are also equal to $\frac{2}{5}$.

This is a geometric sequence with $-1 < r < 1$, so the formula for the sum of an infinite geometric sequence applies.

$$S_\infty = \frac{a_1}{1 - r}$$

$$= \frac{75}{1 - \frac{2}{5}}$$

$$= \frac{75}{\frac{3}{5}}$$

$$S_\infty = 125$$

14. $\displaystyle\sum_{i=1}^{\infty} 54\left(\frac{2}{9}\right)^i$

This is the sum of the infinite geometric sequence with

$a_1 = 54\left(\frac{2}{9}\right)^1 = 12$ and $r = \frac{2}{9}.$

$$\sum_{i=1}^{\infty} 54\left(\frac{2}{9}\right)^i = S_\infty = \frac{a_1}{1 - r}$$

$$= \frac{12}{1 - \frac{2}{9}}$$

$$= \frac{12}{\frac{7}{9}}$$

$$= \frac{9 \cdot 12}{7}$$

$$= \frac{108}{7}$$

15. This problem describes an arithmetic sequence with $a_1 = 50$ and $d = 5$.

The amount deposited after 1 mo (on March 1) is

$$a_2 = a_1 + d = 55.$$

The amount deposited after 2 mo (on April 1) is

$$a_3 = a_1 + 2d = 60.$$

Thus, the amount deposited after 20 mo is

$$a_{21} = a_1 + 20d$$
$$= 50 + 20(5)$$
$$= 150.$$

The total amount deposited after 20 mo will be given by S_{21}.

$$S_n = \frac{n}{2}(a_1 + a_{21})$$

$$= \frac{21}{2}(50 + 150)$$

$$= 2100.$$

Thus, the total amount deposit after 20 mo is $2100.

16. The problem describes a geometric sequence with $a_1 = 50$ and $r = 2$. Since the population doubles every 30 minutes, we have

$$a_2 = 100 \text{ at } 12:30$$
$$a_3 = 200 \text{ at } 1:00, \ldots$$
$$a_{10} = \text{number present at } 4:30.$$

$$a_{10} = a_1 r^9$$
$$= 50 \cdot 2^9$$
$$= 50(512)$$
$$= 25,600$$

There will be 25,600 bacteria present at 4:40 P.M.

17. Prove that

$$8 + 14 + 20 + 26 + \ldots + (6n + 2)$$
$$= 3n^2 + 5n$$

is true for every positive integer n.

Step 1 S_1 is the statement

$$8 = 3(1^2) + 5(1),$$

which is true.

Step 2 Show that if S_k is true, then S_{k+1} is also true. S_k is the statement

$$8 + 14 + 20 + 26 + \ldots + (6k + 2)$$
$$= 3k^2 + 5k,$$

and S_{k+1} is the statement

$$8 + 14 + 20 + 26 + \ldots + (6k + 2) + [6(k + 1) + 2]$$
$$= 3(k + 1)^2 + 5(k + 1).$$

Start with S_k.

$$8 + 14 + 20 + 26 + \ldots + (6k + 2)$$
$$= 3k^2 + 5k$$

Add the $(k+1)$st term, $6(k+1)+2$, to both sides.

$8 + 14 + 20 + 26 + \ldots + (6k + 2)$
$+ [6(k + 1) + 2]$

$= (3k^2 + 5k) + [6(k + 1) + 2]$

$= 3k^2 + 5k + 5(k + 1) + (k + 1) + 2$

$= (3k^2 + 6k + 3) + 5(k + 1)$

$= 3(k^2 + 2k + 1) + 5(k + 1)$

$3 + 6 + 9 + \ldots + 3k + 3(k + 1)$

$= 3(k + 1)^2 + 5(k + 1)$

The final equation is the statement S_{k+1}. Thus, we have shown that if S_k is true, then S_{k+1} is also true. The two steps required for a proof by mathematical induction have been completed, so the statement

$8 + 14 + 20 + 26 + \ldots + (6n + 2)$
$= 3n^2 + 5n$

is true for every positive integer n.

18. $P(11, 3) = \dfrac{11!}{(11 - 3)!}$

$= \dfrac{11!}{8!}$

$= \dfrac{11 \cdot 10 \cdot 9 \cdot 8!}{8!}$

$= 990$

19. $P(7, 7) = \dfrac{7!}{(7 - 7)!}$

$= \dfrac{7!}{0!}$

$= 7!$

$= 5040$

20. $\dbinom{10}{2} = \dfrac{10!}{(10 - 2)!2!}$

$= \dfrac{10!}{8!2!}$

$= \dfrac{10 \cdot 9 \cdot 8!}{8! \cdot 2 \cdot 1}$

$= 45$

21. $\dbinom{21}{0} = \dfrac{12!}{(12 - 0!)0!}$

$= \dfrac{12!}{12!0!}$

$= 1$

22. Use the multiplication principle of counting. The first event, selecting a style, can occur in 4 ways. The second event, selecting a fabric, can occur in 3 ways. The third event, selecting a color, can occur in 5 ways. Thus, there are

$$4 \cdot 3 \cdot 5 = 60 \text{ different coats.}$$

23. Use permutations.

$$P(30, 3) = \dfrac{30!}{(30 - 3)!}$$

$= \dfrac{30!}{27!}$

$= 30 \cdot 29 \cdot 28$

$= 24{,}360$

The offices can be filled in 24,360 ways.

24. Choose 2 of the 14 women and 2 of the 16 men. Since the order of the men and women is unimportant, this is a combination problem. Since choosing the women and choosing the

men are independent events, we can
use the multiplication principle of
counting.

$$\binom{14}{2} \cdot \binom{16}{2}$$

$$= \frac{14!}{(14-2)!2!} \cdot \frac{16!}{(16-2)!2!}$$

$$= \frac{14!}{12!2!} \cdot \frac{16!}{14!2!}$$

$$= \frac{14 \cdot 13 \cdot 12!}{12!2!} \cdot \frac{16 \cdot 15 \cdot 14!}{14!2!}$$

$$= \frac{14 \cdot 13}{2} \cdot \frac{16 \cdot 15}{2}$$

$$= 91 \cdot 120$$

$$= 10,920$$

The four who will attend the con-
ference can be chosen in 10,920
ways.

For Problems 26–29, the sample space is
the set of all cards in a standard deck,
so n(S) = 52.

26. Consider the event E: "drawing a
     red three." There are 2 red threes
     in a deck, so n(E) = 2.

$$P(E) = \frac{n(E)}{n(S)}$$

$$= \frac{2}{52}$$

$$= \frac{1}{26}$$

The probability of drawing a red
three is $\frac{1}{26}$.

27. Consider the event E: "draw a face
     card." Each suit contains 3 face
     cards (jack, queen, and king), so
     the deck contains 12 face cards.

     Thus n(E) = 12

     and $P(E) = \frac{12}{52} = \frac{3}{13}$.

The probability of drawing a card
that is not a face card is

$$P(E') = 1 - P(E)$$

$$= 1 - \frac{3}{13} = \frac{10}{13}.$$

28. The events E: "draw a king" and
     F: "draw a spade" are not mutually
     exclusive, since it is possible
     to draw the king of spades, an out-
     come satisfying both events.

$$P(E \text{ or } F) = P(E \cup F)$$

$$= P(E) + P(F) - P(E \cap F)$$

P(king or spade)

$$= P(\text{king}) + P(\text{spade})$$

$$- P(\text{king and spade})$$

$$= \frac{4}{52} + \frac{13}{52} - \frac{1}{52}$$

$$= \frac{16}{52} = \frac{4}{13}$$

The probability of drawing a king or

a spade is $\frac{4}{13}$.

29. Consider the event E: "draw a face
     card." As shown in the solution
     to Problem 27,

$$P(E) = \frac{3}{13} \quad \text{and} \quad P(E') = \frac{10}{13}.$$

The odds in favor of drawing a face
card are

$$\frac{P(E)}{P(E')} = \frac{\frac{3}{13}}{\frac{10}{13}} = \frac{3}{10} \quad \text{or} \quad 3 \text{ to } 10.$$

30.   "At most 2" means "0 or 1 or 2."

P(at most 2)

= P(0 or 1 or 2)

= P(0) + P(1) + P(2)
    (since the events are mututally
    exclusive)

= .19 + .43 + .30

= .92

The probability that at most 2
filters are defective is .92.

APPENDIX A SETS

1. The elements of the set $\{12, 13, 14, \ldots, 20\}$ are all the natural numbers from 12 to 20 inclusive. These numbers are

$$12, 13, 14, 15, 16, 17, 18, 19, \text{ and } 20.$$

3. The elements of the set $\{1, 1/2, 1/4, \ldots, 1/32\}$ form a geometric sequence with $a_1 = 1$ and $r = 1/2$, that is, the first number is 1, and each number after the first is found by multiplying the preceding number by 1/2. There are 6 elements:

$$1, 1/2, 1/4, 1/8, 1/16, \text{ and } 1/32.$$

5. To find the elements of the set $\{17, 22, 27, \ldots, 47\}$, start with 17 and add 5 to find the next number. The elements of this set form an arithmetic sequence with $a_1 = 17$ and $d = 5$. There are 7 elements.

$$17, 22, 27, 32, 37, 42, \text{ and } 47.$$

7. The elements of the set $\{\text{all natural numbers greater than 7 and less than 15}\}$ are

$$8, 9, 10, 11, 12, 13 \text{ and } 14.$$

9. The set $\{4, 5, 6, \ldots, 15\}$ has a limited number of elements, so it is a finite set.

11. The set $\{1, 1/2, 1/4, 1/8, \ldots\}$ has an unlimited number of elements, so it is an infinite set.

13. The set $\{x \mid x \text{ is a natural number larger than 5}\}$, which can also be written as $\{6, 7, 8, 9, \ldots\}$, has an unlimited number of elements, so it is an infinite set.

15. There are an infinite number of fractions between 0 and 1, so $\{x \mid x \text{ is a fraction between 0 and 1}\}$ is an infinite set.

17. 6 is an element of the set $\{3, 4, 5, 6\}$, so we write $6 \in \{3, 4, 5, 6\}$.

19. -4 is not an element of $\{4, 6, 8, 10\}$, so we write $-4 \notin \{4, 6, 8, 10\}$.

21. 0 is an element of $\{2, 0, 3, 4\}$, so we write $0 \in \{2, 0, 3, 4\}$.

23. $\{3\}$ is a subset of $\{2, 3, 4, 5)$, not an element of $\{2, 3, 4, 5\}$, so we write $\{3\} \notin \{2, 3, 4, 5\}$.

25. $\{0\}$ is a subset of $\{0, 1, 2, 5\}$, not an element of $\{0, 1, 2, 5\}$, so we write $\{0\} \notin \{0, 1, 2, 5\}$.

27. 0 is not an element of $\emptyset$, since the empty set contains no elements. Thus, $0 \notin \emptyset$.

29. $3 \in \{2, 5, 6, 8\}$

3 is not one of the elements in $\{2, 5, 6, 8\}$, so this statement is false.

31. $1 \in \{3, 4, 5, 11, 1\}$

Since 1 is one of the elements of $\{3, 4, 5, 11, 1\}$, the statement is true.

33. $9 \notin \{2, 1, 5, 8\}$

Since 9 is not one of the elements of $\{2, 1, 5, 8\}$, the statement is true.

35. $\{2, 5, 8, 9\} = \{2, 5, 9, 8\}$

This statement is true because both sets contain exactly the same elements.

37. $\{5, 8, 9\} = \{5, 8, 9, 0\}$

These two sets are not equal because $\{5, 8, 9, 0\}$ contains the element 0, which is not an element of $\{5, 8, 9\}$. Therefore, the statement is false.

39. $\{x \mid x$ is a natural number less than $3\} = \{1, 2\}$

Since 1 and 2 are the only natural numbers less than 2, this statement is true.

41. $\{5, 7, 9, 19\} \cap \{7, 9, 11, 15\} = \{7, 9\}$

The symbol "$\cap$" means the intersection of the two sets, which is the set of all elements that belong to both sets. Since 7 and 9 are the only elements belonging to both sets, the statement is true.

43. $\{2,\ 1,\ 7\} \cup \{1,\ 5,\ 9\} = \{1\}$

The symbol "∪" means the union of two sets, which is the set of all elements that belong to either one of the sets or to both sets.

$$\{2,\ 1,\ 7\} \cup \{1,\ 5,\ 9\} = \{1,\ 2,\ 5,\ 7,\ 9\},$$

while

$$\{2,\ 1,\ 7\} \cap \{1,\ 5,\ 9\} = \{1\}.$$

Therefore, the statement is false.

45. $\{3,\ 2,\ 5,\ 9\} \cap \{2,\ 7,\ 8,\ 10\} = \{2\}$

Since 2 is the only element belonging to both sets, the statement is true.

47. $\{3,\ 5,\ 9,\ 10\} \cap \emptyset = \{3,\ 5,\ 9,\ 10\}$

In order to belong to the intersection of two sets, and element must belong to both sets. Since the empty set contains no elements, $\{3,\ 5,\ 9,\ 10\} \cap \emptyset = \emptyset$, so the statement is false.

49. $\{1,\ 2,\ 4\} \cup \{1,\ 2,\ 4\} = \{1,\ 2,\ 4\}$

Since the two sets are equal, their union contains the same elements, 1, 2, and 4. Thus, the statement is true.

51. $\emptyset \cup \emptyset = \emptyset$

Since the empty set contains no elements, the statement is true.

For Exercises 53–63,

$$A = \{2,\ 4,\ 6,\ 8,\ 10,\ 12\},$$
$$B = \{2,\ 4,\ 8,\ 10\},$$
$$C = \{4,\ 10,\ 12\},$$
$$D = \{2,\ 10\},$$
$$\text{and}\ \ U = \{2,\ 4,\ 6,\ 8,\ 10,\ 12,\ 14\}.$$

53. $A \subseteq U$

This statement says "A is a subset of U." Since every element of A is also an element of U, the statement is true,

55. D ⊆ B

Since both elements of D, 2 and 10, are also elements of B, D is a subset of B. The statement is true.

57. A ⊆ B

Set A contains a two elements, 6 and 12, that are not elements of B. Thus, A is not a subset of B. The statement is false.

59. ∅ ⊆ A

The empty set is a subset of every set, so the statement is true.

61. $\{4, 8, 10\}$ ⊆ B

Since 4, 8, and 10 are all elements of B, $\{4, 8, 10\}$ is a subset of B. The statement is true.

63. B ⊆ D

Since B contains two elements, 4 and 8, that are not elements of D, B is not a subset of D. The statement is false.

65. Every element of $\{2, 4, 6\}$ is also an element of $\{3, 2, 5, 4, 6\}$, so $\{2, 4, 6\}$ is a subset of $\{3, 2, 5, 4, 6\}$. We write

$$\{2, 4, 6\} \subseteq \{3, 2, 5, 4, 6\}.$$

67. Since 0 is an element of $\{0, 1, 2\}$, but is not an element of $\{1, 2, 3, 4, 5\}$, $\{0, 1, 2\}$ is not a subset of $\{1, 2, 3, 4, 5\}$. We write

$$\{0, 1, 2\} \nsubseteq \{1, 2, 3, 4, 5\}.$$

69. The empty set is a subset of every set, so ∅ ⊆ $\{1, 4, 6, 8\}$.

For Exercises 71–93,

$$U = \{0, 1, 2, 3, 4, 5, 6, 7, 8, 9, 10, 11, 12, 13\},$$
$$M = \{0, 2, 4, 6, 8\},$$
$$N = \{1, 3, 5, 7, 9, 11, 13\},$$
$$Q = \{0, 2, 4, 6, 8, 10, 12\},$$
and $R = \{0, 1, 2, 3, 4\}.$

71. $M \cap R$

The only elements belonging to both M and R are 0, 2, and 4, so

$$M \cap R = \{0, 2, 4\}.$$

73. $M \cup N$

The union of two sets contains all elements that belong to either set or to both sets.

$$M \cup N = \{0, 1, 2, 3, 4, 5, 6, 7, 8, 9, 11, 13\}$$

75. $M \cap U$

Since $M \subseteq U$, the intersection of M and U will contain the same elements as M.

$$M \cap U = M \quad \text{or} \quad \{0, 2, 4, 6, 8\}$$

77. $N \cup R = \{0, 1, 2, 3, 4, 5, 7, 9, 11, 13\}$

79. N'

The set N' is the complement of set N, which means the set of all elements in the universal set U that do not belong to N.

$$N' = \{0, 2, 4, 6, 8, 10, 12\}$$

81. $M' \cap Q$

First form M', the complement of M. M' contains all elements of U that are not elements of M.

$$M' = \{1, 3, 5, 7, 9, 10, 11, 12, 13\}$$

Now form the intersection of M' and Q.

$$M' \cap Q = \{10, 12\}$$

83. $\emptyset \cap R$

Since the empty set contains no elements, there are no elements belonging to both $\emptyset$ and R. Thus, $\emptyset$ and R are disjoint sets, and $\emptyset \cap R = \emptyset$.

85. $N \cup \emptyset$

Since $\emptyset$ contains no elements, the only elements belonging to N or $\emptyset$ are the elements of N. Thus,

$$N \cup \emptyset = N \quad \text{or} \quad \{1, 3, 5, 7, 9, 11, 13\}.$$

87. $(M \cap N) \cup R$

First form the intersection of M and N. Since M and N have no common elements, $M \cap N = \emptyset$.
Thus,

$$(M \cap N) \cup R = \emptyset \cup R$$
$$= R \quad \text{or} \quad \{0, 1, 2, 3, 4\}.$$

89. $(Q \cap M) \cup R$

First form the intersection of Q and M.

$$Q \cap M = \{0, 2, 4, 6, 8\} = M$$

Now form the union of this set with R.

$$(Q \cap M) \cup R = M \cup R$$
$$= \{0, 1, 2, 3, 4, 6, 8\}$$

91. $(M' \cup Q) \cap R$

First, find M', the complement of M.

$$M' = \{1, 3, 5, 7, 9, 10, 11, 12, 13\}$$

Next, form the union of M' and Q.

$$M' \cup Q = \{0, 1, 2, 3, 4, 5, 6, 7, 8, 9, 10, 11, 12, 13\}$$
$$- U$$

Thus,

$$(M' \cup Q) \cap R = U \cap R$$
$$= R \text{ or } \{1, 2, 3, 4\}.$$

93. $Q' \cap (N' \cap U)$

First, find Q', the complement of Q.

$$Q' = \{1, 3, 5, 7, 9, 11, 13\} = N$$

Now find N', the complement of P.

$$N' = \{0, 2, 4, 6, 8, 10, 12\} = Q$$

Next, form the union of N' and U.

$$N' \cup U = Q \cup U = Q$$

Finally, we have

$$Q' \cap (N' \cap U) = Q' \cap Q = \emptyset$$

Since the intersection of Q' and $(N' \cap U)$ is $\emptyset$, Q' and $(N' \cap U)$ are disjoint sets.

95. M' is the set of all students in this school who are not taking this course.

97. $N \cap P$ is the set of all students in this school who are taking both calculus and history.

99. $M \cup P$ is the set of all students in this school who are taking this course or history or both.

101. The possible outcomes for rolling an honest die are 1, 2, 3, 4, 5, and 6, so the sample space is

$$\{1, 2, 3, 4, 5, 6\}.$$

103. There are eight possible outcomes for tossing a coin three times. The sample space is

$$\{hhh, hht, hth, thh, htt, tht, tth, ttt\}.$$

105. X ∩ Y = ∅

This statement is true if X and Y are disjoint.

107. X ∩ Y = X

This statement will be true if X is a subset of Y, which is written X ⊆ Y.

109. X ∩ ∅ = X

The intersection of any set with the empty set is the empty set, that is, X ∩ ∅ = ∅.

Thus, the given statement can only be true if X = ∅.

NOTES

NOTES

NOTES

NOTES

NOTES